MW00560786

Introduction to Fluid Mechanics

James A. Fay

Introduction to Fluid Mechanics

MIT Press
Cambridge, Massachusetts
London, England

This book was set in Times Roman by TechBooks and was printed and bound in the United States of America.

Library of Congress Cataloging-in-Publication Data

Fay, James A.
Introduction to fluid mechanics / James A. Fay.
p. cm.
Includes bibliographical references and index.
ISBN 0-262-06165-1
ISBN-13 978-0-262-06165-0

1. Fluid mechanics. I. Title.
TA357.F39 1994
532—dc20 93-34567
CIP

10 9 8 7 6 5 4

Contents

Tables

Preface

This text is an outgrowth of the development of the undergraduate subject in fluid mechanics that is required of all mechanical engineering students at MIT. The subject is one semester in length; no additional subject incorporating fluid mechanics is required, although a subsequent heat transfer subject is the option of many students. The subject satisfies a science requirement in that it emphasizes the application of physical laws and the understanding of fluid phenomena. Given the time limitations of one semester, compressible flow cannot be covered. Nevertheless, all the important physical principles are treated in depth, albeit succinctly, including the conservation of energy as it applies to compressible as well as incompressible flows.

The primary objective of this text is to emphasize the important principles of the conservation of mass, momentum and energy as they apply to a fluid. Most applications of these principles are to incompressible flow, although compressible flows are not excluded in the derivation and discussion of the expressions of these principles and are treated separately in chapter 12. Both inviscid and viscous flows are treated extensively, in that order. Two-dimensional irrotational flow is considered in chapter 11 as an example of multidimensional flow analysis.

Students in our fluid mechanics class have taken vector calculus and, usually, thermodynamics before they begin to study fluid mechanics. This text uses vector algebra and calculus extensively in deriving the integral and differential expressions for the conservation of mass, momentum and energy as well as a shorthand for expressing their application to problems, but individual problems and examples seldom require the explicit use of vector expressions per se, scalar variables sufficing. Having invested time and effort to master vector calculus, it is to the student's advantage to profit from the economy and simplicity that it affords in displaying the content and meaning of the principles of fluid mechanics. In this way, the text minimizes the time spent on derivations, which are seldom the ingredients of engineering practice, while solidifying the concepts that are the essential building blocks of design and analysis.

The expression of the laws of the conservation of mass, momentum and energy are developed simultaneously in both integral and differential form, usually in that order. The relationship between the two forms is made quite clear in their derivation through use of the applicable integral theorems of vector calculus. Theses laws are expressed in words as well as equations, and the distinction between scalar and vector expressions are emphasized. The use of control volumes and the integral formulation of the conservation laws applied to them is common in thermodynamics subjects so that most students have had some exposure to such usage prior to beginning fluid mechanics.

Special emphasis has been accorded unsteady inviscid and viscous flows, an important aspect of fluid mechanics often missing from, or inadequately treated in, introductory texts despite the fact that there are many examples of its importance in engineering systems. Enough examples and problems involving unsteady flow are included to give the student an understanding of the importance of not automatically neglecting the unsteady component of the conservation laws.

In many introductory fluid mechanics texts, the conservation of energy is not explained clearly enough. True, it is not easy to reduce the subtleties of effects of viscous dissipation in incompressible flows to simple terms, but the significance of the laws of thermodynamics to fluid flows needs to be elucidated for the student's benefit. In chapter 8, these difficult connections are made explicit by first treating incompressible viscous flows and deriving the expressions for the change in head in such flows. After this discussion, the corresponding development for compressible and nonadiabatic flows follows . It may be desirable to defer the second half of this chapter to the beginning of a treatment of compressible flow, to which it is more appropriate.

The order of presentation of the material in this text has been chosen so as to defer the complexities (such as the viscous force) to later portions while introducing interesting flow problems as early as possible. There is no loss of generality since everything essential is eventually treated properly, but clarity in the student's mind is gained by proceeding from the simple toward the complex.

The order of presentation of the material is not sacrosanct. It is possible to treat irrotational flow (chapter 11) immediately after inviscid flow (chapter 4), although the intervening discussion of viscous flow is necessary for the full appreciation of the limitations that viscous effects have on the practical applications of irrotational flow. Equally well, the material on dimensional analysis and modeling (chapter 11) could be moved forward to follow viscous flow (chapters 6 and 7) without encountering any difficulty. (There is some benefit to the student to deferring this discussion until a wider experience with applications has been reached.)

While we cannot cover compressible flow in our MIT subject for lack of time, chapter 12 has been included for those who require some exposure to compressible flow. It includes the usual topics of an introductory segment on compressible flow together with a beginning treatment of unsteady one-dimensional flow, a part of that subject that is usually overlooked even though it has many practical applications. The major advantage of treating compressible flow at this level is to expose the student to the considerable difference between incompressible and compressible flows while providing some ability to handle the simpler compressible flow applications.

The problems and examples are all illustrative of significant or interesting consequences of the fundamental principles. Although sometimes simplified, they utilize examples of actual engineering devices or the fluid mechanical aspects of their use. In many problems, the student is required to evaluate numerically the analytical expressions used or derived for each problem so as to develop a physical sense about the magnitude of the quantities and an ability to detect errors when the results of the calculations do not appear to be physically reasonable. Since the end result of most engineering designs and analyses is the production of numerical results, it is desirable that students have an appreciation for the integrity and reliability of there numerical evaluations.

The problems are of medium difficulty because mostly they require some analysis rather than the application of a formula. Lengthy algebraic or numerical manipulations are avoided. The emphasis is placed on the selection of the appropriate physical principles and their application to a situation of engineering interest.

Great care has been taken to be both correct and complete without overwhelming the student with little-used qualifications. Footnotes are used to add qualifications without interrupting the important trend of an argument.

This book would not have been written without the stimulation of teaching many MIT students, both undergraduate and graduate. The author is indebted even more to his colleagues in the Fluid Mechanics Laboratory of MIT: Ain Sonin, Ronald Probstein, Ascher Shapiro, C. Forbes Dewey, Roger Kamm, Anthony Patera, Harry Kytomma, Ahmed Ghoniem, James Keck, and John Heywood, who have contributed in various ways, both directly and indirectly, to this book. Of course, the material is solely the responsibility of the author.

Without the constant encouragement and forbearance on the part of my spouse, Gay, authoring this book would not have been the rewarding task that it became.

To the Student

Fluid mechanics is a subject you know a lot about—at least intuitively—through your everyday experiences. Breathing air, drinking water, taking a shower, speaking and hearing all involve utilizing some aspect of a fluid flow. What you don't know yet is how to express this information in quantitative terms or how to design systems that will utilize these phenomena for other purposes. The objective of this text is to help you understand fluid mechanics better than you do now and to equip you to use this new information in a quantitatively useful way.

The origin of fluid mechanics lies in the history of human use of wind, rivers, ponds and ocean for the practical advantages of transportation, irrigation, potable water supply, etc. Archimedes, that astounding mathematician of the ancient world, was the first to show why a ship floats and when it might sink. In the intervening centuries, much more has been learned about fluid mechanics, and today many inventions that utilize this information materially impinge upon our lives. The study of fluid mechanics can help us to appreciate both the complexity of the natural world around us, of which we are an integral part, and the ingenious uses of this understanding that has helped make possible our material world.

Your study begins with the properties of motionless fluids, called hydrostatics. It determines the designs of ships, dams and reservoirs. It explains why ponds form ice on their surface in the winter and why the atmospheric density declines with altitude. You then proceed to inviscid flow — how fluids with negligible viscosity behave. Surprisingly, this explains much of fluid mechanics phenomena, from flow over an airplane wing to cardiac and arterial flow in humans. But there are many other flows, ones that are strongly influenced by viscosity, and these are discussed in the chapters on laminar and turbulent viscous flows. You will find that viscous flows are more difficult to treat than inviscid flows — they are mathematically more complex to describe and embody more varied physical effects. From an engineering perspective, viscous effects can be very important since their existence accounts for a loss in efficiency of a device or system. This is the stuff of precise practical design and analysis. The text goes on to consider some aspects of multidimensional inviscid fluid flow that will give you an opportunity to see how a flow field is analyzed in detail and introduce you to the concepts employed in computational fluid dynamics. Finally, you will be introduced to the mysteries of compressible flow, a wide world of unusual phenomena that is very much tied to engineering applications.

This text employs vector algebra and calculus as a shorthand in describing the three-dimensional flow of a fluid. The theorems of vector calculus are utilized in deriving the differential and integral expressions for the conservation of mass, momentum and energy. These derivations will help you to understand the physical meaning of the terms utilized in the conservation laws and are included for your information (and contemplation!). The more important lesson is to think vectorially, *i.e.*, to use vectors as comfortably as you use scalar quantities. In fact, most example and problems use only scalar components of vector equations so that lack of adeptness with vector algebra and calculus should not be an obstacle to working problems. A section of chapter 1 reviews those aspects of vector algebra and calculus that are employed in the text.

The problems and examples are not mere exercises in plugging numbers into formulae. All of them illustrate an important or interesting application of a fluid mechanical principle, very often an example of a practical engineering design. They are readily solved by finding an analytical answer in terms of the variables and parameters of the problem and then evaluating this answer numerically. This procedure mimics the process by which an engineering design is perfected. Almost every engineering task results in the production of quantitative information, and you should take seriously the obligation to develop quantitatively accurate results in your work. Besides, numerical evaluation often leads to the discovery of mistakes and errors — at least it does for me!

When I first engaged in engineering design and development after obtaining my M.S. degree at MIT, I had to learn on my own all about unsteady compressible flow. It was not then a subject covered in graduate courses. My undergraduate education, in Naval Architecture, was more than ample in hydrostatics but quite inadequate in the dynamics of fluid flow. It took many years of research and teaching before I felt that I had a sound grasp of the fundamentals of fluid mechanics. This shouldn't happen to you. There is no excuse today for an engineering student not to be fully informed about the facts of life — fluid mechanicswise. This text contains all you need to know — but were afraid to ask — about fluid mechanical principles. Once you have mastered it, the more advanced aspects will never trouble you.

Introduction to Fluid Mechanics

1 Introduction

1.1 The World of Fluid Mechanics

We are all experts in fluid mechanics. We sustain our lives by breathing air and drinking water. Those of us who know how to swim have learned to propel ourselves along the free surface of a liquid medium. We observe the changes in the atmosphere accompanying weather patterns, feeling or seeing the changes in wind speed and recognizing the signs of impending precipitation. We see streams and rivers running downhill toward the sea and notice the ocean waves generated by the wind. We are so used to the behavior of the atmosphere, and of lakes, rivers and the ocean when they are nearby, that we do not consciously think of these phenomena as examples of fluid mechanics.

Without fluid mechanics, our bodies could not function. The heart pumps blood throughout the body, supplying all living cells with oxygen and nutrients, carrying away waste products and maintaining a uniform temperature within. The lungs provide for the inhalation of air containing oxygen, which is absorbed in the blood stream, and the exhalation of carbon dioxide that has been evolved from the blood stream. Liquid waste products are stripped from the blood stream in the kidneys and the liquid, urine, is excreted. We are barely conscious of all these processes, but we are certainly aware of the fact that these are vital processes whose failure brings sudden death.

Our principal means of communication to other humans is by speech. Air pumped by our lungs is forced through the larynx, causing vibrations of the vocal chords that modulate the flow rate and produce voice tones. The control of this system, which produces words, both spoken and sung, is indeed complex.

Of course it is the engineered environment that seems the most obvious user of fluid mechanical principles. Our houses and schools are supplied with potable water pumped from the ground or collected in lakes, rivers or reservoirs at some distance and then piped to our doorstep. Liquid and gaseous fuels are delivered for use in furnaces providing hot fluids that convey the heat to living and working spaces. The family refrigerator circulates a fluid that removes heat from the appliance's interior and dumps it into the kitchen; the air conditioner moves interior heat to the outdoors in summer months. For most of us, liquid household wastes are carried away for processing into supposedly harmless form.

Our society consumes remarkable amounts of electric energy, most of which is produced in machines utilizing fluid mechanics to transform the chemical energy of fuels, the gravitational energy of stored water or even the kinetic energy of the wind into electrical power. Because most of this power is produced in giant plants remote from our daily activities, we are not so aware of the amount of power being processed unless we see the smokestack and its plume against the skyline.

Probably the most remarkable of all engineered systems that depend upon the utilization of fluid mechanics are airplanes and ships. These vehicles are designed to sustain themselves in the fluid medium through which they move, consuming the least possible power for the speed that they attain and doing so without experiencing damage or destruction even under adverse weather conditions. Their structures are so shaped that they provide a fluid force to suspend the vehicles in a stable manner against the pull of gravity and against disturbances in the medium through which they travel.

The automobile is the prime example of a complex mechanical, fluid mechanical and electrical system that integrates the various functions of its components into a remarkably sophisticated consumer product. Air and liquid fuel flowing into the cylinders are precisely metered to control the power of the engine, ensure the efficiency of combustion and reduce to a minimum the creation of air pollutants in the exhaust. Hydraulic systems transmit power to operate the brakes and, in some cases, the steering system. Fans circulate fresh air throughout the interior, the air flow having been cooled by an air conditioner or heated by hot fluid from the engine cooling system. A lubricating oil pump circulates oil for both lubrication and cooling of hot engine parts. A fuel pump extracts fuel from the fuel tank and delivers it under pressure to the carburetor or to the fuel injection system. A fan draws cooling air past the radiator, thereby cooling the engine coolant fluid. Even the windshield cleaning fluid is pumped upon demand, spraying the windshield when needed. Air pressure inside the tires maintains a flexible, shock absorbing surface. Fluid filled shock absorbers dissipate the energy imparted by rough roads. A radio's loudspeaker vibrates the air in the interior. Audible safety signals urge one to buckle the seat belt or to remove the ignition key.

It is one thing to observe or experience these fluid flows, but it is another to be able to describe them in quantitative terms, to predict how they would change if circum-

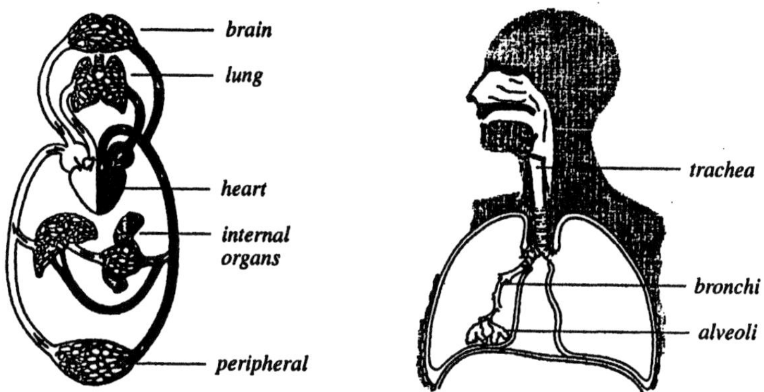

Figure 1.1 *Left*, the human cardiovascular and, *right*, the human pulmonary systems. (Reprinted by permission, from J. F. Green, 1987, *Fundamental Cardiovascular and Pulmonary Physiology*, 2nd ed [Philadelphia: Lea and Febiger.])

stances were different and especially to design new systems utilizing fluid mechanical principles to accomplish results not now attainable. The study of the principles of fluid mechanics is the first step in proceeding beyond the point of awareness and interest toward the development of predictive and analytical skills.

1.1.1 Fluid Mechanics of Human Physiology

Let us look first at the importance of fluid mechanics to the human body. The two principal fluid mechanical systems of the human body are the *cardiovascular* and *pulmonary* systems, as sketched in figure 1.1. The first is a *closed* system in which the fluid (blood) circulates repeatedly within closed loops. The pulmonary system, on the other hand, is an *open* system in which the fluid (air) flows in and out of the lungs in a cyclical manner. Mechanical assistance to either system may be provided during acute medical care.

The cardiovascular flow is generated by the heart, a pump of two stages arranged in series. Oxygenated blood is pumped by the left ventricle and distributed to all parts of the body through a branching system of arteries and capillaries. It is then returned through veins to the heart, having reduced oxygen content and enhanced carbon dioxide content. The right ventricle then pumps the returned blood through the lungs, where oxygen is absorbed and carbon dioxide evolved, and back to the left side of the heart for recirculation. The left heart must supply more pumping work than the right because the blood travels much greater distances in circulating throughout the body than through the lungs.

Under normal conditions, the heart beats seventy-five times per minute, delivering about 5 liters per minute, but its flow can be five times greater under extreme conditions. The peak blood pressure in the heart is about one-sixth of an atmosphere, and the heart develops about two watts of mechanical power. Since the volume of blood in the cardiovascular system is about 6 liters, the average time for the blood to circulate through the system is about a minute. Through a complex control system, oxygenated blood is distributed in different amounts to different parts of the body, with preferential dispatch to the brain and heart. During open heart surgery, a heart-lung machine bypasses the heart and lungs, substituting temporarily for these organs. Despite the seeming simplicity of the heart as a pump, permanent artificial heart machines have not yet proved successful.

Most of the exchanges of oxygen, nutrients and waste products between the cells of the body and the the blood takes place in the capillaries, tubes having a diameter of 5–10 micrometers. The blood flows quite slowly in the capillaries, but there are millions of these tubes connected in parallel so as to reach all the body cells.

Humans respire at a normal rate of 12 breaths per minute, inhaling a volume of about 0.5 liters per breath, or 360 liters per hour. The normal uptake of oxygen in the lungs is about 15 liters per hour of oxygen (about 0.3 kg per day), only 20% of the oxygen available in the air in the lungs. But this uptake rate decreases when the oxygen partial pressure in the lung is reduced, which happens at high altitudes, and when carbon monoxide present in the lung air decreases the oxygen carrying capacity of the blood.

Like the cardiovascular system, the lung volume branches into many small passages, called alveoli. The total lung surface area exposed to the air is about 100 square meters. The lungs are particularly susceptible to damage caused by even small amounts of toxic air pollutants that are easily deposited on the surfaces of the lung.

The mechanical power needed to ensure respiration is provided by the muscles surrounding the lungs. They can provide greater pressure than is ordinarily needed, such as when we cough or blow up a balloon. When inhaling, lung pressure falls below atmospheric. The lung passages must be elastic enough to spring open during inhalation, just like a squeezed-out sponge absorbing water. Emphysema is a degenerative disease that destroys this elasticity and thus the ability to inhale.

These human fluid mechanical systems are very complex compared to most engineering systems, yet they are very resilient. The human heart beats several billion times during a lifetime, and the lungs cycle about a sixth as many times. In recent years, our increased understanding of the fluid mechanical aspects of these flows has aided medical research.

1.1.2 The Atmosphere and the Hydrosphere

Fluid flow external to our bodies is also important. There, air and water are the two fluids we know most about. They are both essential to the maintenance of our lives, providing a hospitable environment for all living things. We have a direct, lifelong experience in observing how they behave and how we utilize them. In the natural environment, we can see, feel or hear examples of almost all the kinds of fluid flows that we will study in this book.

The atmosphere is a layer of gas held to the surface of the earth by gravitational attraction. Most of the mass of the atmosphere is confined to the first 15 kilometers above sea level, yet the small amount above this level is responsible for filtering out the deadly high energy radiation from the sun which would otherwise destroy life. The interaction of the atmosphere with sunlight helps to maintain the earth's surface temperature above that of an airless planet, such as the moon. (This increase in temperature, called the greenhouse effect, threatens to grow in the future because of anthropogenic emissions of heat-absorbing gases.)

The motion of the atmosphere, which we observe around us, is driven by the diurnal pattern of heating by the sun and cooling by radiation to outer space. Over the year, these heating/cooling patterns shift to different latitudes, giving rise to annual climate variations. An intimate part of this process is the evaporation from the earth's surface of water, which forms clouds at high elevations before precipitating back to the surface. This distillation of ocean water, moved to land by winds, provides the fresh water that maintains terrestrial life.

Local weather provides a variety of wind motions. Sometimes the wind speed is quite small, especially at night when radiative cooling stabilizes the atmosphere, but storms driven by precipitation of water vapor, such as thunderstorms and hurricanes, can have very high wind speeds. Cold air is more dense than warm air, so that a cold air mass tends to flow under a warm air mass, forming a cold front. Large-scale weather patterns drift past our locality, bringing changes that are not greatly affected by local conditions. As we can see from watching the daily television weathercast, the main features of the weather pattern extend over many thousands of kilometers, a distance that is hundreds of times the atmospheric height. Yet the changes of pressure, temperature and humidity are much greater in the vertical direction than in the horizontal direction, despite the much greater horizontal size of a weather pattern. The pull of gravity is so strong over large distances that it forces the atmosphere to flow mostly in the horizontal direction.

Because we are so small compared to the vertical and horizontal dimensions of the atmosphere, we can observe how the wind blows in a tiny portion of the atmosphere that is nearby us. It is noticeable that the wind speed and direction are somewhat variable, even over time intervals of less than a minute. These changes are much more rapid than the changes accompanying a weather pattern, which may take a day to pass us by. Fluid flows that exhibit variability over time and length scales which are small compared to that of the overall flow are called turbulent flows. The atmospheric wind is a turbulent flow.[1]

The atmospheric motion is responsible for diluting air pollutants, such as those emitted by power plants and automobiles. When these pollutant streams are marked by smoke, we can readily observe how the smoke intensity decreases as the wind turbulence mixes the pollutant stream with clean air, diluting the strength of the pollutant within the plume (or, if you prefer, dirtying more and more of the atmosphere). Most of these pollutants mix no higher in the atmosphere than a few kilometers and are eventually carried far downwind and deposited back to the earth's surface. Some, however, do not soon return to the earth and instead mix gradually throughout the entire atmosphere, including the stratosphere. Some of these gases lead to the destruction of stratospheric ozone and to increased average surface temperature. The mixing properties of the atmosphere are extremely important in determining the degree of atmospheric pollution in urban areas.

The water in the ocean, lakes and rivers, as well as that in underground aquifers, is called the hydrosphere. The volume of fresh water on the continents is small compared to the oceanic volume, but it is the part of the hydrosphere that is most important to the maintenance of terrestrial life. The management and use of this water for agricultural and other purposes forms an important branch of engineering. However, the ocean is just as important. It provides the source of precipitation over the land, and in its surface layer, an environment for the growth of microscopic plants and animals that form the base of the oceanic food chain. The ocean tends to make the climate more uniform in the latitudinal direction by moving warm tropical waters toward the poles and displacing cold polar water toward the equator.

We are all familiar with the downhill motion of streams and rivers flowing toward the sea. The energy in this flow can be tapped by building dams that force the river flow through turbines to generate electric power. Sometimes this energy is dramatically dissipated as the river plunges over a precipice to form a waterfall, where the violently turbulent motion at the base of the falls converts the river's directed energy into heat. When the river reaches the sea, its fresh water, being lighter than the sea water, floats on top of the sea. However, beyond its mouth, the river gradually mixes with the sea.

Most of the fresh water on the continents is out of sight, below ground. It exists in the pores between mineral deposits and is fed by precipitation that percolates through the ground under the influence of gravity. The fluid velocity in the underground

[1] We treat turbulent flows in some detail in chapter 7.

aquifers is much lower than it is in rivers, the water being impeded by the frictional force of the porous medium through which it flows. Underground water is often the source of potable water, but locating underground water and pumping it from the ground for human use is limited by the characteristics of the underground aquifer.

At the edge of the ocean, we see the ocean waves crashing on the shore. The waves carry to the shore energy generated by the wind blowing over the ocean surface. Of course, the ocean surface doesn't move (on the average) in the direction of the waves, but it oscillates as the wave passes by. Ocean waves are called gravity waves because the pull of gravity at the air-sea interface is responsible for the propagation of these waves, which do not penetrate far below the ocean surface.

The other oceanic motion that we notice at the sea shore is the tidal rise and fall of the sea surface. This twice-a-day cycle is caused by the difference in gravitational pull of the moon (and, to a lesser extent, the sun) on opposite sides of the earth. The differential gravity force gives rise to a bulging of the ocean surface in the direction of the moon, which passes a given location twice in the lunar day of 25 hours. The tidal motion, consisting of both a vertical and horizontal oscillation, may be amplified greatly along the continental coastline, sometimes by a factor of ten above the general oceanic amplitudes.

Oceans may contain localized currents, such as the Gulf stream, that are mighty rivers flowing across a nearly stationary ocean. Earthquakes can generate tidal waves that travel many thousands of kilometers before crashing ashore, sometimes wreaking devastation on lowlying coastlines. Even hurricanes can generate tidal waves that cause coastal flooding. Such tidal waves are not caused by the moon's gravity, as is ordinary tidal motion, but by other disturbances.

The rise and fall of the tide can be utilized to produce mechanical power, but at the present time, using this power is seldom economical as compared to using river power. Mechanisms devised to extract power from ocean waves have also proved to be uneconomical. Still, the forces exerted on ships and wave barriers by ocean waves can be very substantial, and protection from these forces can be very expensive. Thus knowledge of the dynamics of the ocean is important to many of mankind's pursuits.

1.1.3 Engineering Fluid Mechanics

There is a vast array of engineered systems that utilize fluid mechanics as an essential component. Many of these can be found in the home or school (heating and air conditioning, water supply and liquid waste removal, personal computer cooling system, refrigerator, fan, vacuum cleaner and hair dryer) or in the outside environment (road and rail vehicles, airplanes, ships, water reservoirs and aqueducts, municipal waste treatment plants, incinerators and electric generating plants). On the other hand, many of them are components of industrial systems (hydraulic power systems for construction and manufacturing equipment, catalytic crackers in petroleum refineries) that we seldom see but which are vital ingredients of an industrial society. Mechanical, civil, chemical, aeronautical and naval engineers need to be sufficiently versed in fluid mechanics when designing any of these systems.

Probably nothing epitomizes the interrelationships of various fluid mechanical processes as much as the familiar automobile engine, as shown in figure 1.2. The major fluid flow in this engine is that of air into the intake system, where it is mixed with fuel vapor and sucked into the engine cylinder during the intake stroke. After the air/fuel mixture is compressed, ignited and expanded during the power stroke, it flows into the exhaust duct, passing first through a catalytic converter that reduces air pollutants and then through a muffler to damp exhaust noise generated by the high speed flow out of the engine cylinder. The amount of air drawn through the engine in each revolution of the crankshaft is regulated by the throttle valve in the intake system, which reduces the air pressure (and hence air density) when less power is required. (A wide open throttle allows the maximum air and fuel flow into the cylinder and hence the maximum work that can be generated per engine revolution.)

In older vehicles, the carburetor is the device attached to the air intake system which meters and mixes the liquid fuel with the incoming air, adjusting the fuel flow so as to provide the correct proportion of fuel to air that ensures the subsequent ignition of the mixture in the cylinder by a spark source. In newer vehicles with fuel injection, the fuel is sprayed into the air at each intake port rather than being mixed into the common stream in the carburetor. Combustion gas that leaks past the pistons is collected in the crankcase space and piped back into the intake system to prevent its emission to the atmosphere as an air pollutant.

However, there are other fluid flow systems in the automobile engine. A lubricating oil pump draws oil from the crankcase and delivers it to bearings, valves and other moving parts. Liquid fuel stored in the gas tank is transferred by the fuel pump to the carburetor or the fuel injection pump. A coolant circulating pump delivers cooling fluid to the engine and thence to the radiator where it is cooled by air flow and returns to the engine. The cooling air flow through the radiator is maintained by a fan if forward motion of the vehicle is insufficient to do so. The warmed coolant fluid leaving the engine may be diverted to the vehicle air heating system. If the vehicle is equipped with air conditioning, a compressor, driven by the engine, circulates the refrigerant gas through the condenser segment of the radiator, where it is cooled, and then through the evaporator, where heat is absorbed from the interior air of the vehicle.

In the compression ignition (diesel) engine, fuel is introduced directly into the cylinder after the air charge has been compressed. There is no throttle valve in the

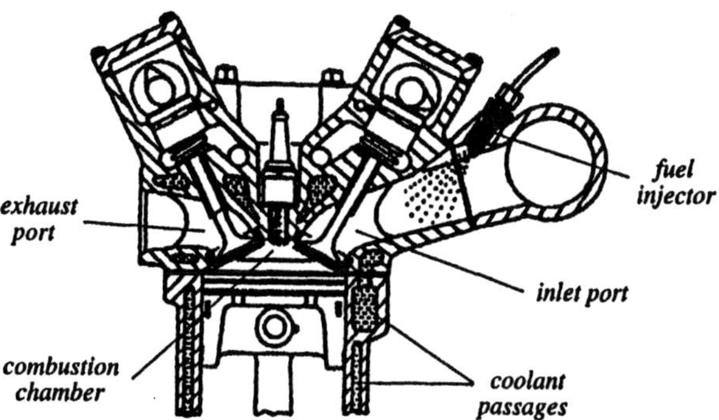

Figure 1.2 Cross-section of an automobile engine showing some of the passages for fluid flow. (Reprinted by permission, from K. Newton, W. Steeds, and T. K. Garrett, 1989, *The Motor Vehicle*, 11 ed [London: Butterworth-Heinemann Ltd.])

intake system and no reduction in pressure of the intake air at low power. In other respects, the fluid flows are similar to that in the spark ignition (gasoline) engine.

The fluid flow in a gas turbine engine is entirely different. Intake air is compressed by rapidly moving compressor blades followed by stationery blades that convert some of the kinetic energy to a rise in air pressure. Fuel is burned in the combustion chamber after compression has been completed, and the hot combustion products pass through the rapidly moving turbine blades, generating mechanical power. On average, the flow is steady and not pulsating as in the reciprocating automotive engine. In the aircraft jet engine, the increase in flow speed of the hot exhaust gas beyond that of the intake air produces a thrust that propels the aircraft.

The high speed flow of steam through a steam turbine in an electric power plant is very similar to that through a gas turbine. But a steam plant is a closed system, with the steam being condensed into water downstream of the turbine and then being pumped to high pressure before passing through the boiler, where it is converted to hot, high pressure steam to feed the turbine.

The flows described above are called *internal flows* because they are contained within mechanical structures. They are predominantly flows involving the transfer of heat or fluids or the generation or utilization of mechanical power. But there are other flows where the fluid is external to the mechanical structures, such as the flow of air around an airplane or that of water around a ship, in which the structure interacts with its fluid environment. These are called *external flows.*

Perhaps the most familiar of the external flows is that surrounding an aircraft. The shaping of the wing structure to produce sufficient lift to balance the pull of gravity while reducing the aerodynamic drag to a low enough value to be matched by the thrust from a propeller or jet engine requires a deep understanding of fluid flow in three dimensions. The design of an aircraft propeller or wind turbine requires equally careful consideration of fluid flow in order to achieve the required efficiencies. For ocean vessels, the shape of the vessel determines both its stability against overturning and the amount of power required to move it, especially the power that is dissipated in forming the ship waves that carry away the energy supplied by the propulsion system. The flow of air over an automobile is now recognized as an important determinant of its energy efficiency, and current vehicle shapes reflect this significance.

It is evident that the design of many engineered systems requires an understanding of the properties of the fluid flows that they utilize. There are many such applications of the principles of fluid flows, and they are very diverse. To participate in the design of such systems requires a sound understanding of the principles of fluid mechanics and an ability to apply these principles to the particular problem at hand.

1.2 The Physics of Fluids

Fluid mechanics is a macroscopic science, *i.e.*, it is concerned with characteristics that can be observed and measured on the laboratory scale. Even though we cannot resolve the motions of individual molecules by measurement in our laboratory, we recognize that it is the aggregation of properties of individual molecules that determines the behavior of a macroscopic element of fluid. Our measuring instruments average the behavior of a very large number of molecules in the vicinity of a point in a fluid, thereby determining the macroscopic fluid properties at that point.

For example, suppose we could measure the velocities of all the molecules within a cube of air that is 1 micrometer (μm) on a side, or 10^{-18} m^3 in volume. Since there are about 2×10^{25} molecules of air in a cubic meter, this tiny volume would still contain more than 10^7 molecules, a very large number. These molecules would be moving in all directions, and their average speed would be about 300 meters per second (m/s). However, if the air is stationary, the average value of the vector sum of the molecular velocities, $\mathbf{V}$, is zero. On the other hand, if the air is moving slowly, this average vector velocity $\mathbf{V}$ will equal that of the bulk motion that we observe. As long as there are a large enough number of molecules in the volume element of interest during the time period of measurement, this averaging process defines macroscopic properties of the fluid that can be ascribed to a point located at the center of the element. These properties are thereby defined as continuous functions of space and time throughout the fluid. It is these average properties that we observe in the laboratory when we measure the velocity of the fluid, or its pressure or density, with suitable instruments.

We usually categorize common fluids as either *gas* or *liquid*. A liter of gas, such as air, has much less mass than a liter of liquid, such as water. The molecules of a gas move about quickly, occasionally colliding with each other but are mostly unrestrained by intermolecular forces. In a liquid, on the other hand, the molecules are always interacting with each other, jiggling about but not moving very far before being repulsed by a neighbor. When a gas is compressed into a smaller volume, molecular collisions occur more frequently and a molecule spends a smaller fraction of its time in free flight. When compressed enough so that its molecules are as close to each other as those in a liquid, a gas is indistinguishable from a liquid. While the distinction between a liquid and its gaseous vapor is important thermodynamically, both states exhibit fluid behavior on a macroscopic scale.

The dominant property of fluids is their propensity to flow spontaneously within their containers. They have to be stored in containers to prevent their moving off to other locations, in contrast to solids which need not be so constrained. The volume of a liquid is conserved when it is moved from one storage vessel to another, but a gas always expands to fill its container since its molecules are free to move about. Fluids have no ability to hold a shape independent of their surroundings. This property of fluids is a direct consequence of the inability of the intermolecular forces to maintain an unchanging angular orientation of the molecules with respect to each other. Fluid molecules that are close to each other at one instant can move far apart with relative ease.

Many fluids are mixtures of several chemical species, such as air which is composed of nitrogen, oxygen and many trace species. Liquids can be solutions of solute species dissolved in a solvent, such as sea water. Mixtures like these still retain the characteristics of fluids, but have the added feature that the proportions of the constituents can be separately measured and may vary from point to point within the fluid.

1.2.1 Physical Properties of a Fluid

The macroscopic properties of a fluid reflect its underlying molecular structure. Some of these properties are very significant in affecting how the fluid reacts to applied forces or motions of the fluid surface and are of prime interest to the understanding of fluid mechanics. Other properties, such as electrical or magnetic properties, may affect a flow only under very unusual circumstances but may be quite consequential to other physical phenomena, such as electromagnetic wave propagation in the fluid. We shall focus here on those properties that are most closely coupled to the motion of a fluid.

Density

The density of a fluid is the ratio of the mass of fluid in a fluid element to its volume. Since we picture a fluid element as a tiny volume, as small as we can possibly measure, density can be considered to be a continuous function of position within the fluid field. We customarily assign the symbol ρ to represent the property density and usually measure it in units of kilograms per cubic meter (kg/m^3).

The density of a fluid affects its flow in two different ways. First of all, it determines the inertia of a unit volume of fluid and hence its acceleration when subject to a given force. Low density fluids, such as gases, accelerate more readily than high density fluids, such as liquids, when subject to the same force per unit volume. Thus a low density fluid like air requires less force per unit volume to accelerate it than does a high density fluid like water. It is for this reason that we find it more difficult to wade through water than to walk through air. Similarly, the gravity force per unit volume is determined by the fluid density. It requires more work to lift a given volume of water than an equal volume of gas.

The density of a liquid is a function of its temperature and pressure. At a fixed pressure, as the temperature of a liquid is increased, its density decreases because a fixed mass of fluid expands with increasing temperature. At a fixed temperature, as the pressure acting on a liquid increases, it is compressed and its density rises. The same variations in density occur for a gas when subject to changes in pressure or temperature, but these changes are relatively much larger for a gas than for a liquid.

Small changes in density $\delta\rho$ that accompany small changes in pressure δp or temperature δT may be expressed in terms of the partial derivatives of ρ with respect to pressure and temperature:[2]

$$\delta\rho = \left(\frac{\partial\rho}{\partial p}\right)_T \delta p + \left(\frac{\partial\rho}{\partial T}\right)_p \delta T$$

Dividing both sides of this expression by ρ, the fractional change in density, $\delta\rho/\rho$, is found to be:

$$\frac{\delta\rho}{\rho} = \left(\frac{\partial \ln\rho}{\partial p}\right)_T \delta p + \left(\frac{\partial \ln\rho}{\partial T}\right)_p \delta T \qquad (1.1)$$

The reciprocal of the coefficient of the first term is called the *bulk modulus E* while the negative of the coefficient of the second term is called the *coefficient of thermal*

[2]The subscript of the partial derivative in parenthesis denotes the variable that is held constant in the differentiation.

expansion β:

$$E \equiv \left(\frac{\partial p}{\partial \ln \rho}\right)_T \tag{1.2}$$

$$\beta \equiv -\left(\frac{\partial \ln \rho}{\partial T}\right)_p \tag{1.3}$$

The values of E for water and air at room temperature and atmospheric pressure are $2.1 \times 10^9\ N/m^2$ and $1.0 \times 10^5\ N/m^2$, respectively. For a given small increase in pressure, the fractional change in air density $\delta\rho/\rho$ is thus 2.1×10^4 times greater for air than for water. This difference is caused by the much smaller spacing between molecules in a liquid than in a gas at atmospheric pressure, making it more difficult to squeeze the liquid into a smaller volume because of the stronger repulsive forces between the liquid molecules.

The ratio of the thermal expansion coefficients of water and air, which are $1.53 \times 10^{-4} K^{-1}$ and $3.5 \times 10^{-3} K^{-1}$, respectively, is smaller than that of the bulk moduli, but air expands more readily than water. Heating of fluids in the earth's gravitational field causes a rising motion of the heated fluid, both for water and air, but the effect is stronger for air than for water because air has a greater value of the thermal expansion coefficient.

The thermal expansion of water is anomalous in the temperature range between 0 °C and 4 °C where water contracts as it is heated. The density of water is a maximum at 4 °C, and heating above or cooling below this temperature causes the water to expand. When the surface of a lake or pond is cooled during the fall of the year, the cooled surface water descends to the bottom as long as its temperature exceeds 4 °C, but when cooled below that value, it floats at the top because it is less dense than the warmer water below. Thus ice forms at the top of a pond in winter, while the bottom fluid stays warmer at 4 °C.

It is sometimes convenient to express the density ρ of a fluid as a ratio to the density ρ_{ref} of water at 4 °C. This ratio is called the *specific gravity SG*:

$$SG \equiv \frac{\rho}{\rho_{ref}} \tag{1.4}$$

$$\rho = (SG)\rho_{ref}$$

It is possible to measure directly the specific gravity of a liquid with a *hydrometer*. The value of the specific gravity may reveal practical information on the state of the fluid. For example, the specific gravity of the sulfuric acid solution in a lead-acid storage battery of an automobile indicates the degree of electrical charge—a higher *SG* denotes a more concentrated solution and thereby a greater battery charge. The specific gravity of milk is used to measure its butter fat content (more fat decreases the SG) and that of maple syrup to measure its sugar content.

Example 1.1

The average temperature of the earth's surface is expected to rise in the next century because of an enhanced greenhouse effect. When the ocean is heated, it will expand, raising the sea level. If the ocean's average depth is 3800 m and its average thermal expansion coefficient is $1.6 \times 10^{-4} K^{-1}$, calculate the rise in sea level when the ocean increases in temperature by 1 K.

Solution

A column of sea water will expand vertically when heated since lateral expansion is restricted by the continents. The rise in sea level δh will be the product of the thermal expansion coefficient times the temperature increase times the depth of the ocean:

$$\begin{aligned} \delta h &= (1.6 \times 10^{-4} K^{-1}) \times (1\,K) \times (3800\,m) \\ &= 0.608\,m \end{aligned}$$

Viscosity

Although it is readily apparent that liquids move under the force of gravity so as to conform to the shape of their containers, some liquids move more slowly than others. When we pour honey from its jar into a dish, it moves less rapidly than does water when subject to the same experimental conditions. A heavy lubricating oil is intermediate between water and honey in its response to being poured. The property of a fluid that measures its resistance to change of shape is called the *viscosity* (denoted by the symbol μ). Thus honey has a greater viscosity than does water.

We need a more quantitative method of determining the relative viscosity of fluids than the simple pouring experiment just described. One way is to place a thin layer of the fluid between two flat parallel plates separated by a constant distance h (see figure 1.3). If we move one plate at a speed V relative to the other (the direction of motion being parallel to the plate surfaces) and measure the force required to accomplish this motion, expressing it as the force per unit area of plate surface τ, then we find that τ is proportional to V and inversely proportional to h for any given fluid, but that the proportionality constant is different for each fluid tested. We call the proportionality constant in this experiment the ***absolute viscosity coefficient***, or

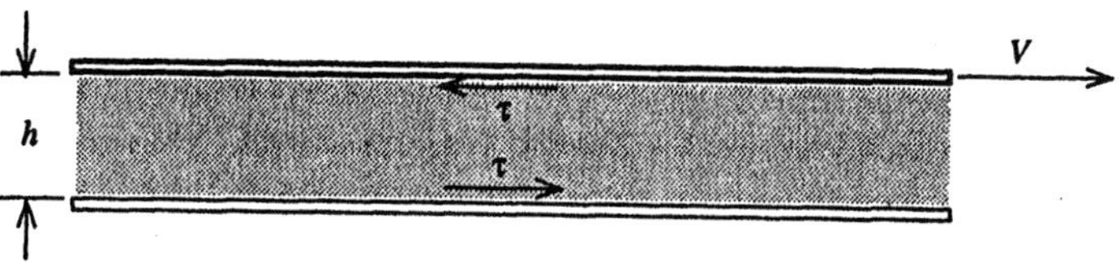

Figure 1.3 A fluid filling the space h between two parallel plates is subject to a simple shearing motion. The upper plate moves with a speed V while the lower is stationary. Both plates are subject to an external force per unit area τ parallel to the plate but in opposite directions.

simply *viscosity* for short:

$$\tau = \mu \left(\frac{V}{h} \right)$$

In this experiment, the force exerted on one plate causing it to move is balanced by an equal but opposite force on the other plate. Both forces are transmitted to the fluid, subjecting the fluid layer to a surface force per unit area τ, called a shear stress, which is the same in magnitude but opposite in direction on the two surfaces of the fluid in contact with the plates. But any thin layer of fluid, of thickness δn, between the two plates must also experience the same stress τ and a corresponding velocity difference δV of such a magnitude that:

$$\tau = \mu \left(\frac{\delta V}{\delta n} \right)$$

In the limit of infinitesimally thin layers, $\delta V / \delta n = \partial V / \partial n$, and we have:

$$\tau = \mu \left(\frac{\partial V}{\partial n} \right) \tag{1.5}$$

Since τ is the same at all points within the fluid in this experiment, it follows that the velocity difference V varies linearly with the distance n normal to the plate surfaces. The flow illustrated in figure 1.3 is called *simple shear.*

Fluids exhibiting the simple behavior of equation 1.5 are called *Newtonian fluids* because Newton first proposed this linear relationship between shear stress and velocity derivative.[3] Most fluids are Newtonian, but some have a more complex relationship between the shear stress and the velocity derivative. In these fluids, called *non-Newtonian*, the different behavior usually arises because the fluid molecules are very large, like polymers or proteins, or they contain non-fluid particles, like blood. It is more difficult to describe the shear stress in such fluids.

[3]Isaac Newton (1642–1727) in his *Principia* also considered sound wave propagation and the kinetic theory of gases. Of course, his laws of motion are the foundation of fluid mechanics.

The viscous stress τ in a fluid undergoing shear flow, as in figure 1.3, is a consequence of the average relative motion of the fluid molecules. In a stationary fluid, individual molecules constantly exchange energy with neighboring molecules, sometimes gaining and at other times losing energy, but over a period of time averaging out to no gain or loss. But, in a shear flow, the molecules tend to gain energy by colliding with other molecules that are, on average, moving toward them with an average speed proportional to the fluid strain rate $\partial V/\partial n$. This energy must be supplied by a force, the shear stress τ, acting on a deforming fluid element, doing work at the rate $\tau(\partial V/\partial n) = \mu(\partial V/\partial n)^2$ per unit volume of fluid. As the energy of the fluid molecules is increased by the work of viscous deformation, the temperature of the fluid will rise because the increased molecular energy is randomly distributed among the molecules, as it is when the fluid is heated. This process is called *viscous dissipation.*

A lubricant is a fluid placed between two moving solid surfaces whose purpose is to prevent the surfaces from coming into direct contact and thereby limiting the shear stress they experience to a low value, $\mu V/h$, rather than the much higher value obtained from solid, sliding contact. The more viscous the lubricant, the higher is this shear stress but the greater is its ability to keep the two surfaces from coming into contact.

The viscosity of a fluid varies markedly with temperature but very little with pressure. Liquid viscosities tend to decrease with increasing temperature. (In cold weather, the greater viscosity of engine oil increases the power required to start an automobile engine.) However, gas viscosities increase with temperature.

Later on we will use a quantity called the *kinematic viscosity*, denoted by the symbol ν. It is defined as the ratio of the viscosity μ to the density ρ:

$$\nu \equiv \frac{\mu}{\rho} \tag{1.6}$$

The kinematic viscosity affects how quickly a shear force exerted on the surface of a fluid penetrates into its interior. For this reason, it is sometimes called the *viscous diffusivity.*

Table 1.1 lists the properties of some common liquids and gases that are important to fluid mechanics: the density ρ, viscosity μ and kinematic viscosity $\nu \equiv \mu/\rho$. Also listed is the rate of change of viscosity μ with increasing temperature, expressed as the derivative $d\ln\mu/dT$. Using this value, we may calculate the viscosity at a temperature $T(°\,C)$ from the viscosity at $20\,°\,C$, μ_{20}, by:

$$\mu = \mu_{20} \exp\left(\frac{d\ln\mu}{dT} \times [T(°\,C) - 20\,°C]\right) \tag{1.7}$$

When a fluid moves along a solid surface, the fluid molecules at the solid surface do not move with the fluid. We say that a fluid sticks to the surface and that the speed of the fluid at the fluid-solid interface is the same as that of the solid. This motion

Table 1.1 Fluid properties at 20 °*C* (68 °*F*) and atmospheric pressure

	ρ kg/m^3	μ $Pa\ s$	ν m^2/s	$d \ln \mu/dT$ K^{-1}
Liquids				
Water	9.982E(2)	1.00E(-3)	1.00E(-6)	-2.84E(-2)
Normal octane	7.02 E(2)	5.42E(-4)	7.72E(-7)	-1.26E(-2)
Ethyl alcohol	7.89 E(2)	1.20E(-3)	1.52E(-6)	-1.95E(-2)
Methyl alcohol	7.92 E(2)	5.84E(-4)	7.37E(-7)	-1.57E(-2)
Benzene	8.79 E(2)	6.52E(-4)	7.42E(-7)	-1.57E(-2)
Ethylene glycol	1.110E(3)	1.99E(-2)	1.79E(-5)	-6.03E(-2)
Glycerine	1.260E(3)	1.49	1.18E(-3)	-9.23E(-2)
Mercury	1.355E(4)	1.55E(-3)	1.14E(-7)	-3.71E(-3)
Perfect gases				
Air	1.204	1.82E(-5)	1.51E(-5)	2.56E(-3)
Hydrogen	8.382E(-2)	8.83E(-6)	1.05E(-4)	3.95E(-3)
Helium	1.664E(-1)	1.95E(-5)	1.17E(-4)	2.15E(-3)
Water vapor	7.498E(-1)	9.57E(-6)	1.28E(-5)	3.67E(-3)
Carbon monoxide	1.165	1.76E(-5)	1.51E(-5)	2.62E(-3)
Nitrogen	1.165	1.76E(-5)	1.51E(-5)	2.50E(-3)
Oxygen	1.330	2.03E(-5)	1.53E(-5)	2.56E(-3)
Argon	1.660	2.25E(-5)	1.36E(-5)	2.68E(-3)
Carbon dioxide	1.830	1.47E(-5)	8.03E(-6)	3.07E(-3)

is in contrast to a solid sliding over another solid, where the solid molecules at the interface move with the same speed as their respective solids, bumping along like a car on a rough road.

Surface Tension

Fluid properties, such as density and viscosity, apply to a fluid element in the interior of a fluid that is surrounded by contiguous fluid elements. These properties are sometimes termed *bulk properties*. But, when a fluid element is located at the boundary of a fluid, where it comes in contact with a dissimilar fluid or a solid, its properties can be different from those of an interior element.

Consider a water droplet surrounded by air, for example. In its interior, each water molecule is subject to attractive and repulsive forces from nearby molecules that,

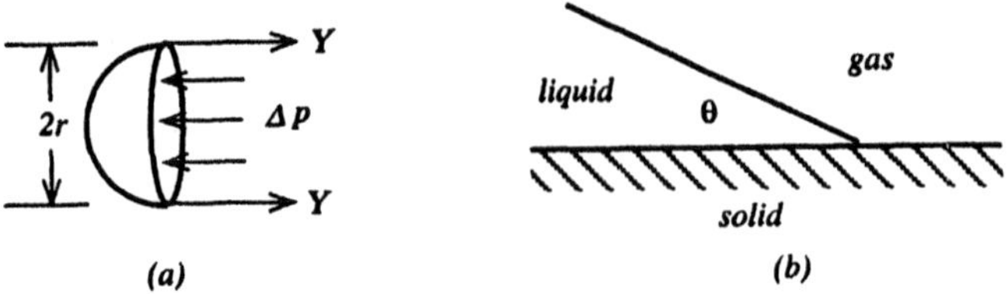

Figure 1.4 Illustrating (*a*) the effects of surface tension on the pressure Δp in a droplet and (*b*) the contact angle θ at a gas/liquid/solid interface.

on average, are distributed equally in all directions about the given molecule. For a surface molecule, however, only half of the surrounding space is occupied by other water molecules so that a new arrangement of the intermolecular forces occurs in order to hold the surface molecule in place. Surface molecules are subject to an attractive force from nearby surface molecules so that the surface is in a state of tension. (For a liquid under positive pressure—liquids in general cannot sustain a negative pressure—the interior molecules experience on average a repulsive force.) This tensile force per unit of length along the surface, called the *surface tension* Υ, is a property of the fluid and its adjacent fluid or solid. It is more properly called the *interfacial tension* because it appears at the interface between two fluids in contact with each other. The unit of surface tension is force per unit length; *e.g.*, newton per meter (N/m).

The surface tension of a fluid drop creates a higher pressure within the drop than exists in the surrounding fluid, much as a balloon compresses the air inside it. This excess pressure Δp inside the drop, acting upon the cross-sectional area πr^2 (where r is the drop radius) is counterbalanced by the surface tension Υ acting on the circumference $2\pi r$, as shown in figure 1.4 (*a*):

$$\Delta p(\pi r^2) = \Upsilon(2\pi r)$$

$$\Delta p = \frac{2\Upsilon}{r} \tag{1.8}$$

By measuring the pressure increment Δp of a drop of radius r, we can calculate the value of the interfacial tension Υ of the liquid in its surrounding fluid.

Note that when an interface is plane so that the radius of curvature r is infinite, the pressure difference Δp is zero, *i.e.*, the pressure is equal on both sides of the plane interface despite the interfacial tension.

The interfacial tension of a liquid is susceptible to change when impurities in the liquid congregate on its surface, altering its surface properties. Surfactants are chemicals deliberately introduced to change the interfacial tension—usually to reduce it. Detergents are surfactants that change the surface properties of oil, grease and dirty particles in contact with water, making it easier to clean them away in a washing process.

The work required to blow up a balloon is stored partly in the stretched balloon material. Similarly, since it also requires work to form a droplet at the end of a capillary tube by pumping liquid through the tube, we can say that this work is stored in the surface of the droplet as surface energy. If we calculate this surface energy using the pressure Δp from the force balance given above, we find that the surface energy per unit area (J/m^2) is equal to the surface tension (N/m), as it should be since the droplet formation process is reversible in the thermodynamic sense.

Both the drop excess pressure Δp and the surface energy per unit drop volume ($4\pi r^2 \Upsilon/[4\pi r^3/3] = 3\Upsilon/r$) become large for very small droplets. This greatly inhibits the formation of rain drops starting from individual water molecules in the atmosphere. Instead, rain drops are usually formed from solid nuclei of sufficient initial size that large pressure differences are not encountered.

Surface tension can be important for fluid flows involving small droplets or bubbles where the pressure excess is comparable to pressure changes in the flow field. Surface tension is important in the formation of liquid droplets in atomizers and fuel spray equipment.

Surface tensions of most liquids in air are less than $0.1\ N/m$. From equation 1.8, a drop of $1\ mm$ radius thereby has an excess pressure Δp less than $200\ N/m^2$, or 2×10^{-3} times atmospheric pressure.

When an interface between two fluids meets a solid surface, the interface forms an angle with respect to the solid surface called the *contact angle*, as shown in figure 1.4 (*b*). This contact angle depends upon the nature of the two fluids and the solid. The values of the angle and the interfacial tension then determine the effects of capillarity, such as the vertical rise height of a fluid in a capillary tube.

Cavitation and Boiling

If the pressure of a liquid is reduced sufficiently far, it will begin to boil, *i.e.*, some portion of it will change from the liquid to the vapor phase. The pressure at which this happens, called the *saturation pressure* and denoted by p_v, depends sensitively upon the fluid temperature and increases rapidly with increasing temperature.[4] In adiabatic liquid flows, such as through pumps, hydroturbines and around ship's propellers, this phase change is called *cavitation* and results from the reduction in pressure of the liquid as it flows along curved surfaces. In flows through heat exchangers, such as a boiler, the vaporization occurs when the saturation pressure increases above the fluid pressure because the fluid is being heated. In both cases, the phase change usually originates at solid surfaces where vaporization nuclei are present.

---

[4] For water, p_v increases from $6.108E(2)\ Pa$ at $0\,^\circ C$ to $1.0131E(5)\ Pa$ at $100\,^\circ C$.

Cavitating flows can damage solid surfaces when the vapor bubbles collapse as the fluid pressure rises above the fluid's saturation pressure. The collapsing bubbles can induce damaging, but localized, stresses in the surface of the solid they contact. In addition, cavitation can change considerably the flow through pump impellers and around ships' propellers, affecting their overall performance. Cavitating flow conditions are to be avoided if at all possible. In contrast, the boiling of liquids is usually a desired process, and the fluid flow is designed to enhance its occurrence.

1.2.2 Thermodynamic Properties

Thermodynamics is a science that tells us how matter behaves when we heat or cool it and when we compress or expand it. It enables us to predict how much work can be produced from matter by the exchange of heat with its surroundings. Most systems that produce mechanical power from a source of heat utilize a fluid such as water or air as the medium that converts the heat to work. The thermodynamic properties of these fluids determine how they behave in the power producing process. To describe properly the detailed motion of these fluids, we need to know their thermodynamic properties.

The laws of thermodynamics define several properties: internal energy $\hat{u}$, absolute temperature T, and entropy $\hat{s}$. These properties are related incrementally to each other by the second law of thermodynamics in the form:

$$T\,d\hat{s} = d\hat{u} + p\,d\left(\frac{1}{\rho}\right) \tag{1.9}$$

From these properties, we may define additional useful ones, among which are the enthalpy $\hat{h} \equiv \hat{u} + p/\rho$, the constant pressure specific heat $\hat{c}_p \equiv (\partial\hat{h}/\partial T)_p$ and the constant volume specific heat $\hat{c}_v \equiv (\partial\hat{u}/\partial T)_\rho$. In chapter 8, we will show how some of these properties may be involved in the flow of a fluid.

Alternatively, we may write the second law of thermodynamics (equation 1.9) in terms of the enthalpy and pressure changes as:

$$T\,d\hat{s} = d\hat{h} - \frac{1}{\rho}\,dp \tag{1.10}$$

The Perfect Gas

Very often we have fluid flows of gases at, or near, atmospheric pressure. In these cases, the changes in pressure p, density ρ and absolute temperature T of a gas particle may be related accurately to each other by the perfect gas law:

$$p = \rho RT \tag{1.11}$$

where R is the gas constant for the particular gas in question. But this information alone is insufficient to explain how the properties of a gas change as it moves. In addition, the laws of thermodynamics, which include equation 1.9, must be invoked. Compressible flows are inherently complicated because the laws of thermodynamics, as well as the laws of fluid mechanics, operate simultaneously.

The perfect gas constant R is related to the *Universal gas constant* $\mathcal{R}$ and the gas molecular weight[5] $\mathcal{M}$ by:

$$R \equiv \frac{\mathcal{R}}{\mathcal{M}} \tag{1.12}$$

For a perfect gas, the internal energy $\hat{e}$, the enthalpy $\hat{h}$ and the specific heats, $\hat{c}_p$ and $\hat{c}_v$, are functions of the absolute temperature T alone and do not depend upon pressure or density. The specific heats are related to the gas constant R by:

$$\hat{c}_p - \hat{c}_v = R \tag{1.13}$$

An important property of a fluid, the speed a of a sound wave, is related to its thermodynamic properties by:

$$a^2 \equiv \left(\frac{\partial p}{\partial \rho}\right)_s \tag{1.14}$$

In the case of a perfect gas, the speed of sound becomes:

$$a^2 = \left(\frac{\hat{c}_p}{\hat{c}_v}\right) RT \tag{1.15}$$

Table 1.2 lists some perfect gas properties of common gases. The viscosity of a perfect gas is a function of temperature alone.

Heat Transfer and Compressible Flows

There are two types of flow in which the fluid thermodynamic properties are important. The first is a flow in which the fluid is heated or cooled—a common process when a fluid is used as a heat transfer medium. The second type, called a *compressible flow*, is one where the internal random motion of the fluid molecules is converted to the directed motion of a fluid particle. Flow through a gas turbine and the propagation of a sound wave are examples of compressible flows. In this book, we will not treat

[5]The molecular weight is the ratio of the mass of a molecule to one-twelfth of the the mass of the carbon-12 atom.

Table 1.2 Perfect gas properties

Gas	Symbol	$\mathcal{M}$	R $J/kg\,K$	$\hat{c}_p/R$ [a]	μ [a] $Pa\,s$
Hydrogen	H_2	2.016	4124	3.46	8.83E(-6)
Helium	He	4.003	2077	2.50	1.95E(-5)
Methane	CH_4	16.04	518.3	4.23	1.11E(-5)
Water vapor	H_2O	18.02	461.5	4.04	9.57E(-6)
Carbon monoxide	CO	28.01	296.8	3.50	1.76E(-5)
Nitrogen	N_2	28.02	296.8	3.50	1.76E(-5)
Air		28.97	287.0	3.50	1.82E(-5)
Oxygen	O_2	32.00	259.8	3.53	2.03E(-5)
Argon	Ar	39.94	208.2	2.50	2.25E(-5)
Carbon dioxide	CO_2	44.01	188.9	4.44	1.47E(-5)

[a] Value at $20\,^\circ C$

flows with heat transfer, but chapter 12 considers some simple compressible flows in perfect gases for which the specific heats are assumed constant.

Incompressible Flow

There are many instances of flows in which the thermodynamic properties of the fluid do not appreciably affect the fluid motion, *i.e.*, the random motion of the fluid molecules is not changed into directed motion of the fluid to any noticeable extent. These flows are called *incompressible flows* because they are not compressible! Every fluid is compressible in that an increase in pressure, no matter how small, is accompanied by an increase in density. In an incompressible flow, a fluid particle is actually compressed or expanded as the pressure acting on it increases or decreases, yet the flow is such that these changes do not couple with the flow, altering its character. In other words, the flow behaves as if the fluid does not undergo any change in density, even though its density changes by a small amount. But it is not sufficient that the density change in a flow be small in order to make it incompressible—the density change in an audible sound wave is minute, yet it is a compressible flow. In section 3.33, we will see how an incompressible flow is defined mathematically. For the time being, we will treat most flows as incompressible and defer dealing with compressible flow until chapters 8 and 12.

1.2.3 Chemical Properties

Sometimes it is desirable to keep track of certain chemical species present in a fluid flow. For example, we may want to determine how air pollutants are carried along by the wind or how toxic chemicals move through an underground aquifer that supplies potable water. In an automobile engine, the fuel is evaporated and mixed with air in precise proportions that are needed to ensure that the mixture will burn rapidly when ignited by the spark. Whatever the reason, we need a simple measure of the amount of each chemical specie present in a fluid element. One such measure is the *mass concentration*, expressed as the mass of constituent *i* per unit volume of fluid, which we designate as ρ_i:

$$\rho_i \equiv \textit{Mass of constituent i per unit volume} \tag{1.16}$$

We might call ρ_i the partial density, in the sense that it would be the density of the specie *i* if it occupied the volume of the fluid element all by itself. Of course, the sum of the mass concentrations of all the constituents must add to the density ρ of the fluid element:

$$\rho = \sum_i \rho_i \tag{1.17}$$

In the case of a mixture of perfect gases, the concentrations may be determined from the partial pressures p_i of the constituents since each specie obeys the perfect gas law equation 1.11:

$$\rho_i = \frac{p_i}{R_i T} \tag{1.18}$$

Note that the partial pressures sum to the mixture pressure:

$$p = \sum_i p_i \tag{1.19}$$

Example 1.2

The maximum oxygen concentration in oxygenated blood is said to be 20%, *i.e.*, the amount of oxygen in one liter of blood is equal to 1/5 of a liter of oxygen at atmospheric pressure and body temperature. Calculate the partial density ρ_{O_2} of oxygen in blood at this concentration.

Solution

First calculate the density of oxygen at one atmosphere and 98.6 $°F$ from the perfect gas law equation 1.11, using $p = 1.0133E(5)\,N/m^2$, $\mathcal{M} = 32$ and $\mathcal{R} = 8.3143E(3)$

$J/kg\,K$:

$$\rho = \frac{p}{RT} = \frac{p\mathcal{M}}{\mathcal{R}T}$$

Inserting numerical values,

$$\rho = \left(\frac{1.0133E(5)Pa \times 32}{8.3143E(3)J/kg\,K \times (273.15 + \frac{5}{9} \times (98.6 - 32))K}\right)$$

$$= 1.2574\,kg/m^3$$

The partial density of blood oxygen is 20% of this amount:

$$\rho_{O_2} = 0.2 \times 1.2574\,kg/m^3 = 0.2515\,kg/m^3$$

1.3 Dimensions and Units of Measurement

Science shows us how to relate quantitatively the things we observe in the world around us or in the laboratory. Our senses readily perceive such things as distance, temperature, the pitch of a sound, the motion of the sun in the sky, the passage of time, the weight of an object and the speed of the wind. But to quantify each of these observables requires the use of an instrument and the definition of a unit of measurement, so that the number of units in the measurement of the observable is its magnitude. The science of measurement enables us to transform our qualitative perceptions into quantitative ones.

We call each of the observables a *dimension* and the unit of measurement the *unit* of that dimension. The water in a glass on a table has such dimensions as volume, mass, temperature, density, pressure, surface area and velocity. Some of these dimensions are related to each other, *e.g.*, the density is the ratio of the mass to the volume. Others, such as mass and temperature, are not related since they can be independently varied. In the former case, the unit of density must equal the ratio of the units of mass and volume, while in the latter, no relationship exists between the units of mass and temperature.

Furthermore, physical laws place additional constraints on the independence of measurements. Physical laws equate observables of one kind to others. For example, Newton's law of motion equates force with the time rate of change of momentum. Each term in the equation representing a physical law must have the same dimension.

Thus force must have the same dimension as time rate of change of momentum. Additionally, the magnitude of the terms in the equation of a physical law must be such as to satisfy the equality expressed—a requirement that is easily satisfied by choosing the unit of measurement for each term to be the same; otherwise we sprinkle arbitrary constants of proportionality throughout Newton's law, Maxwell's equations, the laws of thermodynamics, etc.

SI System

If we examine all of the known physical laws, we find that only five independent dimensions are needed to express all the observables used in these laws, such as length, time, mass, force, temperature, velocity and electric field. By independent dimensions, we mean that none of them can be measured in terms of any combination of the others. By international agreement among scientific groups, these independent dimensions, which we call *fundamental dimensions*, are taken to be those of mass, length, time, temperature and electric current, and the values of the units are defined and labeled (*symbol*) as kilogram (kg), meter (m), second (s), kelvin (K) and ampere (A), respectively. These units are called *base units*. The units of all other dimensions, which we call *derivative dimensions*, are then expressible in terms of these five units of the fundamental dimensions. These latter units are called *derived units*. This system of measurement units is called the *International System of Units*, or SI for short.

It is sometimes convenient to use a nickname for the unit of a derivative dimension. The nickname for the unit of force, which has the dimension $kg\,m/s^2$, is the newton (N). Thus one N equals one $kg\,m/s^2$, and the product $Ns^2/kg\,m$ is unity. Other common nicknames are joule (J) for the unit of energy ($kg\,m^2/s^2 = N\,m$), watt (W) for the unit of power ($kg\,m^2/s^3 = J/s$) and pascal (Pa) for the unit of pressure ($kg/m\,s^2 = N/m^2$). A list of the fundamental and derivative dimensions commonly used in fluid mechanics, including their names, symbols, and equivalent SI units, is contained in table 1.3.

It is also customary to define SI units that are larger or smaller than those of table 1.3 by one or more factors of ten. These units are identified by prefixes, such as *kilo* in *kilojoule* or *milli* in *millimeter*, and by a prefix abbreviation, *e.g.*, *kJ* or *mm*. The factors, their prefixes and abbreviations are listed in table 1.4, together with equivalent word modifiers commonly used for some of them.

Because the SI system was adopted only recently, there exist other units of measurement related to SI units that continue to be used by scientists. These are listed in table 1.5 together with their SI equivalents. They include units of viscosity (*poise*) and kinematic viscosity (*stoke*).

Table 1.3 SI dimensions and units

Quantity	SI unit	Symbol	Equivalent
Mass	kilogram	kg	kg
Length	meter	m	m
Time	second	s	s
Temperature	kelvin	K	K
Force	newton	N	$kg\,m/s^2$
Pressure	pascal	Pa	$kg/m\,s^2 = N/m^2$
Density			kg/m^3
Viscosity			$kg/m\,s = Pa\,s$
Kinematic viscosity			m^2/s
Surface tension			$kg/s^2 = N/m$
Energy, heat, work	joule	J	$kg\,m^2/s^2 = N\,m$
Power	watt	W	$kg\,m^2/s^3 = J/s$
Specific heat, entropy, gas constant			$m^2/s^2K = J/kg\,K$

EES System

Much more significant is the use by engineers in the U.S. of what is termed the English Engineering System (EES) of units. In this system, both the unit of force, the *pound force (lbf)*, and the unit of mass, the *pound mass (lbm)*, as well as the units of length, the *foot (ft)*, and time, the *second (s)*, are considered as fundamental units. The *pound force* is defined as equal to the force of gravity acting on a *pound mass* under standard conditions of gravitational acceleration ($32.174 ft/s^2$). As a consequence, Newton's law of motion contains a constant of proportionality:

$$Force = \frac{(Mass)(Acceleration)}{g_c}$$

where the proportionality constant g_c equals:

$$g_c \equiv 32.174\ lbm\,ft/lbf\,s^2$$

Thus a one pound force applied to a one pound mass will cause an accelereation of $32.174 ft/s^2$. Because the EES uses the *ampere* as the unit of current, the Biot-Savart law for the force acting on a current-carrying conductor must also be modified by a proportionality constant.

While the EES definition of a unit force is suitable for calculating the stresses in a statically loaded structure, it is awkward for use in dynamical systems. Furthermore,

Table 1.4 SI unit prefixes

Factor	Prefix	Symbol	Word modifier
10^{15}	*peta*	*P*	quadrillion
10^{12}	*tera*	*T*	trillion
10^{9}	*giga*	*G*	billion
10^{6}	*mega*	*M*	million
10^{3}	*kilo*	*k*	thousand
10^{2}	*hecto*	*h*	hundred
10^{-1}	*deci*	*d*	
10^{-2}	*centi*	*c*	percent
10^{-3}	*milli*	*m*	
10^{-6}	*micro*	μ	
10^{-9}	*nano*	*n*	
10^{-12}	*pico*	*p*	

Table 1.5 Alternative SI-related units

Dimension	Unit	Symbol	SI unit value
Length	*centimeter*	*cm*	$1.0E(-2)\,m$
Area	*hectare*	*h*	$1.0E(4)\,m^2$
Volume	*litre*	*l*	$1.0E(-3)\,m^3$
Mass	*gram*	*g*	$1.0E(-3)\,kg$
	(metric) ton	*t*	$1.0E(3)\,kg$
Force	*dyne*		$1.0E(-5)\,N$
Pressure	*bar*	*b*	$1.0E(5)\,Pa$
Viscosity	*poise*		$1.0E(-1)\,Pa\,s$
Kinematic viscosity	*stoke*		$1.0E(-4)\,m^2/s$
Energy	*thermochemical calorie*	*cal*	$4.184\,J$
Frequency	*Hertz*	*Hz*	$1/s$
Temperature	*Celsius degree*	*C*	*K*
Permeability	*darcy*		$1.0E(-4)\,m^2$

Table 1.6 Conversion between EES and SI units

Dimension	EES unit	Symbol	SI unit value
Length	*foot*	*ft*	$3.048E(-1)\,m$
	inch	*in*	$2.540E(-2)\,m$
	statute mile		$1.609E(3)\,m$
	nautical mile		$1.852E(3)\,m$
Area	*square foot*	ft^2	$9.290E(-2)\,m^2$
	acre		$4.408E(3)\,m^2$
Volume	*cubic foot*	ft^3	$2.832E(-2)\,m^3$
	cubic inch	in^3	$1.6387E(-5)\,m^3$
	U.S. gallon	*gal*	$3.785E(-3)\,m^3$
Force	*pound force*	*lbf*	$4.448\,N$
Mass	*pound mass*	*lbm*	$4.536E(-1)\,kg$
	(short) ton		$9.072E(2)\,kg$
	(long) ton		$1.0161E(3)\,kg$
Pressure	*pound (force)/sq.in.*	*psi*	$6.895E(3)\,Pa$
Velocity	*miles/hour*	*mph*	$4.470E(-1)\,m/s$
Viscosity	*pound (mass)/foot second*	*lbm/ft.s*	$1.448\,Pa\,s$
Kinematic viscosity	*sq. ft./second*	ft^2/s	$9.294E(-2)\,m^2/s$
Energy	*foot pound (force)*	*ft lbf*	$1.356\,J$
	British thermal unit	*Btu*	$1.0551E(3)\,J$
Power	*horsepower*	*Hp*	$7.457E(2)\,W$
Temperature	*Rankine degree*	*R*	$5/9\,K$

Note: To convert an EES measurement to SI units, multiply its numerical value by the term in the last column. To convert SI to EES, divide.

the EES unit of energy, the *British thermal unit*, is not an integral multiple of one *ft lbf*. These historical choices produce such complications that today most engineering societies require that data be reported in SI units with their corresponding EES units following in parentheses. It is now common practice for engineers to utilize SI units in their calculations, converting, if desired, from EES units at the beginning and to EES units at the end. It is recommended that the engineering student adopt this practice in working out the problems in this text.

To aid in the conversion of EES units to SI units and vice versa, table 1.6 lists many of the commonly used units of the EES and their SI equivalents.

Example 1.3

The unit of water volume used in the distribution of irrigation water is the *acre–foot.* It is a volume of water one foot in depth covering a base area of one acre. How many cubic meters of water are there in one acre-foot?

Solution
Using the unit conversion values shown in table 1.6,

$$1\, acreft = 1\, acreft \times \frac{4.048E(3)\, m^2}{1\, acre} \times \frac{0.3048\, m}{1\, ft}$$
$$= 1234\, m^3$$

Irrespective of the units of measurement, there are several physical constants that fix points on the scales of measurement. Those that are useful in fluid mechanics include the melting point of ice[6], standard atmospheric pressure, standard acceleration of gravity and the universal gas constant. Their values are listed in table 1.7 in terms of SI units.

1.4 Vector Algebra and Calculus

In this book, we make frequent use of vectors in explaining the motion of fluids in physical space. The position, velocity and acceleration of a fluid particle, as well as the forces acting on it, are vector quantities. The conservation of mass, momentum and energy of a fluid particle are more easily expressed by utilizing vector forms. At the same time, numerical or analytic solutions to fluid flow problems may have to be worked out in scalar form. We therefore need to shift from one mode of expression to the other. In this section, we review those aspects of vector algebra and calculus that are needed in the ensuing discussion and use of the principles of fluid mechanics.

In describing the motion of a fluid in physical space, we need to choose a convenient coordinate system. Mostly we use the familiar *cartesian* coordinates x, y, z, but occasionally it is better to use *cylindrical* coordinates r, θ, z when the flow constraints

[6]The temperature of the melting point of ice at one atmosphere pressure, $273.15\, K = 491.67\, R$, marks the zero point on the Celsius scale and $32\,^{\circ}F$ on the Farenheit scale. The SI reference point on the absolute temperature scale is the triple point of water, $273.16\, K$.

Table 1.7 Physical constants

Quantity	SI unit value
Melting point of ice (0 C)	273.15 K
Standard atmospheric pressure	1.0133E(5) Pa
Standard acceleration of gravity	9.8066 m/s^2
Universal gas constant	8.3143E(3) $J/kg\ K$

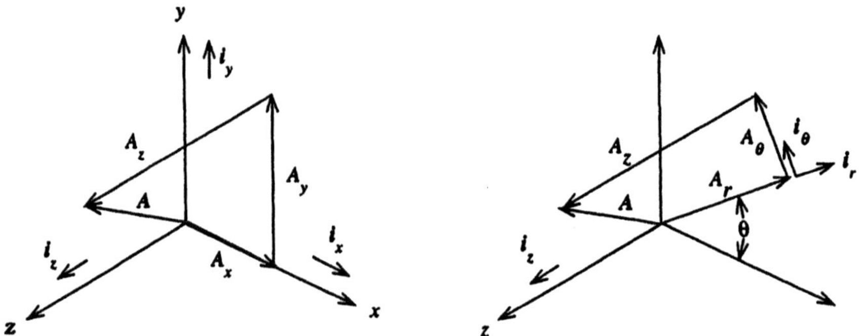

Figure 1.5 Cartesian and cylindrical coordinate systems, showing unit vectors and a vector **A**.

make this an appropriate choice, such as flow in a rotating machine. The advantage of vector algebra and calculus is that their definitions and theorems are independent of the choice of coordinate system. However, their expression in scalar form is different between cartesian and cylindrical coordinates. In what follows, we first use cartesian coordinates in illustrating vector operations and, at the end, give the corresponding forms for cylindrical coordinates.

1.4.1 Vector Algebra

Using **boldface** type to indicate a vector and ordinary type to indicate a scalar, a vector **A** is defined in terms of its three components A_x, A_y, A_z in the x, y, z directions:

$$\mathbf{A} \equiv A_x\mathbf{i}_x + A_y\mathbf{i}_y + A_z\mathbf{i}_z \tag{1.20}$$

in which $\mathbf{i}_x$, $\mathbf{i}_y$, $\mathbf{i}_z$ are the unit vectors in the x, y, z directions, as shown in figure 1.5.

Some examples of vector quantities we will use are the fluid particle velocity **V**, force **F**, acceleration of gravity **g**, torque or couple **T**, angular velocity $\mathbf{\Omega}$ and vorticity $\boldsymbol{\omega}$.

The *scalar product* $\mathbf{A} \cdot \mathbf{B}$ of the vectors $\mathbf{A}$ and $\mathbf{B}$ is defined as:

$$\mathbf{A} \cdot \mathbf{B} \equiv A_x B_x + A_y B_y + A_z B_z \tag{1.21}$$

The magnitude A of a vector $\mathbf{A}$ is found from the scalar product of $\mathbf{A}$ with itself:

$$A^2 \equiv \mathbf{A} \cdot \mathbf{A} = A_x^2 + A_y^2 + A_z^2$$

$$A = \sqrt{A_x^2 + A_y^2 + A_z^2} \tag{1.22}$$

The scalar product $\mathbf{A} \cdot \mathbf{B}$ is the product of the component of $\mathbf{A}$ in the direction of $\mathbf{B}$ times the magnitude of $\mathbf{B}$, and vice versa. If ϕ is the angle between the vectors $\mathbf{A}$ and $\mathbf{B}$, then

$$\mathbf{A} \cdot \mathbf{B} = AB \cos \phi \tag{1.23}$$

For example, the product $\mathbf{V} \cdot \mathbf{i}_x$ is the magnitude of the component of $\mathbf{V}$ in the direction of $\mathbf{i}_x$, or V_x. If $\mathbf{V}$ is the velocity of a particle on which the force $\mathbf{F}$ acts, then $\mathbf{F} \cdot \mathbf{V}$ is the rate of doing work, or power expended, moving the particle. Likewise, if $\mathbf{T}$ is the torque applied to a structure and $\mathbf{\Omega}$ is its angular velocity, then $\mathbf{\Omega} \cdot \mathbf{T}$ is the power expended in rotating it.

The *position vector* $\mathbf{R}$ is the distance from the origin to a point in space:

$$\mathbf{R} \equiv x\mathbf{i}_x + y\mathbf{i}_y + z\mathbf{i}_z$$

$$R = \sqrt{x^2 + y^2 + z^2} \tag{1.24}$$

The *vector product* $\mathbf{A} \times \mathbf{B}$ of the vectors $\mathbf{A}$ and $\mathbf{B}$ is defined as:

$$\mathbf{A} \times \mathbf{B} \equiv \mathbf{i}_x (A_y B_z - A_z B_y) + \mathbf{i}_y (A_z B_x - A_x B_z) + \mathbf{i}_z (A_x B_y - A_y B_x) \tag{1.25}$$

The magnitude of $\mathbf{A} \times \mathbf{B}$ is $AB \sin \phi$, and its direction is perpendicular to both $\mathbf{A}$ and $\mathbf{B}$, *i.e.*, it is normal to the plane that contains $\mathbf{A}$ and $\mathbf{B}$. Note that the vector product $\mathbf{A} \times \mathbf{A}$ is zero identically:

$$\mathbf{A} \times \mathbf{A} \equiv 0 \tag{1.26}$$

In fluid mechanics, a common use of the vector product is that for the moment $\mathbf{R} \times \mathbf{F}$ of a force $\mathbf{F}$ acting at a point located at $\mathbf{R}$.

1.4.2 Vector Calculus

The vector differential operator *del*, denoted by the symbol ∇, is defined as :

$$\nabla \equiv \mathbf{i}_x \frac{\partial}{\partial x} + \mathbf{i}_y \frac{\partial}{\partial y} + \mathbf{i}_z \frac{\partial}{\partial z} \tag{1.27}$$

If ∇ operates on a scalar function of x, y and z, say $a\{x, y, z\}$, then the resulting vector is called the *gradient* of a and is denoted by ∇a:

$$\nabla a \equiv \mathbf{i}_x \frac{\partial a}{\partial x} + \mathbf{i}_y \frac{\partial a}{\partial y} + \mathbf{i}_z \frac{\partial a}{\partial z} \tag{1.28}$$

The gradient of $a\{x, y, z\}$, ∇a, has a simple geometric interpretation. The equation $a\{x, y, z\} = k$, where k is a constant, describes a surface in x, y, z-space on each point of which the value of the function $a\{x, y, z\}$ is the same, namely, k. The gradient of a is perpendicular to this surface at any point on it and has the direction of an increase in the value of the constant k. The magnitude of ∇a is the spatial rate at which k increases in the normal direction.

In fluid mechanics, the pressure gradient, ∇p, is an example of a gradient of a scalar function, the pressure p, and appears prominently in the equation of motion of a fluid.

The scalar product of ∇ and a vector $\mathbf{A}$, denoted by $\nabla \cdot \mathbf{A}$, is called the *divergence* of $\mathbf{A}$:

$$\nabla \cdot \mathbf{A} \equiv \frac{\partial A_x}{\partial x} + \frac{\partial A_y}{\partial y} + \frac{\partial A_z}{\partial z} \tag{1.29}$$

The reverse of this scalar product, $\mathbf{A} \cdot \nabla$, is a scalar differential operator:

$$\mathbf{A} \cdot \nabla = A_x \frac{\partial}{\partial x} + A_y \frac{\partial}{\partial y} + A_z \frac{\partial}{\partial z} \tag{1.30}$$

It is common to enclose this operator in parenthesis, $(\mathbf{A} \cdot \nabla)$, to avoid confusion when operating on a scalar or vector. For example,

$$(\mathbf{A} \cdot \nabla)a = A_x \frac{\partial a}{\partial x} + A_y \frac{\partial a}{\partial y} + A_x \frac{\partial a}{\partial z}$$

$$(\mathbf{A} \cdot \nabla)\mathbf{B} = A_x \frac{\partial \mathbf{B}}{\partial x} + A_y \frac{\partial \mathbf{B}}{\partial y} + A_x \frac{\partial \mathbf{B}}{\partial z} \tag{1.31}$$

The scalar product of ∇ with itself, $\nabla \cdot \nabla$, is called the *Laplacian* operator, and is denoted by the symbol ∇^2:

$$\nabla^2 \equiv \frac{\partial^2}{\partial x^2} + \frac{\partial^2}{\partial y^2} + \frac{\partial^2}{\partial z^2} \tag{1.32}$$

∇^2 can operate on either a scalar a or a vector $\mathbf{A}$, yielding a scalar or vector, respectively:

$$\nabla^2 a \equiv \frac{\partial^2 a}{\partial x^2} + \frac{\partial^2 a}{\partial y^2} + \frac{\partial^2 a}{\partial z^2} \tag{1.33}$$

$$\nabla^2 \mathbf{A} \equiv \frac{\partial^2 \mathbf{A}}{\partial x^2} + \frac{\partial^2 \mathbf{A}}{\partial y^2} + \frac{\partial^2 \mathbf{A}}{\partial z^2} \tag{1.34}$$

The vector product of ∇ and a vector $\mathbf{A}$, $\nabla \times \mathbf{A}$, is called the *curl* of $\mathbf{A}$:

$$\nabla \times \mathbf{A} \equiv \mathbf{i}_x \left(\frac{\partial A_z}{\partial y} - \frac{\partial A_y}{\partial z} \right) + \mathbf{i}_y \left(\frac{\partial A_x}{\partial z} - \frac{\partial A_z}{\partial x} \right) + \mathbf{i}_z \left(\frac{\partial A_y}{\partial x} - \frac{\partial A_x}{\partial y} \right) \tag{1.35}$$

There are two products involving the curl and divergence of a vector that are identically zero:

$$\nabla \times (\nabla a) \equiv 0 \tag{1.36}$$

$$\nabla \cdot (\nabla \times \mathbf{A}) \equiv 0 \tag{1.37}$$

where a is a scalar and $\mathbf{A}$ a vector. In addition, there are several instances where we will need to use expressions for vector operations involving products, as in the following:

$$\nabla \cdot (a\mathbf{A}) = a(\nabla \cdot \mathbf{A}) + (\mathbf{A} \cdot \nabla)a \tag{1.38}$$

$$\nabla \times (a\mathbf{A}) = a(\nabla \times \mathbf{A}) + (\nabla a) \times \mathbf{A} \tag{1.39}$$

$$(\mathbf{A} \cdot \nabla)\mathbf{A} = \nabla \left(\frac{A^2}{2} \right) - \mathbf{A} \times (\nabla \times \mathbf{A}) \tag{1.40}$$

$$\nabla \times (\mathbf{A} \times \mathbf{B}) = \mathbf{A}(\nabla \cdot \mathbf{B}) + (\mathbf{B} \cdot \nabla)\mathbf{A} - \mathbf{B}(\nabla \cdot \mathbf{A}) - (\mathbf{A} \cdot \nabla)\mathbf{B} \tag{1.41}$$

$$\nabla \times (\nabla \times \mathbf{A}) = \nabla(\nabla \cdot \mathbf{A}) - \nabla^2 \mathbf{A} \tag{1.42}$$

1.4.3 Vector Integrals

In fluid mechanics, it is often convenient to express the conservation laws in integral form. There are three types of integrals used for this purpose: line, area and volume integrals. When these integrals are complete, *i.e.* the line is a closed curve marking the edge of an area it encloses, or the area completely encloses a volume, then there are relations between line, surface and volume integrals of certain vector quantities.

The first of these is a vector equation involving a scalar function a and is known as *Gauss' theorem*:

$$\iiint_{\mathcal{V}} \nabla a \, d\mathcal{V} = \iint_{S} a \, \mathbf{n} \, dS \tag{1.43}$$

The volume element is designated by $d\mathcal{V}$ (*e.g.*, $dx\,dy\,dz$) and the surface element by

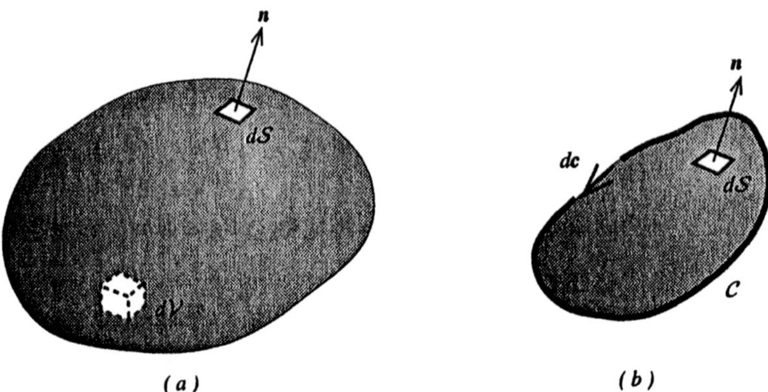

Figure 1.6 Elements of volume, surface and line integrals.

dS. The multiple integral signs indicate the volume and surface integrals, for which the limits of integration are such that *the surface S completely encloses the volume $\mathcal{V}$*. The vector $\mathbf{n}$ is the unit vector normal to the surface S, *being defined as positive when pointing outward from the enclosed volume $\mathcal{V}$*. Figure 1.6 (*a*) illustrates the volume and surface integrals involved in equation 1.43.

The second relationship, known as the *divergence theorem*, is a scalar equation:

$$\iiint_{\mathcal{V}} \nabla \cdot \mathbf{A}\, d\mathcal{V} = \iint_S \mathbf{A} \cdot \mathbf{n}\, dS \tag{1.44}$$

Stokes' theorem is a scalar equation relating line and surface integrals:

$$\iint_S \mathbf{n} \cdot (\nabla \times \mathbf{A})\, dS = \int_C \mathbf{A} \cdot d\mathbf{c} \tag{1.45}$$

Here $d\mathbf{c}$ is the vector line element of the *closed curve* C, and $\mathbf{n}$ is the unit normal to the surface S enclosed by the curve C, as shown in figure 1.6 (*b*). The direction of positive $\mathbf{n}$ is given by the right hand rule: if the index finger of the right hand indicates the direction of the line element $d\mathbf{c}$, then the thumb indicates the direction of $\mathbf{n}$.

We will have occasion later on to utilize line integrals that do not form a closed curve, as in figure 1.6 (*b*) but only join two points, say 1 and 2, along the curve C. When the vector integrand is the gradient of a scalar, *e.g.*, ∇a, the line integral can be integrated:

$$\int_1^2 (\nabla a) \cdot d\mathbf{c} = \int_1^2 \left(\frac{\partial a}{\partial x} dx + \frac{\partial a}{\partial y} dy + \frac{\partial a}{\partial z} dz \right)$$

$$= \int_1^2 da$$

$$= a_2 - a_1 \tag{1.46}$$

In fluid mechanics, some of the integrals in equations 1.43–1.45 represent physical quantities of interest. For example, the net force acting on a fluid volume caused by the pressure p is $\iiint(-p)\mathbf{n}\,dS$, and the net outflow of fluid across an imaginary surface in the flow is $\iint \mathbf{V}\cdot\mathbf{n}\,dS$, where $\mathbf{V}$ is the fluid velocity. In particle mechanics, the work required to move a particle around a closed path C in space is $\int \mathbf{F}\cdot d\mathbf{c}$, where $\mathbf{F}$ is the force applied to the particle.

1.4.4 Cylindrical Coordinates

For some fluid flows, the use of cylindrical coordinates r, θ, z, illustrated in figure 1.5, can be more convenient than employing cartesian coordinates. But, because the unit vectors $\mathbf{i}_r$ and $\mathbf{i}_\theta$ vary their direction (but not their magnitude) with a change in θ, it is necessary to include this variation when differentiating the r and θ components of a vector with respect to θ. The derivatives of $\mathbf{i}_r$ and $\mathbf{i}_\theta$ are:

$$\frac{\partial \mathbf{i}_r}{\partial\theta} = \mathbf{i}_\theta; \quad \frac{\partial \mathbf{i}_\theta}{\partial\theta} = -\mathbf{i}_r \tag{1.47}$$

so that the θ-derivatives of the θ- and r-components of $\mathbf{A}$ become:

$$\frac{\partial}{\partial\theta}(A_\theta \mathbf{i}_\theta) = \mathbf{i}_\theta \frac{\partial A_\theta}{\partial\theta} - A_\theta \mathbf{i}_r$$

$$\frac{\partial}{\partial\theta}(A_r \mathbf{i}_r) = \mathbf{i}_r \frac{\partial A_r}{\partial\theta} + A_r \mathbf{i}_\theta \tag{1.48}$$

For reference later on, we list here the forms of vector algebra and calculus used above, expressed in cylindrical coordinates:

$$\mathbf{A} = A_r\mathbf{i}_r + A_\theta\mathbf{i}_\theta + A_z\mathbf{i}_z \tag{1.49}$$

$$\mathbf{A}\cdot\mathbf{B} = A_rB_r + A_\theta B_\theta + A_zB_z \tag{1.50}$$

$$\mathbf{A}\times\mathbf{B} = \mathbf{i}_r(A_\theta B_z - A_zB_\theta) + \mathbf{i}_\theta(A_zB_r - A_rB_z) + \mathbf{i}_z(A_rB_\theta - A_\theta B_r) \tag{1.51}$$

$$\nabla = \mathbf{i}_r\frac{\partial}{\partial r} + \mathbf{i}_\theta\frac{1}{r}\frac{\partial}{\partial\theta} + \mathbf{i}_z\frac{\partial}{\partial z} \tag{1.52}$$

$$\nabla\cdot\mathbf{A} = \frac{1}{r}\frac{\partial}{\partial r}(rA_r) + \frac{1}{r}\frac{\partial A_\theta}{\partial\theta} + \frac{\partial A_z}{\partial z} \tag{1.53}$$

$$\mathbf{A}\cdot\nabla = A_r\frac{\partial}{\partial r} + A_\theta\frac{1}{r}\frac{\partial}{\partial\theta} + A_z\frac{\partial}{\partial z} \tag{1.54}$$

$$\nabla\times\mathbf{A} = \mathbf{i}_r\left(\frac{1}{r}\frac{\partial A_z}{\partial\theta} - \frac{\partial A_\theta}{\partial z}\right) + \mathbf{i}_\theta\left(\frac{\partial A_r}{\partial z} - \frac{\partial A_z}{\partial r}\right) + \mathbf{i}_z\left(\frac{1}{r}\frac{\partial}{\partial r}(rA_\theta) - \frac{1}{r}\frac{\partial A_r}{\partial\theta}\right) \tag{1.55}$$

$$\nabla^2 a = \frac{\partial^2 a}{\partial r^2} + \frac{1}{r}\frac{\partial a}{\partial r} + \frac{1}{r^2}\frac{\partial^2 a}{\partial\theta^2} + \frac{\partial^2 a}{\partial z^2} \tag{1.56}$$

$$\nabla^2\mathbf{A} = \mathbf{i}_r\left[\nabla^2 A_r - \frac{A_r}{r^2} - \frac{2}{r^2}\left(\frac{\partial A_\theta}{\partial\theta}\right)\right] + \mathbf{i}_\theta\left[\nabla^2 A_\theta - \frac{A_\theta}{r^2} + \frac{2}{r^2}\left(\frac{\partial A_r}{\partial\theta}\right)\right] + \mathbf{i}_z\nabla^2 A_z \tag{1.57}$$

1.5 Problems

Problem 1.1

Sea water has a density of $1.02478E(3)\,kg/m^3$ at $20°\,C$ and one atmosphere pressure. At the same pressure and temperature, its thermal expansion coefficient β is $2.57E(-4)\,K^{-1}$ and its bulk modulus E is $2.147E(9)\,Pa$. Calculate the density of a sea water at $10°\,C$ and a pressure 100 times atmospheric pressure.

Problem 1.2

A droplet of volume $V = 4\pi r^3/3$ and surface area $S = 4\pi r^2$ has an internal pressure $\Delta p = 2\Upsilon/r$ greater than the surrounding atmosphere. Show that the work $(\Delta p)\delta V$ required to increase the volume by a small increment δV is equal to the increment in surface energy $\Upsilon\delta A$ associated with the surface area increment δA.

Problem 1.3

In the experiment illustrated in figure 1.3, calculate the shear stress τ when $V = 1\,m/s$ and $h = 1\,mm$, for air and water at $20°\,C$.

Problem 1.4

Show that the thermal expansion coefficient β and the bulk modulus E of a perfect gas are $1/T$ and p, respectively.

Problem 1.5

Standard (dry) air is a perfect gas mixture of four gases: nitrogen, oxygen, argon and carbon dioxide. Their partial pressures are 0.7809, 0.2095, $9.3E(-3)$ and $3E(-4)$ times the total pressure, and their molecular weights are 28, 32, 40 and 44, respectively. (a) Calculate the partial densities of the four species for air at one atmosphere

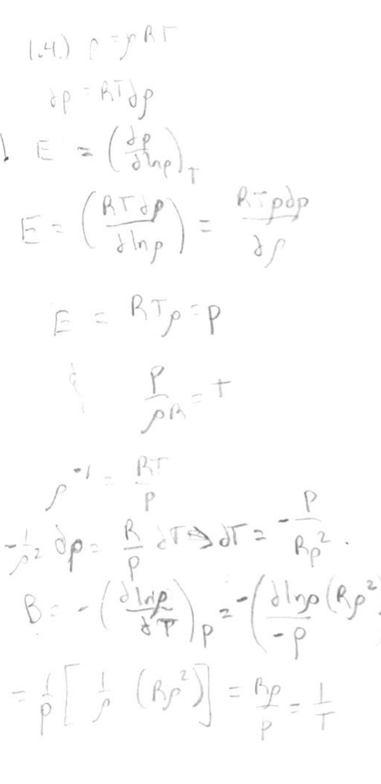

pressure and 0° C. (b) What is the density of air at this condition? (c) What is the (average) molecular weight of air?

Problem 1.6

Calculate the speed of sound in air at 20° C.

Problem 1.7

An automobile tire has the shape of a torus of major diameter $D = 60\,cm$ and minor diameter $d = 15\,cm$. The volume V_t of the tire is $\pi D(\pi d^2/4) = \pi^2 D d^2/4$. The tire contains air at an absolute pressure of $p_t = 50\,psia$ and temperature of 20° C. Calculate the mass $M_t(kg)$ of air in the tire.

Problem 1.8

An automobile engine consumes 4 pounds of fuel per hour while it generates 10 horsepower. The fuel has a heating value of 18,500 *Btu* per pound of fuel. (a) Express the fuel consumption rate, the engine power and the heating value of the fuel in *SI* units. (b) Calculate the engine thermal efficiency, which is the ratio of the engine power to the rate of consumption of fuel heating value.

Problem 1.9

Convert the entries in table 1.7 to EES units. (Express the universal gas constant in two alternative forms, using *ft lbf* and *Btu* as the alternative energy units.)

Problem 1.10

The velocity field of a flow is:

$$\mathbf{V}\{x, y, z\} = Ax\mathbf{i}_x + By\mathbf{i}_y + Cz\mathbf{i}_z$$

where A, B and C are constants. Derive expressions for $\nabla \cdot \mathbf{V}$ and $\nabla \times \mathbf{V}$.

Problem 1.11

The gradient of the fluid pressure, $p\{x, y\}$, is given by:

$$\nabla p = -A(x\mathbf{i}_x + y\mathbf{i}_y)$$

where A is a constant. Derive an expression for $p\{x, y\} - p\{0, 0\}$.

Problem 1.12

The pressure field of an irrotational flow over a circular cylinder of radius a, expressed in cylindrical coordinates, is:

$$p\{r, \theta, z\} = p_\infty - \frac{\rho U}{2}\left(\frac{a^2}{r^2}\right)\left[\left(\frac{a^2}{r^2}\right) - 2\cos(2\theta)\right]$$

where p_∞ and U are the pressure and flow speed far from the cylinder. Derive an expression for ∇p as a function of r, θ and z.

Bibliography

Green, Jerry Franklin. 1987. *Fundamental Cardiovascular and Pulmonary Physiology*, 2nd ed. Philadelphia: Lea and Febiger .

Neumann, Gerhard and Willard J. Pierson, Jr. 1966. *Principles of Physical Oceanography*, Englewood Cliffs, N.J.: Prentice-Hall.

Sabersky, Rolf H., Allan J. Acosta, and Edward G. Hauptman. 1989. *Fluid Flow, A First Course in Fluid Mechanics*. 3rd ed. New York: Macmillan Publishing Co.

U.S. Department of Commerce. 1955. *Tables of Thermal Properties of Gases*. National Bureau of Standards Circular 564. Washington, D.C.: U.S. Government Printing Office.

U. S. Department of Commerce. 1981. *The International System of Units (SI)*. National Bureau of Standards Special Publication 330. Washington, D.C.: U.S. Government Printing Office.

2 Fluid Statics

A static fluid is one which does not move—its velocity and acceleration, as measured in an inertial reference frame, are everywhere zero. Water stored in a reservoir, gasoline stored in an automobile tank and liquid propane stored under pressure in a gas tank are examples of static fluids. The forces acting on these fluids are in balance, and no motion ensues. In this chapter, we explore how this force balance is maintained and what the observable consequences of this force equilibrium are on the structures containing, or surrounded by, the fluid.

By expressing the force balance on a static particle, we can find how the pressure in a static fluid varies with position, an interesting example of which is the earth's atmosphere. This information makes it possible to measure pressure with manometers and barometers. We can calculate the force exerted by a static fluid on a solid surface with which it is in contact, including bodies floating on or submerged in water, and we can determine the stability of the latter. As long as a fluid is stationary, we can find out much practical information about how it interacts with its surroundings with relatively little difficulty.

2.1 Forces on a Fluid Particle

A fluid particle is subject to two quite different types of forces. The surface of the particle experiences a force per unit area called the *stress*. This is the force acting between molecules on the surface and those outside the particle in the surrounding medium but still close to the surface. The intermolecular force that gives rise to the surface stress is short-range, *i.e.*, it is appreciable only when the molecules are closer than than about $1E(-10)\,m$ apart. On the other hand, there are long-range forces that can be exerted throughout the volume of the fluid particle. These are called *body forces* because they act on the entire particle, not just the surface.

In a fluid, the surface stress depends upon both the relative *position* of the molecules near the surface and the relative average *motion* of these molecules. It is convenient to subdivide the surface stress into these two categories, the first of which we will call the *pressure* p and the second which we call the *viscous stress* τ. When a fluid has no relative motion, the only stress is the first kind, the pressure. We shall find out that this stress component is always normal to the surface of the particle. However, when the fluid moves, there will be a viscous stress component whose direction and magnitude depends upon the rate of distortion of the moving fluid element. For nearly all fluids, this viscous stress is proportional to the fluid viscosity μ. (We will treat the viscous stress in greater detail in chapter 6.)

The most common body force affecting the motion of fluids is the gravity force exerted on a fluid particle by the entire earth. Although the gravity force between molecules of a fluid particle is extremely weak, that due to the cumulative effect of very large masses at large distances produces a noticeable force.[1] The strength and direction of the gravity force is conveniently given by the product of the particle mass times the local acceleration of gravity $\mathbf{g}$. The other possible body forces acting on fluids are the electrostatic force between electrically charged fluid particles and surrounding charged fluid or solids and the electromagnetic force induced by a fluid particle conducting electric current in the presence of a magnetic field. The treatment of fluids subject to these forces is called electrohydrodynamics and magnetohydrodynamics. We will not consider such flows in this book.

2.2 Stress in a Fluid

When we compress a fluid, forcing the molecules closer together, we apply an inward force on the fluid boundaries. This force is transmitted throughout the fluid as the molecules, being pushed closer together, repel each other more strongly, on average. This average intermolecular force is called an internal stress.

Let us be more precise. Consider a finite volume $\mathcal{V}$ of fluid, completely enclosed by a surface $\mathcal{S}$, as shown in figure 2.1, that is completely surrounded by the same kind of fluid. At a point on this surface, the fluid molecules on or close to the surface experience a force per unit area of surface, called the stress σ, caused by the molecules outside the surface. In general, the stress vector σ has a component normal to the surface (called a normal stress) and one parallel to the surface (called a shear stress) which are respectively parallel and perpendicular to the unit outward normal $\mathbf{n}$. By

[1] Ocean tides are caused by the differential gravity force of the moon and the sun acting upon the ocean fluid on opposite sides of the earth.

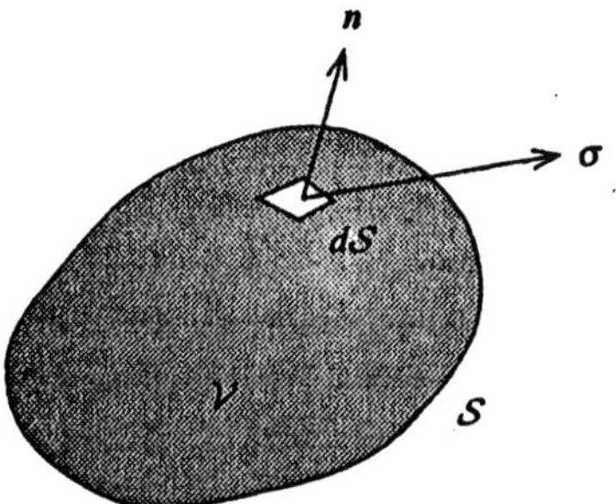

Figure 2.1 The fluid inside a volume $\mathcal{V}$ is enclosed by the surface $\mathcal{S}$. At a point on the surface, where the unit outward normal is $\mathbf{n}$, the stress is $\boldsymbol{\sigma}$.

Newton's law of action and reaction, the stress acting on the fluid inside $\mathcal{S}$ is equal and opposite to the stress at this point acting on the fluid outside $\mathcal{S}$.[2]

The stress at a point within a fluid is more complex than is implied by figure 2.1 because the stress vector $\boldsymbol{\sigma}$ depends upon the direction of the surface normal $\mathbf{n}$. In order to describe the state of stress at a point P in a fluid (or a solid, for that matter), we must specify the three components of the stress vector for each of three mutually perpendicular orientations of a surface passing through the point. These nine quantities form what is called the *stress tensor*. In a fluid, some of the components of the stress tensor depend upon the rate at which a fluid element is being distorted, and these components are proportional to the fluid viscosity. In chapter 6, where we treat viscous flows, we will have to take these components into account. For now we can ignore them because we are considering a case of no flow and therefore no distortion of a fluid element. Only the remaining components of the stress tensor that are not dependent upon the distorting motion need to be taken into account.

2.3 Pressure in a Static Fluid

2.3.1 Pascal's Law

In a static fluid, where there is no motion, the stress vector $\boldsymbol{\sigma}$ cannot be different for different orientations of the surface normal because there is no preferred direction

[2]If part, or all, of the surface $\mathcal{S}$ is surrounded by a solid or a different kind of fluid, an interfacial tension would exist, as explained in the section on surface tension on page 17. There would then be a difference between the normal stress acting on the interior fluid and that acting on the exterior fluid or solid. For the time being, we'll ignore this effect.

in the fluid, which is isotropic in structure.[3] *At a point P in a static fluid, $\boldsymbol{\sigma}$ must therefore have the direction of* **n** *and have the same magnitude for all directions of* **n**. This startling result is called *Pascal's Law*.[4]

To prove that the magnitude of $\boldsymbol{\sigma}$ is independent of the direction of the surface normal **n**, consider a force balance on the fluid element as pictured in figure 2.2. The stresses acting perpendicular to two faces of area $\delta y\, \delta z$ and $\delta y\,(\delta z/\sin\phi)$ are designated by σ_x and σ, respectively. Since the fluid element is motionless, the sum of the forces in the x direction must add to zero:[5]

$$-\sigma_x\, \delta y\, \delta z + (\sigma \sin\phi)\left(\delta y \frac{\delta z}{\sin\phi}\right) = 0$$

$$(-\sigma_x + \sigma)\, \delta y\, \delta z = 0$$

$$\sigma = \sigma_x$$

As the fluid element shrinks to zero, the normal stresses σ and σ_x are colocated at the point P. However, the magnitude of $\boldsymbol{\sigma}$ must be the same for any direction since we could have chosen z or y for the direction of x in figure 2.2. Because we know that fluids can sustain only a compressive stress, or pressure, we identify the magnitude of $\boldsymbol{\sigma}$ with the pressure p and its direction as opposite to **n**:

$$\boldsymbol{\sigma} = (-p)\, \mathbf{n} \tag{2.1}$$

2.3.2 The Pressure Force

Using Pascal's Law, equation 2.1, we can determine the force exerted on a volume $\mathcal{V}$ of fluid by the stress acting on its surface $\mathcal{S}$, as illustrated in figure 2.1, if we integrate the pressure force per unit area, $-p\mathbf{n}$, over the surface to find the **pressure force** vector:

---

[3]If the stress $\boldsymbol{\sigma}$ had a shear component tangential to the surface $\mathcal{S}$, that component would have a direction. But if the fluid is stationary, there is no local property of the fluid, such as the velocity, that can give a direction. Thus this shear stress must be zero.

[4]Blaise Pascal (1632-1662) described this principle in *Traits de l'equilibre des liqeurs*, published posthumously. He was the first to suggest experiments to show the variation of atmospheric pressure with altitude. The SI unit of pressure, the *pascal*, is named in his honor.

[5]We do not include the force of gravity in this force balance because it is proportional to $\delta x\, \delta y\, \delta z$ which is a smaller quantity than the areas $\delta y\, \delta z$, etc., that enter into the forces due to stresses. As the size of the element shrinks to zero, volume forces become negligible compared to surface forces.

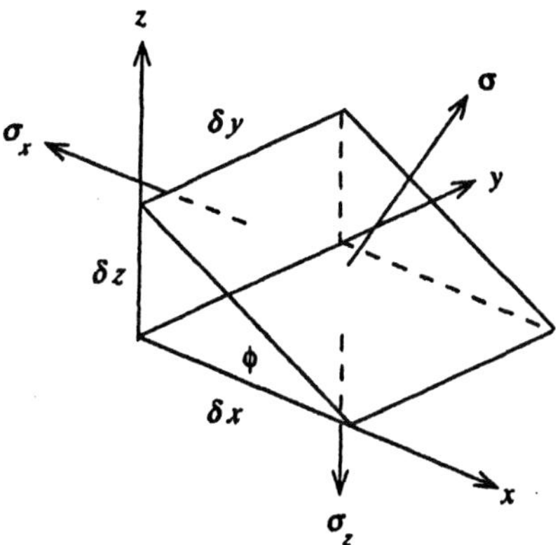

Figure 2.2 The forces acting on an element of static fluid.

$$\textbf{pressure force} = \iint_S (-p\mathbf{n})\, dS \tag{2.2}$$

But it is also possible to express this pressure force as a volume integral by using Gauss' theorem, equation 1.43, to convert the surface integral to a volume integral:

$$\textbf{pressure force} = \iiint_{\mathcal{V}} (-\nabla p)\, d\mathcal{V} \tag{2.3}$$

Since the total pressure force exerted on the volume $\mathcal{V}$ is the quantity $-\nabla p$ integrated over the volume, the integrand $-\nabla p$ must be the **pressure force** *per unit volume* at any point in the fluid:

$$\textbf{pressure force}\ \textit{per unit volume} = -\nabla p \tag{2.4}$$

Any small volume element $\delta x\, \delta y\, \delta z$ will be subject to a pressure force caused by the difference in pressure on the faces of the volume element. For example, the pressure difference in the x-direction is $(\partial p/\partial x)\delta x$ and the corresponding force difference is $-(\partial p/\partial x)\delta x(\delta y\, \delta z)\mathbf{i}_x$, so that the total pressure force is $-\nabla p(\delta x\, \delta y\, \delta z)$. This force acts in the direction of decreasing pressure, *i.e.*, opposite to the direction of the gradient of pressure, ∇p.

Example 2.1

Near the surface of the earth, the atmospheric pressure p decreases with increasing altitude z above sea level approximately as:

$$p = p_0 \exp(-\alpha z)$$

where p_0 is the sea level pressure of $1.0133E(5)\,Pa$ and $\alpha = 1.2E(-4)\,m^{-1}$. Calculate the pressure force per unit volume at $z = 0$ and $z = 5\,km$.

Solution

$$-\nabla p = -\frac{dp}{dz}\,\mathbf{i}_z$$

$$= \alpha p_0 \exp(-\alpha z)\,\mathbf{i}_z$$

At $z = 0$,

$$-\nabla p = (1.2E(-4)\,m^{-1} \times 1.033E(5)\,Pa)\,\mathbf{i}_z$$

$$= (12.40\,N/m^3)\,\mathbf{i}_z$$

At $z = 5\,km$,

$$-\nabla p = (1.2E(-4)\,m^{-1} \times 1.033E(5)\,Pa \times \exp[-1.2E(-4) \times 5E(3)])\,\mathbf{i}_z$$

$$= (6.805\,N/m^3)\,\mathbf{i}_z$$

2.3.3 Pressure in a Fluid in a Gravitational Field

When we dive deeply into a swimming pool or a pond, we feel an increasing pressure on our ears as we descend. Conversely, when we ascend in altitude, as in a rising elevator, we experience a decreasing pressure. The pressure in a static fluid appears to increase in the direction of the pull of gravity. Why does fluid pressure change in this manner?

A fluid element in a pond or swimming pool remains motionless because there is a balance of forces acting in the vertical direction: the pull of gravity acting downwards is balanced by the upward force of pressure. The gravitational force is a body force whose magnitude is the product of the mass of a fluid element times the gravitational acceleration. For a unit volume of fluid, the mass is ρ so that the gravitational force per unit volume becomes:

$$\textbf{gravitational force } \textit{per unit volume} = \rho\,\mathbf{g} \tag{2.5}$$

In order for a fluid to remain motionless when subject to a gravitational force field, the sum of the pressure force per unit volume and the gravitational force per unit volume must add to zero:

$$-\nabla p + \rho \mathbf{g} = 0 \tag{2.6}$$

This equation of hydrostatic equilibrium expresses the force balance at each point in a stationary fluid. It shows that the pressure p must increase in the direction of **g** (since $\nabla p = \rho \mathbf{g}$) and that the magnitude of the pressure gradient is ρg. The denser the fluid, the greater is the increase of pressure with depth. Furthermore, any horizontal plane in the fluid is a surface of constant pressure, because ∇p has no component in the horizontal direction.

The differential equation of hydrostatic equilibrium, equation 2.6, may be integrated explicitly along any line lying entirely within the fluid and connecting two points, 1 and 2, *whenever the density* ρ *is constant within the fluid.* Denoting the line element by $d\mathbf{c}$, the line integral of equation 2.6 becomes:

$$-\int_1^2 \nabla p \cdot d\mathbf{c} + \int_1^2 \rho \mathbf{g} \cdot d\mathbf{c} = 0$$

The first of these terms is easily integrated using equation 1.46:

$$\int_1^2 \nabla p \cdot d\mathbf{c} = p_2 - p_1$$

The second term may also be integrated if we note that the acceleration of gravity can be expressed as the gradient of the scalar product of **g** and the position vector **R**:[6]

$$\begin{aligned}
\nabla(\mathbf{g} \cdot \mathbf{R}) &= \nabla(g_x x + g_y y + g_z z) \\
&= \mathbf{i}_x \frac{\partial}{\partial x}(g_x x) + \mathbf{i}_y \frac{\partial}{\partial y}(g_y y) + \mathbf{i}_z \frac{\partial}{\partial z}(g_z z) \\
&= \mathbf{i}_x g_x + \mathbf{i}_y g_y + \mathbf{i}_z g_z \\
&= \mathbf{g}
\end{aligned} \tag{2.7}$$

Noting that ρ is constant, we now can evaluate the second integral using equation 1.46 to find:

$$\begin{aligned}
\int_1^2 \rho \mathbf{g} \cdot d\mathbf{c} &= \rho \int_1^2 \nabla(\mathbf{g} \cdot \mathbf{R}) \cdot d\mathbf{c} \\
&= \rho \mathbf{g} \cdot \mathbf{R}_2 - \rho \mathbf{g} \cdot \mathbf{R}_1
\end{aligned} \tag{2.8}$$

---

[6]We are assuming here that the acceleration of gravity does not change with position **R**. This is acceptable as long as the vertical dimension of the fluid is small compared to the radius of the earth.

Finally, combining these two integrals, the condition of hydrostatic equilibrium becomes:

$$-(p_2 - p_1) + \rho\mathbf{g}\cdot\mathbf{R}_2 - \rho\mathbf{g}\cdot\mathbf{R}_1 = 0$$

$$p_1 - \rho\mathbf{g}\cdot\mathbf{R}_1 = p_2 - \rho\mathbf{g}\cdot\mathbf{R}_2 \tag{2.9}$$

It is customary to choose Cartesian coordinates for which the z-axis points upward, opposite to the direction of the gravitational acceleration **g**. In this convention,

$$\mathbf{g} = -g\mathbf{i}_z$$

$$\mathbf{g}\cdot\mathbf{R} = -gz \tag{2.10}$$

and the general condition of hydrostatic equilibrium, Equation 2.9, assumes the form:

$$p_1 + \rho g z_1 = p_2 + \rho g z_2 \tag{2.11}$$

There is a simple way to understand the relationship of equation 2.11. Suppose we isolate a vertical cylindrical column of fluid of base area A and height $z_2 - z_1$. The pressure p_1 at the bottom of this column exceeds that (p_2) at the top by an amount $p_1 - p_2$, so that there is an upward force of amount $(p_1 - p_2)A$ due to the pressure acting on the top and bottom of the column. Since the sides of the column are vertical, there is no vertical component of the pressure force acting on the sides. This upward force must be balanced by the downward pull of gravity on the fluid inside the column, which equals the product of g times the mass $\rho(z_2 - z_1)A$. Thus the force balance requires that $(p_1 - p_2)A = \rho g(z_2 - z_1)A$, which is equivalent to equation 2.11. The pressure distribution of equation 2.11 provides for this vertical force balance on all columns of liquid.

The integral equation of hydrostatic equilibrium, equation 2.11, is more general than it seems. This relation holds not only between any two points in a fluid for which the density is everywhere constant but also between all points in the fluid. In other words, the sum $p + \rho g z$ has the same value at all points in the fluid that can be connected to each other by a line that always lies in the fluid. This conclusion can be represented by the equation:

$$p + \rho g z = constant \tag{2.12}$$

where the constant can be evaluated from knowledge of p and z at one point in the fluid, say 1. The pressure $p\{z\}$ as a function of height z can then be found from:

$$p\{z\} = p_1 + \rho g z_1 - \rho g z \tag{2.13}$$

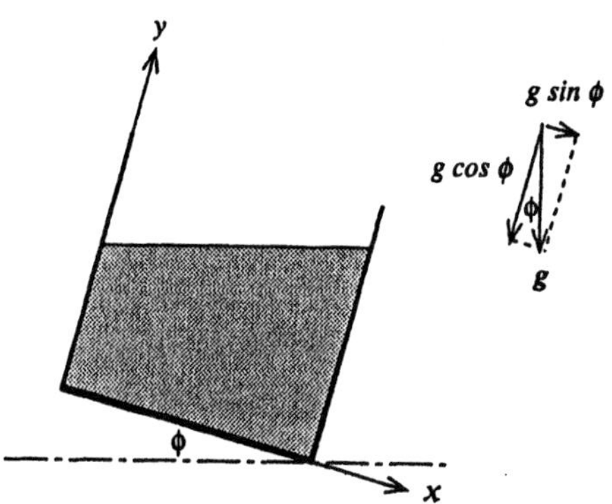

Figure E 2.2

Example 2.2

A rectangular water tank is to be installed at an angle ϕ to the horizontal, as shown in figure E 2.2. In order to calculate the stresses in the tank, the designer needs to know the fluid pressure as a function of the distances x and y from the corner of the tank (see figure E 2.2). Derive an expression for $p\{x, y\}$ equivalent to equation 2.13.

Solution

The gravitational acceleration **g** has components in the x and y directions:

$$\mathbf{g} = g \sin\phi\, \mathbf{i}_x - g\cos\phi\, \mathbf{i}_y$$

and the invariant $p - \rho\mathbf{g}\cdot\mathbf{R}$ becomes:

$$\begin{aligned} p - \rho\mathbf{g}\cdot\mathbf{R} &= p - \rho(g\sin\phi\,\mathbf{i}_x - g\cos\phi\,\mathbf{i}_y)\cdot(x\,\mathbf{i}_x + y\,\mathbf{i}_y + z\,\mathbf{i}_z) \\ &= p - \rho g(x\sin\phi - y\cos\phi) \end{aligned}$$

The equivalent of equation 2.13 is:

$$p\{x, y\} = p_1 - \rho g(x_1\sin\phi - y_1\cos\phi) + \rho g(x\sin\phi - y\cos\phi)$$

For this distribution of pressure, the horizontal lines, $y = x\tan\phi + constant$, are lines of constant pressure.

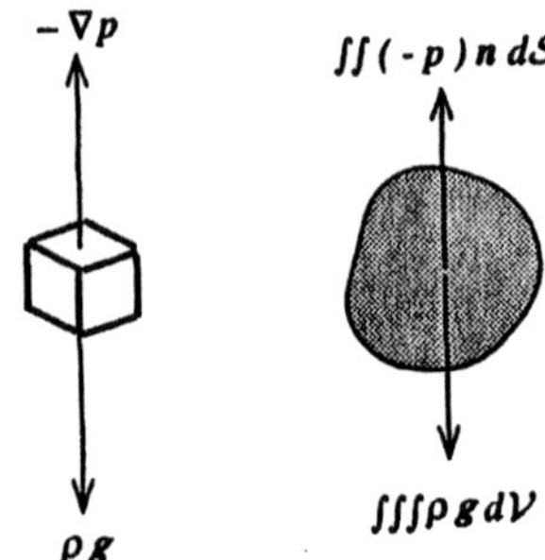

Figure 2.3 The hydrostatic force balance on (*left*) a unit volume of fluid and (*right*) a finite volume of fluid are alternate differential and integral forms, respectively, of the equation for static equilibrium in a fluid.

There is another integral of the hydrostatic equilibrium equation, equation 2.6. If we multiply it by a volume element $d\mathcal{V}$ and integrate over a finite volume $\mathcal{V}$, we find:

$$\iiint_{\mathcal{V}} (-\nabla p)\,d\mathcal{V} + \iiint_{\mathcal{V}} \rho \mathbf{g}\,d\mathcal{V} = 0$$

$$\iint_{S} (-p)\mathbf{n}\,dS + \iiint_{\mathcal{V}} \rho \mathbf{g}\,d\mathcal{V} = 0 \tag{2.14}$$

using, as we did before, Gauss' theorem (equation 1.43) to convert the volume integral to a surface integral. Equation 2.14 states that the pressure force acting on the surface of a volume of fluid plus the gravity force acting on the fluid inside the volume add to zero.

The force balance on a unit volume of fluid, equation 2.6, and that on a finite volume of fluid, equation 2.14, is illustrated in figure 2.3. These equations are the differential and integral forms of the hydrostatic force balance.

2.3.4 Pressure Measurement

Measuring Atmospheric Pressure

The standard instrument for measuring atmospheric pressure is the mercury barometer.[7] As shown in figure 2.4, it consists of glass tube about a meter in length and closed at one end. The tube, after being filled with mercury, is inverted and the open end placed

[7]The mercury barometer was invented by Evangelista Torricelli (1608–1647). Chemists often use as a unit of pressure the *torr*, which equals 1 *mm* of change in the height of the mercury column of a barometer or manometer.

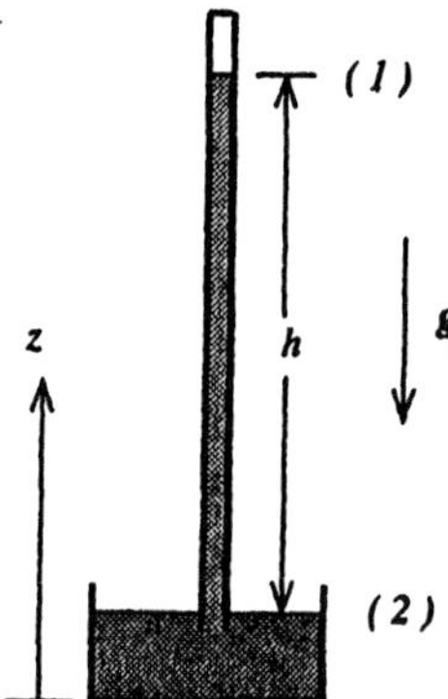

Figure 2.4 A sketch of a mercury barometer. Fluid heights are measured positively in the z-direction, gravitational acceleration is downward, and the points 1 and 2 identify the free surfaces of the mercury column and pool, respectively.

below the free surface of a pool of mercury. The mercury in the column falls below the upper end of the glass tube, leaving a space which is nearly a perfect vacuum.[8] The pressure of the atmosphere at the level of the free surface of the mercury pool is calculated from a measurement of the vertical distance h between the mercury surface in the column (1) and in the pool (2), as shown in figure 2.4, utilizing the equation of hydrostatic equilibrium, equation 2.11:

$$p_1 + \rho g z_1 = p_2 + \rho g z_2$$

$$p_2 = p_1 + \rho g(z_1 - z_2)$$

$$= \rho g h$$

where we have set $p_1 = 0$, the pressure of a vacuum. The density of mercury at $0\,°C$ is $1.360E(4)\ kg/m^3$, and the product ρg is $1.3337E(5)\ Pa/m$ at the standard value of $g = 9.8066\ m/s^2$. If the atmospheric pressure equals its standard value, $1.0133E(5)\ Pa$, then $h = 0.760\ m = 760\ mm = 29.92\ in$. Local weather forecasters often quote atmospheric pressure in inches of mercury, although the official meteorological unit of pressure is the *bar*, which equals $1.0E(5)\ Pa$, with local atmospheric pressure being quoted in *millibars*.

We think of atmospheric pressure as being constant, yet it varies with altitude just like the pressure in the mercury column of the barometer. In air, the pressure gradient, $\partial p/\partial z = -\rho g$, is much smaller in magnitude than in mercury because the density of

[8]This space is filled with mercury vapor, whose pressure at room temperature is exceedingly low compared to atmospheric pressure.

air is smaller than that of mercury by a factor of about 10^4. For differences in altitude of the order of a meter, air pressure changes by about 10^{-4} atmospheres, an amount so small as to be negligible for most engineering purposes. It is generally permissible to assume that, on a laboratory scale, the pressure of the atmosphere is constant.

The mercury barometer measures the absolute pressure of the atmosphere. The *aneroid barometer* is a mechanical instrument that measures changes in atmospheric pressure. It consists of a cylindrical container that is evacuated to a very small pressure but one end of which is flexible. As the atmospheric pressure rises or falls, the flexible end moves inward or outward in proportion, moving a dial that indicates the amount of pressure change along a linear scale that is calibrated against a mercury barometer. This same principle of pressure differences causing proportionate deformation of elastic structures is used in most pressure measuring devices.

The Manometer

The principal of the barometer can be extended to the measurement of pressure in closed containers by use of the *manometer*, as illustrated in figure 2.5. A U-shaped glass tube is partially filled with a liquid, such as water or mercury. One end of the tube is open to the atmosphere, while the other is connected to the container whose pressure is to be measured. Applying equation 2.11 to the manometer fluid (whose density is ρ_m) between the fluid surfaces at 2 and 1:

$$p_2 + \rho_m g z_2 = p_1 + \rho_m g z_1$$

$$p_2 = p_1 + \rho_m g(z_1 - z_2)$$

$$= p_{at} + \rho_m g(z_1 - z_2)$$

where we have noted that the pressure p_1 equals the atmospheric pressure p_{at}. The pressure p_2 is not necessarily the same as the pressure p_3 of the fluid at the center of the container if the fluid is a liquid. To account for this difference, we apply equation 2.11 to the container fluid (of density ρ_c) between points 3 and 2:

$$p_3 + \rho_c g z_3 = p_2 + \rho_c g z_2$$

$$p_3 = p_2 - \rho_c g(z_3 - z_2)$$

$$= p_{at} + \rho_m g(z_1 - z_2) - \rho_c g(z_3 - z_2)$$

If the container fluid is a gas, then its density ρ_c is much less than that of the manometer liquid ρ_m, and p_3 and p_2 are substantially equal. On the other hand, if the container fluid is a liquid then the pressure difference $p_2 - p_3 = \rho_c(z_3 - z_2)$ may contribute significantly to the determination of the container pressure and should not be disregarded.

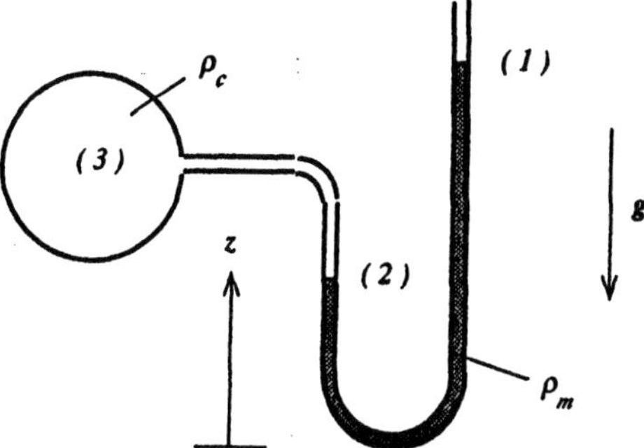

Figure 2.5 A U-tube manometer used to measure the pressure of a fluid in a container.

A manometer is commonly used in the measurement of human blood pressure. In this procedure, a flexible cuff containing an inflatable chamber is wrapped around the upper arm. A hand pump is used to inflate the cuff, which applies a positive pressure to the arm. The cuff pressure in excess of atmospheric pressure is measured by a mercury-filled manometer. When the cuff pressure is higher than the blood pressure in the artery of the arm, the artery collapses and no blood flows into the lower arm. The flow of blood through the artery can be detected by listening with a stethoscope applied to the lower arm. The cuff pressure at which blood first begins to flow to the lower arm, called the systolic pressure, is recorded as well as the lower pressure at which the blood flow to the lower arm is first impeded by the cuff inflation, called the diastolic pressure. These pressures are reported in millimeters of mercury (*mm Hg*). Normal values in healthy adults are 140/90. These are blood pressures in excess of atmospheric pressure.

Gage and Vacuum Pressures

Use of the manometer shown in figure 2.5 enables us to calculate the difference in pressure between the fluid in the container and atmospheric pressure. Unless we measure the atmospheric pressure with a barometer, we won't be able to calculate the absolute pressure of the container fluid. (The latter might be important if the container fluid is a gas and we want to calculate its density from a knowledge of its absolute pressure and temperature.) Like the manometer, most pressure measuring devices measure the difference between the absolute pressure of the pressurized fluid and atmospheric pressure. When this difference is positive, the pressure reading is called the *gage pressure*:

$$\textit{gage pressure} \equiv \textit{absolute pressure} - \textit{atmospheric pressure} \qquad (2.15)$$

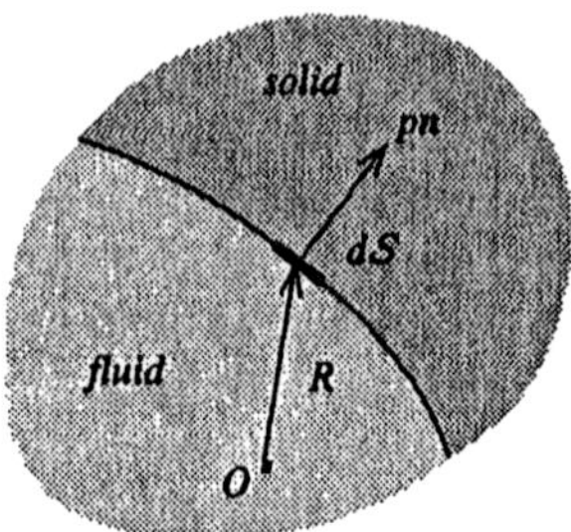

Figure 2.6 The pressure force per unit area acting on an element dS of the surface S of a solid is $p\mathbf{n}$, where $\mathbf{n}$ is the unit normal pointing outward from the fluid. $\mathbf{R}$ is the position vector of the surface element dS measured from the origin O of the coordinate system.

On the other hand, when this difference is negative, its magnitude is called the *vacuum pressure*:

$$\textit{vacuum pressure} \equiv \textit{atmospheric pressure} - \textit{absolute pressure} \tag{2.16}$$

Vacuum pressure is always positive, by definition, but cannot exceed the atmospheric pressure in magnitude.

The convention of reporting gage and vacuum pressures stems from the construction of pressure measuring devices. The most common mechanical pressure gage is the *Bourdon gage*, which consists of a somewhat flattened, long metal tube closed at one end and coiled in the form of a spiral. When the interior of the tube is connected to the pressurized fluid, the tube straightens slightly in an amount proportional to the difference between the internal pressure and atmospheric pressure. The small displacement of the tube is amplified mechanically to rotate a dial from which the pressure is read. *Electromechanical pressure gages* generate a voltage signal that is proportional to the deflection of an elastic structure responding to the pressure difference between the fluid and the atmosphere. (This is the type of gage used to record the lubricating oil pressure in an automobile.) The zero position on all these gages corresponds to an equal (atmospheric) pressure both internally and externally.

2.4 Pressure Forces on Solid Surfaces

There are many examples of structures that are stressed by pressure forces acting on them. Pressurized gas storage tanks, dams, ships' hulls, aircraft wings, and skyscrapers exemplify the necessity of designing structures to withstand the forces from fluids they contact. An important part of fluid mechanics is the determination of the pressure forces that such structures have to withstand in order to function properly. In this section, we consider the pressure forces exerted by static fluids on portions of solid structures.

In many cases, it is sufficient to know the total force **F** and moment **T** acting on a segment of a structure and caused by the pressure of the fluid in contact with the surface. We may calculate **F** and **T** by integration of the differential force $d\mathbf{F}$ and its moment $d\mathbf{T}$ that act on an element dS of the solid surface. Referring to figure 2.6, the increments in force $d\mathbf{F}$ and moment $d\mathbf{T}$ caused by the pressure p acting on the surface element dS located at the distance **R** from the origin O of the coordinate system are:

$$d\mathbf{F} = p\mathbf{n}\,dS$$

$$d\mathbf{T} = \mathbf{R} \times p\mathbf{n}\,dS$$

Note that the unit normal **n** points outward from the fluid (and into the solid). Integrating over the surface S,

$$\mathbf{F} = \iint p\mathbf{n}\,dS \tag{2.17}$$

$$\mathbf{T} = \iint (\mathbf{R} \times p\mathbf{n})\,dS \tag{2.18}$$

In general, these surface integrals do not completely enclose a volume but apply only to a portion of the structure acted upon by the fluid.

The pressure forces acting on the surface may be replaced by a single force **F** acting at a point $\mathbf{R}_{cp}$, called the *center of pressure*, located so as to give the same moment **T** as the pressure forces:

$$\mathbf{R}_{cp} \times \mathbf{F} \equiv \mathbf{T} \tag{2.19}$$

Note that the moment of the pressure forces about the center of pressure is zero:

$$\begin{aligned}
\iint (\mathbf{R} - \mathbf{R}_{cp}) \times p\mathbf{n}\,dS &= \iint (\mathbf{R} \times p\mathbf{n})\,dS - \mathbf{R}_{cp} \times \iint p\mathbf{n}\,dS \\
&= \mathbf{T} - \mathbf{R}_{cp} \times \mathbf{F} \\
&= 0
\end{aligned}$$

After **F** and **T** have been found from equations 2.17-2.18, equation 2.19 may be solved for $\mathbf{R}_{cp}$.

In calculating the pressure forces on structures that are completely or partially surrounded by the atmosphere, it is sufficient to replace the absolute pressure p by the gage pressure $p - p_{at}$ since the total pressure force on the structure is independent

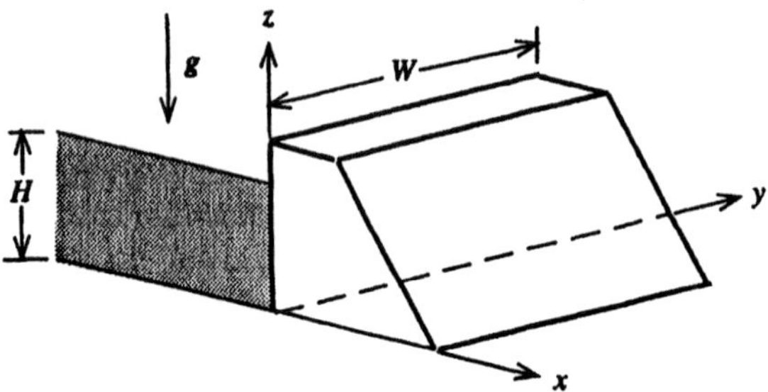

Figure 2.7 A perspective sketch of a dam of width W retaining a reservoir of water of depth H. The coordinate axes are centered at the left corner of the base of the dam.

of the magnitude of atmospheric pressure:

$$\begin{aligned}\mathbf{F} &= \iint p\mathbf{n}\,d\mathcal{S} \\ &= \iint (p - p_{at})\mathbf{n}\,d\mathcal{S} + \iint p_{at}\mathbf{n}\,d\mathcal{S} \\ &= \iint (p - p_{at})\mathbf{n}\,d\mathcal{S} - \iiint \nabla p_{at}\,d\mathcal{V} \\ &= \iint (p - p_{at})\mathbf{n}\,d\mathcal{S} + 0\end{aligned}$$

where the surface integral of atmospheric pressure acting on the entire volume of the structure is found to be zero from Gauss' theorem, equation 1.43, because $\nabla p_{at} = 0$. *Thus a uniform pressure applied to the surface of a structure produces no net force or moment on the structure.*

As an example of pressure forces and moments, consider the dam sketched in figure 2.7, which holds in place a reservoir of water of depth H and width W. To calculate the pressure force of the water on the dam, first determine the gage pressure on the dam surface at an elevation z above the dam base by applying the condition of hydrostatic equilibrium, equation 2.13:

$$\begin{aligned}p(z) &= p_1 + \rho g z_1 - \rho g z \\ &= \rho g H - \rho g z\end{aligned}$$

where we have taken the point 1 to be at the surface of the reservoir and noted that the gage pressure of the atmosphere is zero. We next determine $\mathbf{F}$ by substituting this expression for p into equation 2.17:

$$\mathbf{F} = \iint p\mathbf{n}\, dS$$

$$= \int_0^W \int_0^H \rho g(H - z)\mathbf{i}_x\, dy\, dz$$

$$= W\rho g \left(Hz - \frac{z^2}{2}\right)\Big|_0^H \mathbf{i}_x$$

$$= \rho g \left(\frac{WH^2}{2}\right) \mathbf{i}_x$$

Note that the average pressure experienced by the dam is $\rho g H/2$ and that the force magnitude F is the product of the average pressure times the area WH.

The moment $\mathbf{T}$ is determined by substituting the expression for the pressure into equation 2.18:

$$\mathbf{T} = \iint (\mathbf{R} \times p\mathbf{n})\, dS$$

$$= \int_0^H \int_0^W (x\mathbf{i}_x + y\mathbf{i}_y + z\mathbf{i}_z) \times \rho g(H - z)\mathbf{i}_x dy\, dz$$

$$= \rho g \int_0^H \int_0^W (z\mathbf{i}_y - y\mathbf{i}_z)(H - z) dy\, dz$$

$$= \rho g \int_0^H \left(Wz\mathbf{i}_y - \frac{W^2}{2}\mathbf{i}_z\right)(H - z) dz$$

$$= \rho g W \left| \frac{Hz^2}{2} - \frac{z^3}{3} \right|_0^H \mathbf{i}_y - \rho g \frac{W^2}{2} \left| Hz - \frac{z^2}{2} \right|_0^H \mathbf{i}_z$$

$$= \rho g \left(\frac{WH^3}{6}\right) \mathbf{i}_y - \rho g \left(\frac{W^2H^2}{4}\right) \mathbf{i}_z$$

The moment $\mathbf{T}$ of the pressure force of the water has two components. The component T_y in the y direction has the magnitude of the force F times a distance $H/3$:

$$T_y = \rho g \left(\frac{WH^3}{6}\right) = \rho g \left(\frac{WH^2}{2}\right)\left(\frac{H}{3}\right) = F\left(\frac{H}{3}\right)$$

while the z component T_z has the magnitude of the force F times a distance $W/2$:

$$T_z = -\rho g \left(\frac{W^2H^2}{4}\right) = -\rho g \left(\frac{WH^2}{2}\right)\left(\frac{W}{2}\right) = -F\left(\frac{W}{2}\right)$$

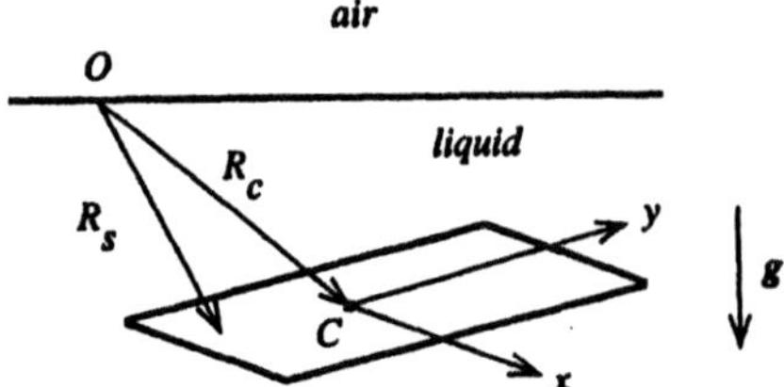

Figure 2.8 The coordinate system for determining forces and moments on a plane surface in contact with a fluid.

The distance to the center of pressure needed to produce the moment **T** is thus:

$$\mathbf{R}_{cp} = \left(\frac{H}{3}\right)\mathbf{i}_z + \left(\frac{W}{2}\right)\mathbf{i}_y$$

which satisfies equation 2.19. The center of pressure is located halfway along the width of the dam and one-third of the water depth above the dam base. To overcome both the force **F** and the moment **T**, which tend to displace and topple the dam, the thickness and mass of the dam need to be made sufficiently great that it will not translate or rotate by deforming the soil beneath.

2.4.1 Pressure Force on a Plane Surface

A simplification of equations 2.17 and 2.18 for the force **F** and moment **T** on a solid surface is possible whenever the surface is a plane. Such a situation is illustrated in figure 2.8, showing a plane surface of arbitrary shape located beneath a liquid surface whose pressure is p_a. Choosing a cartesian coordinate system fixed in the plate, with origin at the plate centroid C and axes x and y lying in the plane of the plate, the position vector $\mathbf{R}_s$ of a point on the plate surface is:

$$\mathbf{R}_s = \mathbf{R}_C + x\mathbf{i}_x + y\mathbf{i}_y \tag{2.20}$$

where $\mathbf{R}_C$ is the position vector of the centroid C as measured from the origin O on the liquid surface (see figure 2.8). The centroid position $\mathbf{R}_C$ is the area-averaged position of the surface area S:

$$\mathbf{R}_C \equiv \frac{1}{S}\iint \mathbf{R}_s \, dS \tag{2.21}$$

From equation 2.9, noting that $p = p_a$ and $\mathbf{g} \cdot \mathbf{R} = 0$ on the liquid surface, the pressure p_s on the plane surface is:

$$p_s - \rho\mathbf{g}\cdot\mathbf{R}_s = p_a$$

$$p_s = p_a + \rho\mathbf{g}\cdot\mathbf{R}_s = p_a + \rho\mathbf{g}\cdot\mathbf{R}_C + \rho(g_x x + g_y y)$$

$$= p_C + \rho(g_x x + g_y y) \tag{2.22}$$

where $p_C \equiv p_a + \rho\mathbf{g}\cdot\mathbf{R}_C$ is the pressure at the centroid C, and g_x and g_y are the components of $\mathbf{g}$ in the x- and y-directions.

If C is the centroid of the plane surface, then the moments about the x- and y-axes of a uniformly distributed unit pressure on the surface must be zero:

$$\iint x\,dS = 0; \quad \iint y\,dS = 0 \tag{2.23}$$

Using this relation, we can determine the force $\mathbf{F}$ from equation 2.17 to be:

$$\mathbf{F} = \mathbf{n}\iint p_s\,dS$$

$$= \mathbf{n}\iint p_C\,dS + \rho\mathbf{n}\iint (g_x x + g_y y)\,dS$$

$$= (p_c A)\,\mathbf{n} \tag{2.24}$$

where A is the plate area. Thus the force $\mathbf{F}$ acting on a plane surface having any orientation equals the product of the pressure at the centroid C times the plate area A and of course acts in a direction normal to the plate.

For a regular figure, such as a square, rectangle, circle, ellipse or triangle, the centroid location is easily found by symmetry. For irregular figures, it can be obtained by simple integration so as to satisfy equation 2.23.

The *center of pressure cp* is the point on the plane surface about which the moment of the pressure force is zero:

$$\iint p(x - x_{cp})\,dS = 0; \quad \iint p(y - y_{cp})\,dS = 0 \tag{2.25}$$

where $x_{cp}\mathbf{i}_x + y_{cp}\mathbf{i}_y$ is the distance from the centroid C to the center of pressure cp. To satisfy this condition, substitute equation 2.22 into 2.25 and simplify by using 2.23:

$$\iint [p_C(x - x_{cp}) + \rho(g_x x + g_y y)(x - x_{cp})]\,dS = 0$$

$$-p_C x_{cp} A + \rho g_x \iint x^2\,S + \rho g_y \iint xy\,dS = 0$$

$$-p_C x_{cp} A + \rho(g_x I_{xx} + g_y I_{xy}) = 0$$

where $I_{xx} \equiv \iint x^2\,dS$ and $I_{xy} \equiv \iint xy\,dS$ are the moments of inertia of the plane surface about the centroid C. Solving for x_{cp} and similarly for y_{cp}, we find:

$$x_{cp} = \frac{\rho(g_x I_{xx} + g_y I_{xy})}{p_C A}$$

$$y_{cp} = \frac{\rho(g_y I_{yy} + g_x I_{xy})}{p_C A} \qquad (2.26)$$

The moment $\mathbf{T}$ of the pressure force about the origin O is simply the moment of the force $\mathbf{F}$ (equation 2.24) acting through the center of pressure cp, as in 2.19:

$$\mathbf{T} = (\mathbf{R}_C + x_{cp}\mathbf{i}_x + y_{cp}\mathbf{i}_y) \times (p_c A)\,\mathbf{n} \qquad (2.27)$$

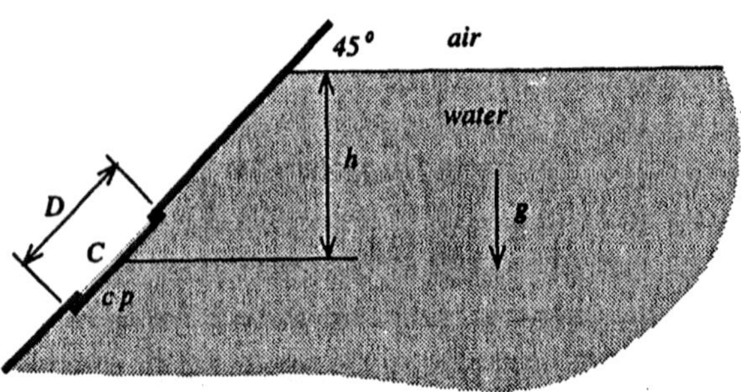

Figure E 2.3

Example 2.3

A circular flat plate of diameter $D = 1\,m$ closes off an opening in a ship's hull at a distance $h = 3\,m$ below the water surface. The plane of the plate is 45° from the vertical, as shown in figure E 2.3. If the water density, $\rho = 1E(3)\,kg/m^3$, calculate the total force exerted on the plate by the water and the distance between the center of pressure cp and the centroid of the circular plate. (For a circle, $I_{yy} = \pi D^4/64$.)

Solution

The gage pressure p_C at the plate centroid is:

$$p_C = \rho g h = (1E(3)\,kg/m^3)(9.807\,m/s^2)(3\,m) = 2.942E(4)\,Pa$$

so that the force F becomes:

$$F = p_C A = (2.942E(4)\,Pa)\frac{\pi(1\,m^2)}{4} = 2.311E(4)\,N$$

Taking y upward along the plate and x horizontal, $g_x = 0$ and $g_y = -g/\sqrt{2}$. From equation 2.26,

$$y_{cp} = \frac{\rho(-g/\sqrt{2})I_{yy}}{p_c A} = -\frac{\rho g D^2}{16\sqrt{2}p_c}$$

$$= -\frac{(1E(3)\,kg/m^3)(9.807\,m/s^2)(1\,m^2)}{16\sqrt{2}(2.942E(4)\,Pa)} = -1.473E(-2)\,m$$

2.4.2 Pressure Force on a Curved Surface

For curved surfaces, there is no general simplification of the expressions for the force **F** and moment **T** corresponding to those for plane surfaces. For curved surfaces that are portions of regular shapes, such as spheres, cylinders or cones, it may be possible to define a coordinate system that makes it easy to determine the surface area element dS and the normal **n** so that the integrals of equations 2.17 and 2.18 can be evaluated. Alternatively, it may be possible to form an imaginary closed surface $\mathcal{S}$, of which the curved surface is part but of which the remaining parts are simple plane or cylindrical surfaces for which it is easy to compute the forces and moments exerted by the surrounding fluid. By performing a force and moment balance on this imaginary surface $\mathcal{S}$ enclosing a volume of fluid, it will be possible to solve for the unknown force and moment acting on the given curved surface.

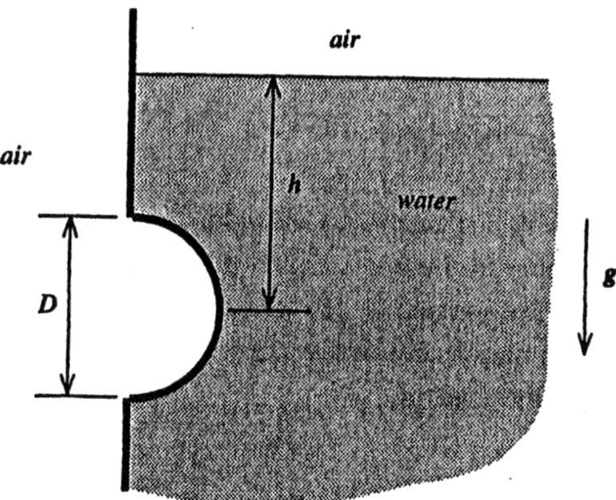

Figure E 2.4

Example 2.4

In the flat side of a tank holding water there is a hemispherical surface of diameter D at a distance h below the water surface, as shown in figure E 2.4. Derive expressions for the horizontal force F_h and upward force F_v exerted on the hemispherical surface by the surrounding water.

Solution
Replace the hemispherical surface by a closed hemispherical volume completely surrounded by water and filled with water. Such a volume would be in equilibrium with the pressure forces on its surfaces. In the horizontal direction, the pressure force (F_h) on the hemispherical portion of the surface would be balanced by that on the plane vertical circular surface of area $\pi D^2/4$ and centroid pressure $p_C = p_a + \rho g h$, so that:

$$F_h = (p_a + \rho g h)\left(\frac{\pi D^2}{4}\right)$$

In the vertical direction, the upward pressure force (F_v) on the hemispherical surface (there is no upward component of the plane portion) must balance the pull of gravity on the fluid inside the volume. This volume is $(1/2)(4\pi/3)(D/2)^3 = \pi D^3/12$, so that:

$$F_v = \frac{\pi \rho g D^3}{12}$$

2.5 Pressure Forces on Bodies Immersed in Fluids

2.5.1 Archimedes' Principle

If a body completely enclosed by a solid surface is immersed in a fluid, the total pressure force acting on the body, called the *buoyant force* and denoted by $\mathbf{F}_b$, is readily calculated from equation 2.17 because the pressure force integral over the surface of the body may be converted to a volume integral over the volume $\mathcal{V}$ of the fluid displaced by the structure:

$$\mathbf{F}_b = \iint_S p\mathbf{n}\, dS$$

$$= -\iiint_{\mathcal{V}} \nabla p \, d\mathcal{V}$$

$$= -\iiint_{\mathcal{V}} \rho \mathbf{g} d\mathcal{V}$$

$$= -\rho \mathbf{g} \mathcal{V} \tag{2.28}$$

Here we have used equation 1.43 to form the volume integral (noting that **n** in equation 2.17 is an inward normal for the displaced volume $\mathcal{V}$) and equation 2.6 to replace the pressure gradient by the gravitational force per unit volume. Equation 2.28 states that *the pressure force (buoyant force) on an immersed body is equal in magnitude but opposite in direction to the force of gravity acting on the displaced fluid.* This is *Archimedes' principle.*[9]

The position $\mathbf{R}_b$ of the *center of buoyancy* of a floating or submerged object is defined as that of the mass center of the displaced fluid:

$$\mathbf{R}_b \equiv \frac{1}{\mathcal{V}} \iiint_{\mathcal{V}} \mathbf{R} \, d\mathcal{V} \tag{2.29}$$

The buoyant force $\mathbf{F}_b$ can be considered to act through the center of buoyancy since the moment $\mathbf{T}_b$ of the gravity force on the displaced fluid is $\mathbf{R}_b \times \mathbf{F}_b$:

$$\mathbf{T}_b = -\iiint_{\mathcal{V}} (\mathbf{R} \times \rho \mathbf{g}) \, d\mathcal{V} = \iiint_{\mathcal{V}} \mathbf{R} \, d\mathcal{V} \times (-\rho \mathbf{g})$$

$$= \mathbf{R}_b \mathcal{V} \times (\mathbf{F}_b/\mathcal{V}) = \mathbf{R}_b \times \mathbf{F}_b \tag{2.30}$$

where we have used equations 2.28 and 2.29. The moment of the pressure force about the center of buoyancy is zero.

When a body floats at the interface between two fluids, such as a boat floating on water, each fluid will contribute to the total buoyant force an amount $\rho g \mathcal{V}$ equal to the force of gravity acting on the displaced volume of the respective fluids. However, in the case of a body floating on water, the density of air is so much less than that of water that we can neglect the buoyant force of air compared to that of water and consider only the gravitational force of the displaced volume of water when calculating the buoyant force $\mathbf{F}_b$.

---

[9]Archimedes (287(?)-212 B.C.) was the most important mathematician of his millenium. He contributed to statics and dynamics as well as the hydrostatics of floating and submerged bodies.

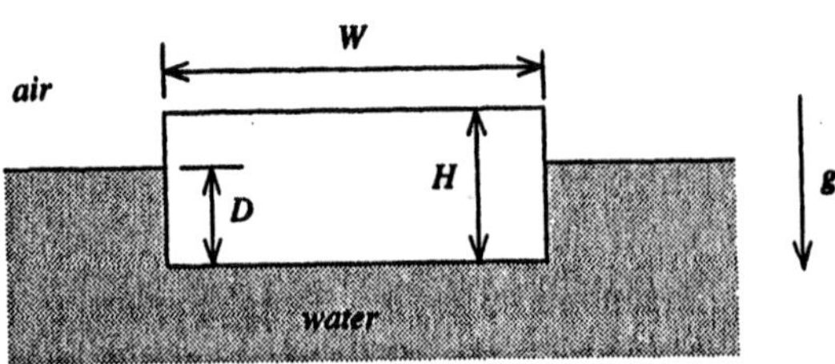

Figure E 2.5

Example 2.5

A bar of soap floats on the surface of water, its bottom surface a distance D below the surface of the water, as shown in figure E 2.5. The bar has a width W, a thickness H and a length L normal to the plane of figure E 2.5. What is the specific gravity SG of the soap?

Solution
The density of the soap is $(SG)\rho$, and the gravity force acting on the soap is $(SG)\rho g WHL$. According to Archimedes principle, equation 2.28, this is equal in magnitude to the gravity force acting on the displaced fluid mass, ρDWL:

$$(SG)\rho g WHL = \rho g DWL$$

$$SG = \frac{D}{H}$$

2.5.2 Equilibrium of Immersed Bodies

Static Equilibrium

A body of mass M immersed in a fluid will not move if the forces (and their moments) acting on the body add to zero. The forces consist of the buoyant force $\mathbf{F}_b$, the gravitational force $M\mathbf{g}$ and any external force $\mathbf{F}_{ex}$ that might be present, such as the tension in a balloon tether or an anchor line. Utilizing equation 2.28, the force balance may be written as:

$$M\mathbf{g} - \rho \mathcal{V} \mathbf{g} + \mathbf{F}_{ex} = 0 \qquad (2.31)$$

In the absence of an external force, a body will remain stationary only if its mass M is equal to $\rho\mathcal{V}$, the mass of the displaced fluid. (A submarine maintains this balance by adjusting the volume of water in its ballast tanks.) A body floating at an air-liquid interface sinks to a draft that just provides the displaced volume necessary to balance the mass of the body. (When a ship takes on cargo, it floats deeper in the water by an amount needed to offset the gravitational force acting on the cargo.)

It is also necessary that the moments of these forces add to zero:

$$\mathbf{R}_g \times M\mathbf{g} - \mathbf{R}_b \times \rho\mathcal{V}\mathbf{g} + \mathbf{R}_{ex} \times \mathbf{F}_{ex} = 0 \tag{2.32}$$

where $\mathbf{R}_{ex}$ is the point of application of the external force $\mathbf{F}_{ex}$ and $\mathbf{R}_g$ is the position of the center of gravity of the body. If there is no external force, then the moment balance, together with the force balance, requires that:

$$(\mathbf{R}_g - \mathbf{R}_b) \times \mathbf{g} = 0 \tag{2.33}$$

Thus the center of gravity and the center of buoyancy lie on the same vertical line, making $\mathbf{R}_g - \mathbf{R}_b$ parallel to $\mathbf{g}$ and their vector product zero.

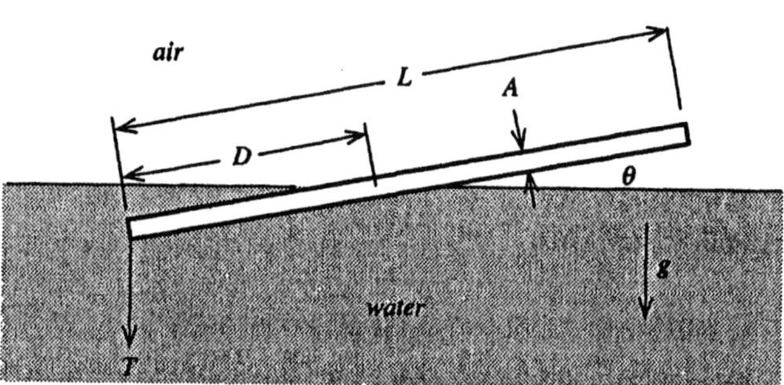

Figure E 2.6

Example 2.6

A slender pole floating in a pond is anchored by a line tied to one end, as shown in figure E 2.6, so that the tethered end is submerged and the other extends into the air. The pole lies at an angle θ with the horizontal. The pole has a length L, a uniform cross-sectional area A and a density ρ_p that is less than the water density ρ_w. Derive expressions for the length D of the submerged portion of the pole and the tension T in the anchor line, in terms of the parameters ρ_p, ρ_w, A and L.

Solution

Taking the moments of the gravity force $\rho_p gAL$ and the buoyant force $\rho_w gAD$ about the point of attachment of the anchor line, the condition of rotational equilibrium, equation 2.32, is:

$$\left(\frac{L}{2}\cos\theta\right)\rho_p gAL - \left(\frac{D}{2}\cos\theta\right)\rho_w gAD = 0$$

$$D^2\rho_w = L^2\rho_p$$

$$D = \left(\frac{\rho_p}{\rho_w}\right)^{\frac{1}{2}} L$$

Applying Archimedes' principle, equation 2.28, the equilibrium of vertical forces gives:

$$T + \rho_p gAL = \rho_w gAD$$

$$T = \rho_w gA\left(\frac{\rho_p}{\rho_w}\right)^{\frac{1}{2}} L - \rho_p gAL$$

$$= \rho_p gAL\left[\left(\frac{\rho_w}{\rho_p}\right)^{\frac{1}{2}} - 1\right]$$

Note that D and T are independent of the angle θ.

Stable Equilibrium

The balance of forces and moments is necessary for an enclosed body immersed in a fluid to remain stationary. However, this equilibrium might be precarious, like that of a pin balanced on its point. Stable equilibrium requires that, when disturbed slightly from its equilibrium position, the body will tend to return to that position.

Let us examine how this principle might be applied to the case of a body completely surrounded by a fluid. In figure 2.9, the buoyant force $\mathbf{F}_b$ acts through the center of buoyancy B while the gravity force $\rho g\mathcal{V} = -\mathbf{F}_b$ acts through the center of mass G. As long as B and G are aligned vertically, the body is in static equilibrium. But suppose the body is rotated clockwise through a small angle ϵ, as shown in figure 2.9. If G lies below B, then there will be a restoring couple of magnitude $\rho g\mathcal{V}[BG]\epsilon$, where $[BG]$ is the distance between B and G. On the other hand, if G lies above B, the couple will tend to increase the angle of rotation. In that case, the body will eventually turn upside down, so that the center of gravity will lie below the center of buoyancy. We

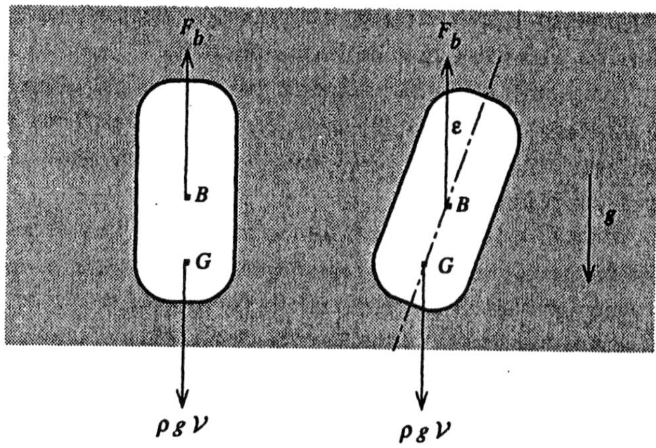

Figure 2.9 Illustrating the stability of equilibrium of a body submerged in a fluid.

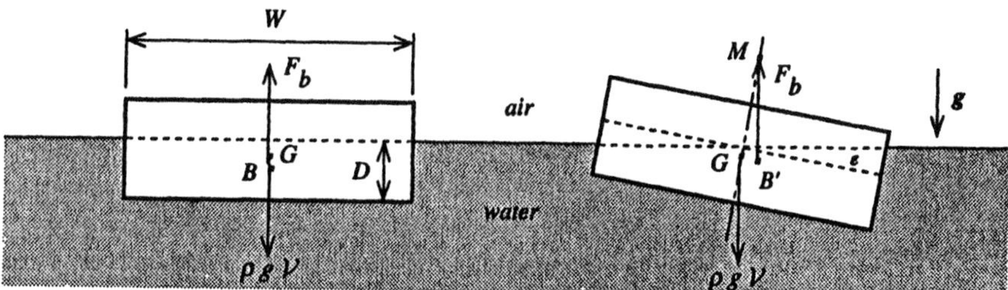

Figure 2.10 The stable equilibrium of a floating body results from the movement of the center of buoyancy when the body is tilted through a small angle.

can conclude that the stability of a submerged body requires that the center of gravity lie below the center of buoyancy.

Does the same principle apply to a body floating on the surface of a liquid? Consider the case of a bar of soap floating freely on water, as shown in figure 2.10. The center of gravity is located halfway between the upper and lower surfaces of the bar. However, the center of buoyancy is located halfway between the lower surface and the water line and hence is below the center of gravity. Yet we know that the bar is stable because it will return to this position when perturbed.

Consider how this differs from the case of a submerged body. The behavior of a floating bar of soap when rotated clockwise through a small angle ε is also illustrated in figure 2.10. While the displaced volume of water has not changed, the center of buoyancy has shifted to the right because more of the right side of the bar is submerged than the left. If the new center of buoyancy, B', lies to the right of the center of gravity, then the bar will return to its original position.

The point M on the bar centerline that lies above B' when the bar is tilted through a small angle ε is called the *metacenter*. If the metacenter M lies above the center of gravity G, as in figure 2.10, then the bar will be stable and return to its original position when disturbed.

To locate the metacenter M for the bar of soap, we will first find an expression for the moment of the displaced volume about the center of buoyancy B'. The displaced volume consists of two components, the original displaced volume DHL (L is the length of the bar of soap) plus a wedge-shaped piece of volume $(1/2)(W/2)(\varepsilon W/2)L = \varepsilon W^2L/8$ removed from the left side and added to the right side, its center of gravity having been displaced a distance of $2W/3$. The sum of the moments of both these components about the point B' must add to zero:

$$-DHL \times [BB'] + \varepsilon\left(\frac{W^2L}{8}\right)\left(\frac{2W}{3}\right) = 0$$

$$[BB'] = \varepsilon\left(\frac{W^2}{12D}\right) \tag{2.34}$$

However, the distance $[BB']$ is equal to $\varepsilon\,[BM]$, so that:

$$[BM] = \frac{W^2}{12D} \tag{2.35}$$

We can conclude that the stability of a floating body is improved by making the width W large and the draft D small and by keeping the center of mass as low as possible (thereby increasing $[GM]$). This explains why a rowboat is more stable than a canoe and why it is destabilizing to stand up in a canoe! It also explains why a bar of soap will not float stably on its edge, *i.e.*, with the smallest dimension in the horizontal direction.

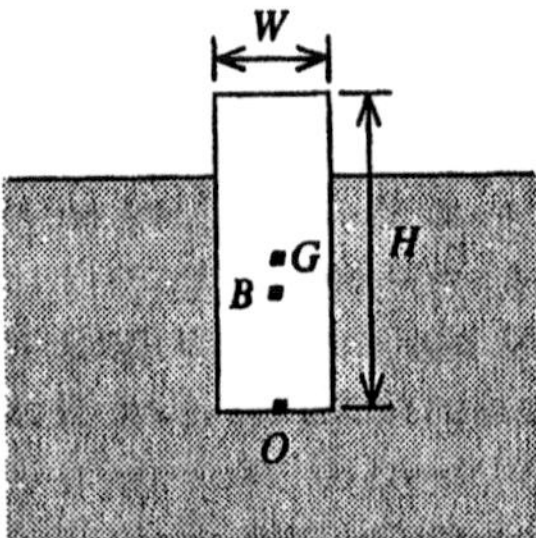

Figure E 2.7

Example 2.7

A rectangular block of wood floats on the surface of water, as illustrated in figure E 2.7. The block has a width W, a height H and a specific gravity SG. Find the minimum ratio of width to height W/H that will ensure stability, *i.e.*, for which $GM = 0$.

Solution

Denote the point at the bottom center of the block by O, the center of gravity by G and the center of buoyancy by B, as shown in figure E 2.7. The distance $[OG]$ is half of H because the density of the block is uniform and $[OB]$ is half of the distance $SG \times H$ of submergence of the block. Thus the distance $[BG]$ is:

$$[BG] = (1 - (SG))\frac{H}{2}$$

The distance $[BM]$ from B to the metacenter M can be evaluated from equation 2.35 by noting that the distance $D = SG \times H$:

$$[BM] = \frac{W^2}{12(SG)H}$$

Setting $[BM] = [BG]$ gives the condition for marginal stability, $[GM] = 0$:

$$\frac{W^2}{12(SG)H} = (1 - (SG))\frac{H}{2}$$

$$\left(\frac{W}{H}\right) = \sqrt{6(SG)(1 - SG)}$$

For a cube ($W = H$) to float upright, SG must be smaller than or greater than the smaller or larger root of $6(SG)(1 - SG) = 1$:

$$SG \leq \frac{3 - \sqrt{3}}{6}$$

$$\leq 0.2113$$

$$SG \geq \frac{3 + \sqrt{3}}{6}$$

$$\geq 0.7887$$

Ice cubes and styrofoam cubes will float upright, but not soap cubes!

2.6 Stratified Fluids

Sometimes the density of a static fluid in a gravitational field is not uniform. In the atmosphere, the air density declines precipitously with altitude. In the deep ocean, the fluid density at the bottom is greater than that at the surface. When fluids are immiscible, such as water and air or gasoline and water, the denser fluid settles to the bottom of a container enclosing the two fluids. In all these instances of ***stratified fluids***, the integral relation of equation 2.13, expressing the pressure as a function of height z in the gravitational field, is no longer true because the density of the fluid does not have a constant value independent of z.

What will be the pressure in a stratified fluid? Certainly, the pressure will decrease as z increases. However, the density of the fluid will also tend to decrease as the pressure decreases, so that the rate of decrease of pressure, which is proportional to density (equation 2.6), will be smaller in magnitude as z increases. This coupling of the density with pressure is important in the atmosphere and the ocean over vertical distances of several kilometers but is insignificant for most engineering systems. In the latter case, density differences arise principally from heating or cooling, from differences in chemical composition or from phase differences of the fluid layers.

2.6.1 Stability of Stratified Fluids

Before determining the pressure distribution in a static stratified fluid, let us first determine the conditions that are needed to ensure that a fluid of variable density will remain motionless in a gravitational field. Taking the curl of the hydrostatic force balance equation 2.6, we find:

$$-\nabla \times (\nabla p) + \nabla \times (\rho \mathbf{g}) = 0$$

$$0 + \rho(\nabla \times \mathbf{g}) + (\nabla \rho) \times \mathbf{g} = 0$$

$$(\nabla \rho) \times \mathbf{g} = 0 \tag{2.36}$$

where we have used vector calculus equations 1.36 and 1.39 to evaluate the curl products and have noted that $\nabla \times \mathbf{g}$ is zero since $\mathbf{g}$ is a constant. equation 2.36 requires that $\nabla \rho$ is either zero or is a vector having the same direction as $\mathbf{g}$. Since $\mathbf{g}$ has the vertical direction, *ρ must be a function of vertical height alone* in order to ensure that $\nabla \rho$ has only one component, in the direction of $\mathbf{g}$. This is the condition of *static stability*.

This certainly agrees with our everyday experience. The surface of water in a glass is horizontal, the density of the fluid changing abrubtly from that of water to that of

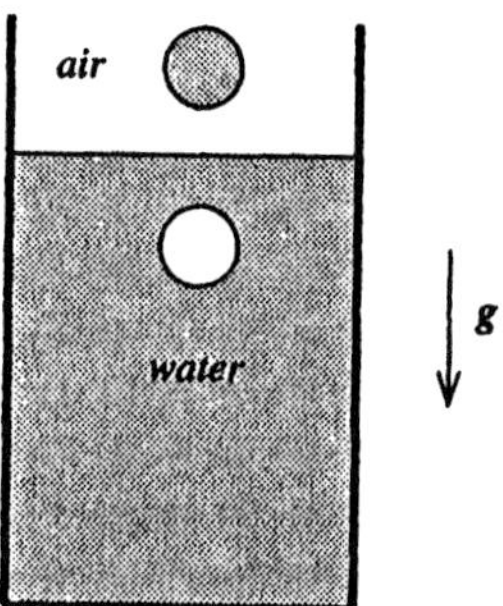

Figure 2.11 The dynamic stability of a dense fluid beneath a less dense fluid depends upon the tendency of fluid disturbances to return to the undisturbed state.

air at the height of this surface irrespective of horizontal location within the glass. A lighter fluid poured into a container of denser fluid will float on top, forming a horizontal interface between the two fluids.

It is also obvious from daily experience that the denser fluid always underlies the lighter. If we invert a container half filled with water and air, the water always flows to the bottom. But to have a layer of water above the layer of air does not violate the condition of static stability provided the interface remains horizontal. We never observe such upside-down fluids because the interface of a dense fluid above a less dense fluid is dynamically unstable, like a pin balanced on its point.

To illustrate the principle of dynamic stability, consider the container half filled with water and air shown in figure 2.11. Suppose we interchange a small droplet of water with an equal volume of air, resulting, as shown, in an air bubble below the interface and a droplet suspended above it. This will require some work, not only to raise the drop through the air and to push the bubble into the liquid, but also to form the air/water interface on the surface of the bubble and droplet where surface tension exists. If we release the droplet and the bubble, the former will drop, while the latter will rise to the air-water interface, annihilating each other. Since the system returns spontaneously to its original configuration after being disturbed in this manner, we say that it is *dynamically stable.*

Now turn your textbook upside down and study figure 2.11 for the case where the water layer lies above the air. Upon release of the bubble and the droplet, the bubble will rise further into the water and the droplet will descend into the air. The upside-down configuration is definitely unstable since it departs further from its original state. We call this *dynamic instability*. Note also that the work to displace the droplet and bubble is negative, although positive work is required to form the interfaces against the force of surface tension. For very small samples of water, such as a drop suspended from the end of an eyedropper, it is possible to have a dynamically stable interface with water above air, as is found at the bottom of such a drop, because of the stabilizing influence of surface tension which requires positive work when the interface is deformed.

2.6.2 Pressure in Stratified Fluids

When the fluid density is a known function of the height z in the gravitational field, say $\rho\{z\}$, the pressure $p\{z\}$ as a function of height can be found by integrating the equation of hydrostatic force balance, equation 2.6. Taking the vertical (z) component of this equation, we can find a total differential equation for the pressure:

$$-(\nabla p)\cdot \mathbf{i}_z + \mathbf{g}\cdot \mathbf{i}_z\,\rho\{z\} = 0$$

$$-\left(\frac{dp}{dz}\right) - g\,\rho\{z\} = 0$$

$$dp + g\,\rho\{z\}\,dz = 0 \qquad (2.37)$$

Integrating between a point z_0 where the pressure p_0 is known and any point z within the fluid, we find:

$$\int_{p_0}^{p\{z\}} dp + g\int_{z_0}^{z} \rho\{z\}\,dz = 0$$

$$p\{z\} = p_0 - g\int_{z_0}^{z} \rho\{z\}\,dz \qquad (2.38)$$

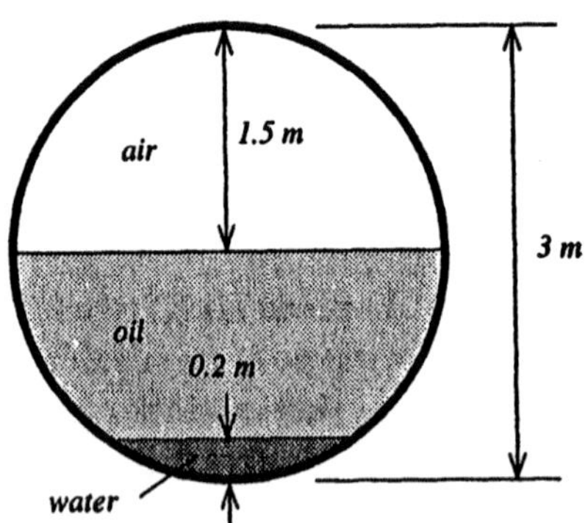

Figure E 2.8

Example 2.8

A horizontal cylindrical fuel oil storage tank of internal diameter 3 m is half full of liquid. The liquid consists of a layer of fuel oil ($SG = 0.87$) above a layer of water whose thickness is 0.2 m, as shown in figure E 2.8. The upper half of the tank is vented to the atmosphere. Calculate the gage pressure at the bottom of the tank.

Solution

Measuring the height z from the oil-air interface, the bottom of the tank is at $z = -1.5\,m$, and the water-oil interface is at $z = -1.3\,m$. The pressure p_b at the tank bottom may be determined using equation 2.38:

$$p_b = p_0 - g\int_0^{-1.3\,m} \rho_o dz - g\int_{-1.3m}^{-1.5m} \rho_w dz$$

$$\begin{aligned} p_b - p_0 &= \rho_w g(SG)(1.3\,m) + \rho_w g(0.2\,m) \\ &= 1.0E(3) kg/m^3 \times 9.807\,m/s^2 (0.87 \times 1.3\,m + 0.2\,m) \\ &= 1.3053E(4)\,Pa \end{aligned}$$

2.6.3 The Earth's Atmosphere

The earth's atmosphere is a comparatively thin layer of gas held to the earth's surface by gravitational attraction. Its density and pressure are greatest at the earth's surface, and both decline approximately exponentially with altitude. Most of the mass of the atmosphere is contained within the first ten kilometers of altitude. The atmospheric mass is not so large as to be unaffected on a global scale by human activity. In recent decades, gases mixed into the atmosphere have resulted in a decrease in the amount of ozone and an increase in surface temperature because of the interaction of the atmosphere with sunlight.

For the purpose of calculating the pressure and density of the atmosphere, we can regard air as a perfect gas obeying the perfect gas law equation 1.11, where the gas constant R does not vary with altitude.[10] Substituting the perfect gas law into the differential equation of force balance, equation 2.37, and integrating, we find an

---

[10] At extremely high altitudes, sunlight dissociates and ionizes the air molecules, causing an increase in the gas constant.

expression for the pressure $p\{z\}$:

$$dp + g\left[\frac{p}{RT\{z\}}\right]dz = 0$$

$$\left(\frac{dp}{p}\right) + \left(\frac{g}{R}\right)\frac{dz}{T\{z\}} = 0$$

$$\int_{p_0}^{p(z)}\left(\frac{dp}{p}\right) + \left(\frac{g}{R}\right)\int_0^z \frac{dz}{T\{z\}} = 0$$

$$\ln\left[\frac{p\{z\}}{p_0}\right] + \left(\frac{g}{R}\right)\int_0^z \frac{dz}{T\{z\}} = 0$$

$$p\{z\} = p_0 \exp\left[-\left(\frac{g}{R}\right)\int_0^z \frac{dz}{T\{z\}}\right] \qquad (2.39)$$

where p_0 is the atmospheric pressure at the earth's surface, $z = 0$. The density $\rho\{z\}$ may be found readily by dividing equation 2.39 by $RT\{z\}$. Note that the atmospheric absolute temperature $T\{z\}$ must be known as a function of altitude in order to evaluate the integral of equation 2.39.

The Isothermal Atmosphere

A close approximation to the earth's atmospheric pressure distribution may be obtained from equation 2.39 if we assume that the atmospheric temperature is everywhere the same, say T_0. In that case the integral may be evaluated to give:

$$p\{z\} = p_0 \exp\left[-\left(\frac{g}{RT_0}\right)z\right] \qquad (2.40)$$

The pressure decreases exponentially with altitude, declining by a factor of e for each increment of RT_0/g in altitude. The length RT_0/g, called the *scale height* of the atmosphere, has the value of 8.433 km when $T_0 = 288.15\,K$ (15 C).[11]

The U.S. Standard Atmosphere

The temperature of the atmosphere varies somewhat with altitude depending upon the location on the earth and the time of the year. Averaged over the continental U.S. and over many years, the temperature profile $T\{z\}$ may be represented by a piecewise

[11]For air, $R = 287.0\,J/kg\,K$ (see Table 1.2) and $R/g = 29.26\,m/K$.

Table 2.1 Properties of the U.S. Standard Atmosphere

i	z_i (km)	T_i (K)	$\left(\frac{dT}{dz}\right)_{i,i+1}$ (K/km)	p_i (Pa)	ρ_i (kg/m^3)
0	0	288.15		1.0133E(5)	1.225
			−6.5		
1	11	216.65		2.263E(4)	3.639E(-1)
			0.0		
2	20	216.65		5.472E(3)	8.800E(-2)
			1.0		
3	32	228.65		8.673E(2)	1.322E(-2)

linear function:

$$T_{i,i+1}\{z\} = T_i + \left(\frac{dT}{dz}\right)_{i,i+1}(z - z_i), \qquad z_i \leq z \leq z_{i+1} \tag{2.41}$$

where T_i is the temperature at the lower edge of each interval i and $(dT/dz)_{i,i+1}$ is the constant temperature gradient within the interval. When this expression is substituted into equation 2.40 and the integral evaluated, the pressure distribution in each linear segment is found to be:

$$p_{i,i+1}\{z\} = p_i \left[\frac{T_i + (dT/dz)_{i,i+1}(z - z_i)}{T_i}\right]^{-(g/R)(dz/dT)_{i,i+1}}, \qquad z_i \leq z \leq z_{i+1} \tag{2.42}$$

The values of z_i, T_i, $(dT/dz)_{i,i+1}$, p_i and ρ_i are exhibited in table 2.1 for altitudes that include the troposphere and stratosphere. Values of T, p and ρ at intermediate altitudes may be found by use of equations 2.40–2.42 and the perfect gas law.

Example 2.9

Calculate the temperature, pressure and density of the U.S. Standard Atmosphere at an altitude of 5 *km*.

Solution

From table 2.1, the temperature gradient from 0 to 5 *km* is $-6.5\,K/km$ so that the temperature T is:

$$T = 288.15\,K + (-6.5\,K/km) \times 5\,km$$

$$= 255.7\,K$$

The pressure p may be calculated using equation 2.42:

$$p = 1.0133E(5)\,Pa\left(\frac{255.7\,K}{288.15\,K}\right)^{-1/(-6.5E(-3)\,K/m\times 29.26\,m/K)}$$
$$= 5.4065E(4)Pa$$

Finally, the density ρ may now be calculated from the perfect gas law, equation 1.11:

$$\rho = \frac{p}{RT}$$
$$= \frac{5.4065E(4)\,Pa}{287\,J/kg\,K \times 255.7\,K}$$
$$= 0.7367 kg/m^3$$

Atmospheric Stability

If a stratified fluid is to be dynamically stable, it must resist small displacements in the vertical direction. For a stratified liquid having a density that decreases in the vertical direction, a fluid particle moved vertically would be surrounded by less dense fluid and therefore would return spontaneously to its original position. Thus successively less dense layers of liquid stacked vertically will be dynamically stable. We may express this requirement by noting that the density gradient should be negative for dynamic stability:

$$\left(\frac{d\rho\{z\}}{dz}\right) < 0 \tag{2.43}$$

In the case of a stratified gas, however, we must take into account the fact that a gas particle moved vertically will experience a reduction in density because it moves to a region of lower pressure and expands in volume. Thus the density of the surrounding fluid must not only decrease with height but must decrease more rapidly than does that of the displaced particle. This condition of dynamic stability may be expressed as:

$$\left(\frac{d\rho\{z\}}{dz}\right) < \left(\frac{\partial \rho}{\partial p}\right)_{\mathfrak{s}} \frac{dp\{z\}}{dz}$$
$$< \left(\frac{\partial \rho}{\partial p}\right)_{\mathfrak{s}} (-\rho g) \tag{2.44}$$

where the density change with pressure of the displaced gas particle occurs at constant entropy $\mathfrak{s}$ because it is a reversible adiabatic change. For the atmosphere, we may use the perfect gas properties of equations 1.11–1.15 to express this condition in the form:

$$\left(\frac{dT\{z\}}{dz}\right) > -\frac{g}{\hat{c}_p} \tag{2.45}$$

The ratio $g/\hat{c}_p$ has the value $9.762\,K/km$ for air, and the corresponding temperature gradient $dT\{z\}/dz$ of $-9.762\,K/km$ is called the *adiabatic lapse rate.* Note that the U.S. Standard Atmosphere (table 2.1) is a stable atmosphere since the temperature gradients always exceed the adiabatic lapse rate. Generally speaking, the earth's atmosphere is stable except in the lower troposphere when it is heated by the earth, usually during daylight hours.

2.7 Surface Tension and Capillarity

Small droplets or bubbles of one fluid immersed in another fluid have a spherical shape because of the interfacial tension, or surface tension, between the two fluids. Whenever the fluids have different densities—which is almost always the case—the droplets or bubbles move downward or upward because the buoyant force is not in balance with the gravity force. However, droplets or bubbles can remain stationary in a gravitational field when they come into contact with a solid surface provided the surface tension force exceeds the buoyant force. Since the former is proportional to the drop radius while the latter is proportional to the cube of the drop radius, a sufficiently small drop or bubble can defy gravity and remain attached to a solid surface. The bubble attached to the side of a glass of carbonated beverage and the rain drop sticking to the window pane are two common examples of this phenomenon.

Example 2.10

Estimate the approximate maximum diameter d of a bubble of air in water that can remain attached to a wall of a water container.

Solution

The buoyant force on a spherical air bubble of diameter d is $\rho_w g \pi d^3/6$, where $\rho_w = 10^3\,kg/m^3$ is the density of water. The maximum force that surface tension could exert to hold the bubble in place is $\pi d \Upsilon$, where $\Upsilon = 7.3E(-2)\,N/m$ is the air/water interfacial tension. Equating these two approximately,

$$\rho_w g \frac{\pi d^3}{6} \approx \pi d \Upsilon$$

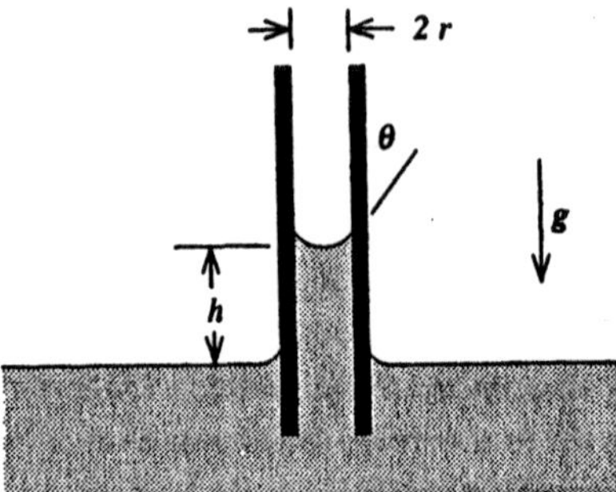

Figure 2.12 Liquid rises in a capillary to an equilibrium position determined by the balance of the surface tension force and the gravity force acting on the column of elevated fluid.

$$d^2 \approx \left(\frac{6\Upsilon}{\rho_w g}\right)$$

$$\approx \frac{6 \times 7.3E(-2)\,N/m}{1.0E(3)\,kg/m^3 \times 9.8066\,m/s^2}$$

$$d \approx 6.7\,mm$$

Since we have overestimated the surface tension force, bubbles of only a couple of millimeters in diameter will break free.

As shown in figure 1.4 (b), when the interface between two fluids comes in contact with a solid surface, the interface makes an angle θ with the solid surface, called the *contact angle*, which is a property of the two fluids and the solid. (By convention, the angle is that subtended by the solid surface and fluid interface in the denser fluid.) This angle reflects the balance of interfacial tensions between the solid and the fluids at the point where the fluid interface meets the solid surface. In the case of a liquid/gas interface having a contact angle less than $\pi/2$, the liquid tends to creep along the surface, displacing the gas, and the liquid is said to wet the surface. On the other hand, when the contact angle exceeds $\pi/2$, the liquid moves away from the gas and is said to be nonwetting. A water-air interface in contact with clean glass is wetting, whereas when it is in contact with paraffin it is nonwetting.

Another example of the effect of surface tension acting in a static fluid is the rise of a liquid in a small diameter tube, or capillary tube, extending vertically above the surface of a liquid exposed to air, as shown in figure 2.12. The average rise height h of the liquid in the tube above the air/liquid interface may be computed from the balance of the surface tension force component in the vertical direction, $2\pi r\Upsilon\cos\theta$, and the gravity force $\rho g(\pi r^2 h)$ acting on the elevated column of liquid:[12]

$$\rho g(\pi r^2 h) = 2\pi r \Upsilon \cos\theta$$

$$h = \left(\frac{2\Upsilon \cos\theta}{\rho g r}\right) \tag{2.46}$$

Note that the rise height will increase as the radius r of the tube is made smaller. For nonwetting liquids, such as mercury, the contact angle θ is greater than $\pi/2$, and the height h is negative ($\cos\theta < 1$), *i.e.*, the fluid moves downward in the capillary tube rather than upward.

Capillary action can be important when liquids infiltrate into porous materials. For example, the upward flow of liquid by capillary action is essential to the maintenance of a candle flame. Radiant heat from the candle flame melts the wax at the base of the candle wick. The melted wax is drawn upward in the wick by capillary action, evaporating as it rises because of heat transfer from the flame surrounding the wick. The vapor then diffuses into the flame, burning by mixing with the surrounding air and feeding back some heat to maintain the process of melting and evaporation.

2.8 Hydraulic Force Transmission

Pressurized fluids can be used to exert forces on pistons that move mechanical components of systems. Because the fluid pressure can be transmitted through long tubes or pipes, the force may be induced at a distance from the source of the pressurized fluid. For example, the hydraulic brake system of an automobile actuates the brakes of all four wheels when the master brake cylinder pressurizes the brake fluid. Because the same pressure is applied to all brake cylinders, an equal brake-setting force is applied to each wheel, providing even braking for the vehicle. In systems of this type, the pressure p is much greater than the product of ρg times the difference in elevation between points within the system so that the pressure is substantially constant throughout the system. The fluid transmits an equal pressure throughout the system just like an electric wire transmits an equal voltage.

A property of hydraulic force transmission systems is that they can amplify the magnitude of the actuating force, just as a mechanical lever does. This principle is illustrated in figure 2.13 which shows two piston-cylinder combinations connected

[12]Note that the fluid pressure on the bottom surface of the column is atmospheric pressure since it has the same elevation as the air/water interface. Thus the atmospheric pressure force acting on the upper surface of the column is balanced by the pressure force on the bottom surface of the column.

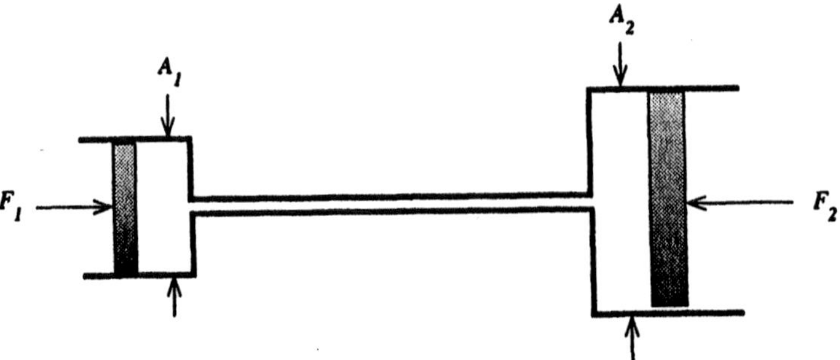

Figure 2.13 A sketch of a fluid system that increases the force F_1 applied to piston 1 to a greater value, F_2, at the larger piston 2.

by a tube containing oil. The smaller piston, of area A_1, is subject to a force F_1 that pressurizes the fluid in the cylinder to a pressure $p_1 = F_1/A_1$. If there is no flow, there must be a restraining force F_2 applied to piston 2 of such amount that the pressure $p_2 = p_1$. Thus,

$$\frac{F_1}{A_1} = p_1 = p_2 = \frac{F_2}{A_2}$$

$$F_2 = \left(\frac{A_2}{A_1}\right) F_1 \tag{2.47}$$

By making $A_2 > A_1$, the applied force F_1 exerts a greater force F_2 on the mechanism to be moved. This is the principle of operation of the hydraulic jack.

The hydraulic lever illustrated in figure 2.13 does not produce more work than it consumes if the pistons move. For a small displacement δx_1 of the first piston, a volume of fluid $A_1\delta x_1$ will be moved into the second cylinder, causing it to be displaced an amount δx_2 such that:

$$A_2\delta x_2 = A_1\delta x_1$$

Combining this with equation 2.47,

$$F_1\delta x_1 = F_2\delta x_2 \tag{2.48}$$

Thus the work done on piston 1 is equal to the work done by piston 2 even though the forces are unequal.

The relations of equations 2.47 and 2.48 are not dependent upon the length or volume of the transmission line connecting the two cylinders. The diameter of the line is chosen to maintain a uniform pressure within the system if the pistons undergo some movement. Hydraulic systems can transmit considerable amounts of power, but such transmission requires the flow of fluid such that the static pressure distribution is no longer achieved and fluid dynamics may become important.

2.9 Problems

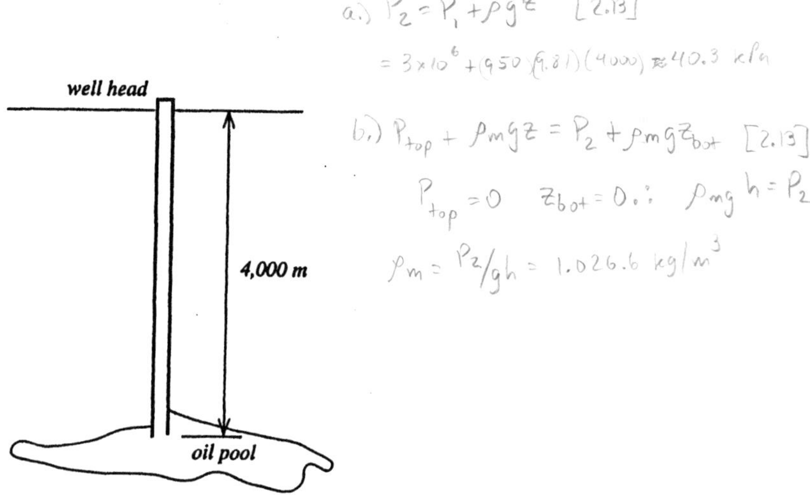

Figure P 2.1

Problem 2.1

A capped oil well drilled to a depth h of $4,000\,m$ terminates in a pool of oil that has filled the well completely to the surface, where the well head gage pressure p_1 of the oil is measured to be $3E(6)\,Pa$. (a) If the oil has a density $\rho_o = 950\,kg/m^3$, what is the oil gage pressure p_2 at the bottom of the well? (b) Drilling mud is a fluid with a density ρ_m that can be adjusted to any value greater than that of water. In order to continue drilling below $4,000\,m$, the drilling crew will pump drilling mud into the well to displace the oil. Calculate the density of mud needed if the well head gage pressure is to be reduced to zero when the well is completely filled with mud.

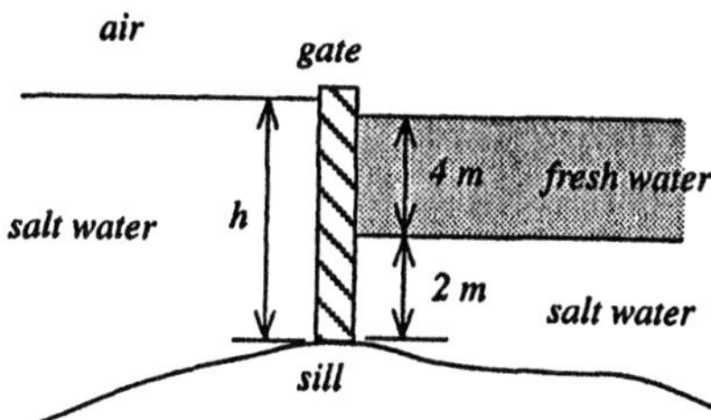

Figure P 2.2

Problem 2.2

A lock gate separates the water in the Charles River basin from the sea water in Boston harbor. As shown in figure P 2.2, the water in the basin consists of an upper fresh water layer of density $\rho_f = 1.0E(3)\, kg/m^3$ and a depth of $4\, m$ and a lower salt water layer of density $\rho_s = 1.03E(3)\, kg/m^3$. The fresh/salt water interface is $2\, m$ above the sill of the gate. On the harbor side, the salt water depth h, measured above the level of the sill, rises and falls with the tide. (a) At some water level h on the harbor side, the total horizontal force of the sea water exerted on the lock gate will be balanced by the total horizontal force of the basin water acting against the other side of the gate. Calculate the value of h at this equilibrium. (b) Under this condition, if there is a leak in the lock gate seal at the sill, will water leak into or out of the basin?

Problem 2.3

The density ρ of sea water increases with the gage pressure p according to the relation:

$$\rho = \rho_0(1 + Kp)$$

The density ρ_0 at sea level ($p = 0$) is $1.03E(3)\, kg/m^3$, and the compressibility constant K is $5E(-10)\, Pa^{-1}$. (a) If sea water were incompressible ($K = 0$) so that its density were everywhere ρ_0, calculate the numerical value of the pressure at a depth of $5,000\, m$. (b) Because sea water density increases with pressure, the pressure at $5,000\, m$ will be slightly greater than that calculated in part (a). Derive the differential equation for the pressure in sea water, integrate it and evaluate the integral to find the pressure at a depth of $5,000\, m$.

Problem 2.4

Assuming that the earth is a sphere of radius $R_E = 6.35E(6)\, m$, calculate the mass M_{at} of the earth's atmosphere. (Hint: The mass per unit of surface area is $\int_0^\infty \rho\, dz$.)

Problem 2.5

A water barometer is constructed, as in figure 2.4, but using water rather than mercury as the barometer fluid. At $20\,°C$, the vapor pressure p_v of water at the top of the water column is $2.337E(3)\, Pa$ and the water density $\rho_w = 9.982E(2)\, kg/m^3$. Calculate the height h of the water column if the atmospheric pressure is $p_a = 1.0133E(5)\, Pa$.

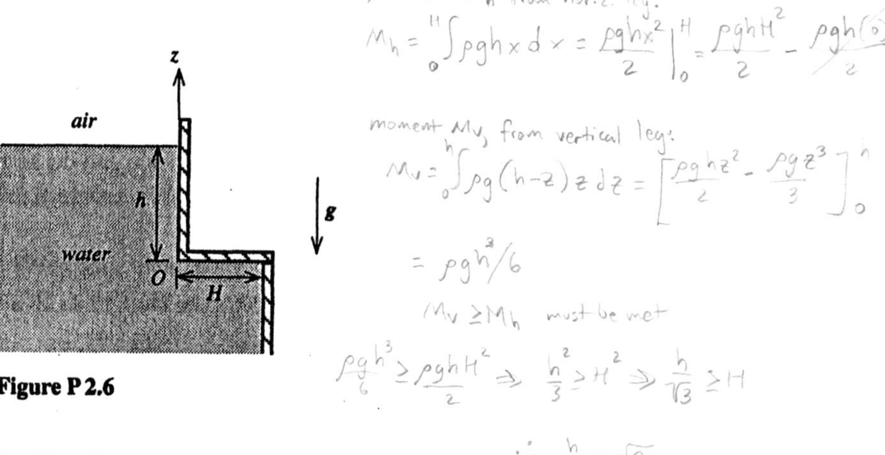

Figure P 2.6

Problem 2.6

A reservoir of water is closed at one end by a barrier. At the top of the barrier is an L-shaped channel that further restrains the water when its level is higher than the barrier top by the amount h, as shown in figure P 2.6. The channel is hinged at the corner point O, so that it can rotate in a counterclockwise direction, but not in a clockwise direction. If the water level h is high enough, the pressure force on the vertical face of the L-shaped channel will maintain a clockwise moment that will exceed the counterclockwise moment of the pressure force on the horizontal leg of the channel, and the channel will remain upright and prevent water leaking beneath it. Calculate the minimum value of the ratio h/H that will just ensure against rotation of the channel.

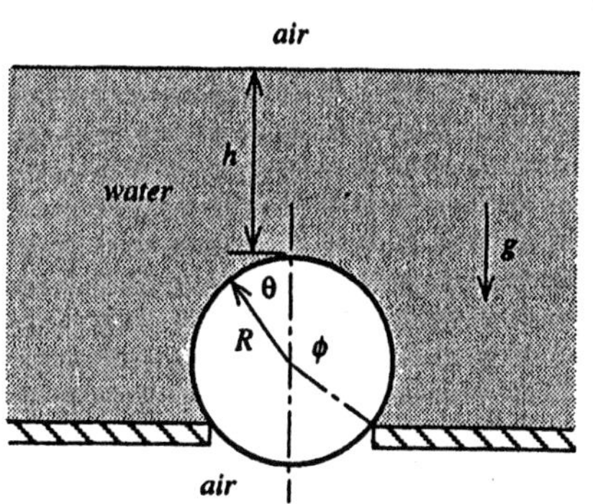

Figure P 2.7

Problem 2.7

The valve at the bottom of a toilet tank consists of a sphere of radius R and negligible mass that closes off a circular opening in the bottom of the tank, as shown in figure P 2.7. The line of contact between the sphere and the opening subtends an angle ϕ from the vertical (see figure P 2.7). The air/water interface at the top of the tank is a distance h above the top of the sphere.

The water pressure on the upper surface of the sphere will hold the sphere in place unless the opening is too small, *i.e.*, ϕ is too close to π. (If $\phi = \pi$, there would be an upward force equal to ρg times the sphere volume.)

(a) Derive an expression for the gage pressure $p\{\theta\}$ on the surface of the sphere as a function of the angle θ measured from the vertical, in terms of the parameters R, h and the water density ρ. (b) Derive an expression for the net downward force F due to water pressure on the sphere as a function of the angle ϕ. (c) Calculate the minimum value of h/R that will make $F \geq 0$ when $\phi = 3\pi/4$.

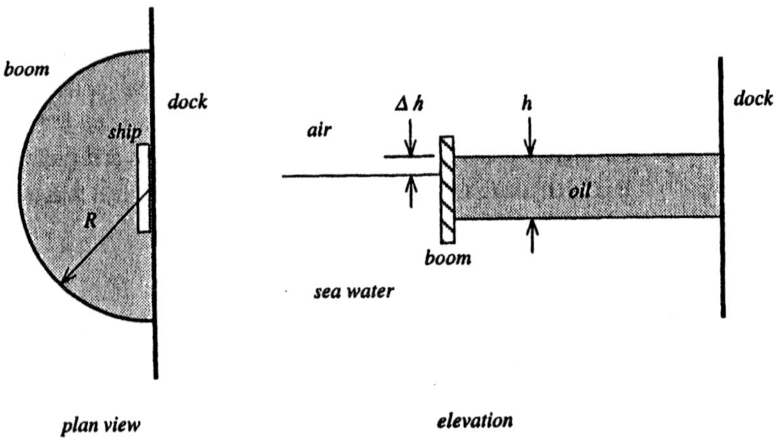

Figure P 2.8

Problem 2.8

Oil spills of various magnitudes may occur in ports where oil tankers are loaded or unloaded. The density ρ_o of oil is less than that of sea water ρ_w and the two fluids are immiscible so that, if a spill of oil occurs, the oil will float on top of the water and spread sideways. To contain any possible spills and minimize pollution, a semicircular oil boom of radius R is deployed around the dock where the loading takes place, as shown in figure P 2.8. The boom is a vertical fence that floats in the water, its bottom being submerged in the water and its top extending above the water

surface. It prevents the spread of oil beyond the barrier when the oil layer does not extend above the top or below the bottom of the fence.

Suppose a volume V of oil is spilled inside the boom. After sufficient time for hydrostatic equilibrium to be reached, a layer of oil of uniform thickness h will fill the interior of the boom. Because of the lighter density of the oil, it will float with its upper surface slightly higher than that of the surface of the sea water by an amount Δh, as shown in figure P2.8. Since the oil layer inside is higher than that of the sea outside, there will be a net outward force F per unit length of boom. In terms of the quantities V, ρ_o, ρ_w, R and g, derive expressions for (a) h, (b) Δh and (c) F. (d) Calculate the numerical values of h, Δh and F for the case of $V = 1.0E(4)\,m^3$, $R = 100\,m$, $\rho_w = 1.03E(3)\,kg/m^3$ and $\rho_o = 9.5E(2)\,kg/m^3$.

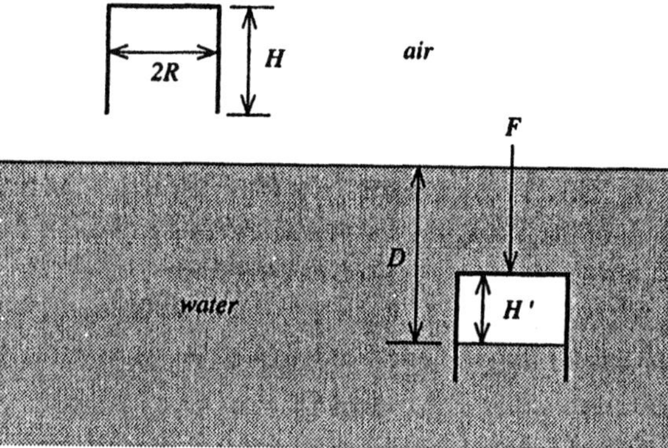

Figure P2.9

Problem 2.9

A cylindrical container of radius R and height H, open at the bottom, is pushed down through the surface of a pool of water. As it pierces the surface, it traps a volume of air, $V = \pi R^2 H$, at atmospheric pressure p_{at} and density ρ_{at}. As the cylinder is pushed beneath the surface, the height of the trapped air column H' decreases, as shown in figure P2.9, because the pressure of the air has increased. In this process, the product of the air pressure times its volume, pV, remains constant.

(a) If the depth of the air/water interface is D, derive expressions for the height H' and for the pressure force F exerted on the cylinder by the fluids in terms of V, H, p_{at}, ρ_{at}, D and the water density ρ_w. (Neglect the volume of the cylinder material.) (b) Calculate the values of H' and F if $V = 1.0\,m^3$, $H = 1.0\,m$, $p_{at} = 1.0E(5)\,Pa$, $\rho_{at} = 1.23\,kg/m^3$, $\rho_w = 1.0E(3)\,kg/m^3$ and $D = 10\,m$.

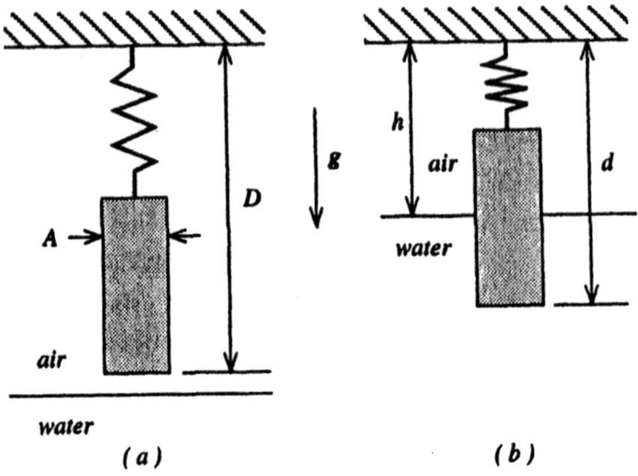

Figure P 2.10

Problem 2.10

In a laboratory experiment a wooden block of cross-sectional area A is suspended above a pan of water by a linear spring of stiffness k, as shown in figure P 2.10(a). The bottom of the block is a distance D below the level of the spring support. When the water level in the pan is raised to a distance h below the support level, as shown in figure P 2.10(b), the block floats upward to a distance d below the support. Derive an expression for the distance d in terms of the parameters A, D, k, h and the water density ρ.

Problem 2.11

Prior to launch, a high altitude balloon is filled with a mass $M_{He} = 100\,kg$ of helium whose pressure and temperature equal those of the atmosphere, $1.033E(5)\,Pa$ and $288.15\,K$, respectively. (a) Calculate the maximum mass of balloon material plus its payload of scientific instruments that can be lifted by this amount of helium. (b) The balloon ascends to an altitude of $11\,km$ where the atmospheric pressure and temperature are $2.263E(4)Pa$ and $216.65\,K$. If the helium pressure and temperature match those of the atmosphere, calculate the net force exerted on the balloon material and its payload at this altitude. (The gas constants for helium and air are $R_{He} = 2077\,J/kg\,K$ and $R_a = 287.0\,J/kg\,K$, respectively.)

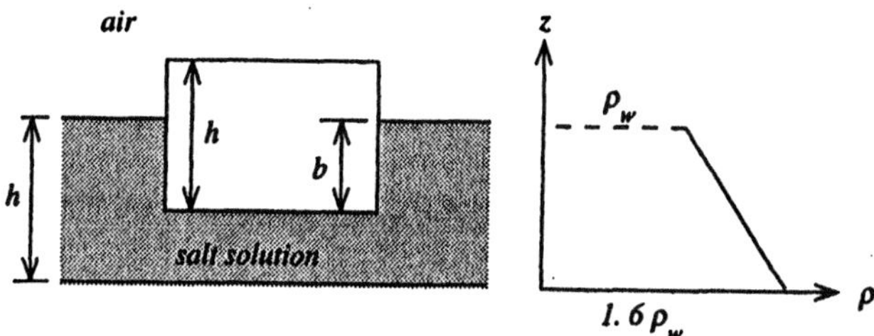

Figure P 2.12

Problem 2.12

A tank of density-stratified salt solution, having a depth $h = 0.1\,m$, is prepared in the laboratory. The density of the liquid increases linearly with depth, as indicated in figure P 2.12, starting from a value of pure water (ρ_w) at the upper surface and increasing to a value of $1.6\,\rho_w$ at the bottom of the tank. A rectangular wood block of height $h = 0.1\,m$ and density $\rho_b = 0.5\,\rho_w$ is placed in the tank. Calculate the depth b below the surface of the liquid to which the bottom surface of the block sinks when it is floating at equilibrium.

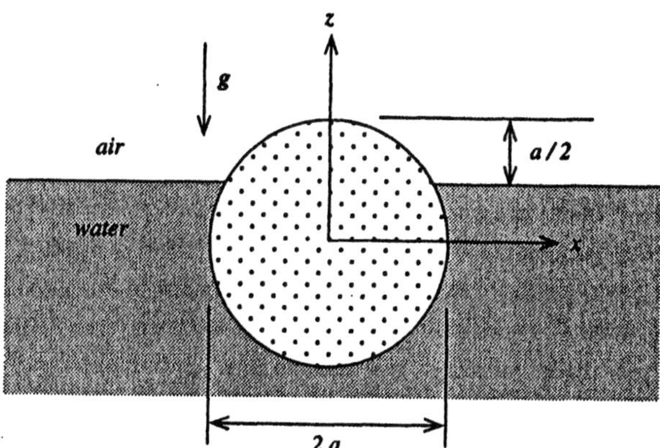

Figure P 2.13

Problem 2.13

As sketched in figure P 2.13, a wooden log of circular cross-section floats on the surface of a pond with a vertical height $a/2$ exposed to the air, where a is the log radius. (a) Using the integral relation:

$$2\int \sqrt{a^2 - z^2}\, dz = z\sqrt{a^2 - z^2} + a^2 \sin^{-1}\left(\frac{z}{a}\right)$$

calculate the specific gravity of the log. (b) Show that the log is neutrally stable, *i.e.*, $[GM] = 0$.

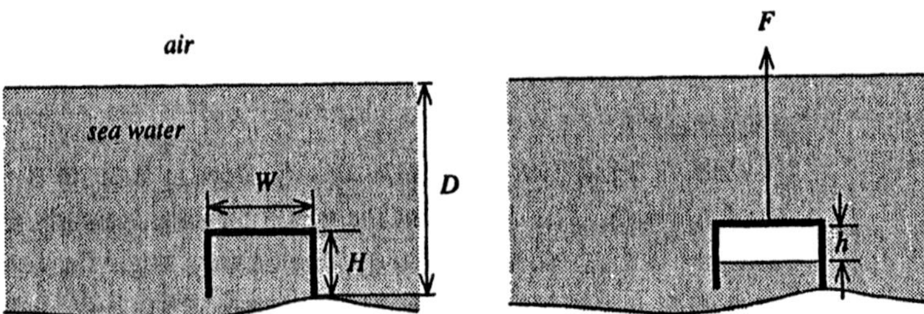

Figure P 2.14

Problem 2.14

A barge has overturned and sunk to the bottom of the sea, as shown in figure P 2.14. It lies on the bottom in water of depth $D = 30\,m$. The barge has a length $L = 30\,m$, a depth $H = 10\,m$, a width $W = 10\,m$ and a mass M of 300 metric tons (1 metric ton = $1.0E(3)\,kg$). A marine crane is to be used to raise the barge from the sea floor. However, the crane has a lifting capacity of only 100 metric tons. It is therefore planned to displace sea water from inside the barge by pumping air into the overturned barge until the force F required to lift the barge from the bottom is equal to that needed to raise a 100 metric ton mass in the air, as shown in figure P 2.14.

(a) Calculate the volume V_b of sea water that must be displaced to meet the required lifting force F. (In calculating V_b, neglect the volume of water displaced by the steel structure of the barge.) Sea water density $\rho_w = 1.03E(3) kg/m^3$. (b) Calculate the gage pressure p_b of the air in the overturned barge.

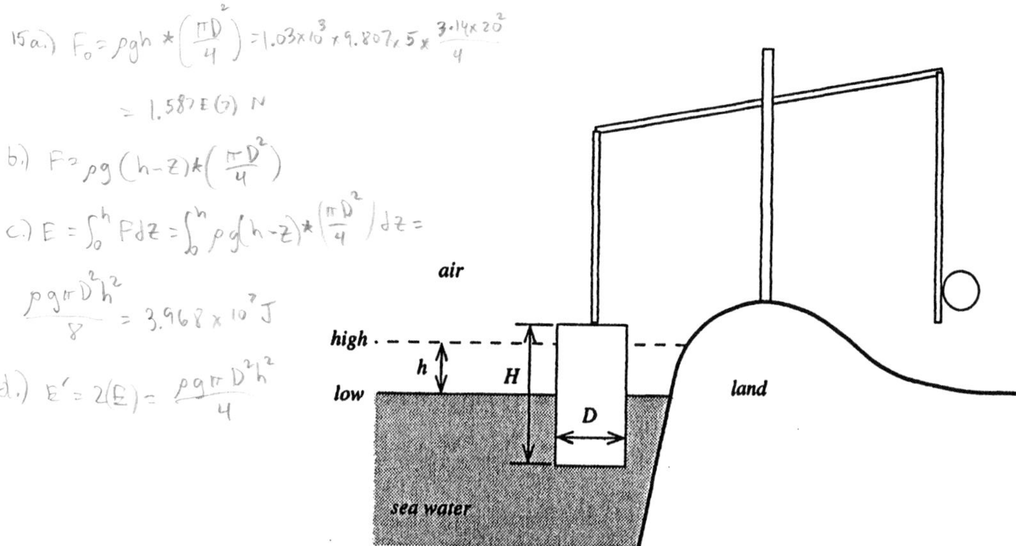

Figure P 2.15

Problem 2.15

An inventor proposes to generate power from the tidal rise and fall of the ocean surface by linking a float to an electric generator. The device is sketched in figure P 2.15, which shows a cylindrical tank of diameter $D = 20\,m$ and height $H = 15\,m$ attached to a lever that causes the armature of an electric generator to rotate as the tank rises and falls with the level of the ocean. (The density of ocean water is $\rho = 1.03E(3)\,kg/m^3$.)

The device will be operated in the following manner. At low tide, the float and lever will be locked at its equilibrium position, where the buoyant force of the water is balanced by the gravity force on the float (neglecting the gravity force on the lever), the float being half submerged as shown in the sketch above. Later on, at high tide, the sea having risen a distance $h = 5\,m$, the device will be unlocked. The extra buoyant force acting on the float will push it upward, actuating the electric generator and generating power. When the float has risen the distance h, the extra buoyant force will have returned to zero, and no more power will be generated. At this point, the apparatus will be locked in place until the tide has returned to its low tide level when the second half of the power generation cycle will be completed.

(a) Calculate the force F_0 exerted by the float on the lever at high tide just before the lever is released. (b) Derive an expression for the force F exerted by the float when it has risen slowly a distance z from its initial position. (c) Calculate the energy E generated by the electric generator during the process of the float rising through the distance h. (d) Derive an expression for the total energy E_t produced in one tidal cycle.

Problem 2.16

A linear weighing scale is calibrated to read $100\,kg$ when a metal weight of specific gravity $SG_m = 4$ and volume $\mathcal{V}_m = 2.5E(-2)\,m^3$ is placed upon it. A person of mass $M_p = 100\,kg$ and specific gravity $SG_p = 1$ steps on the scale. Calculate the reading on the scale if the air density $\rho_a = 1.2\,kg/m^3$.

Problem 2.17

Two parallel glass plates separated by a small distance d are placed vertically in a dish of water. Because of surface tension, water rises in the gap between the plates a distance h above the surface of the water in the dish. Derive an expression for the rise distance h in terms of the surface tension Υ, the contact angle θ, the water density ρ and the plate spacing d.

Problem 2.18

A vehicular tunnel under a river bed is constructed of sections in the form of circular cylindrical steel shells of outside diameter $D = 10\,m$. When equipped with temporary end closures, the sections will be lowered to the river floor, connected with watertight seals and the end closures removed to form a continuous tunnel. To keep the hollow tunnel on the river bottom, a layer of concrete of thickness h will be added outside the exterior of the steel tunnel sections in sufficient amount that the gravity force on the concrete alone will just counteract the buoyant force on the tunnel. If the density of concrete is $\rho_c = 2.3E(3)\,kg/m^3$ and that of the water is $\rho_w = 1E(3)\,kg/m^3$, calculate the thickness h of concrete needed for this tunnel.

Bibliography

Olson, Reuben M., and Steven J. Wright. 1990. *Essentials of Engineering Fluid Mechanics.* New York: Harper and Row.

3 Conservation of Mass

The conservation of mass is fundamental to all sciences. In fluid mechanics, it implies constraints on how a fluid can move. If we squeeze a balloon or toothpaste tube at one place, it bulges out at another. We need to express this intuitively obvious physical constraint in quantitative, mathematical terms if we are to advance toward the goal of describing and predicting fluid flow in general. The application of the principle of mass conservation to a fluid flow problem can tell us some, but not necessarily all, of the things we want to know about how a fluid behaves.

In this chapter, we derive expressions for the conservation of mass as it applies to fluid flows. However, before we can do this, we must understand the conventions used in describing how a fluid moves, called fluid kinematics.

3.1 Kinematics of Fluid Flow

Kinematics is the study of how things move. For example, in an automobile engine, there is a relationship between the rotary motion of the crankshaft and the motions of the piston, connecting rod, valves, timing gear, etc. They all are related by the constraints imposed by the linkages among the components. In a fluid, there is no equivalent relationship between the motions of distant particles because fluid particles are not rigidly connected to each other like the parts of an automobile engine. Nevertheless, not every kind of fluid motion is possible, and there are constraints on fluid motion that can be expressed in terms of a kinematical description of the fluid flow.

3.1.1 Lagrangian and Eulerian Descriptions

In describing the motion of a rigid body, such as a ship or airplane, we use the three spatial coordinates of the center of mass and the three angular coordinates of rotation about the mass center to describe the position and angular orientation of the body at each instant of time. Knowing these six coordinates as functions of time, we can find their time derivatives to determine the components of the velocity of the mass center and the angular velocity about the mass center. An additional differentiation will determine the linear and angular accelerations of the body.[1] This mode of describing the motion of an identified body of finite mass is called the *Lagrangian* description.

Can we apply the same method of description to a fluid? Since a fluid is not rigid, but easily deformable, we need to take into account the motion of the fluid relative to the mass center, which is certainly not a simple rotation as it is in a rigid body. One possible way around this difficulty is to subdivide the fluid into a very large number, N, of individual material particles and then employ a Langrangian description for each of these N particles. Needless to say, this imposes great difficulties, except in a few simple flows.

The solution to this dilemma is to abandon the requirement of following the fate of each fluid particle and focus on the velocity $\mathbf{V}$ of the fluid particle that occupies a point in space, $\mathbf{R}$, at the time t. An *Eulerian* description of the flow field consists of the specification of the velocity $\mathbf{V}$ as a function of position $\mathbf{R}$ and time t:

$$\mathbf{V} = \mathbf{V}\{\mathbf{R}, t\}$$

By a suitable differentiation with respect to time, we can find the acceleration of the fluid particle at any position $\mathbf{R}$ and time t and, if necessary, the displacement of the particle from its position at an earlier time by integration.

In the Eulerian description of fluid motion, other physical variables of interest, such as pressure p and density ρ, are also considered to be functions of position $\mathbf{R}$ and time t. Mathematically speaking, $\mathbf{V}, p, \rho,$ *etc.* are considered to be the dependent variables of the flow that are functions of the independent variables $\mathbf{R}$ and t.[2] This general description is termed a field description, and the dependent variables are called field variables. Thus we use the terms such as velocity field and pressure field to imply the dependence of velocity and pressure on the independent variables $\mathbf{R}$ and t.

In this text, we use two coordinate systems to describe the position vector $\mathbf{R}$, cartesian and cylindrical coordinates:

$$\mathbf{R} \equiv x\,\mathbf{i}_x + y\,\mathbf{i}_y + z\,\mathbf{i}_z = r\,\mathbf{i}_r + z\,\mathbf{i}_z$$

---

[1] The equations of motion of the body are derived by equating the linear and angular acceleration times the mass and mass moment of inertia to the sum of the forces and their moments about the mass center, respectively.

[2] $\mathbf{R}$ has three spatial dimensions so that there are four independent scalar variables, *e.g.*, x, y, z and t. The number of dependent variables includes all physical properties of interest.

In the cylindrical coordinate system, the direction of the unit vectors $\mathbf{i}_r$ and $\mathbf{i}_\theta$ are functions of the angular coordinate θ, whereas $\mathbf{i}_x$, $\mathbf{i}_y$ and $\mathbf{i}_z$ have fixed directions. Thus, implicitly, $\mathbf{R}$ is a function of θ as well as explicitly a function of r and z. Furthermore, in performing differentiation of vectors expressed in cylindrical coordinates, the unit vectors $\mathbf{i}_r$ and $\mathbf{i}_\theta$ have nonzero derivatives:

$$\frac{\partial}{\partial \theta}(\mathbf{i}_r) = \mathbf{i}_\theta; \quad \frac{\partial}{\partial \theta}(\mathbf{i}_\theta) = -\mathbf{i}_r$$

We generally will use cartesian coordinates to describe fluid flows, except in cases where axial symmetry of the flow makes using cylindrical coordinates more convenient.

The spatial derivatives of vector and scalar fields, expressed in cartesian and cylindrical coordinates, were reviewed in section 1.4.2.

Example 3.1

A fluid moves in the x direction only so that its velocity $\mathbf{V} = u\mathbf{i}_x$. The Lagrangian description of the flow gives the position x of the fluid particle that was located at x_0 at time t_0 as:[3]

$$x = \frac{x_0}{t_0}t$$

provided $t \geq t_0 \geq 0$. Derive the Eulerian description of this flow, $u\{x, t\}$.

Solution

The velocity u of the fluid particle that was located at x_0 at time t_0 is found by differentiating x with respect to t:

$$u = \frac{dx}{dt} = \frac{x_0}{t_0}$$

But at any x, t,

$$\frac{x_0}{t_0} = \frac{x}{t}$$

---

[3]This is the one-dimensional, nonrelativistic analog of the big bang model of the expanding universe. As an exercise, show that the relative speed of a particle of fluid a distance d from the particle on which an observer is located depends upon d and t but not the position x of the observer.

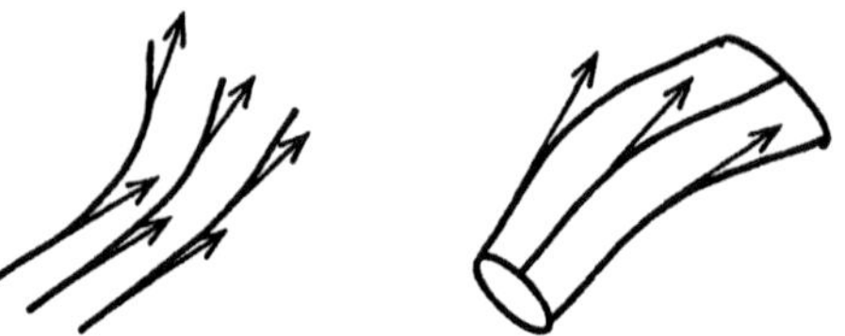

Figure 3.1 Above, *left*, streamlines and, *right*, a streamtube in a flow field.

Thus

$$u\{x,t\} = \frac{x}{t}$$

3.1.2 Streamlines and Pathlines

In colloquial usage, the word streamline connotes a smoothed flow, as in streamlined body. In fluid mechanics, it has a precise meaning. A *streamline* is a line in a flow field that is everywhere tangent to the velocity vector **V** at each point along the streamline for any instant of time t (see figure 3.1.) We can think of the flow field being filled with streamlines, much as a pot is filled with cooked spaghetti. Like spaghetti, the streamlines never intersect each other because, at any point, there can be only one direction of the velocity. Later on, we will show how to calculate the streamlines for some multidimensional flows. For the simple flow of example 3.1, the streamlines are straight lines parallel to the x axis.

A *streamtube* is a surface in the flow formed from streamlines and closed upon itself to form a tube of variable cross-section, as shown in figure 3.1. The use of the streamtube is primarily a conceptual one, helping us to visualize mentally how a flow field might be subdivided into streamtubes that entirely fill the flow field. Since the fluid does not cross through the surface of the stream tube, we may think of the stream tubes as flexible, moveable pipes containing the flow inside them.

Both streamlines and streamtubes are instantaneous snapshots of lines and surfaces in the flow field. As time progresses, these lines and surfaces will move to different locations in space unless the flow is steady, *i.e.*, does not depend upon time. In steady flow, any snapshot of the flow is identical to every other one, and the streamlines and stream surfaces are fixed in space.

The useful line to define in the Lagrangian description of a flow field is the *pathline*, which is the path followed over later times of a particular particle identified at an initial time and location. We can observe a pathline in an experiment by marking a fluid particle with a puff of dye or smoke and taking a time-exposed picture of the marked fluid. (A common analog is the light path of auto headlights in a nighttime time-exposed photograph.) A pathline is the trajectory of a single fluid particle.

If a flow is steady, then a streamline and pathline passing through the same point in space are identical because the velocity field depends only upon position and not time. Experimentally, this is convenient for the observation of steady streamlines because the set of pathlines formed by a steady stream (*i.e.*, a closely spaced series of puffs) of dye or smoke follow the same trajectory through space as the streamline. Even in unsteady flows, the use of dye or smoke markers can be helpful in visualizing the flow behavior despite the fact that the marked fluid does not denote either a streamline or a pathline.

3.1.3 The Material Derivative

The standard forms of Newton's law of motion and the laws of thermodynamics apply to a fixed mass of identified matter whose properties change as time progresses. The natural mode for expressing these laws is the Lagrangian description of motion because it directly describes the history of an identified particle. Since we use the Eulerian description for a moving fluid, we need to establish the Eulerian expression of the rate of change of any property of a fluid particle as it moves through the flow field. The time rate of change of a fluid property, as measured by an observer moving with the particle, is called the *material derivative* of that property.

As an example, consider the rate of change of density ρ of a fluid particle that is located at position $\mathbf{R}$ at time t. During the time interval dt, the particle moves an amount $d\mathbf{R} = \mathbf{V}\,dt$. The total increment $d\rho$ in density is the sum of the part due to the time increment dt and that due to the spatial increment $d\mathbf{R}$. Using cartesian coordinates to express the amount of $d\rho$,

$$\begin{aligned}
d\rho &= \frac{\partial \rho}{\partial t}dt + \frac{\partial \rho}{\partial x}dx + \frac{\partial \rho}{\partial y}dy + \frac{\partial \rho}{\partial z}dz \\
&= \frac{\partial \rho}{\partial t}dt + \left(\frac{\partial \rho}{\partial x}\mathbf{i}_x + \frac{\partial \rho}{\partial y}\mathbf{i}_y + \frac{\partial \rho}{\partial z}\mathbf{i}_z\right) \cdot (\mathbf{i}_x dx + \mathbf{i}_y dy + \mathbf{i}_z dz) \\
&= \frac{\partial \rho}{\partial t}dt + \nabla\rho \cdot d\mathbf{R} = \frac{\partial \rho}{\partial t}dt + d\mathbf{R} \cdot \nabla\rho \\
&= \frac{\partial \rho}{\partial t}dt + \mathbf{V}dt \cdot \nabla\rho \\
&= \left(\frac{\partial \rho}{\partial t} + \mathbf{V} \cdot \nabla\rho\right) dt
\end{aligned}$$

$$\frac{d\rho}{dt} = \left(\frac{\partial}{\partial t} + \mathbf{V} \cdot \nabla\right)\rho \tag{3.1}$$

To emphasize that the material time derivative includes both spatial and time partial derivatives, and is not simply the partial time derivative, we will denote it by D/Dt:

$$\frac{D}{Dt} \equiv \left(\frac{\partial}{\partial t} + \mathbf{V} \cdot \nabla\right) \tag{3.2}$$

Note that D/Dt is a scalar operator so that the material derivative of a scalar variable, such as density ρ, for example, is a scalar quantity:

$$\frac{D\rho}{Dt} = \frac{\partial \rho}{\partial t} + (\mathbf{V} \cdot \nabla)\rho \tag{3.3}$$

Equation 3.3 may be expressed in terms of cartesian coordinates as :

$$\begin{aligned}\frac{D\rho}{Dt} &= \frac{\partial \rho}{\partial t} + \left[(u\mathbf{i}_x + v\mathbf{i}_y + w\mathbf{i}_z) \cdot (\mathbf{i}_x\frac{\partial}{\partial x} + \mathbf{i}_y\frac{\partial}{\partial y} + \mathbf{i}_z\frac{\partial}{\partial z})\right]\rho \\ &= \frac{\partial \rho}{\partial t} + u\frac{\partial \rho}{\partial x} + v\frac{\partial \rho}{\partial y} + w\frac{\partial \rho}{\partial z}\end{aligned} \tag{3.4}$$

and in terms of cylindrical coordinates as:

$$\begin{aligned}\frac{D\rho}{Dt} &= \frac{\partial \rho}{\partial t} + \left[(V_r\mathbf{i}_r + V_\theta\mathbf{i}_\theta + V_z\mathbf{i}_z) \cdot (\mathbf{i}_r\frac{\partial}{\partial r} + \mathbf{i}_\theta\frac{\partial}{r\partial \theta} + \mathbf{i}_z\frac{\partial}{\partial z})\right]\rho \\ &= \frac{\partial \rho}{\partial t} + V_r\frac{\partial \rho}{\partial r} + \frac{V_\theta}{r}\frac{\partial \rho}{\partial \theta} + V_z\frac{\partial \rho}{\partial z}\end{aligned} \tag{3.5}$$

Expressions for the material derivative of a vector property, such as the velocity $\mathbf{V}$, will be treated in section 4.2.

Example 3.2

A velocity field and density field in cartesian space is given as:

$$\mathbf{V} = \frac{L}{t}\mathbf{i}_x$$

$$\rho = Kte^{-x/L}$$

where L and K are constants having the dimensions of length and density divided by time, respectively. Find the material derivative of the density ρ.

Solution

Substituting into equation 3.3,

$$\frac{D\rho}{Dt} = \frac{\partial}{\partial t}\left(Kte^{-x/L}\right) + \frac{L}{t}\frac{\partial}{\partial x}\left(Kte^{-x/L}\right)$$

$$= Ke^{-x/L} + \frac{L}{t}\left(\frac{-Kt}{L}e^{-x/L}\right)$$

$$= 0$$

In this example, the density of a given fluid particle does not change with time, even though the density of the fluid at a given location increases linearly with time.

3.1.4 Kinematical Flow Classifications

We have seen that the flow variables of interest (*e.g.*, $\mathbf{V}, p, \rho$, *etc.*) depend in general upon four independent scalar variables, the three components of **R** and t. However, in many cases, the flow variables depend only upon some of these four variables and not upon the remaining ones. Such flows are much easier to deal with both analytically and numerically because less information is needed to describe all that needs to be known about the flow. We use special names for flows that have this simplifying dependence upon space and time.

Steady flows are ones for which there is no variation of the flow variables as time progresses. If we sit in an automobile travelling at a constant speed along an empty highway, the air flow relative to us and the vehicle is a steady flow. The partial time derivative of any flow variable is zero, and these variables depend only upon the distance **R** measured from a fixed point in the vehicle. On the other hand, if the vehicle were accelerating or if it were to pass another vehicle on the road, then the flow observed by the vehicle's occupant would be *unsteady*.

Sometimes an unsteady flow can be described as a steady flow by suitable choice of a reference frame. The flow observed by a pedestrian at the side of a highway who is passed by a bus moving at constant speed is an unsteady flow, but the same flow observed by a passenger in the bus is a steady flow.

A *quasi-steady* flow is one for which the change with time is so small that it can be disregarded when describing the flow at a particular time. For example, if a vehicle accelerates from 0 to 60 *mph* in several minutes, the flow around the vehicle when it reaches 30 *mph* is indistinguishable from the flow that exists when the vehicle moves steadily at that speed. However, before you can disregard a flow unsteadiness, you must prove that its effects are small enough to ignore.

Example 3.3

An automobile of length $L = 4m$ accelerates uniformly from 0 to $100 m/s$ over a time period of T seconds. It is suggested that the vehicle drag encountered during the acceleration period is negligibly different from that of a steady speed vehicle moving at the corresponding instantaneous speed of the acceleration period. How long should the time period T be in order for this suggestion to be acceptable?

Solution

The flow around the automobile can be considered steady if the time derivative term of equation 3.3 is much smaller than the spatial derivative:

$$\frac{\partial \rho}{\partial t} \ll (\mathbf{V} \cdot \nabla)\rho$$

The magnitude of these terms can be estimated using the time and length scales T and L:

$$\frac{\partial \rho}{\partial t} \simeq \frac{\rho}{T}$$

$$(\mathbf{V} \cdot \nabla)\rho \simeq \frac{V\rho}{L}$$

Inserting these approximate values in the required inequality:

$$\frac{\rho}{T} \ll \frac{V\rho}{L}$$

$$T \gg \frac{L}{V}$$

The average speed during the acceleration period is $50 m/s$ so that a time period T much greater than $4/50 \simeq 0.1$ seconds, say 10 seconds, would be acceptable. Yet it is evident that the unsteady term is not negligible at the start when $V = 0$. However, the drag is so small at low speeds that it need not be estimated accurately at low speeds for the purpose of analyzing the dynamics of accelerating vehicles.

The number of spatial variables upon which all the flow variables depend is called the *dimensionality* of the flow. In a *one-dimensional* flow, for example, the variables might depend upon x alone so that the partial derivatives with respect to y and z would be zero. A two-dimensional flow, which in cylindrical coordinates does not depend upon the azimuthal variable θ but only upon the radial and axial coordinates r and z, is called an *axisymmetric* flow.

A *quasi-one-dimensional* flow is one which depends upon two spatial variables, but the dependence upon one is so small that it can be neglected in describing the flow changes in the direction of the other dimension. An example of a quasi-one-dimensional flow is that of a fluid in a circular pipe whose diameter varies only gradually with distance along the pipe so that the derivates with respect to the axial distance z are much smaller than those with respect to the radial distance r.

The *directionality* of the flow refers to the number of nonzero components of the velocity vector $\mathbf{V}$. For example, fluid flowing steadily through a long pipe and having a velocity that is everywhere parallel to the pipe axis is called a *unidirectional* flow. A *bidirectional* flow has velocity components in two directions and is commonly called a *plane flow* because the velocity vector lies in the plane of the two directions.

3.2 Control Volumes and Surfaces

In deriving Archimedes' principle, we invoked the concept of a closed surface within a fluid that enclosed the volume displaced by a submerged or floating body. In hydrostatics, such volumes are motionless and their enclosing surfaces are impervious to the movement of fluid. We might say that the volume and its enclosing surface is "frozen" in the fluid. We find the total force acting on the fluid in the volume by visualizing the fluid as a rigid body filling the volume.

We will find that the concept of defining a volume of fluid in a flow field, surrounded by an imaginary enclosing surface, is extremely useful in fluid mechanics. We call such a volume a *control volume* and its surrounding surface a *control surface.* In this book, we will use only control volumes and their surfaces that are fixed in the reference frame of the flow field.[4] Control volumes fixed in space are the natural choice in an Eulerian flow description.

Control volumes and surfaces can be very useful in applying the laws of conservation of mass, momentum and energy to fluid flows. In many applications, there is incomplete information regarding the entire flow field but sufficient information about the fluid flow at the control surface to enable us to find total forces and moments on structures and the quantities of heat and work involved, as well as other useful information. The expressions of the physical laws governing fluid flow in integral form utilize control volumes and surfaces.

---

[4]While it is possible to utilize control volumes that deform and/or move with respect to the reference frame of the flow field, their applicability is limited and the expressions of the conservation laws are different from what is given in this book (see F. M. White, 1986 *Fluid Mechanics*, 2nd ed. [New York: McGraw-Hill Book Co.]).

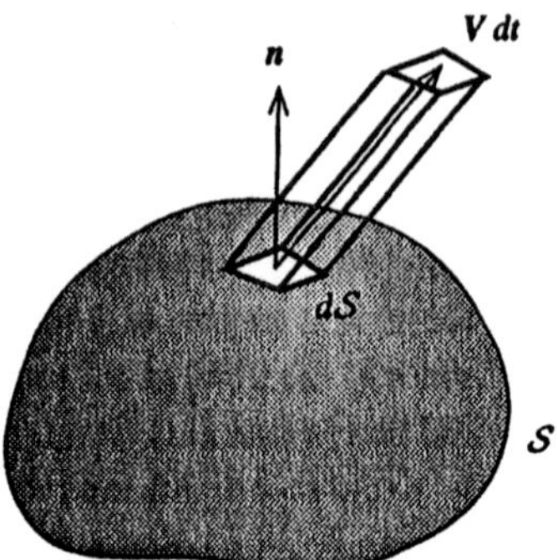

Figure 3.2 The volume of fluid flowing outward across an element dS of a control surface during the time interval dt is $\mathbf{V} \cdot \mathbf{n}\, dt\, dS$, where $\mathbf{n}$ is the unit normal to the surface S pointing outward from the control volume.

3.2.1 Flow across a Control Surface

A control surface is an imaginary surface fixed in the flow reference frame, chosen for the convenience it affords in analyzing a flow field. It may coincide in part, or in whole, with a real solid boundary of the flow, but in general, fluid will flow across this imaginary control surface. How fast does the fluid flow across the control surface?

Let us calculate the amount of fluid that crosses a small element of area dS of a control surface in a time interval dt. Figure 3.2 shows the outward unit normal $\mathbf{n}$ at the surface element dS of the control surface S surrounding the control volume $\mathcal{V}$. Fluid that was located in the surface element dS at the beginning of the time interval dt moves a distance $\mathbf{V}dt$ during this time interval. This moving fluid sweeps out a volume in the shape of a cylinder of base area dS and slant height $\mathbf{V}dt$, a volume filled with fluid that has passed through the surface element dS during the time interval dt. The volume of the cylinder is the product of the base area dS times the component of the slant height in the direction of $\mathbf{n}$, $\mathbf{V}dt \cdot \mathbf{n} = \mathbf{V} \cdot \mathbf{n}\, dt$, giving a volume of $\mathbf{V} \cdot \mathbf{n}\, dt\, dS$. The *volume flow rate, i.e.*, the volume of fluid flowing per unit time across the control surface, is denoted by Q:

$$dQ \equiv \mathbf{V} \cdot \mathbf{n}\, dS$$

$$Q \equiv \iint \mathbf{V} \cdot \mathbf{n}\, dS \tag{3.6}$$

The volume flow rate per unit area of control surface is $\mathbf{V} \cdot \mathbf{n}$.

Example 3.4

A fluid flowing steadily in a circular pipe of radius a has a velocity $\mathbf{V}$ that is everywhere parallel to the pipe axis (z-axis), that is a maximum U at the center and that is zero at the pipe wall:

$$\mathbf{V} = U\left(1 - \frac{r^2}{a^2}\right)\mathbf{i}_z$$

where r is the radial distance from the pipe axis. Derive an expression for the volume flow rate Q in the pipe.

Solution

Using equation 3.6 and choosing the surface element dS to be $2\pi r\,dr$, Q becomes:

$$\begin{aligned} Q &= \iint \mathbf{V}\cdot\mathbf{n}\,dS \\ &= \int_0^a U\left(1 - \frac{r^2}{a^2}\right)(2\pi r)\,dr \\ &= 2\pi U\left(\frac{r^2}{2} - \frac{r^4}{4a^2}\right)\Big|_0^a \\ &= \frac{1}{2}U(\pi a^2) \end{aligned}$$

U/2 is the volume-averaged velocity, *i.e.*, the velocity which, when multiplied by the flow area πa^2, equals the volume flow rate Q.

Volume flow rate is a common measure used in the handling of fluids. Liquid pumps are usually rated in gallons per minute (GPM) and ventilation fans in cubic feet per minute (cfm). While volume flow rates may be useful for practical applications, they do not suffice for expressing the laws governing fluid flow. Instead, we focus on the mass of a fluid sample or other physical properties proportionate to mass, such as momentum or energy, when describing the quantities that enter into the expression for these laws.

The first flow constraint we will treat is the conservation of mass. To express this conservation law, we need to determine the rate at which fluid mass crosses a control surface. As we found above, and as is illustrated in figure 3.2, the volume of fluid passing through the surface element dS in the time dt is $\mathbf{V}\cdot\mathbf{n}dS\,dt$. Since the fluid mass per unit volume is ρ, the mass of the fluid in this volume is $\rho\mathbf{V}\cdot\mathbf{n}dS\,dt$. The *mass flow rate* across the control surface is denoted by $\dot{m}$:

$$d\dot{m} \equiv \rho\mathbf{V}\cdot\mathbf{n}\,dS$$

$$\dot{m} \equiv \iint \rho\mathbf{V}\cdot\mathbf{n}\,dS \tag{3.7}$$

Note that $\rho\mathbf{V}\cdot\mathbf{n}$ is the mass flow rate per unit area of control surface. Also, note that if the fluid density ρ is the same at all points on the control surface, then:

$$\begin{aligned}\dot{m} &= \iint \rho\mathbf{V}\cdot\mathbf{n}\,dS\\ &= \rho\iint \mathbf{V}\cdot\mathbf{n}\,dS\\ &= \rho Q \qquad (\rho = \mathit{constant})\end{aligned} \tag{3.8}$$

3.3 Conservation of Mass

3.3.1 Integral Form of Mass Conservation

The conservation of mass in a fluid flow requires that the accumulation of mass inside a control volume is accounted for by the net flow of mass across the control surface because mass cannot be created or destroyed within the control volume. It is analogous to the balancing of a bank account: the increase in value of the principle is equal to the deposits plus interest payments minus the withdrawals plus service charges. Flow into the control volume is the equivalent of a "credit" while outflow is a "debit" to the mass account.

Consider a flow through a control volume during a small time interval dt. The net outflow of mass from the control volume during this time interval would be the mass flow rate times the increment, $\dot{m}\,dt = (\iint \rho\mathbf{V}\cdot\mathbf{n}\,dS)\,dt$. The increment in mass inside the control volume, $d(\iiint \rho d\mathcal{V})$, must equal the negative of the outflow mass:

$$d\Big(\iiint_{\mathcal{V}} \rho d\mathcal{V}\Big) = -\Big(\iint_S \rho\mathbf{V}\cdot\mathbf{n}S\Big)\,dt$$

Dividing by the time increment dt, we obtain an expression for the time rate of increase of mass inside the control volume:

$$\frac{d}{dt}\iiint_{\mathcal{V}} \rho\,d\mathcal{V} = -\iint_S \rho\mathbf{V}\cdot\mathbf{n}\,dS$$

$$\frac{d}{dt}\iiint_{\mathcal{V}} \rho\, d\mathcal{V} + \iint_{\mathcal{S}} \rho \mathbf{V} \cdot \mathbf{n}\, d\mathcal{S} = 0 \tag{3.9}$$

This equation expresses the conservation of mass applied to a fixed control volume, namely, that *the rate of accumulation of mass inside the control volume plus the net rate of outflow of mass across the control surface add to zero.*

This equation of mass conservation is a scalar equation and is not sufficient by itself to determine the velocity field **V**. Additional information, in the form of an equation of motion, is also required.[5] Nevertheless, the conservation of mass imposes some constraints on how the velocity field may behave. For example, if the control volume $\mathcal{V}$ is always filled with fluid of fixed density ρ, then the first term of equation 3.9 is zero, and thus the second term as well so that the velocity **V** cannot have only an outward (or inward) component everywhere on the control surface $\mathcal{S}$. Sometimes the fluid mass conservation equation is called the *continuity equation* in recognition of the constraint it imposes on the velocity field.

In many engineering applications, it may be easier to apply equation 3.9 by considering the inflow and outflow portions of the mass flow across the control surface as separate items, $\dot{m}_{in}$ and $\dot{m}_{out}$, respectively:

$$\frac{d}{dt}\iiint_{\mathcal{V}} \rho\, d\mathcal{V} + \iint_{\mathcal{S}} \rho \mathbf{V} \cdot \mathbf{n}\, d\mathcal{S} = 0$$

$$\frac{d}{dt}\iiint_{\mathcal{V}} \rho\, d\mathcal{V} + \dot{m}_{out} - \dot{m}_{in} = 0$$

$$\frac{d}{dt}\iiint_{\mathcal{V}} \rho\, d\mathcal{V} = \dot{m}_{in} - \dot{m}_{out} \tag{3.10}$$

where the inflow and outflow mass flow rates $\dot{m}_{in}$ and $\dot{m}_{out}$ are the *magnitudes* of $\iint \rho \mathbf{V} \cdot \mathbf{n}\, d\mathcal{S}$ summed over all inflow and outflow streams, respectively.

For a flow process that commences at an initial time t_i and lasts until a final time t_f, equation 3.10 may be integrated on time from t_i to t_f to obtain:

$$\iiint_{\mathcal{V}} \rho\{t_f\}\, d\mathcal{V} - \iiint_{\mathcal{V}} \rho\{t_i\}\, d\mathcal{V} = \int_{t_i}^{t_f} [\dot{m}_{in}\{t\} - \dot{m}_{out}\{t\}]\, dt \tag{3.11}$$

Thus the difference between the final and the initial masses within the control volume equals the difference between the total inflow and outflow masses during the time interval.

---

[5]For compressible flows, an equation of energy conservation must also be used.

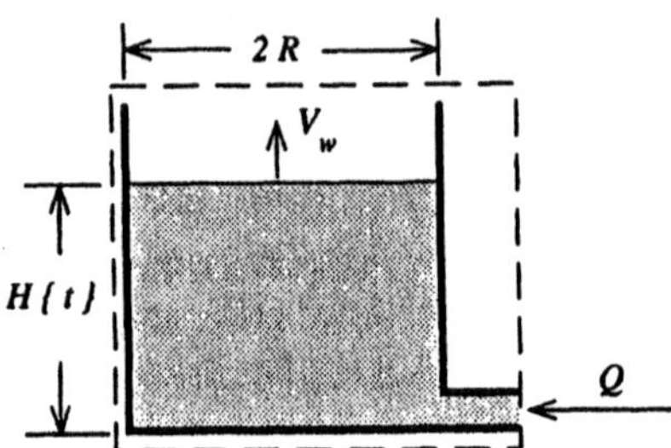

Figure E 3.5

Example 3.5

A cylindrical tank of radius $R = 1\,m$ is being filled with water by a pump. The tank axis is vertical, as shown in figure E 3.5, and the level of the water in the tank is observed to be rising at a rate of $V_w = 1\,mm/s$. Calculate the volume flow rate Q of water through the pump.

Solution

Choose a control volume that encloses the tank, as shown by the dashed line in figure E 3.5. Denoting the height of water in the tank by $H\{t\}$, the mass of water in the tank is $\rho_w(\pi R^2 H\{t\})$, and mass conservation, equation 3.10, requires that:

$$\frac{d}{dt}\iiint_{\mathcal{V}} \rho\, d\mathcal{V} = \dot{m}_{in} - \dot{m}_{out}$$

$$\frac{d}{dt}\left[\rho_w(\pi R^2 H\{t\})\right] = \rho_w Q$$

$$\pi R^2 \frac{dH\{t\}}{dt} = Q$$

$$Q = \pi R^2 V_w$$

$$= \pi \times 1.0\,m^2 \times 1.0E(-3)\,m/s$$

$$= 3.142E(-3)\,m^3/s$$

Note that we have applied the conservation of mass only to the water in the tank and have not included the air stored at the top of the tank nor its outflow across the control surface.

In engineering problems, quite often a control surface is chosen that passes through a duct or pipe normal to its axis. If the density and velocity are the same at all points within the duct, the magnitude of the mass flow rate $\dot{m}$ is the product of the density ρ times the velocity V times the cross-sectional area A of the duct or pipe:

$$|\dot{m}| = \rho V A \tag{3.12}$$

On the other hand, where the velocity **V** is not normal to the control surface, the normal component of the velocity, $\mathbf{V} \cdot \mathbf{n}$, must be used to evaluate the mass flow rate.

Example 3.6

A fire hose is supplied by a fire truck with a water volume flow rate Q of 1,000 gallons per minute. The exit nozzle of the fire hose has a diameter D of 2 *in*. Calculate the velocity V of the water leaving the nozzle.

Solution

Since the flow is steady, the mass flow out of the nozzle equals the mass flow into the nozzle:

$$\dot{m}_{out} = \dot{m}_{in}$$

$$\rho V \left(\frac{\pi D^2}{4}\right) = \rho Q$$

$$V = \frac{4Q}{\pi D^2}$$

$$= \frac{4(1.0E(3)\, gal/min)(3.785E(-3)\, m^3/gal)(min/60\, sec)}{\pi(2\, in)(2.54E(-2)\, m/in)}$$

$$= 31.12\, m/s$$

We have used table 1.6 to convert gallons to cubic meters.

3.3.2 Differential Form of Mass Conservation

We have derived the equation of mass conservation by applying the mass conservation principle to the fluid contained in a control volume of finite size. If we shrink the control volume to the size of a differential volume element $d\mathcal{V} = dx\,dy\,dz$, the mass conservation law becomes a differential equation. To derive the form of this equation, rewrite equation 3.9 as:

$$\iiint_{\mathcal{V}} \left(\frac{\partial \rho}{\partial t}\right) d\mathcal{V} + \iint_{S} \rho \mathbf{V} \cdot \mathbf{n}\, dS = 0$$

by bringing the time differentiation inside the volume integral.[6] Now apply the divergence theorem, equation 1.44, to the surface integral, giving:

$$\iiint_{\mathcal{V}} \left(\frac{\partial \rho}{\partial t}\right) d\mathcal{V} + \iiint_{\mathcal{V}} \nabla \cdot (\rho \mathbf{V}) d\mathcal{V} = 0$$

$$\iiint_{\mathcal{V}} \left(\frac{\partial \rho}{\partial t} + \nabla \cdot (\rho \mathbf{V})\right) d\mathcal{V} = 0$$

Because this holds for all control volumes, the integrand must be zero identically:

$$\frac{\partial \rho}{\partial t} + \nabla \cdot (\rho \mathbf{V}) = 0 \tag{3.13}$$

The differential mass conservation equation (also called the continuity equation), equation 3.13, expresses the same physical constraint as does the integral form, equation 3.9. The first term is the rate of accumulation of mass per unit volume of an infinitesimal fluid element and the second is its net mass outflow rate per unit volume. In cartesian coordinates, equation 3.13 has the form:

$$\frac{\partial \rho}{\partial t} + \frac{\partial}{\partial x}(\rho u) + \frac{\partial}{\partial y}(\rho v) + \frac{\partial}{\partial z}(\rho w) = 0 \tag{3.14}$$

while in cylindrical coordinates it has the form (see equation 1.53):

$$\frac{\partial \rho}{\partial t} + \frac{1}{r}\frac{\partial}{\partial r}(r\rho V_r) + \frac{1}{r}\frac{\partial}{\partial \theta}(\rho V_\theta) + \frac{\partial}{\partial z}(\rho V_z) = 0 \tag{3.15}$$

Example 3.7

The fluid flow of example 3.1 has a velocity field, $\mathbf{V} = (x/t)\,\mathbf{i}_x$. Assuming that the density depends only upon time, find $\rho\{t\}$.

---

[6]Since the control volume is fixed in space, the limits of integration are not functions of time so the time differentiation may be included in the integrand.

Solution

Since the velocity has only one component, in the x direction, mass conservation equation 3.13 becomes:

$$\frac{\partial \rho}{\partial t} + \frac{\partial}{\partial x}(\rho u) = 0$$

$$\frac{\partial \rho}{\partial t} + \frac{\partial}{\partial x}(\rho \frac{x}{t}) = 0$$

$$\frac{\partial \rho}{\partial t} + \frac{\rho}{t} + \frac{x}{t}\frac{\partial \rho}{\partial x} = 0$$

If ρ does not depend upon x, the third term is zero and:

$$\frac{d\rho}{dt} + \frac{\rho}{t} = 0$$

$$\rho\, dt + t\, d\rho = 0$$

$$\rho t = constant$$

The density decreases inversely with time because the fluid particles are being stretched in the x direction in proportion to time.

3.3.3 Incompressible Flow

Mass conservation equation 3.13 may be written in a different form by expanding the divergence term using equation 1.38:

$$\frac{\partial \rho}{\partial t} + \nabla \cdot (\rho \mathbf{V}) = 0$$

$$\frac{\partial \rho}{\partial t} + (\mathbf{V} \cdot \nabla)\rho + \rho \nabla \cdot \mathbf{V} = 0$$

$$\frac{D\rho}{Dt} + \rho \nabla \cdot \mathbf{V} = 0 \tag{3.16}$$

where we have used equation 3.3 to identify the material derivative of ρ.

An *incompressible flow* is one for which the rate of density change of a fluid particle, $D\rho/Dt$, is negligible compared with the component terms of $\rho\nabla \cdot \mathbf{V}$. Using cartesian coordinates, we may write this condition as:

$$\left|\frac{D\rho}{Dt}\right| \ll \rho \left(\left|\frac{\partial u}{\partial x}\right| + \left|\frac{\partial v}{\partial y}\right| + \left|\frac{\partial w}{\partial z}\right| \right)$$

When this condition is satisfied, the material derivative term in equation 3.16 may be dropped and the equation of mass conservation becomes:

$$\nabla \cdot \mathbf{V} = 0 \qquad (incompressible) \tag{3.17}$$

This condition for incompressible flow may be expressed in a different manner. Since equation 3.16 is always true, substituting equation 3.17 into it leads to the conclusion that:

$$\frac{D\rho}{Dt} = 0 \qquad (incompressible) \tag{3.18}$$

This form expresses the incompressibility of the flow by stating that the density of a fluid particle does not change as it moves through the flow field.

It is important to emphasize that an incompressible flow does not require that the density have the same constant value throughout the flow field but only that the density be unchanging along a particle path. Different fluid particles may have different densities, yet each can have an unchanging density in an incompressible flow. If the density is the same at all locations at a particular time, it will remain so at subsequent times if the flow is incompressible. We call such a flow a *constant-density* flow. However, not all incompressible flows are constant-density flows.[7]

Unless specifically noted otherwise, we will assume that all flows are incompressible. However, there are three conditions for which this assumption would be incorrect: (a) when the density of the fluid is changed by a significant amount (such as compressing a gas in an engine cylinder), (b) when the flow velocity is not small compared to the speed of sound and (c), in the case of unsteady flow, when the time for the velocity to change appreciably is not long compared to the time for a sound wave to traverse the flow field. A nearly sonic or supersonic flow is a *compressible* flow, as is sound wave propagation. We will treat compressible flows in chapters 8 and 12.

Example 3.8

A plane flow has velocity components $u = x/T$, $v = -y/T$ and $w = 0$, where T is a constant having the dimension of time. Is this flow incompressible?

---

[7] Some authors call a constant-density flow a *homogeneous flow*, meaning that the density is everywhere and always the same.

Solution

Evaluating $\nabla \cdot \mathbf{V}$,

$$\begin{aligned}\nabla \cdot \mathbf{V} &= \frac{\partial u}{\partial x} + \frac{\partial v}{\partial y} + \frac{\partial w}{\partial z} \\ &= \frac{\partial}{\partial x}\left(\frac{x}{T}\right) + \frac{\partial}{\partial y}\left(-\frac{y}{T}\right) + 0 \\ &= \frac{1}{T} - \frac{1}{T} \\ &= 0\end{aligned}$$

The flow is incompressible.

3.4 Conservation of Chemical Species

Sometimes we need to keep track of individual chemical constituents of a mixture of fluids. In an automobile engine, liquid fuel droplets are mixed with air in the carburetor in definite proportions that are needed to ensure ignition and burning in the engine cylinder. The mass flow rate of automobile exhaust pollutants is measured to determine whether the vehicle conforms to pollution control regulations. We may need to express the conservation of the mass of each chemical species involved in a fluid flow.

The mass concentration of a chemical specie i, denoted by ρ_i, is the mass per unit volume of that constituent in a mixture. The expression for the mass conservation of any constituent i, in the absence of any chemical reaction that might create or remove it, must be identical to that for the fluid as a whole, equation 3.9, except that ρ_i replaces ρ:[8]

$$\frac{d}{dt}\iiint_{\mathcal{V}} \rho_i \, d\mathcal{V} + \iint_S \rho_i \mathbf{V} \cdot \mathbf{n} \, dS = 0 \tag{3.19}$$

Note that, if we sum the mass conservation equations for all of the species in a mixture,

[8]We are assuming here that each specie moves with the same velocity $\mathbf{V}$. In flows where the density ratio ρ_i/ρ varies in space, there is diffusion of species and equation 3.19 must be modified to take this into account.

then we will obtain equation 3.9 because the sum of the partial densities ρ_i equals ρ:

$$\sum_i \frac{d}{dt}\iiint_{\mathcal{V}} \rho_i \, d\mathcal{V} + \sum_i \iint_S \rho_i \mathbf{V}\cdot\mathbf{n}\, dS = 0$$

$$\frac{d}{dt}\iiint_{\mathcal{V}} \sum_i \rho_i \, d\mathcal{V} + \iint_S \sum_i \rho_i \mathbf{V}\cdot\mathbf{n}\, dS = 0$$

$$\frac{d}{dt}\iiint_{\mathcal{V}} \rho \, d\mathcal{V} + \iint_S \rho \mathbf{V}\cdot\mathbf{n}\, dS = 0$$

The mass flow rate of species i across the control surface will be designated by $\dot{m}_i$:

$$\dot{m}_i \equiv \iint_S \rho_i \mathbf{V}\cdot\mathbf{n}\, dS \tag{3.20}$$

and the equivalent of equation 3.10 has the form:

$$\frac{d}{dt}\iiint_{\mathcal{V}} \rho_i \, d\mathcal{V} = (\dot{m}_i)_{in} - (\dot{m}_i)_{out} \tag{3.21}$$

The differential form of mass conservation is the same as equation 3.13 except that ρ is replaced by ρ_i:

$$\frac{\partial \rho_i}{\partial t} + \nabla\cdot(\rho_i \mathbf{V}) = 0 \tag{3.22}$$

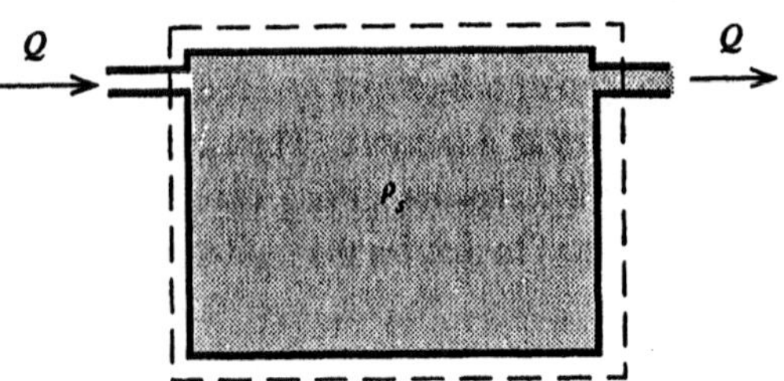

Figure E 3.9

Example 3.9

A tank of volume $\mathcal{V} = 10\,m^3$ is full of a salt solution having an initial salt density $\rho_{s0} = 3.0\,kg/m^3$. At time $t = 0$ fresh water is pumped into the tank at a volume flow rate $Q = 0.01\,m^3/s$, displacing salt solution through an overflow line at an equal volume flow rate. The fluid in the tank is well stirred so that each increment of fresh water dilutes the salt solution uniformly throughout the tank.

(a) Derive and solve a differential equation for the time dependence of the salt density $\rho_s\{t\}$ in the tank. (b) Calculate the volume of fresh water that must be pumped into the tank to reduce the salt concentration by a factor of two from its initial value ρ_{s0}.

Solution

(a) Applying equation 3.21 to the tank volume, and noting that the salt density of the outflow is the same as that in the tank, $\rho_s\{t\}$:

$$\frac{d}{dt}\iiint_{\mathcal{V}} \rho_i\, d\mathcal{V} = (\dot{m}_i)_{in} - (\dot{m}_i)_{out}$$

$$\left(\frac{d\rho_s}{dt}\right)\mathcal{V} = 0 - \rho_s Q$$

$$\frac{d\rho_s}{\rho_s} = -\frac{Q}{\mathcal{V}}dt$$

$$\rho_s = \rho_{s0}\exp\left(-\frac{Qt}{\mathcal{V}}\right)$$

$$= (3.0\,kg/m^3)\exp\left(-\frac{t}{1.0E(3)\,s}\right)$$

(b) At the time when $\rho_s = \rho_{s0}/2$,

$$\exp\left(\frac{Qt}{\mathcal{V}}\right) = 2$$

$$Qt = \mathcal{V}\ln 2$$

$$= 10\,m^3 \times 0.6931$$

$$= 6.931\,m^3$$

3.4.1 Chemical Reactions

When a chemical specie undergoes a chemical reaction, such as fuel reacting with air to form combustion products, the mass conservation equation for each of the reacting species must include the possibility of the production (or loss) of these species caused by the chemical reaction. To account for the effects of chemical rections, we include a term on the right side of equation 3.19 to allow for the production of specie *i* because of chemical reactions:

$$\frac{d}{dt}\iiint_{\mathcal{V}} \rho_i \, d\mathcal{V} + \iint_S \rho_i \mathbf{V} \cdot \hat{\mathbf{n}} \, dS = \iiint_{\mathcal{V}} \left(\frac{\partial \rho_i}{\partial t}\right)_{chem} d\mathcal{V} \tag{3.23}$$

where $(\partial \rho_i / \partial t)_{chem}$ is the net mass rate of production of specie i per unit volume inside the control volume due to all reactions that produce or consume specie i.

The corresponding form of the differential equation of mass conservation is:

$$\frac{\partial \rho_i}{\partial t} + \nabla \cdot (\rho_i \mathbf{V}) = \left(\frac{\partial \rho_i}{\partial t}\right)_{chem} \tag{3.24}$$

Note again that summing equation 3.23 or 3.24 over all species will give equation 3.19 or 3.22, respectively, because the sum over all chemical reactions will not result in any change in the total mass of the species involved.

3.5 Two-Phase Flow

Some flows involve a mixture of two physical phases. Solid particles in a gas or liquid (dust storm or silt in a river), liquid droplets in a gas (fog) and gas bubbles in a liquid (boiling water) are examples of flows in which the fluid components are not mixed thoroughly down to the molecular level as they are in chemical solutions and mixtures. Such flows are called *two-phase* flows.

The concept of the partial density of each phase of a two-phase mixture is a useful one for analyzing such flows. The partial density ρ_i of a phase is simply the ratio of the mass of the phase in a sample of fluid to the volume of that sample. The sample volume must be large enough to contain a large number of particles (or bubbles or droplets) but small enough to include only a small fraction of the flow field. In this sense, the phase density is directly analogous to the partial density of a chemical specie in a mixture. Sometimes the amount of the phase can be determined as the fraction of the volume of the fluid that is occupied by the phase. Note that the phase partial density is quite distinct from the actual density of the material in the particle or bubble because it is averaged over a much larger volume.

When the minor component of a heterogeneous mixture (*e.g.*, dust particle or bubble) is very small in size, it tends to be carried along at the speed of the surrounding fluid. In such cases, the phase may be treated just like a chemical specie and each component of the two-phase fluid would obey a mass conservation law having the form of equation 3.22. On the other hand, when the size of the components are not small enough, as in a rainstorm or boiling water, each phase may have a different velocity, and the mass conservation law would take the form:

$$\frac{\partial \rho_i}{\partial t} + \nabla \cdot (\rho_i \mathbf{V}_i) = 0 \tag{3.25}$$

where $\mathbf{V}_i$ is the velocity of phase i. The difference between the velocities of each phase is determined by the details of the dynamics of the flow around each particle or bubble—a quite complicated situation that is not easily handled by the methods we use in this book.

3.6 Measuring Volume and Volume Flow Rate

The amounts of most fluids we use daily are measured by their volumes. In baking a cake, we measure a cup of milk and a teaspoon of vanilla. A cup of coffee is generally a standard volume. The volume of water used to flush a toilet is fixed by the operation of the float valve in the supply tank. The size of the container holding a fluid is the simplest method for measuring the volume of fluid.

However, many fluids are delivered through pipes or hoses—gasoline to our automobile fuel tank, water and natural gas to our home. Flowmeters measure the volumes of these fluids which are sold in commerce. A typical flowmeter consists of a rotor that rotates in proportion to the volume of fluid that passes through the meter. The rotor motion operates a counter calibrated to the volume of fluid as measured in suitable units. (In gasoline pumps, the meter also actuates the price counter.)

Sometimes it is necessary to know or adjust the volume rate of flow. In an automobile carburetor, the volume rates of flow of liquid fuel and air must be held to the same proportions over the entire speed and power range by suitable mechanical and electronic controls. (In diesel engines, the volume of fuel injected per cycle is the controlled variable.) Some measurement of volume flow rates is accomplished in meters that utilize the dynamical principles to be discussed in later chapters of this book.

3.7 Problems

Problem 3.1

The velocity field of a fluid flow is given as:

$$\mathbf{V} = f\{t\}\,\mathbf{i}_x$$

where $f\{t\}$ is a function of time alone. Prove that the flow is incompressible if the density ρ is a function h of the argument $x - ft$:

$$\rho = h\{x - ft\}$$

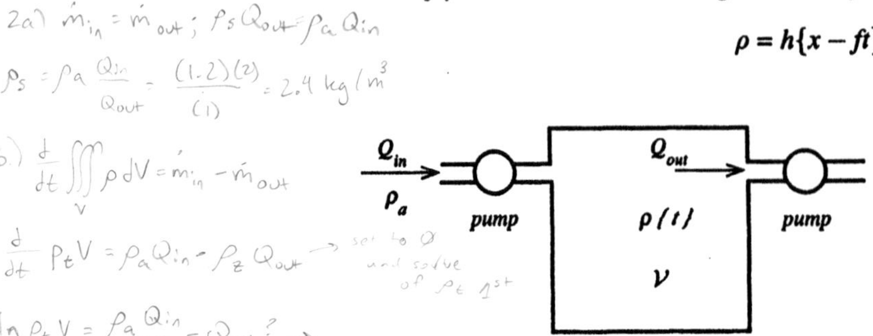

Figure P 3.2

Problem 3.2

A tank of volume $\mathcal{V} = 100\,m^3$ is fitted with two pumps, one at the entrance and the other at the exit of the tank, as shown in figure P 3.2. Each pump is a positive displacement pump, *i.e.*, it pumps at a fixed volume flow rate of the fluid entering the pump. The inflow pump sucks air from the atmosphere and delivers it to the tank, while the outflow pump sucks air from the tank and delivers it to a pipeline.

Initially, neither pump is working and the vessel is filled with air at atmospheric density $\rho_a = 1.2\,kg/m^3$. At $t = 0$ both pumps are switched on, instantaneously drawing a volume flow rate $Q_{in} = 2\,m^3/s$ into the inflow pump and $Q_{out} = 1\,m^3/s$ into the outflow pump. Because the mass inflow and outflow rates are different initially, the density $\rho\{t\}$ of the air in the tank changes with time. After a long time, however, the air density inside the tank reaches a steady value.

(a) Calculate the value of the air density ρ_s in the tank after the steady state has been reached. (b) Derive and solve a differential equation for $\rho\{t\}$ for any time $t \geq 0$.

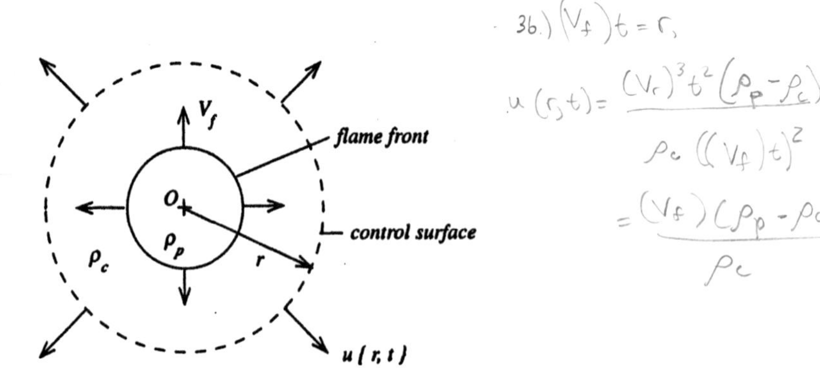

Figure P 3.3

Problem 3.3

In figure P 3.3, a stationary combustible mixture of air and fuel is ignited at the point O at time $t = 0$ by a spark, which initiates a spherical flame front moving outward from the point O at a *constant* speed, V_f. The combustible mixture ahead of the flame front, which has a constant density ρ_c, is converted in the very thin flame front to hot products of combustion that has a much lower density ρ_p. Because the volume of a fluid element enveloped by the flame front is increased, the combustible mixture is pushed radially outward by the expanding flame front, although the combustion products remain stationary. Thus the gas outside the flame front is moving radially outward while that inside is stationary.

(a) Selecting a spherical control volume of fixed radius $r > V_f t$ surrounding the flame front, apply the equation of mass conservation and derive an expression for the radial outflow velocity $u\{r, t\}$ of the combustible mixture at r for any time t prior to the arrival of the flame front at r, *i.e.*, $t < r/V_f$. (b) Derive an expression for the velocity u just in advance of the flame front as a function of ρ_c, ρ_p and V_f.

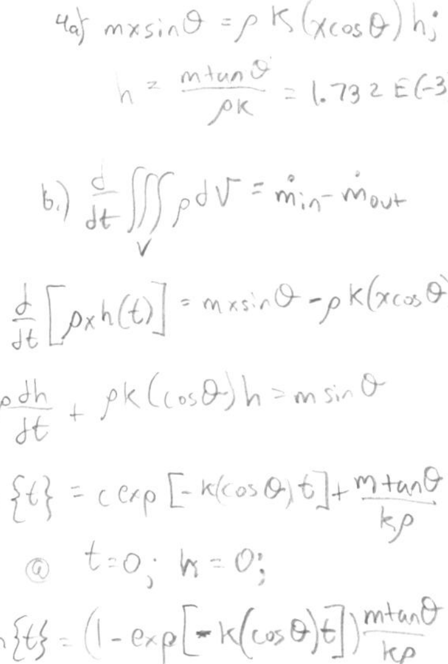

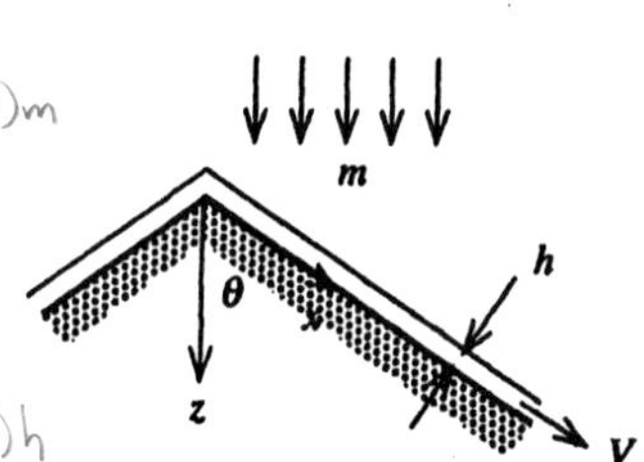

Figure P 3.4

Problem 3.4

Rain falls on a houseroof that is inclined at an angle $\theta = 60°$ to the vertical, as shown in figure P 3.4. The mass flow rate of rain *per unit horizontal area* is $m = 1E(-3)\, kg/m^2s$. A very thin film of water forms on the roof, flowing parallel to the roof surface in the direction of the x axis as shown in figure P 3.4. The flow speed V in the water is uniform across the thickness h of the film and depends only upon the vertical distance z from the apex of the roof, varying as:

$$V = kz$$

where k is a constant having the value $1E(-3)\, s^{-1}$.

(a) Show that in a steady flow the thickness h of the film does not depend upon the distance x, and calculate its value. (b) If the rain starts to fall at time $t = 0$ at a steady rate of m when the roof is dry, derive and solve a differential equation for the time-dependent thickness $h\{t\}$, assuming that, at any instant of time, h is independent of x and that $V = kz$. (c) Show that your solution gives the same value for h as in (a) when $t \rightarrow \infty$.

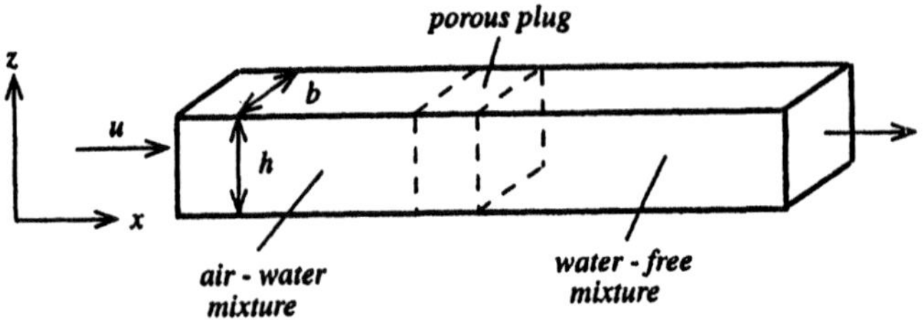

Figure P 3.5

Problem 3.5

Air containing a suspension of very fine water droplets flows unsteadily into a rectangular duct of breadth $b = 0.3\,m$ and height $h = 0.2\,m$, as shown in figure P 3.5. The inlet flow velocity u is not uniform but varies with the vertical z direction and time $t \geq 0$ according to:

$$u\{z\} = at\left(1 - \frac{z}{h}\right)$$

where $a = 0.1\,m/s^2$ is a constant. The fraction of the volume of the incoming mixture of air and water droplets that is occupied by the water droplets is $\epsilon = 0.01$, and the droplet water density ρ_w is $1E(3)\,kg/m^3$.

(a) Calculate the volume flow rate $Q\{t\}$ of mixture flowing into the duct. (b) Calculate the mass flow rate of water into the duct. (c) A porous plug located in the duct absorbs all the water droplets, retaining this water in the plug as shown in figure P 3.5. At what time t will the mass $M\{t\}$ of water retained in the plug equal $10\,kg$?

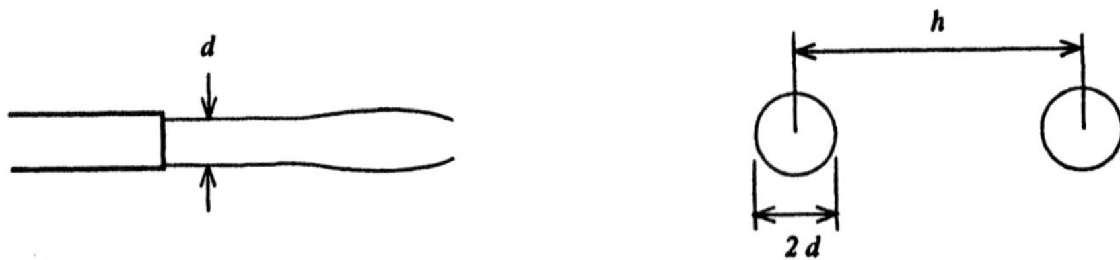

Figure P 3.6

Problem 3.6

A stream of liquid flows from a pipe into the atmosphere. The fluid stream has a diameter d and a uniform velocity. At some distance downstream, the stream breaks up into spherical drops of diameter $2d$ that are spaced a distance h apart, as shown in figure P 3.6. Calculate the ratio h/d.

Problem 3.7

A meteorological balloon is being inflated by a constant volume flow rate of helium $Q_{in} = 0.1\, m^3/s$. The pressure inside the balloon remains at atmospheric pressure until the balloon volume $\mathcal{V}$ reaches a value $\mathcal{V}_0 = 1\, m^3$, after which the balloon's internal gage pressure p increases with the balloon volume as the balloon material stretches, according to the relation:

$$p = k(\mathcal{V} - \mathcal{V}_0)$$

where the constant $k = 1.0E(5)\, Pa/m^3$. However, there is a small leak in the balloon, having a volume flow rate Q_{out} that is dependent upon the gage pressure in the balloon according to the relation:

$$Q_{out} = c\sqrt{p}$$

where the constant $c = 1.0E(-3)\, m^3/Pa^{1/2}s$.

(a) Assuming that the helium is an incompressible fluid in this process, calculate the steady state volume $\mathcal{V}_\infty$ of the balloon and the corresponding gage pressure p_∞. (b) Derive a differential equation for the volume $\mathcal{V}\{t\}$ whose solution will give its time dependence when $\mathcal{V} \geq \mathcal{V}_0$.

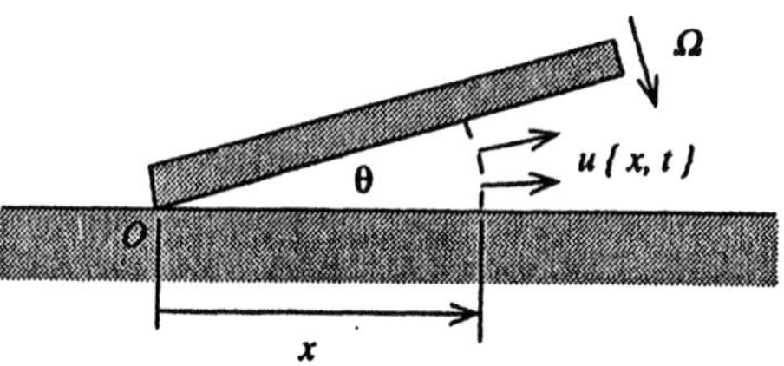

Figure P 3.8

Problem 3.8

As shown in figure P 3.8, a flat plate is hinged at point O to a horizontal surface. The hinged plate is slowly rotated toward the horizontal at a clockwise angular velocity $\Omega = 0.1\, s^{-1}$, squeezing out the air between the two plates. Calculate the radial velocity $u\{x\}$ at $x = 0.1\, m$ when $\theta = 0.1$. (Assume that the flow is incompressible and that the dimension of the plate in a direction normal to the plane of figure P 3.8 is so great that the fluid flows only in the plane of the figure.)

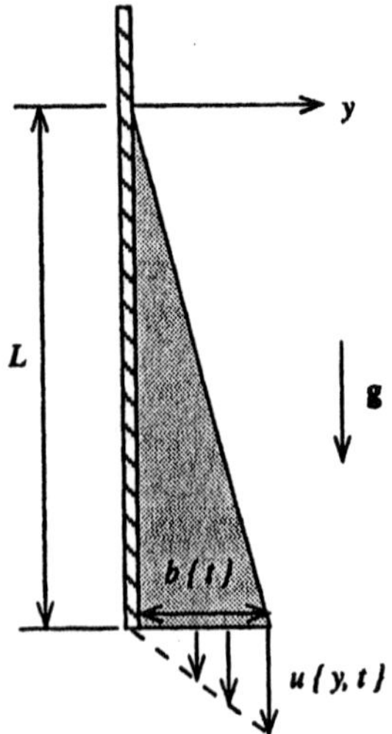

Figure P 3.9

Problem 3.9

A thin flat plate is dipped into honey to a depth $L = 0.1\,m$ and then lifted out, being held stationary while the honey drains off. The draining flow is a plane flow in the plane of figure P 3.9. The width $b\{t\}$ of the honey layer at the bottom of the plate decreases with time t, but the layer of honey retains its triangular shape, as shown in figure P 3.9. The honey velocity at the bottom of the plate, $u\{y\}$, varies linearly with distance from the plate:

$$u\{y\} = \frac{Uy}{b\{t\}}$$

where $U = 1.0E(-3)\,m/s$ is a constant. If the initial value of b at time $t = 0$ is $b_0 = 1.0E(-2)\,m$, derive an expression for $b\{t\}$.

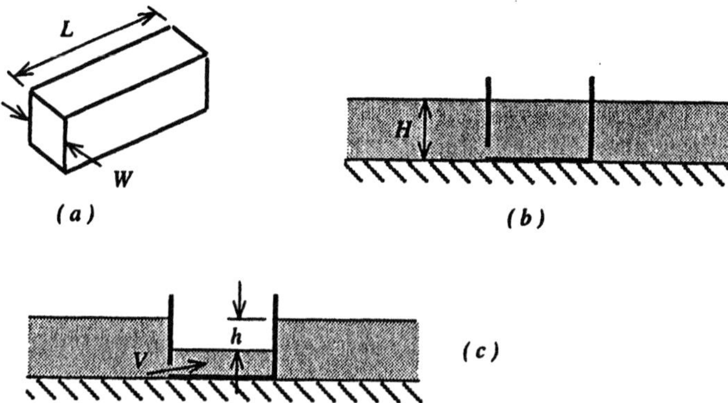

Figure P3.10

Problem 3.10

An empty rectangular barge of width $W = 10\,m$ and length $L = 30\,m$ has sprung a leak, filled with water and settled to the bottom of the sea in water of depth $H = 4\,m$, as shown in figure P3.10 (a)-(b). The barge owner wishes to refloat the barge by pumping it out with a pump having a volume flow rate $Q = 0.5\,m^3/s$. When the pump is operated, water flows into the barge through the leak in the bottom of area $A = 0.1\,m^2$ with a speed V given by:

$$V = \sqrt{2gh}$$

in which h is the difference in level between the air/water interface inside the barge and that in the sea outside (see figure P3.10 [c]).

(a) Derive a differential equation which describes the rate at which the water level difference h increases with time t after the pump is turned on. (b) After operating the pump for some time, the water level inside the barge no longer decreased while the barge remained firmly resting on the bottom. What was the value of h at this conditon? (c) The owner knows that the mass M of the empty barge is $1.0E(6)\,kg$. He installs a pump of much larger capacity than was used in (b), and while the water level was still decreasing, the barge began to refloat. Assuming the density ρ_{sw} of sea water is $1030\,kg/m^3$, at what value of h did the barge begin to refloat? (d) Calculate the smallest value of pump capacity Q for which it is possible to refloat the barge.

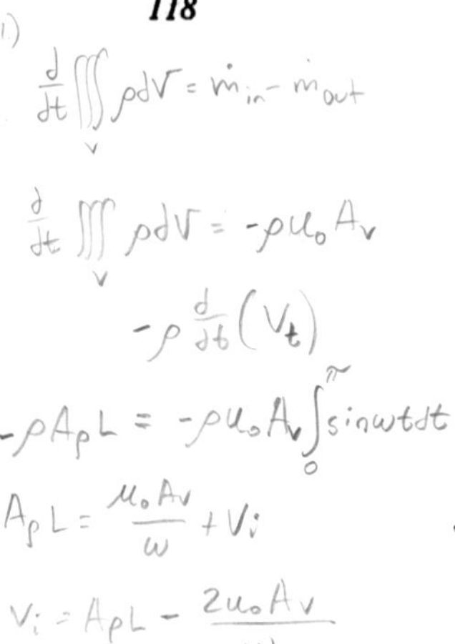

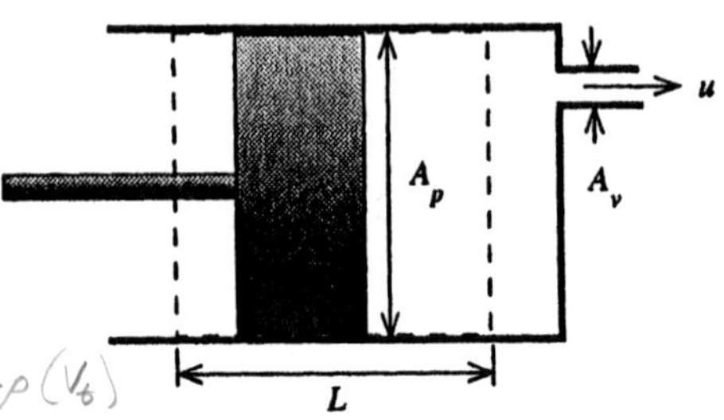

Figure P3.11

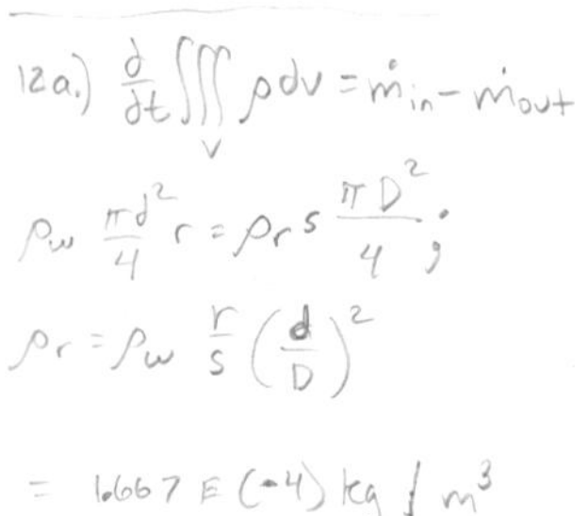

Problem 3.11

During the exhaust process of an internal combustion engine, the gas is pushed out of the cylinder through the exhaust valve by the motion of the piston, which moves from left to right through the distance $L = 0.1\,m$, as shown in figure P3.11. If the piston rings are worn, some gas escapes through the clearance space between the piston and the cylinder walls. To find out how much gas escapes, the gas velocity u at the exhaust valve was measured and found to be:

$$u = u_0 \sin \omega t$$

where $u_0 = 70\,m/s$ and $\omega = 300\,s^{-1}$ are constants and t is the time. The product ωt varies between 0 and π when the piston moves through the distance L. The areas of the piston and valve are $A_p = 1.0E(-2)\,m^2$ and $A_v = 2.0E(-3)\,m^2$, respectively. Assuming that the gas is incompressible, calculate the volume $\mathcal{V}_l$ of gas that leaked past the piston during the exhaust process.

Figure P3.12

Problem 3.12

A rain gage consists of a funnel of diameter $D = 10\,cm$ that collects rain, depositing the liquid water of density $\rho = 1E(3)\,kg/m^3$ into a cylinder of diameter $d = 1\,cm$. The top surface of the column of liquid collected is observed to rise at the rate $r = 1\,cm/min$. (a) If the raindrops are falling at a speed $s = 10\,m/s$ into the funnel, calculate the mass of liquid water in a cubic meter of the atmosphere, ρ_r. (b) If the raindrops have a radius $R = 1\,mm$, calculate the number of raindrops in a cubic meter of the atmosphere, n.

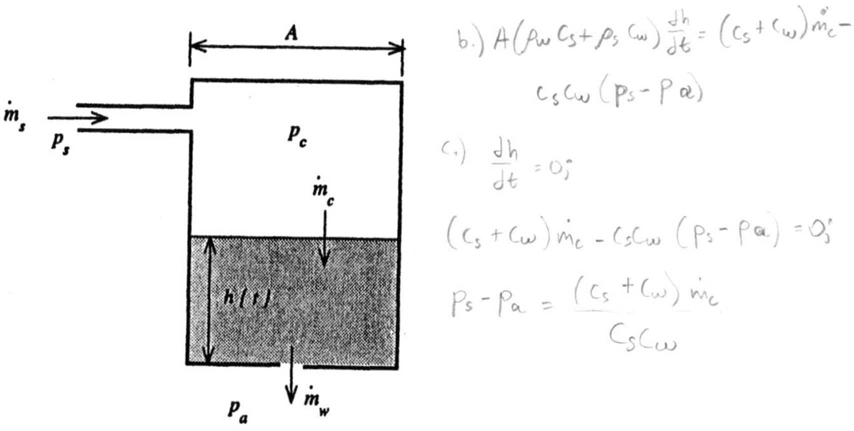

Figure P 3.13

Problem 3.13

A condenser in a steam system consists of a cylindrical tank of cross-sectional area A, the lower part of which is filled with water having a depth $h\{t\}$ that varies with time. The upper part is filled with steam supplied at a mass flow rate $\dot{m}_s$ that is proportional to the difference in pressure between the steam supply pressure p_s and the condenser pressure p_c:

$$\dot{m}_s = c_s(p_s - p_c)$$

where c_s is a constant. The steam condenses to water at a constant mass flow rate $\dot{m}_c$. A small orifice in the bottom of the tank allows the water to drain out at a mass flow rate $\dot{m}_w$ that is proportional to the difference between the condenser pressure and atmospheric pressure p_a:

$$\dot{m}_w = c_w(p_c - p_a)$$

where c_w is a constant. The steam and water densities ρ_s and ρ_w are constant in this process.

(a) Write two conservation of mass equations, one for steam and the other for water, in the form:

$$\frac{dh}{dt} = f\{\dot{m}, p, etc.\}$$

(b) Derive an equation for dh/dt that does *not* contain the unknown condenser pressure p_c. (c) What should be the pressure difference across the condenser, $p_s - p_a$, that would ensure that h would not change with time?

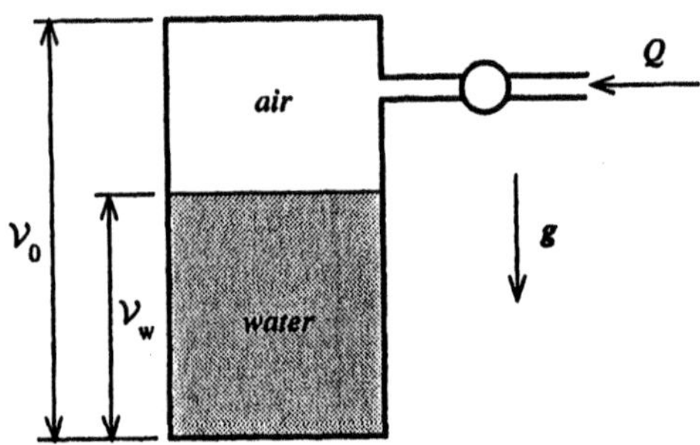

Figure P 3.14

Problem 3.14

A tank of volume $\mathcal{V}_0$ is filled initially with air at atmospheric density ρ_{at}. Beginning at $t = 0$, a pump delivers to the tank a mixture of air and water. The volumetric inflow rate to the pump is Q, half of which is water having a density ρ_w and the other half is air at the density ρ_{at}. Once inside the tank, the water falls to the bottom, occupying a volume $\mathcal{V}_w\{t\}$ that increases with time (see figure P 3.14). The air rises to the top of the tank, forming a volume with uniform density $\rho_a\{t\}$.

(a) In terms of the known quantities $\mathcal{V}_0$, ρ_{at}, Q and ρ_w, derive expressions for $\mathcal{V}_w\{t\}$ and $\rho_a\{t\}$ as functions of time t. (b) The pressure p of the air in the tank is related to its density ρ_a and the atmospheric pressure p_{at} by:

$$p = p_{at}\left(\frac{\rho_a}{\rho_{at}}\right)$$

Calculate the ratio p/p_{at} for the time when the tank is half full of water.

Problem 3.15

A HIVOL sampler is an instrument for measuring the mass concentration of solid particles in atmospheric air. It consists of a fan that sucks air at a constant volume flow rate $Q = 10\,m^3/min$ through a filter which traps the solid particles in the air stream. The filter is weighed before and after the sampler is run for a period of time, and the increase in filter mass is equal to the mass of particles in the ambient air that passed through the filter during the time period.

A HIVOL is run for a period $\Delta t = 24\,h$, and the mass M_p of the collected particles is measured to be $1.34g$. Calculate the ambient particle density ρ_p, using units of micrograms per cubic meter ($\mu g/m^3$).

Problem 3.16

Carbon dioxide is accumulating in the earth's atmosphere because the mass rate of inflow of CO_2 currently exceeds the mass rate of outflow during an annual cycle. The concentration of CO_2, now about 250 parts per million (*i.e.*, 250 of each million air molecules are CO_2 molecules), is increasing at the rate of about $1.5\,ppm$ per year. If the atmospheric mass is $5.2E(18)\,kg$, calculate the mass rate of increase of CO_2 in the atmosphere, measured in metric tons per year. (The molecular weight of CO_2 is 44, while that of air is 28.95.)

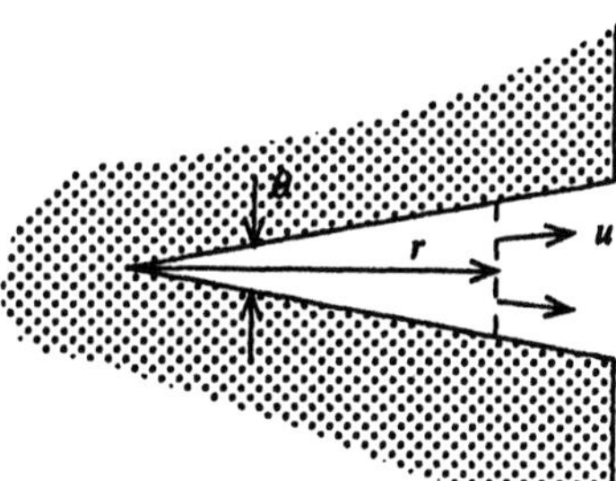

Figure P 3.17

Problem 3.17

As shown in figure P3.17, a tidal estuary, viewed from above, is wedge shaped. Because it is open to the sea at the right, the depth h of water in the estuary is the same as that in the sea, varying with the tidal motion of the sea:

$$h = a + b \sin \omega t$$

where $\omega = 1.396E(-4)\,s^{-1}$ is the tidal frequency, t is the time and where $a = 10\,m$ and $b = 2\,m$ are constants of the local daily tide. Calculate the value of the tidal velocity $u\{r,t\}$ at a position $r = 10\,km$ from the source of the estuary when $\omega t = 0$.

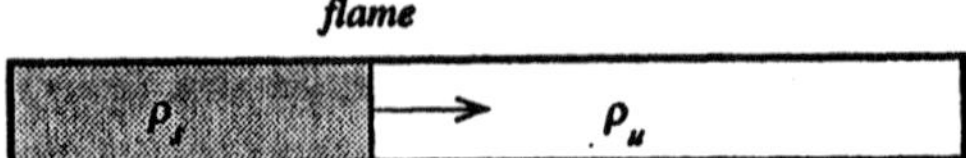

Figure P 3.18

Problem 3.18

A tube closed at both ends is filled initially with a combustible mixture of uniform density ρ_0 and pressure p_0. The mixture is ignited at the left end, and a plane flame propagates through the mixture from left to right, as shown in figure P 3.18. In this process, the pressure p is uniform in space but increases as the flame moves down the tube. Ahead of the flame, in the unburned mixture, the density ρ_u is uniform but increases with pressure according to the relation:

$$\rho_u = \rho_0 \left(\frac{p}{p_0}\right)^n$$

where the constant $n = 0.8$. Behind the flame, in the burned gases, the density ρ_b is also uniform, varying with pressure:

$$\rho_b = \alpha \rho_0 \left(\frac{p}{p_0}\right)^n$$

where the constant $\alpha = 0.25$.

(a) Calculate the pressure ratio p/p_0 when the flame has completely consumed the unburned gas. (b) When the flame has reached the midpoint of the tube, calculate the percent of the mass of the original mixture that has burned.

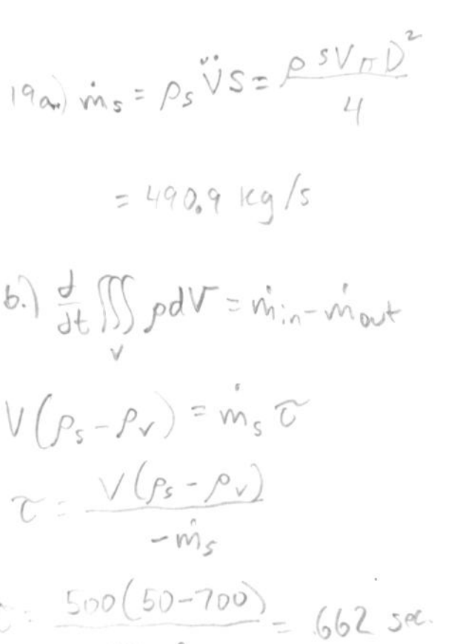

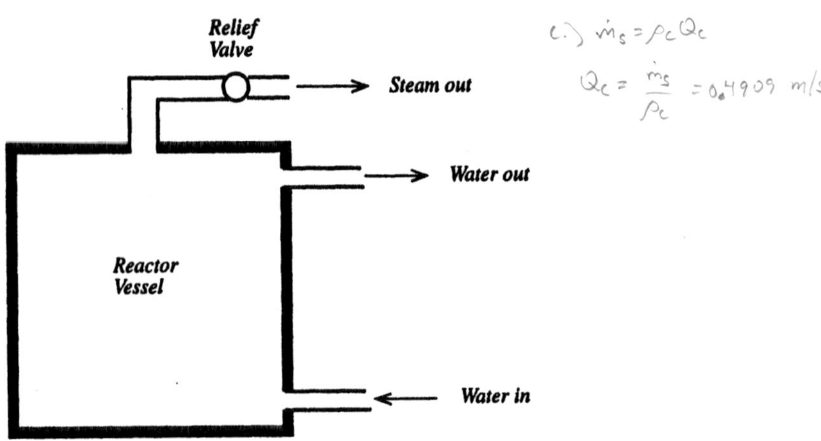

Figure P 3.19

Problem 3.19

A nuclear reactor vessel of volume $\mathcal{V} = 500\,m^3$ is filled completely with water under high pressure. During regular operation, a volume flow rate of $Q = 10\,m^3/s$ of water is pumped into the bottom of the reactor vessel and an equal volume flow of water leaves the top. The average density of the water in the vessel is $\rho_v = 700\,kg/m^3$.

Due to a feed water pump failure, the inflow and outflow of water ceases abruptly. The water remaining in the reactor begins to boil because of heating by the reactor core, and steam escapes through a pressure relief valve into the containment building. The steam in the reactor and flowing through the relief valve has a constant density of $\rho_s = 50\,kg/m^3$. The escaping steam flows at a constant mean speed $\bar{V} = 50\,m/s$ through a circular pipe of diameter $D = 0.5\,m$.

(a) Calculate the mass flow rate $\dot{m}_s$ of steam through the relief valve. (b) Calculate the length of time τ that the steam escapes until all the water in the reactor vessel is converted to steam. (c) An emergency core cooling system has been provided that will inject cold water of density $\rho_c = 1E(3)\,kg/m^3$ into the reactor at a mass flow rate just sufficient to counterbalance the mass rate of steam generation described in (a). Calculate the volume flow rate Q_c of water that is supplied by this system.

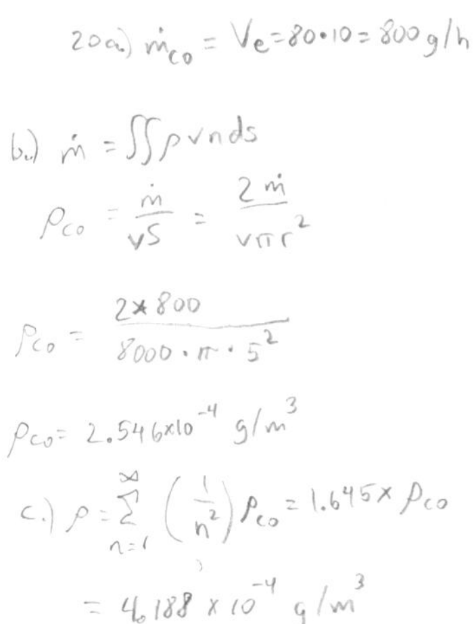

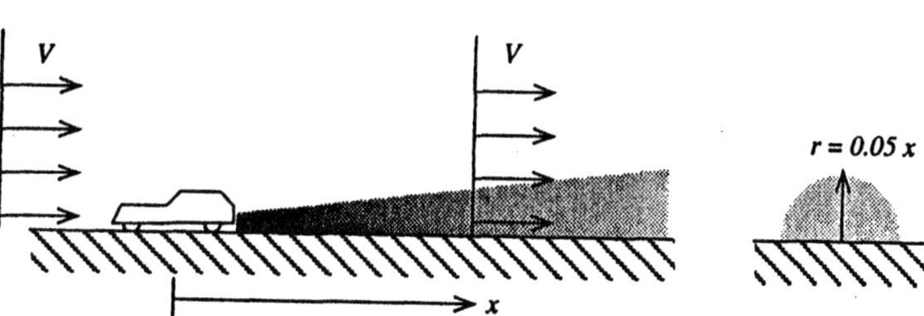

Figure P 3.20

Problem 3.20

An automobile moves along a highway at a speed $V = 80\,km/h$ while emitting carbon monoxide (CO) in its exhaust at a rate e of 10 g per kilometer of travel ($e = 10\,g/km$). In a reference frame attached to the automobile, a wake region extends behind the automobile as shown in the sketch. Within this region the fluid speed equals the vehicle speed V and the emitted CO is uniformly mixed within the semicircular cross-section of the wake, whose radius r equals $0.05\,x$, where x is the distance behind the vehicle. Calculate (a) the mass rate of emission of CO, $\dot{m}_{CO}$ (g/h), and (b) the mass density of CO, ρ_{CO} (g/m^3) in the vehicle wake at a distance $x = 100\,m$ behind the vehicle. (c) The vehicle is preceded by a long line of vehicles spaced $100\,m$ apart, moving at the same speed and emitting CO at the same rate. At any location, each of these vehicles makes an additive contribution to the total value of ρ_{CO}, depending upon its distance away. Calculate the mass density of CO at $x = 100\,m$ behind the vehicle shown in the sketch, due to the cumulative effect of all the preceding vehicles. (Note: The sum $\sum_1^\infty (1/n^2) = \pi^2/6 = 1.645$.)

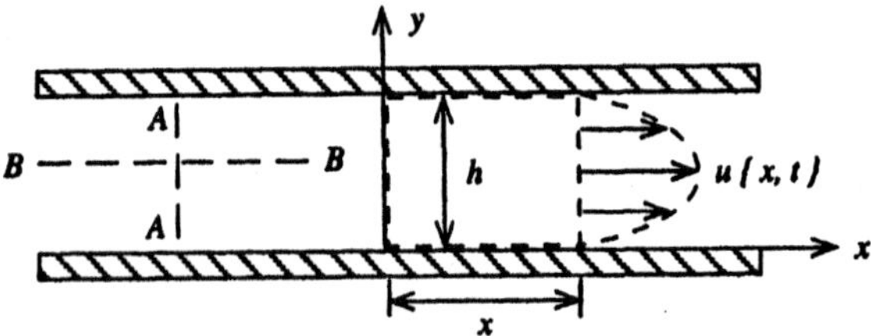

Figure P 3.21

Problem 3.21

A viscous, incompressible fluid is squeezed out of the gap between two parallel plates when the upper plate moves downward at a speed V, the lower plate remaining stationary. Because the plate dimension is very large in the direction z normal to the plane of figure P 3.21, the fluid velocity lies entirely in the x, y plane, and there is no dependence upon the z direction. The u component of the velocity has a parabolic distribution as shown in figure P 3.21:

$$u = f\{x\}\left[\frac{y}{h} - \frac{y^2}{h^2}\right]$$

where $f\{x\}$ is a function of x only. (Note that u is zero at both the lower ($y = 0$) and the upper ($y = h$) surfaces.) The y-axis is a plane of symmetry of the flow, the flow to the left of it being a mirror image of that to the right, and thus $u\{0, y\} = 0$.

(a) Using the control volume indicated by the dashed line in figure P 3.21, determine the function $f\{x\}$ in terms of the variables x, V and h. (b) Utilizing the differential form of the mass conservation law, derive an expression for the vertical velocity component $v\{x, y\}$ as a function of x, y, V and h. (c) Sketch the variation of v as a function of y for $h \geq y \geq 0$. (d) At a particular instant of time, dye lines are inserted in the flow in the form of a cross, $A - A$ and $B - B$, as shown by the dashed lines in figure P 3.21. Sketch the distorted shape that the cross assumes a short time later.

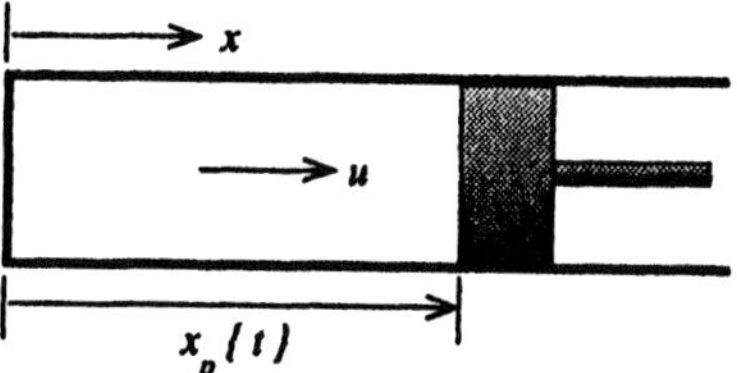

Figure P 3.22

Problem 3.22

A piston moves slowly inside a cylinder closed at one end, as shown in figure P 3.22. Its position $x_p\{t\}$ is a prescribed function of time t. Because the motion is slow, the gas density $\rho\{t\}$ is uniform within the cylinder but varies with time t. However, the gas velocity $u\{x, t\}$ depends upon both x and t. Prove that a solution to the differential form of the mass conservation law for this motion is:

$$\rho\{t\} = \frac{k}{x_p\{t\}}$$

$$u\{x, t\} = \frac{x}{x_p}\frac{dx_p}{dt}$$

where k is a constant.

Bibliography

Batchelor, G. K. 1967. *An Introduction to Fluid Dynamics.* Cambridge: Cambridge University Press.

Olson, Reuben M., and Steven J. Wright. 1990. *Essentials of Engineering Fluid Mechanics.* New York: Harper and Row.

White, Frank M. 1986. *Fluid Mechanics.* 2nd ed. New York: McGraw-Hill Book Co.

4 Inviscid Flow

In this chapter, we consider a kind of flow called *inviscid flow* that occurs in special, although not uncommon, circumstances. In such flows, the effect of fluid viscosity is so small as to be negligible, and the resultant flow is much easier to treat analytically than is the case when viscous effects cannot be ignored. More importantly, it is possible to determine readily significant properties of the flow, such as the pressure and velocity fields, or even to estimate such properties by use of simple algebraic relations. It is easy to develop physical intuition about how an inviscid flow behaves.

We first derive the vector differential equation of motion of an inviscid fluid, called *Euler's equation.* Then we find a scalar integral of this equation, called *Bernoulli's equation*, that provides an algebraic relation between pressure, velocity and position in the earth's gravitational field. While Bernoulli's equation does not tell us everything about the flow field, it may provide us with enough information to find what we need to know in order to solve a particular practical problem.

Later, in chapter 6, we will consider the more difficult problem of how to describe a flow when viscous effects cannot be neglected. For the time being, by treating only inviscid flows, we can develop some familiarity with fluid dynamical principles that will be helpful when we consider the more general case of viscous flows.

4.1 Criterion for Inviscid Flow

If a fluid were to have zero viscosity, then it could not sustain a shear stress, and its flow would be inviscid exactly. However, no fluid has zero viscosity.[1] For a flow

[1]Liquid helium-4 at temperatures below 4.2 K flows without friction through small tubes and channels. This flow is a macroscopic quantum motion of the fluid and is not describable in its entirety by Euler's equation. Under these conditions, helium-4 is called a *superfluid.*

to be regarded as inviscid, the effects of the shear stresses on the motion must be sufficiently small as compared to other influences that they can be ignored as being negligible. A *necessary*, but not *sufficient*, condition for negligible viscous effects is that a dimensionless parameter characterizing the flow, called the *Reynolds number* and denoted by *Re*, is very large. For the steady flow of a fluid of density ρ and viscosity μ over (or through) an object of dimension L at a speed V, the Reynolds number is:

$$Re \equiv \frac{\rho VL}{\mu} = \frac{VL}{\nu} \tag{4.1}$$

where we have used equation 1.6 to replace ρ/μ by $1/\nu$. Thus the necessary condition for inviscid flow is:

$$Re >> 1$$

If the Reynolds number of a flow is not large, then the flow is viscous and cannot be treated as an inviscid flow. However, it is possible that a large Reynolds number flow can be greatly affected by viscous effects under some circumstances, such as when the flow comes in contact with solid boundaries. We cannot always predict when such flows should be regarded as viscous and must be guided by experimental observation and experience. For the time being, we will note in this chapter when a flow cannot be treated as inviscid.

4.2 Acceleration of a Fluid Particle

If we wish to express Newton's law of motion for a fluid particle in the form, *mass × acceleration = force*, we need first an expression for the acceleration of a fluid particle. Since the acceleration of a fluid particle is the time rate of change of its velocity **V**,

$$\text{acceleration} = \frac{D\mathbf{V}}{Dt} \tag{4.2}$$

where we have used the material time derivative D/Dt of equation 3.2 because we need the time rate of change following the fluid particle. In cartesian coordinates, the acceleration $D\mathbf{V}/Dt$ may be written in component form as:

$$\begin{aligned}\frac{D\mathbf{V}}{Dt} &= \frac{\partial \mathbf{V}}{\partial t} + (\mathbf{V}\cdot\nabla)\mathbf{V} \\ &= \left(\frac{\partial u}{\partial t} + u\frac{\partial u}{\partial x} + v\frac{\partial u}{\partial y} + w\frac{\partial u}{\partial z}\right)\mathbf{i}_x\end{aligned}$$

$$+\left(\frac{\partial v}{\partial t}+u\frac{\partial v}{\partial x}+v\frac{\partial v}{\partial y}+w\frac{\partial v}{\partial z}\right)\mathbf{i}_y$$

$$+\left(\frac{\partial w}{\partial t}+u\frac{\partial w}{\partial x}+v\frac{\partial w}{\partial y}+w\frac{\partial w}{\partial z}\right)\mathbf{i}_z \tag{4.3}$$

while in cylindrical coordinates the acceleration becomes:

$$\frac{D\mathbf{V}}{Dt}=\left(\frac{\partial V_r}{\partial t}+V_r\frac{\partial V_r}{\partial r}+\frac{V_\theta}{r}\frac{\partial V_r}{\partial \theta}+V_z\frac{\partial V_r}{\partial z}-\frac{V_\theta^2}{r}\right)\mathbf{i}_r$$

$$+\left(\frac{\partial V_\theta}{\partial t}+V_r\frac{\partial V_\theta}{\partial r}+\frac{V_\theta}{r}\frac{\partial V_\theta}{\partial \theta}+V_z\frac{\partial V_\theta}{\partial z}+\frac{V_r V_\theta}{r}\right)\mathbf{i}_\theta$$

$$+\left(\frac{\partial V_z}{\partial t}+V_r\frac{\partial V_z}{\partial r}+\frac{V_\theta}{r}\frac{\partial V_z}{\partial \theta}+V_z\frac{\partial V_z}{\partial z}\right)\mathbf{i}_z \tag{4.4}$$

Each of the three components of the acceleration vector requires one time derivative and three spatial derivatives of a velocity component in its expression or a total of twelve derivatives needed for determining the acceleration of a fluid particle. When expressed in cylindrical coordinates (equation 4.4), there are two additional, non-derivative terms: the centrifugal acceleration, $-V_\theta^2/r$, in the radial direction and the Coriolis acceleration, $V_r V_\theta/r$, in the tangential direction. Since it would be awkward to write out the complete form of the acceleration in terms of these derivatives every time we need to use it in an equation, we shall use the shorthand notation $D\mathbf{V}/Dt$ or $\partial\mathbf{V}/\partial t+(\mathbf{V}\cdot\nabla)\mathbf{V}$ to indicate the acceleration of a fluid particle.

Example 4.1

The velocity field of a steady incompressible inviscid flow, expressed in cylindrical coordinates, is:

$$V_r=\frac{k_1}{r};\quad V_\theta=\frac{k_2}{r};\quad V_z=0$$

where k_1 and k_2 are constants having the dimension of *velocity* × *length*. Derive expressions for the components of acceleration in the radial, tangential and axial directions.

Solution

Substituting the velocity components into equation 4.4,

$$\frac{D\mathbf{V}}{Dt}\cdot\mathbf{i}_r=0+\frac{k_1}{r}\left(-\frac{k_1}{r^2}\right)+0+0-\frac{1}{r}\left(\frac{k_2}{r}\right)^2=-\frac{k_1^2+k_2^2}{r^3}$$

$$\frac{D\mathbf{V}}{Dt}\cdot\mathbf{i}_\theta = 0 + \frac{k_1}{r}\left(-\frac{k_2}{r^2}\right) + 0 + 0 + \frac{k_1 k_2}{r^3} = 0$$

$$\frac{D\mathbf{V}}{Dt}\cdot\mathbf{i}_z = 0$$

4.3 Euler's Equation

We are now equipped to write Newton's law of motion for a fluid particle. Select a volume element $\delta\mathcal{V}$ of fluid having a mass $\rho\,\delta\mathcal{V}$. Because $-\nabla p$ is the pressure force per unit volume of fluid, this volume element is subject to a pressure force $(-\nabla p)\,\delta\mathcal{V}$. It is also acted upon by a gravity force $(\rho\,\delta\mathcal{V})\,\mathbf{g}$. Equating the product of mass $\rho\,\delta\mathcal{V}$ times the acceleration of a fluid particle to the sum of the pressure force $(-\nabla p)\,\delta\mathcal{V}$ and the gravity force $(\rho\,\delta\mathcal{V})\,\mathbf{g}$ acting on the particle, we write the equation of motion as:

$$(\rho\,\delta\mathcal{V})\left(\frac{\partial\mathbf{V}}{\partial t} + (\mathbf{V}\cdot\nabla)\mathbf{V}\right) = (-\nabla p)\,\delta\mathcal{V} + (\rho\,\delta\mathcal{V})\,\mathbf{g}$$

Dividing by $\rho\,\delta\mathcal{V}$, we obtain *Euler's equation:*[2]

$$\frac{\partial\mathbf{V}}{\partial t} + (\mathbf{V}\cdot\nabla)\mathbf{V} = -\frac{1}{\rho}\nabla p + \mathbf{g} \tag{4.5}$$

The left side of Euler's equation is the acceleration of a fluid particle while the right side is the sum of the forces per unit mass of fluid. Note that the fluid density appears only in the denominator of the pressure force term. For a given amount of acceleration, a high density liquid fluid particle requires a much greater pressure gradient ∇p than does a low density gas particle. On the other hand, a droplet of water and one of mercury would fall freely in a vacuum ($\nabla p = 0$) with the same acceleration $\mathbf{g}$ despite their different densities. In the absence of any motion ($\mathbf{V} = 0$), Euler's equation reduces to the equation for static equilibrium, equation 2.6.

For use in working problems and examples utilizing cartesian coordinates, we write here the x, y and z components of Euler's equation, found by evaluating the scalar product of equation 4.5 times $\mathbf{i}_x$, $\mathbf{i}_y$ and $\mathbf{i}_z$, respectively:

$$\frac{\partial u}{\partial t} + u\frac{\partial u}{\partial x} + v\frac{\partial u}{\partial y} + w\frac{\partial u}{\partial z} = -\frac{1}{\rho}\frac{\partial p}{\partial x} + \mathbf{g}\cdot\mathbf{i}_x$$

[2]Leonhard Euler (1707-1783) was one of the most prolific mathematicians of all time. He made many contributions to mechanics, dynamics and hydrodynamics. He proved that Bernoulli's equation is an integral of Euler's equation.

$$\frac{\partial v}{\partial t}+u\frac{\partial v}{\partial x}+v\frac{\partial v}{\partial y}+w\frac{\partial v}{\partial z}=-\frac{1}{\rho}\frac{\partial p}{\partial y}+\mathbf{g}\cdot\mathbf{i}_y$$

$$\frac{\partial w}{\partial t}+u\frac{\partial w}{\partial x}+v\frac{\partial w}{\partial y}+w\frac{\partial w}{\partial z}=-\frac{1}{\rho}\frac{\partial p}{\partial z}+\mathbf{g}\cdot\mathbf{i}_z \qquad (4.6)$$

Example 4.2

A steady inviscid incompressible flow has a velocity field:

$$u=fx; \qquad v=-fy; \qquad w=0$$

where f is a constant having the dimensions of s^{-1}. Derive an expression for the pressure field $p\{x,y,z\}$ if the pressure $p\{0,0,0)\}=p_0$ and $\mathbf{g}=-g\mathbf{i}_z$.

Solution

Substituting the values of the velocity components into Euler's equation 4.6, we find:

$$fx\frac{\partial(fx)}{\partial x}-fy\frac{\partial(fx)}{\partial y}=-\frac{1}{\rho}\frac{\partial p}{\partial x}; \qquad or\ f^2x=-\frac{1}{\rho}\frac{\partial p}{\partial x}$$

$$fx\frac{\partial(-fy)}{\partial x}-fy\frac{\partial(-fy)}{\partial y}=-\frac{1}{\rho}\frac{\partial p}{\partial y}; \qquad or\ f^2y=-\frac{1}{\rho}\frac{\partial p}{\partial y}$$

$$0=-\frac{1}{\rho}\frac{\partial p}{\partial z}-g$$

Integrating the first of these equations,

$$\frac{p}{\rho}=-\frac{(fx)^2}{2}+h\{y,z\}$$

and substituting into the second and integrating:

$$f^2y=-\frac{\partial h}{\partial y}$$

$$h=-\frac{(fy)^2}{2}+k\{z\}$$

The expression for p now becomes:

$$\frac{p}{\rho}=-\frac{f^2}{2}(x^2+y^2)+k\{z\}$$

Substituting this expression for p into the third of Euler's equations and integrating:

$$0 = -\frac{\partial k}{\partial z} - g$$

$$k\{z\} = -gz + c$$

Substituting $k\{z\}$ into the expression for p gives:

$$\frac{p}{\rho} = -\frac{f^2}{2}(x^2 + y^2) - gz + c$$

Finally, we choose $c = p_0/\rho$ so as to satisfy the condition that $p\{0,0,0\} = p_0$:

$$p = p_0 - \rho g z - \rho \frac{f^2}{2}(x^2 + y^2)$$

4.3.1 Constant-Density Flow

If the fluid density is constant throughout the flow field and unvarying with time—an instance of incompressible flow—it is possible to simplify the form of Euler's equation by defining a new independent variable p^*:

$$p^* \equiv p - \rho \mathbf{g} \cdot \mathbf{R} \tag{4.7}$$

so that Euler's equation 4.5 takes the form:

$$\frac{\partial \mathbf{V}}{\partial t} + (\mathbf{V} \cdot \nabla)\mathbf{V} = -\frac{1}{\rho}\nabla(p^* + \rho \mathbf{g} \cdot \mathbf{R}) + \mathbf{g}$$

$$= -\frac{1}{\rho}\nabla p^* \qquad (\textit{if} \;\; \nabla \rho = 0) \tag{4.8}$$

where we have used equation 2.7 to eliminate the gravity term. The variable p^* is a measure of the amount by which the pressure p differs from a hydrostatic pressure distribution, and ∇p^* is the net force per unit volume available to accelerate the flow.

By introducing the variable p^*, we have eliminated gravity explicitly from Euler's equation. Once we have solved Euler's equation and obtained $p^*\{\mathbf{R}, t\}$, we can then determine $p\{\mathbf{R}, t\}$ from equation 4.7.

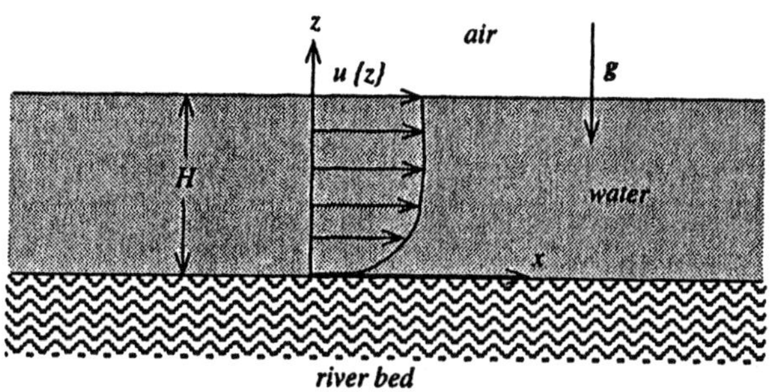

Figure E 4.3

Example 4.3

A river of depth H flows steadily with a horizontal velocity $\mathbf{V} = u\{z\}\,\mathbf{i}_x$ that varies with the height z above the river bed, as shown in figure E 4.3. Derive an expression for the pressure distribution $p\{z\}$ in the river.

Solution

Noting that $D\mathbf{V}/Dt = 0$ since u does not depend upon x, p^* must be a constant to satisfy Euler's equation 4.8. Evaluating this constant at the air-water interface,

$$p\{z\} + \rho g z = p_a + \rho g H$$

$$p\{z\} = p_a + \rho g(H - z)$$

This is the same pressure distribution that would exist if the water in the river were not moving.

To ensure that the river continues to flow downstream in the presence of friction on the bottom (a viscous flow), the bottom surface would have to be rotated clockwise very slightly in figure E 4.3. This would scarcely change the pressure distribution calculated above.

4.4 Bernoulli's Equation

To solve an inviscid fluid flow problem utilizing cartesian coordinates, we must integrate Euler's equation to find the four dependent scalar variables u, v, w and p as functions of the independent variables x, y, z and t (assuming ρ is a constant). Since we need four scalar equations to find the four dependent variables, we must append to the three scalar components of Euler's equation, equation 4.6, the equation of mass conservation of an incompressible fluid, equation 3.17:

$$\nabla \cdot \mathbf{V} = 0$$
$$\frac{\partial u}{\partial x} + \frac{\partial v}{\partial y} + \frac{\partial w}{\partial z} = 0 \tag{4.9}$$

Solving such a complex set of equations is indeed a formidable problem. No general integral of these equations has yet been found. Even a numerical solution for an arbitrary three-dimensional unsteady motion requires the use of a supercomputer. However, if we limit our attention to flows of simple geometry and initial and boundary conditions, we can find some analytical solutions that are practically useful. Examples are given in chapter 11.

It was the genius of Bernoulli[3] to have derived what subsequently proved to be a single scalar integral of Euler's equation, one that applies to any inviscid flow provided that the fluid density does not vary arbitrarily but only in a prescribed manner. This integral, called *Bernoulli's equation*, is often directly applicable to many engineering problems, providing useful, although not complete, information about the fluid flow. (For a complete description of the fluid flow, we would need four scalar integrals of Euler's equation and the mass conservation equation.)

To derive Bernoulli's equation, we begin by using the vector identity of equation 1.40 to replace the term $(\mathbf{V} \cdot \nabla)\mathbf{V}$ in equation 4.5 and rearranging the terms to obtain the following form of Euler's equation:

$$\frac{\partial \mathbf{V}}{\partial t} + \nabla\left(\frac{V^2}{2}\right) + \frac{1}{\rho}\nabla p - \mathbf{g} = \mathbf{V} \times (\nabla \times \mathbf{V})$$

Next we integrate this form of Euler's equation along a line C in space, whose element of length is $d\mathbf{c}$, by forming the scalar product of Euler's equation and $d\mathbf{c}$, then integrating between the points 1 and 2 along the line C:

$$\int_1^2 \frac{\partial \mathbf{V}}{\partial t} \cdot d\mathbf{c} + \int_1^2 \nabla\left(\frac{V^2}{2}\right) \cdot d\mathbf{c} + \int_1^2 \left(\frac{1}{\rho}\nabla p\right) \cdot d\mathbf{c} - \int_1^2 \mathbf{g} \cdot d\mathbf{c}$$
$$= \int_1^2 \mathbf{V} \times (\nabla \times \mathbf{V}) \cdot d\mathbf{c} \tag{4.10}$$

[3]Daniel Bernoulli (1700-1782) was both a mathematician, hydrodynamicist and physician. His *Hydrodynamica* (1738) explains his equation but does not specifically derive it from first principles. The derivation was subsequently given by Euler.

Two of the terms in this equation can be integrated directly using equation 1.46 and equation 2.7:

$$\int_1^2 \nabla\left(\frac{V^2}{2}\right)\cdot d\mathbf{c} = \frac{V_2^2}{2} - \frac{V_1^2}{2}$$

$$\int_1^2 \mathbf{g}\cdot d\mathbf{c} = \int_1^2 \nabla(\mathbf{g}\cdot\mathbf{R})\cdot d\mathbf{c}$$
$$= \mathbf{g}\cdot\mathbf{R}_2 - \mathbf{g}\cdot\mathbf{R}_1$$

To integrate other terms, choose the line C to be a streamline, *i.e.*, $d\mathbf{c}$ is parallel to $\mathbf{V}$ at each point along the line. To emphasize this choice, we denote the streamline element by $d\mathbf{s}$. By this choice, the integral on the right side of equation 4.10 is zero because its integrand is perpendicular to $\mathbf{V}$ and therefore the scalar product of the integrand with $d\mathbf{s}$ is identically zero:

$$\int_1^2 \mathbf{V}\times(\nabla\times\mathbf{V})\cdot d\mathbf{s} = 0 \qquad (\textit{if}\;\; d\mathbf{s}\times\mathbf{V}=0) \tag{4.11}$$

where the condition that $d\mathbf{s}$ is parallel to $\mathbf{V}$ can be expressed as $d\mathbf{s}\times\mathbf{V}=0$. The pressure gradient integral can be evaluated easily if the density does not change along the streamline: [4]

$$\int_1^2 \left(\frac{1}{\rho}\nabla p\right)\cdot d\mathbf{s} = \frac{1}{\rho}\int_1^2 \nabla p\cdot d\mathbf{s}$$
$$= \frac{1}{\rho}(p_2 - p_1) \qquad (\textit{if}\;\; \mathbf{V}\cdot(\nabla\rho)=0) \tag{4.12}$$

where the constancy of density along a streamline is ensured by the condition that the density gradient $\nabla\rho$ is perpendicular to $\mathbf{V}$, or $\mathbf{V}\cdot(\nabla\rho)=0$. Inserting these values for the integrals into equation 4.10,

$$\int_1^2 \frac{\partial\mathbf{V}}{\partial t}\cdot d\mathbf{s} + \left(\frac{V_2^2}{2} + \frac{p_2}{\rho} - \mathbf{g}\cdot\mathbf{R}_2\right) - \left(\frac{V_1^2}{2} + \frac{p_1}{\rho} - \mathbf{g}\cdot\mathbf{R}_1\right) = 0$$
$$(\textit{if}\;\; \mathbf{V}\cdot(\nabla\rho)=0;\; d\mathbf{s}\times\mathbf{V}=0) \tag{4.13}$$

This is the form of *Bernoulli's equation* for the case of constant density along a streamline.

---

[4] It is not sufficient that the flow be incompressible ($D\rho/Dt = 0$) for the density to be constant along a streamline, unless the flow is also steady. However, it is sufficient, but not necessary, for the density to be constant everywhere in the flow field in order to satisfy its constancy along a streamline.

In unsteady flow, the integrand of the first term of Bernoulli's equation, equation 4.13, must be known at all points along the instantaneous streamline in order for the integral to be evaluated. If the flow is such that the streamlines do not change with time but the velocity $\mathbf{V}$ does, then $\partial\mathbf{V}/\partial t$ has the direction of $\mathbf{V}$ and ds, and the integral may be easy to evaluate.

In a steady flow, the first term of equation 4.13 is absent, and the sum, $\mathbf{V}^2/2+p/\rho-\mathbf{g}\cdot\mathbf{R}$, has the same value at all points along the same streamline but not necessarily the same value as points along a different streamline.

In our study of hydrostatics, we made use of a convention for a cartesian coordinate system that the direction of the z-axis is vertical and opposite to the direction of $\mathbf{g}$. In this convention, $\mathbf{g}\cdot\mathbf{R}=-gz$ and Bernoulli's equation takes the form:

$$\int_1^2 \frac{\partial \mathbf{V}}{\partial t}\cdot d\mathbf{s} + \left(\frac{\mathbf{V}_2^2}{2}+\frac{p_2}{\rho}+gz_2\right) - \left(\frac{\mathbf{V}_1^2}{2}+\frac{p_1}{\rho}+gz_1\right) = 0$$

$$(\textit{if}\ \ \mathbf{V}\cdot(\nabla\rho)=0;\ d\mathbf{s}\times\mathbf{V}=0;\ \mathbf{g}=-g\mathbf{i}_z) \tag{4.14}$$

This form of Bernoulli's equation is convenient for working most problems.

4.4.1 Applications of Bernoulli's Equation

Bernoulli's equation is very useful in enabling us to understand the behavior of many engineering flows. In this section, we give examples of both steady and unsteady flows of incompressible fluids that demonstrate the application of Bernoulli's equation to the flow of fluids.

Fluid Streams

One of the simplest inviscid fluid flows is that of a stream of fluid (such as water) flowing through a stationary fluid with which it does not mix (such as air). Consider the case of water flowing from a tap, as illustrated in figure 4.1. The water leaves the tap with a speed V_1 as a circular stream of diameter D_1. As it falls, it speeds up and contracts in diameter. Ultimately, the stream becomes so thin that surface tension forces break it up into droplets, but before this happens, the flow can be described by applying Bernoulli's equation 4.14 to the central streamline of the water stream:

$$\frac{p_2}{\rho}+\frac{V_2^2}{2}+gz_2=\frac{p_1}{\rho}+\frac{V_1^2}{2}+gz_1$$

On the central streamline, the pressure of the water will be the same as that in the atmosphere at the same height z because the radial acceleration of the water stream is negligible. (We also assume here that the surface tension is negligible so that the

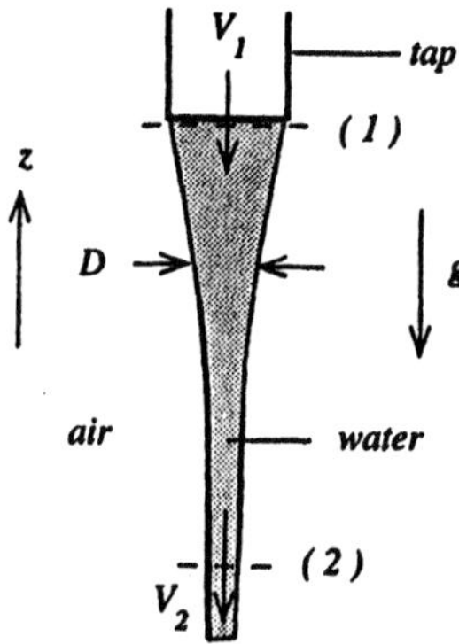

Figure 4.1 Water leaving a tap increases in speed as it falls downward through the stationary air, as described by Bernoulli's equation.

pressure inside the water column is the same as that outside.) The water pressure at 1 and 2 are thus related by the hydrostatic pressure distribution in the atmosphere:

$$p_2 + \rho_a g z_2 = p_1 + \rho_a g z_1$$

where ρ_a is the density of the ambient fluid (air) and is assumed to be constant. Substituting this expression into the previous one and solving for V_2^2 we find:

$$V_2^2 = V_1^2 + 2\left(1 - \frac{\rho_a}{\rho}\right) g(z_1 - z_2)$$

Bernoulli's equation demonstrates how the water velocity increases as the distance from the tap $z_1 - z_2$ increases. Because the density ρ_a of air is only about 10^{-3} times that of water, $\rho_a/\rho \ll 1$, and we may write Bernoulli's equation for this case as:

$$V_2^2 = V_1^2 + 2g(z_1 - z_2)$$

This is equivalent to assuming that the ambient air pressure is a constant and thus $p_1 = p_2$. However, this approximation would not be justifiable if the ambient fluid density were the same order of magnitude as that of the moving fluid stream, as would be the case of an oil stream injected into water or heated air rising into a colder atmosphere.[5] In such instances, the vertical acceleration of the stream is smaller than g, namely, $(1 - \rho_a/\rho)g$.

[5]In the latter case, $1 - \rho_a/\rho$ is negative and $z_2 > z_1$, *i.e.*, the fluid stream flows upward.

The diameter of the flowing stream may next be found by applying mass conservation to the steady flow between 1 and 2:

$$\rho V_2 \left(\frac{\pi D_2^2}{4}\right) = \rho V_1 \left(\frac{\pi D_1^2}{4}\right)$$

$$D_2 = D_1 \sqrt{\frac{V_1}{V_2}}$$

$$= D_1 \left(\frac{V_1^2}{V_1^2 + 2g(z_1 - z_2)}\right)^{1/4}$$

Note how the diameter decreases quite slowly with increasing $z_1 - z_2$.

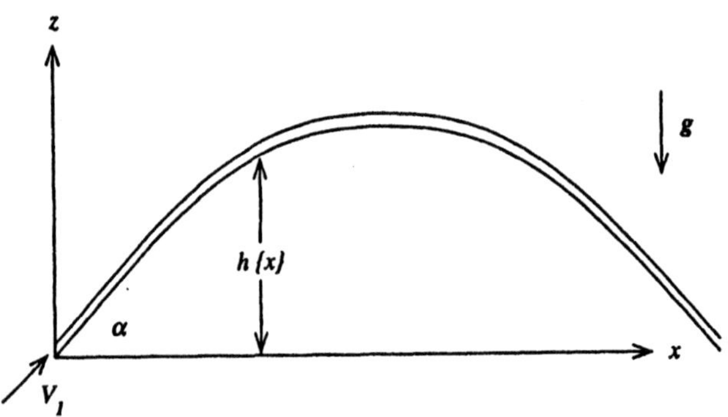

Figure E 4.4

Example 4.4

A fire hose directs a stream of water of velocity V_1 at an angle α above the horizontal, as illustrated in figure E 4.4. The stream rises initially but then eventually falls to the ground.

(a) Derive an expression for the height $h\{x\}$ of the stream above the hose nozzle as a function of the horizontal distance x from the nozzle. (b) Calculate the maximum value of h if $V_1 = 50\,m/s$ and $\alpha = 45°$.

Solution

(a) In terms of the vertical and horizontal components of **V**, w and u, Bernoulli's equation 4.14 is:

$$\frac{u_1^2 + w_1^2}{2} + gz_1 = \frac{u^2 + w^2}{2} + gz$$

assuming constant atmospheric pressure. The horizontal component of Euler's equation 4.6 is $Du/Dt = 0$; therefore u does not change along the fluid stream, and $u = u_1$. Solving Bernoulli's equation for w:

$$w = \sqrt{w_1^2 - 2gh}$$

where $h = z - z_1$. The slope of the fluid stream, dh/dx, must equal the ratio w/u:

$$\frac{dh}{dx} = \frac{w}{u} = \frac{\sqrt{w_1^2 - 2gh}}{u_1}$$

Integrating this differential equation for h from $x = 0$ to x:

$$\int_0^h \frac{dh}{\sqrt{w_1^2 - 2gh}} = \frac{1}{u_1} \int_0^x dx$$

$$-\left| \frac{\sqrt{w_1^2 - 2gh}}{g} \right|_0^h = \frac{x}{u_1}$$

$$w_1 - \sqrt{w_1^2 - 2gh} = \frac{gx}{u_1}$$

$$h\{x\} = \frac{w_1 x}{u_1} - \frac{gx^2}{2u_1^2}$$

where $u_1 = V\cos\alpha$ and $w_1 = V\sin\alpha$. Note that the fluid stream reaches the ground at $x = 2u_1 w_1/g$.

(b) The maximum value of h occurs when $dh/dx = 0$. From the differential equation for h, this occurs when $h = w_1^2/2g$. Calculating h:

$$h = \frac{(50\,m/s \times \sin 45°)^2}{2 \times 9.807\,m/s^2} = 63.73\,m$$

Flow Through an Orifice

Fluids may flow into or out of tanks or chambers through an opening that limits the rate of flow. Inviscid flow through such orifices moves at a speed that depends upon the pressure difference between the fluid inside and outside of the vessel. In figure 4.2, we show some streamlines of the steady inviscid incompressible flow of fluid passing through an orifice of area A from a chamber (on the left), having a pressure p_1, to a

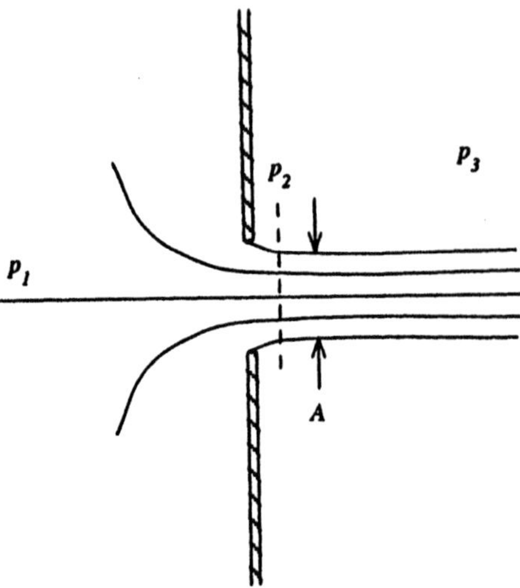

Figure 4.2 The inviscid incompressible flow of fluid through an orifice from a higher pressure chamber on the left to a lower pressure chamber on the right follows the sketched streamlines. The orifice pressure p_2 equals the receiving chamber pressure p_3.

chamber (on the right), having a lower pressure p_3. The pressure p_2 at the exit of the orifice is less than that in the chamber at the left so that the fluid accelerates as it flows toward the orifice. On the other hand, as the fluid emerges into the chamber at the right, it continues moving from left to right at an unchanging speed because there is no further change in pressure between the orifice exit (pressure p_2) and the stationary fluid in the chamber at the right (pressure p_3).[6] Writing Bernoulli's equation 4.14 for steady flow along the central streamline of the flow between a point far from the orifice, where $V_1 = 0$, and a point at the exit of the orifice, we find:

$$\frac{V_2^2}{2} + \frac{p_2}{\rho} + g z_2 = \frac{V_1^2}{2} + \frac{p_1}{\rho} + g z_1$$

$$\frac{V_2^2}{2} = \frac{p_1}{\rho} - \frac{p_2}{\rho}$$

$$V_2 = \sqrt{\frac{2(p_1 - p_3)}{\rho}} \tag{4.15}$$

where we have used the fact that $p_2 = p_3$. The volume flow rate Q and mass flow rate

[6] In a supersonic compressible flow, the pressure p_2 can be greater than p_3.

$\dot{m}$ of fluid through the orifice become:[7]

$$Q = A\sqrt{\frac{2(p_1 - p_3)}{\rho}}$$

$$\dot{m} = A\sqrt{2\rho(p_1 - p_3)} \tag{4.16}$$

The constant-speed flow into the receiving chamber does not continue indefinitely in the downstream direction. Viscous forces between the moving and stationary fluid in the chamber cause the fluid jet to become unsteady, break up into eddies and dissipate its kinetic energy. Needless to say, this process is not an inviscid flow, and therefore Bernoulli's equation cannot be applied to it. Nevertheless, the flow upstream of the orifice can be accurately represented as inviscid.[8]

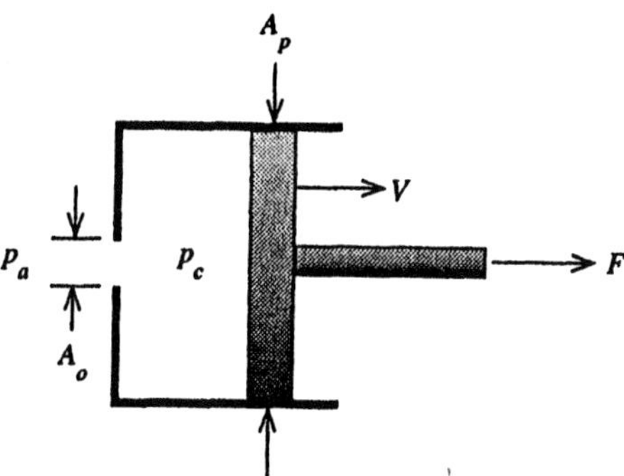

Figure E 4.5

Example 4.5

A circular cylinder of area A_p is fitted with a piston that is retracted at a constant speed V by a force F, as shown in figure E 4.5. As it is retracted, atmospheric air

---

[7]Because of the viscous shear stress exerted on the fluid by the walls of the orifice, the actual volume and mass flow rates are slightly less than those of equation 4.16 for inviscid flow.

[8]It is possible to find a solution to Euler's equations yielding a flow with streamlines that are symmetric about the plane of the orifice. Such flows are not observed in practice, the fluid flow following the streamlines sketched in figure 4.2 instead. This asymmetry is caused by the viscous flow effects near the surface of the orifice that are not taken into account by Euler's equation.

of density ρ flows through an orifice of area A_o into the cylinder where the air has a lower pressure p_c than atmospheric pressure p_a.

(a) Assuming incompressible flow, derive expressions for the pressure difference $p_a - p_c$, the force F and the power P required to move the piston at the speed V. (b) Assuming that figure E4.5 is a reasonable model for the flow of air into the human lung, calculate the power required to inhale 0.5 l of air in 2 s through a trachea "orifice" of area $A_o = 1.0\,cm^2$ when the air density $\rho = 1.225 kg/m^3$.

Solution

(a) By mass conservation, the volumetric flow rate into the cylinder given by equation 4.16 must equal the rate at which the cylinder volume is increasing, $A_p V$:

$$A_0\sqrt{\frac{2\,(p_a - p_c)}{\rho}} = A_p V$$

$$p_a - p_c = \frac{\rho}{2}\left(\frac{A_p V}{A_o}\right)^2$$

The force F applied to the piston is equal to the pressure difference $p_a - p_c$ across the piston faces times the piston area A_p:

$$F = (p_a - p_c)A = \frac{\rho}{2}\left(\frac{A_p V}{A_o}\right)^2 A_p$$

and the power P is the product of the force F times the piston velocity V:

$$P = FV = \frac{\rho}{2}\left(\frac{(A_p V)^3}{A_o^2}\right)$$

(b) For the human lung, the volumetric flow rate $A_p V = 0.5\,l/2\,s = 2.5E(-4)\,m^3/s$. Thus the power P expended in inhaling (or exhaling) is:

$$P = \frac{1.225\,kg/m^3 \times (2.5E(-4)\,m^3/s)^3}{2 \times (1.0E(-4)\,m^2)^2} = 9.57E(-4)W$$

Flow from Pressurized Tanks

Fluids are often stored in containers under pressure greater than atmospheric: domestic water in home storage tanks, propane in home or camp fuel systems, compressed air in workshops, steam in power plant boilers, etc. Flow out of these storage vessels is usually regulated by valves delivering the fluid at a pressure lower than that at which the fluid is stored. Sometimes the fluid leaks to the atmosphere through a small hole

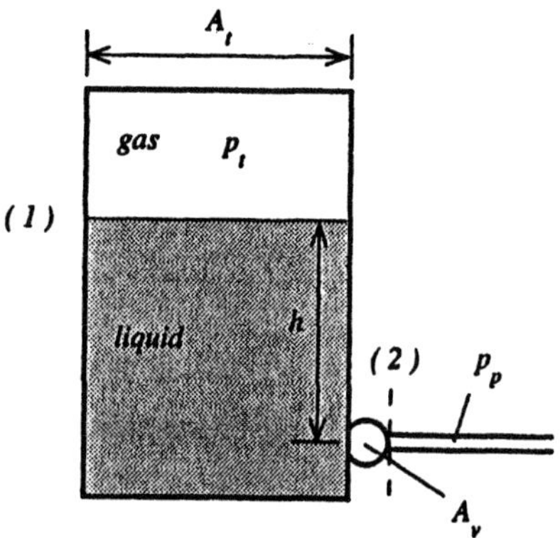

Figure 4.3 A tank holding liquid under pressure from a gas supplies a flow through a valve to a pipe of liquid at a lower pressure.

or crack. The rate of discharge of the fluid under these conditions may be found by applying Bernoulli's equation to the particular flow in question.

Consider a liquid stored in a tank under a pressure p_t maintained by a layer of gas at the top of the tank, as shown in Figure 4.3. A valve at the bottom of the tank regulates the outflow of liquid to a pipe conveying away the liquid, where the pressure is p_p. The speed of flow through the valve, V_v, may be found by applying Bernoulli's equation 4.14 to a streamline connecting the liquid/gas interface in the tank to the valve orifice:

$$\frac{V_2^2}{2} + \frac{p_2}{\rho} + g\,z_2 = \frac{V_1^2}{2} + \frac{p_1}{\rho} + g\,z_1$$

Usually, the area A_t of the free surface in the tank is much greater than the orifice area A_v of the valve. By mass conservation, the speed V_1 of the gas/liquid interface is much smaller than that of the liquid flowing through the valve:

$$A_t V_1 = A_v V_2$$

$$\frac{V_1}{V_2} = \frac{A_v}{A_t} \ll 1$$

Neglecting V_1 compared to V_2, Bernoulli's equation becomes:

$$\frac{p_t}{\rho} + g\,z_1 = \frac{V_v^2}{2} + \frac{p_p}{\rho} + g\,z_2$$

$$V_v = \sqrt{2gh + \frac{2(p_t - p_p)}{\rho}}$$

where we have replaced p_1 by p_t and $z_1 - z_2$ by the height h of the liquid/gas interface above the valve location. Note that the flow velocity through the valve is determined by the difference in the hydrostatic pressure $p_t + \rho g h$ at the level of the valve and the pressure p_p in the pipe downstream.

Example 4.6

An oil storage tank having a diameter $D_t = 30\,m$ is filled with oil to a depth $H = 5\,m$. The space above the oil is vented to the atmosphere. A pipe of inside diameter $D_p = 5\,cm$ leading from the base of the tank is accidentally broken, allowing the oil to spill onto the ground. Calculate how long it will take for the oil to drain completely from the tank.

Solution

Applying Bernoulli's equation, while noting that the pressure of the oil leaving the pipe and the pressure at the surface of the oil in the tank are both equal to atmospheric pressure and assuming that the velocity dh/dt at the oil surface is negligible, the speed V_p of flow out of the pipe is:

$$V_p = \sqrt{2gh\{t\}}$$

where $h\{t\}$ is the height of the oil surface above the bottom of the tank. By mass conservation, the rate of change of height dh/dt is related to V_p by:

$$\frac{d}{dt}\left(\frac{\pi D_t^2}{4}h\right) = -\frac{\pi D_p^2}{4}V_p$$

$$\frac{dh}{dt} = -\left(\frac{D_p}{D_t}\right)^2 V_p = -\left(\frac{D_p}{D_t}\right)^2 \sqrt{2gh}$$

Integrating this differential equation for h from $h = H$ to $h = 0$, we find:

$$\int_H^0 \frac{dh}{\sqrt{h}} = -\int_0^t \sqrt{2g}\left(\frac{D_p}{D_t}\right)^2 dt'$$

$$\left|2\sqrt{h}\right|_H^0 = -\sqrt{2g}\left(\frac{D_p}{D_t}\right)^2 |t'|_0^t$$

$$t = \left(\frac{D_t}{D_p}\right)^2 \sqrt{\frac{2H}{g}}$$

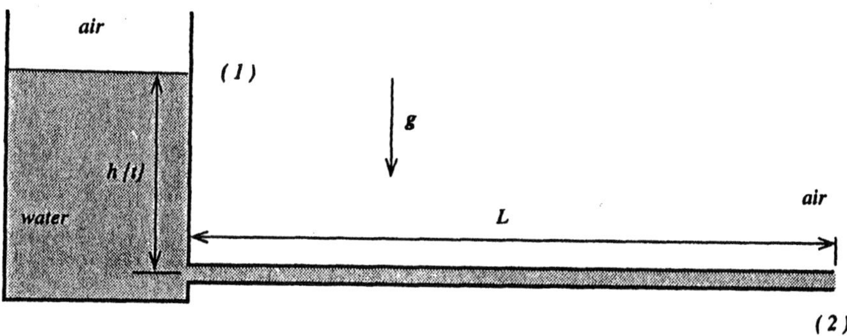

Figure 4.4 Water filling a tank and a long pipe is accelerated from rest when the end of the pipe is suddenly opened to the atmosphere.

$$= \left(\frac{30}{0.05}\right)^2 \sqrt{\frac{2 \times 5\,m}{9.807\,m/s^2}} = 3.635E(5)\,s = 101.0\,hr$$

Unsteady Flow

The examples considered so far are steady flows. When flows are started from rest or altered with time, the unsteady term in Bernoulli's equation 4.13 may be significant.

For example, consider the case of the starting of flow in a long pipe of length L. Supplied with water from a tank and discharging into the atmosphere through a valve at its end, the pipe is suddenly opened wide at time $t = 0$, as shown in figure 4.4. Writing Bernoulli's equation between the air/water interface in the tank and the exit of the pipe,

$$\int_1^2 \frac{\partial V}{\partial t} ds + \frac{V_2^2}{2} + \frac{p_2}{\rho} + g z_2 = \frac{V_1^2}{2} + \frac{p_1}{\rho} + g z_1$$

$$\int_1^2 \frac{\partial V}{\partial t} ds = g h - \frac{V_2^2}{2}$$

since $p_1 = p_2 = p_{atm}$. The only significant contribution to the integral comes from the fluid in the pipe whose velocity $V_2\{t\}$ varies with time. Evaluating this integral over the length L of the pipe,

$$L \frac{dV_2}{dt} = g h - \frac{V_2^2}{2}$$

$$\frac{dV_2}{2gh - V_2^2} = \frac{dt}{2L}$$

Integrating from $V_2 = 0$ at $t = 0$,

$$\tanh^{-1}\left\{\frac{V_2}{\sqrt{2gh}}\right\} = \frac{\sqrt{gh}\,t}{\sqrt{2}L}$$

$$V_2 = \sqrt{2gh}\tanh\left\{\frac{\sqrt{gh}\,t}{\sqrt{2}L}\right\}$$

At the very beginning, at times t that are very small compared with $L/\sqrt{gh}$ so that $\tanh x = x$, the velocity $V_2 = ght/L$ in agreement with the differential equation when $V_2 = 0$. During this period, the fluid acceleration is gh/L. For very long times $t \gg L/\sqrt{gh}$, V_2 approaches its steady value of $\sqrt{2gh}$. The magnitude of the time required to reach a steady flow is the time $L/\sqrt{2gh}$ required for a fluid particle to move the length of the pipe at the steady flow speed.

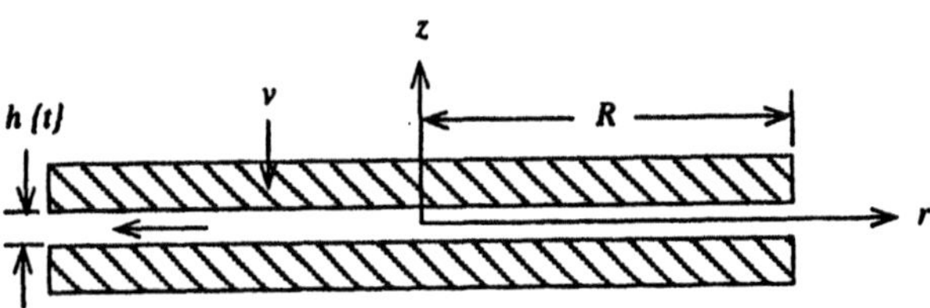

Figure E 4.7

Example 4.7

A pair of circular discs of radius R encloses a thin layer of liquid between their parallel faces, the thickness $h\{t\}$ of this layer decreasing with time at a constant speed v, *i.e.*, $dh/dt = -v$ (see figure E 4.7). The liquid is expelled radially outward at a speed V_r that varies with radius r and time t, but not with axial distance z. The liquid pressure at the exit radius R is atmospheric pressure p_a. Assuming an incompressible flow, derive expressions for (a) the radial velocity V_r, (b) the pressure p and (c) the force F that must be applied to each disc to maintain the given motion.

Solution

(a) Applying mass conservation equation 3.10 to a control volume enclosing the liquid layer out to a radius r,[9]

$$\frac{d}{dt}(\pi r^2 h) = -2\pi r h V_r$$

$$V_r = -\frac{r}{2h}\frac{dh}{dt} = \frac{\nu r}{2h}$$

(b) Applying Bernoulli's equation 4.14 to a horizontal streamline ($z = 0$) between the axis ($r = 0$) and a point at a radius r:

$$\int_0^r \frac{\partial V_r}{\partial t} ds + \frac{V_r^2}{2} + \frac{p}{\rho} - \frac{p_0}{\rho} = 0$$

where p_0 is the pressure on the axis. Noting that:

$$\frac{\partial V_r}{\partial t} = \frac{d}{dt}\left(\frac{\nu r}{2h}\right) = \frac{\nu r}{2}\left(-\frac{1}{h^2}\frac{dh}{dt}\right) = \frac{\nu^2 r}{2h^2}$$

then:

$$\int_0^r \frac{\partial V_r}{\partial t} dr = \frac{\nu^2 r^2}{4h^2}$$

Substituting this into Bernoulli's equation, we find:

$$p = p_0 - \rho\frac{V_r^2}{2} - \int_0^r \frac{\partial V_r}{\partial t} dr = p_0 - \frac{3}{8}\left(\frac{\nu r}{h}\right)^2 \rho$$

If we choose p_0 so that $p = p_a$ at $r = R$, then:

$$p = p_a + \frac{3\nu^2}{8h^2}(R^2 - r^2)\rho$$

(c) The force F is the integral of $p - p_a$ over the area of a disc:

$$F = \int_0^R \frac{3\nu^2}{8h^2}(R^2 - r^2)\rho\,(2\pi r)dr = \frac{3\pi}{16}\left(\frac{\nu R^2}{h}\right)^2 \rho$$

Metering Flow

It is sometimes desirable to be able to measure the rate at which fluids are flowing through a pipe or duct. By forcing the fluid to flow through a constriction inside the pipe or duct while measuring the pressure change accompanying this squeezing down

[9]This value of V_r satisfies the mass conservation equation 3.15 in cylindrical coordinates when $V_z = -\nu z/h$.

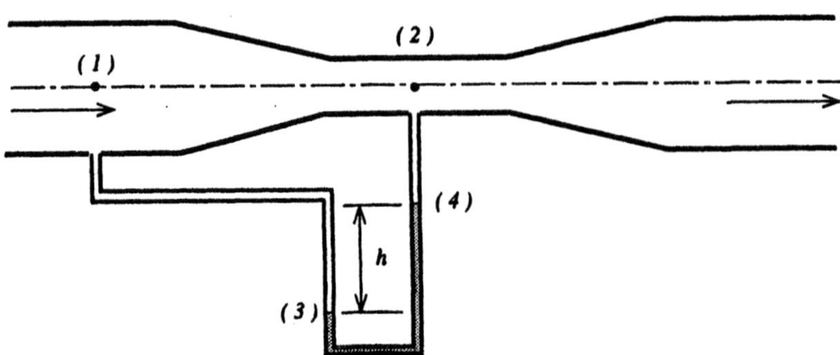

Figure 4.5 A venturi meter constricts the flow of a fluid so as to create a pressure difference that is related to the volumetric and mass flow rates. The manometer shown is used to measure the pressure difference.

of the flow, it is possible to compute the volumetric and mass flow rates of the flow.

A *venturi meter* is sketched in figure 4.5. It consists of a section of pipe that gradually reduces the flow area from A_1 upstream to A_2 at the point of minimum area. If the pressures are measured at these locations, then Bernoulli's equation 4.14 applied to the central streamline of the flow may be solved for the upstream flow speed in terms of the pressure change $p_1 - p_2$ and the velocity ratio V_2/V_1:

$$\frac{V_2^2}{2} + \frac{p_2}{\rho} + g z_2 = \frac{V_1^2}{2} + \frac{p_1}{\rho} + g z_1$$

$$V_1 = \sqrt{\frac{2(p_1 - p_2)}{\rho\left((V_2/V_1)^2 - 1\right)}}$$

since $z_1 = z_2$. Mass conservation requires that $\rho V_1 A_1 = \rho V_2 A_2$, with the result that:

$$V_1 = \sqrt{\frac{2(p_1 - p_2)}{\rho\left((A_1/A_2)^2 - 1\right)}}; \qquad Q = A_1\sqrt{\frac{2(p_1 - p_2)}{\rho\left((A_1/A_2)^2 - 1\right)}};$$

$$\dot{m} = A_1\sqrt{\frac{2\rho(p_1 - p_2)}{\left((A_1/A_2)^2 - 1\right)}} \qquad (4.17)$$

The pressure difference may be measured by use of a manometer connected to the venturi meter as shown in figure 4.5 and filled with a liquid whose density ρ_m is greater than that of the working fluid. Because the moving fluid is not accelerating

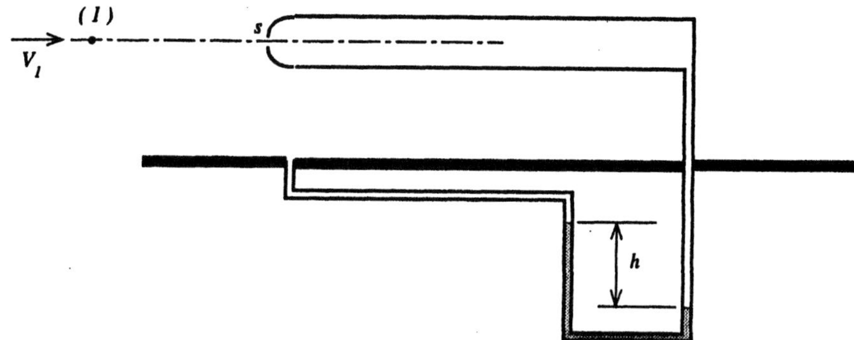

Figure 4.6 A pitot tube measures the flow speed directly upwind by means of the difference in pressure inside and outside the tube.

at 1 and 2, the pressure distribution is hydrostatic between 1 – 3 and 2 – 4:

$$p_1 + \rho g z_1 = p_3 + \rho g z_3$$

$$p_2 + \rho g z_2 = p_4 + \rho g z_4$$

By subtraction,

$$p_1 - p_2 = p_3 - p_4 + \rho g(z_3 - z_4)$$

since $z_1 = z_2$. For the manometer fluid,

$$p_3 + \rho_m g z_3 = p_4 + \rho_m g z_4$$

$$p_3 - p_4 = \rho_m g(z_4 - z_3)$$

Thus the pressure difference $p_1 - p_2$ becomes:

$$p_1 - p_2 = (\rho_m - \rho)gh$$

An instrument that can measure the velocity at a point in the flow, called the *pitot tube*, is illustrated in figure 4.6. It consists of a hollow tube aligned with the oncoming flow and closed at one end with a rounded plug containing a tiny hole at the tube centerline. The fluid inside the pitot tube is stationary, while the oncoming fluid flows around it. A fluid particle moving along the streamline that is coincident with the pitot tube axis comes to rest as it approaches the tip of the pitot tube (designated by s) because it must split and pass on either side of the tube. As it comes to rest momentarily, its pressure rises to a value p_s, called the *stagnation pressure*, that is related to the upstream flow speed V_1 by Bernoulli's equation:

$$\frac{V_1^2}{2} + \frac{p_1}{\rho} + g z_1 = \frac{V_s^2}{2} + \frac{p_s}{\rho} + g z_s$$

$$p_s = p_1 + \rho \frac{V_1^2}{2}$$

$$V_1 = \sqrt{\frac{2(p_s - p_1)}{\rho}}$$

because $V_s = 0$ and $z_1 = z_s$. The pressure of the stationary fluid inside the pitot tube equals the stagnation pressure of the external flow, with which it is in contact through the tiny hole at the *stagnation point s* of the tube. The pressure difference $p_s - p_1$ may be measured by a manometer arranged as shown in Figure 4.6. Following the analysis given above for the venturi meter, the pressure difference is $(\rho_m - \rho)gh$ and the flow speed of the oncoming flow is:

$$V_1 = \sqrt{\frac{2(\rho_m - \rho)gh}{\rho}}$$

To measure the volumetric flow rate Q in a pipe or duct, the pitot tube can be moved to all locations within the cross-section of the flow, and the velocity measurements integrated:

$$Q = \int\int V\{x, y\}\, dx\, dy$$

This type of measurement is often necessary when the fluid velocity varies noticeable within a pipe or duct.

4.5 Euler's Equation in Streamline Coordinates

It is sometimes convenient to choose an orthogonal coordinate system whose local directions are defined by a streamline of the flow. Called streamline coordinates, the three mutually perpendicular directions at a point in the flow are determined by the directions of the tangent, normal and binormal to the streamline passing through the point. This is illustrated in figure 4.7 for a point P on a streamline, where the unit vectors lying in these three directions are labeled $\mathbf{i}_s$, $\mathbf{i}_n$ and $\mathbf{i}_b$, respectively. The unit normal $\mathbf{i}_n$ points in the direction of the center of curvature O of the streamline at the point P, the distance OP being the radius of curvature R. To develop Euler's equation for this coordinate system, we will embed a cylindrical coordinate system with center

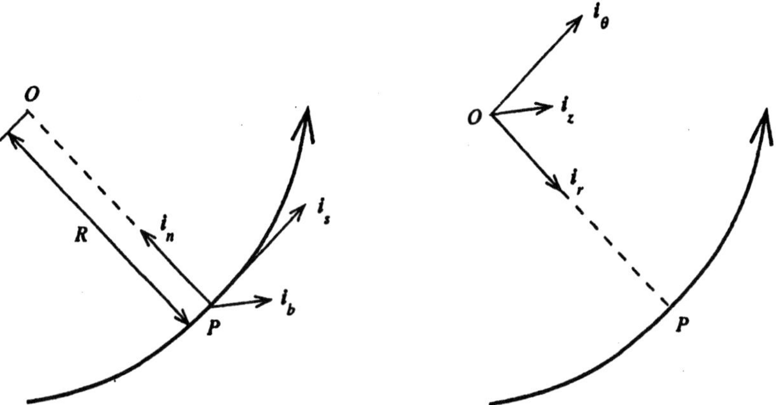

Figure 4.7 Unit vectors for streamline coordinates at a point P on a streamline whose center of curvature is O.

at O and with the z-axis in the direction of the binormal $\mathbf{i}_b$. The point P lies at a radius $r = R$. The unit vectors of the streamline and cylindrical coordinates are related by $\mathbf{i}_r = -\mathbf{i}_n$, $\mathbf{i}_\theta = \mathbf{i}_s$, $\mathbf{i}_z = \mathbf{i}_b$, the magnitude of the components of the velocity vector by $V_r = V_n$, $V_\theta = V$, $V_z = V_b$, and the spatial derivatives by $\partial/\partial r = -\partial/\partial n$, $\partial/r\partial\theta = \partial/\partial s$, $\partial/\partial z = \partial/\partial b$. Substituting these values into equation 4.4, and noting that $V_n = 0$ and $V_b = 0$ at any point on a streamline, the acceleration of a fluid particle in streamline coordinates is:[10]

$$\frac{D\mathbf{V}}{Dt} = \left(-\frac{\partial V_n}{\partial t} + \frac{V^2}{R}\right)\mathbf{i}_n + \left(\frac{\partial V}{\partial t} + V\frac{\partial V}{\partial s}\right)\mathbf{i}_s + \left(\frac{\partial V_b}{\partial t}\right)\mathbf{i}_b \tag{4.18}$$

In the case of *steady flow* in streamline coordinates, Euler's equation has an especially simple form:

$$\begin{aligned}
\left(\frac{V^2}{R}\right) &= -\frac{1}{\rho}\frac{\partial p}{\partial n} + \mathbf{g}\cdot\mathbf{i}_n = -\frac{1}{\rho}\frac{\partial p^*}{\partial n} \\
V\frac{\partial V}{\partial s} &= -\frac{1}{\rho}\frac{\partial p}{\partial s} + \mathbf{g}\cdot\mathbf{i}_s = -\frac{1}{\rho}\frac{\partial p^*}{\partial s} \\
0 &= -\frac{1}{\rho}\frac{\partial p}{\partial b} + \mathbf{g}\cdot\mathbf{i}_b = -\frac{1}{\rho}\frac{\partial p^*}{\partial b} \qquad (\textit{steady flow})
\end{aligned} \tag{4.19}$$

[10]Although V_n and V_b are zero on the instantaneous streamline passing through P, their time derivatives are not zero in general, unless the flow is steady. If we expand any component of the velocity in a Taylor series in time, we can see that the first and higher derivatives in time are not necessarily zero when the first term is zero.

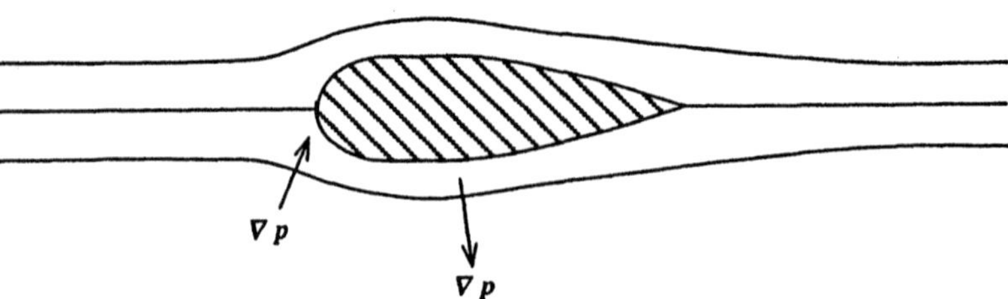

Figure 4.8 Sketch of streamlines near a body shows how the direction of the pressure gradient depends upon the curvature of the streamline.

There are several aspects of these equations that deserve notice. The first of the equations of 4.19, which expresses the motion in the normal direction, shows that the net force in the direction $\mathbf{i}_n$ causes the centrifugal acceleration V^2/R. The second equation, for motion along the streamline, can be integrated to give Bernoulli's equation for steady flow. The third equation of motion in the direction of the binormal gives a hydrostatic pressure distribution in this direction because the acceleration in this direction is zero.

The equation of motion in the normal direction enables us to determine how the pressure varies in a steady flow if we know the shape of the streamlines. If gravity has no component in the normal direction ($\mathbf{g} \cdot \mathbf{i}_n = 0$), then *the pressure decreases in the direction of the center of curvature of the streamline.* For example, consider the steady flow past a streamlined body sketched in figure 4.8. At the front of the body, the streamlines curve away from the body, and so the pressure there must be higher than the uniform pressure far from the body. On the other hand, at the side of the body, the streamlines follow the convex shape of the body, and the pressure at the side must be less than that far away. While such a description of the pressure distribution in this flow is qualitatively correct, a quantitative calculation of the pressure in the flow cannot be obtained easily from the streamline form of Euler's equation, equation 4.19, except for very simple flows.

4.6 Inviscid Flow in Noninertial Reference Frames

In analyzing fluid flows it is sometimes convenient to use a coordinate system tied to a reference frame that is not inertial, *i.e.*, one in which the acceleration measured by an observer in the noninertial reference frame is different from that measured by an observer in the inertial reference frame. Typical examples of such noninertial reference frames would be a reference frame fixed in a booster rocket that is accelerating upward from its launching pad or a rotating reference frame fixed to the rotor of a turbomachine. In such instances, the expressions for Euler's and Bernoulli's equations need to be modified to take into account the motion of the noninertial reference frame.

We commonly regard the laboratory reference frame as an inertial reference frame when describing fluid flow in laboratory experiments. However, because the earth rotates about its axis, the laboratory reference frame is not strictly inertial, although it can be regarded as such for flows having small length and time scales typical of engineering systems. For large scale flows that change slowly with time, such as that of the atmosphere or the ocean, it is necessary to take into account the earth's rotational speed when using a reference frame fixed to the earth. If the angular velocity of a rotating system is sufficiently high, the use of noninertial reference frames may require corrections that are important even for laboratory scale flows.

In this section, we develop the modifications to the equations of inviscid flow that are required in the use of noninertial reference frames for two simple cases. The first of these is that of a translating (but not rotating) reference frame that is accelerating. The second is that of a reference frame that is rotating at a steady angular speed about an axis whose direction is fixed but is otherwise not being translated with respect to the inertial reference frame. These two examples suffice to cover most applications of engineering importance.

4.6.1 Translating, Accelerating Reference Frame

Consider a noninertial reference frame that is translating with respect to the inertial reference frame. (Translation means that the coordinate system axes do not change direction in inertial space.) Denote the position and velocity of a fluid particle in the noninertial reference frame by $\tilde{\mathbf{R}}$ and $\tilde{\mathbf{V}}$, respectively, to distinguish them from the position $\mathbf{R}$ and velocity $\mathbf{V}$ in the inertial reference frame. (Because we are dealing with nonrelativistic velocities, time t will be the same in both reference frames.) Denoting the velocity of the noninertial reference frame (as measured in the inertial reference frame) by $\mathbf{V}_{ni}\{t\}$, the fluid particle velocities in the two reference frames are related by:

$$\mathbf{V}\{\mathbf{R}, t\} = \mathbf{V}_{ni}\{t\} + \tilde{\mathbf{V}}\{\tilde{\mathbf{R}}, t\} \tag{4.20}$$

i.e., the inertial frame particle velocity is the sum of the velocity of the noninertial reference frame and the particle velocity measured relative to the noninertial reference frame. To relate the accelerations of a fluid particle as measured in the two reference frames, we differentiate equation 4.20 following the fluid particle:

$$\frac{D\mathbf{V}}{Dt} = \frac{d\mathbf{V}_{ni}\{t\}}{dt} + \left(\frac{\partial}{\partial t} + \tilde{\mathbf{V}} \cdot \tilde{\nabla}\right) \tilde{\mathbf{V}}\{\tilde{\mathbf{R}}, t\}$$

$$= \mathbf{a}_{ni}\{t\} + \frac{\tilde{D}\tilde{\mathbf{V}}}{\tilde{D}t} \tag{4.21}$$

where $\mathbf{a}_{ni}\{t\}$ is the acceleration of the translating noninertial reference frame, as measured in the inertial reference frame and where the material derivative $\tilde{D}/\tilde{D}t$ in a cartesian noninertial reference frame (for example) is:

$$\frac{\tilde{D}}{\tilde{D}t} = \frac{\partial}{\partial t} + \tilde{u}\frac{\partial}{\partial \tilde{x}} + \tilde{v}\frac{\partial}{\partial \tilde{y}} + \tilde{w}\frac{\partial}{\partial \tilde{z}} \tag{4.22}$$

Euler's Equation

Euler's equation, equation 4.5, may now be written in terms of the moving reference frame by substituting equation 4.21 into 4.5:

$$\mathbf{a}_{ni}\{t\} + \frac{\tilde{D}\tilde{\mathbf{V}}}{\tilde{D}t} = -\frac{1}{\rho}\tilde{\nabla}p + \mathbf{g}$$

$$\frac{\tilde{D}\tilde{\mathbf{V}}}{\tilde{D}t} = -\frac{1}{\rho}\tilde{\nabla}p + \mathbf{g} - \mathbf{a}_{ni}\{t\} \tag{4.23}$$

The net effect of the accelerating reference frame is to make it appear that the acceleration of gravity $\mathbf{g}$ is altered by an amount $-\mathbf{a}_{ni}\{t\}$ that is a function of time in general. It is this increment that we sense when, in an elevator, we feel the increase or decrease in our "weight" as the elevator accelerates upward or downward after a stop. If a glass of water is allowed to drop freely, the acceleration of the glass $\mathbf{a}_{ni}\{t\}$ equals $\mathbf{g}$, and since there is no motion of the water relative to the glass ($\tilde{\mathbf{V}} = 0$), Euler's equation is satisfied if the pressure inside the water is uniform ($\tilde{\nabla}p = 0$).

Bernoulli's Equation

To derive Bernoulli's equation for the noninertial reference frame we integrate Euler's equation along a streamline of the flow in the noninertial reference frame.[11] For the particular case of constant density along a streamline, the equivalent of equation 4.13 becomes:

$$\int_1^2 \frac{\partial \tilde{\mathbf{V}}}{\partial t} \cdot d\tilde{\mathbf{s}} + \left(\frac{\tilde{V}_2^2}{2} + \frac{p_2}{\rho} - (\mathbf{g} - \mathbf{a}_{ni}) \cdot \tilde{\mathbf{R}}_2\right) - \left(\frac{\tilde{V}_1^2}{2} + \frac{p_1}{\rho} - (\mathbf{g} - \mathbf{a}_{ni}) \cdot \tilde{\mathbf{R}}_1\right) = 0$$

$$(\textit{if}\ \tilde{\mathbf{V}} \cdot \tilde{\nabla}\rho = 0) \tag{4.24}$$

---

[11] A streamline passing through a point P in the inertial reference frame *is not* the same as the streamline in the noninertial reference plane passing through the same point because $\mathbf{V} \neq \tilde{\mathbf{V}}$.

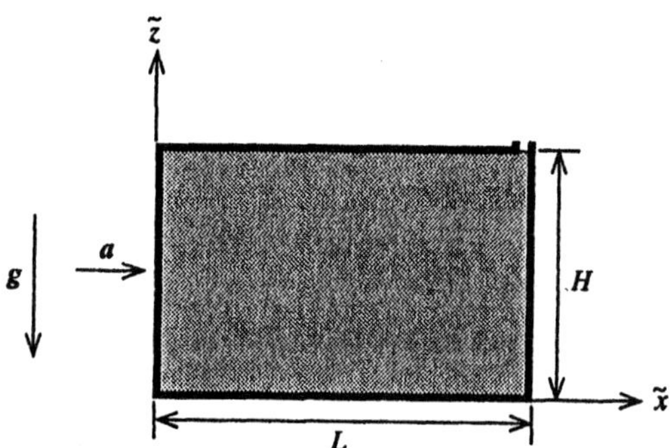

Figure E 4.8

Example 4.8

An automobile, accidentally hit from behind, experiences a forward acceleration of magnitude $2g$. Its fuel tank, filled with gasoline of density $\rho = 8.6E(2)\, kg/m^3$, is open to the atmosphere at its upper forward corner, as shown in figure E 4.8. The tank length $L = 1.0\, m$ and height $H = 0.6\, m$. Calculate the location and value of the maximum pressure in the tank.

Solution

Since there is no relative motion of the fluid in the tank ($\tilde{V} = 0$), Bernoulli's equation 4.24 for an accelerating reference frame becomes:

$$\frac{p}{\rho} + (\mathbf{g} - \mathbf{a}_{in}) \cdot \tilde{\mathbf{R}} = \frac{p_a}{\rho} - (\mathbf{g} - \mathbf{a}_{in}) \cdot \tilde{\mathbf{R}}_1$$

$$p = p_a - \rho(\mathbf{g} - \mathbf{a}_{in}) \cdot (\tilde{\mathbf{R}}_1 - \tilde{\mathbf{R}})$$

where $\tilde{\mathbf{R}}_1 = L\mathbf{i}_x + H\mathbf{i}_z$ is the location of the tank opening to the atmosphere. The pressure will be a maximum when $\tilde{\mathbf{R}} = 0$, *i.e.*, at the lower left corner of the tank ($\tilde{x} = 0$, $\tilde{z} = 0$). Noting that $\mathbf{g} - \mathbf{a}_{in} = -g\mathbf{i}_z - 2\,g\mathbf{i}_x$, the maximum pressure is:

$$\begin{aligned} p &= p_a - \rho(-g\mathbf{i}_z - 2\,g\mathbf{i}_x) \cdot (L\mathbf{i}_x + H\mathbf{i}_z) \\ &= p_a + \rho\,(gH + 2gL) \\ &= p_a + (8.6E(2)\, kg/m^3)(9.807\, m/s^2)(0.6 + 2 \times 1.0)\, m \\ &= p_a + 2.207E(4)\, Pa \end{aligned}$$

4.6.2 Steadily Rotating Reference Frame

Because of the applications for which it is used, we choose a cylindrical coordinate system (r, θ, z) fixed in the inertial reference frame and another cylindrical coordinate system $(\tilde{r}, \tilde{\theta}, \tilde{z})$ fixed to the rotating noninertial reference frame. The relations between these coordinates are:

$$r = \tilde{r}; \ \theta = \tilde{\theta} + \Omega t; \ z = \tilde{z}$$

$$V_r = \tilde{V}_r; \ V_\theta = \tilde{V}_\theta + \Omega \tilde{r}; \ V_z = \tilde{V}_z \qquad (or \ \mathbf{V} = \tilde{\mathbf{V}} + \boldsymbol{\Omega} \times \tilde{\mathbf{r}}) \tag{4.25}$$

where Ω is the magnitude of the angular velocity vector $\boldsymbol{\Omega}$ $(= \Omega\, \mathbf{i}_z)$ of rotation of the noninertial reference frame about the z-axis.

It is not a simple matter to obtain an expression for the acceleration $D\mathbf{V}/Dt$ of a fluid particle in terms of the flow quantities in the noninertial reference frame. If we use the relationships of equation 4.25 in equation 4.4, we can find the following expression for the acceleration:

$$\frac{D\mathbf{V}}{Dt} = \frac{\tilde{D}\tilde{\mathbf{V}}}{\tilde{D}t} - \Omega^2 \tilde{r}\, \mathbf{i}_r + 2\,\boldsymbol{\Omega} \times \tilde{\mathbf{V}} \tag{4.26}$$

The second term on the right of equation 4.26 is the familiar centrifugal acceleration of a particle moving at a tangential speed $\Omega \tilde{r}$ at a radius $\tilde{r}$ about the $\tilde{z}$-axis. The third term is the *Coriolis acceleration*, whose direction is perpendicular to both the velocity $\tilde{\mathbf{V}}$ and the angular velocity $\boldsymbol{\Omega}$.

Euler's Equation

Substituting equation 4.26 into equation 4.5, we find an expression for Euler's equation in terms of the noninertial coordinates and motion:

$$\frac{\tilde{D}\tilde{\mathbf{V}}}{\tilde{D}t} = -\frac{1}{\rho}\tilde{\nabla} p + \mathbf{g} + \Omega^2 \tilde{r}\, \mathbf{i}_r - 2\,\boldsymbol{\Omega} \times \tilde{\mathbf{V}} \tag{4.27}$$

The last two terms on the right, having the dimensions of acceleration, add to the pressure force and gravity force per unit mass ($\tilde{\nabla} p/\rho$ and $\mathbf{g}$) in producing the relative acceleration $\tilde{D}\tilde{\mathbf{V}}/\tilde{D}t$ as measured in the noninertial reference frame. When multiplied by ρ, they are sometimes called the centrifugal force and Coriolis force per unit volume, but of course they are not true forces.

Bernoulli's Equation

It is possible to integrate Euler's equation 4.27 along a streamline of the flow in the noninertial reference frame by noting that the term $2\,\boldsymbol{\Omega} \times \tilde{\mathbf{V}}$ makes no contribution to the integral because it is perpendicular to $d\tilde{\mathbf{s}}$ and that the centrifugal acceleration term $\Omega^2\tilde{r}\,\mathbf{i}_r \cdot d\tilde{\mathbf{s}} = \Omega^2\tilde{r}\,d\tilde{r}$. For the case of constant density along a streamline, Bernoulli's equation has the form:

$$\int_1^2 \frac{\partial \tilde{\mathbf{V}}}{\partial t} \cdot d\tilde{\mathbf{s}} + \left(\frac{\tilde{V}_2^2}{2} + \frac{p_2}{\rho} - \mathbf{g} \cdot \tilde{\mathbf{R}}_2 - \frac{(\Omega\,\tilde{r}_2)^2}{2}\right) - \left(\frac{\tilde{V}_1^2}{2} + \frac{p_1}{\rho} - \mathbf{g} \cdot \tilde{\mathbf{R}}_1 - \frac{(\Omega\,\tilde{r}_1)^2}{2}\right) = 0$$

$$(\textit{if}\;\; \tilde{\mathbf{V}} \cdot \tilde{\nabla}\rho = 0) \qquad (4.28)$$

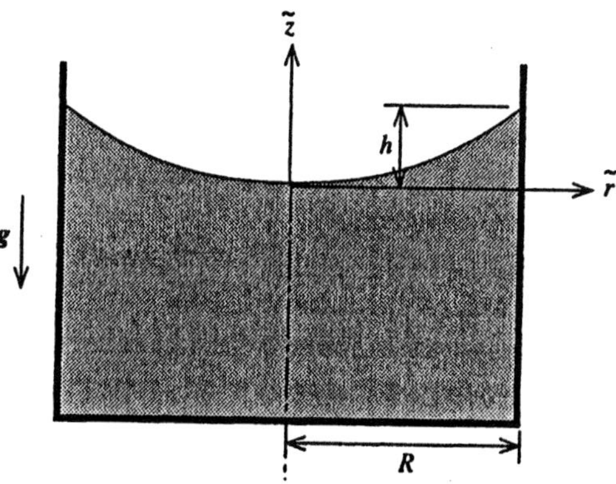

Figure E 4.9

Example 4.9

A jar of radius $R = 5\,cm$, partially filled with water, is placed on a record turntable. After the turntable has been revolving at a steady speed of 33 revolutions per minute for a long time, the water is motionless ($\tilde{V} = 0$) in the rotating reference frame and the water surface has the shape of a paraboloid of revolution, as shown in figure E 4.9, with the water level at the outside edge higher than that at the centerline by an amount h. Calculate the height h.

Solution

Using Bernoulli's equation 4.28 and taking point 2 to be $\tilde{r} = 0$, $\tilde{z} = 0$ and the point 1 to be $\tilde{r} = R$, $\tilde{z} = h$, while noting that $\tilde{V} = 0$ and both p_1 and p_2 equal atmospheric pressure, we find:

$$g\tilde{z}_2 - \left(g\tilde{z}_1 - \frac{(\Omega R)^2}{2}\right) = 0$$

$$h = \frac{(\Omega R)^2}{2g}$$

$$= \frac{((33 \times 2\pi \div 60\,s) \times 5.0E(-2)\,m)^2}{2 \times 9.807\,m/s^2}$$

$$= 1.522E(-3)\,m$$

4.6.3 Inviscid Flow in Rotating Machines

Increasing the pressure of a fluid is commonly accomplished in rotating or reciprocating machines. For the former, if the rotational speed is high enough, the centrifugal acceleration can give rise to a substantial pressure increase. This principle is used in the *centrifugal pump* or *compressor*.

A pump, compressor or turbine consists of two mechanical components, the *rotor* (which rotates about a fixed axis) and the *stator* (which remains stationary). In the laboratory coordinate system, which is fixed to the stator, the flow is unsteady because the rotor blades are moving with respect to this system and pressures and velocities change with time, albeit in a periodic manner. There is no simple solution to Euler's or Bernoulli's equation for this unsteady flow. Instead, by noting that the flow relative to the rotor is steady, we use the noninertial rotating reference frame fixed in the rotor to determine the flow changes from inlet to outlet.

Consider the simple example of a centrifugal pump as shown in figure 4.9. The inlet velocity $\mathbf{V}_1$ is purely axial so that the inlet velocities in the inertial and noninertial reference frames are related by:

$$V_{r1} = 0;\ V_{\theta 1} = 0;\ V_{z1} = -V_1$$

$$\tilde{V}_{r1} = 0;\ \tilde{V}_{\theta 1} = -\Omega\,\tilde{r}_1;\ \tilde{V}_{z1} = -V_1$$

while at the exit from the rotor the relative flow in the noninertial reference frame is purely radial at a speed $\tilde{V}_2$, giving:

$$\tilde{V}_{r2} = \tilde{V}_2;\ \tilde{V}_{\theta 2} = 0;\ \tilde{V}_z = 0$$

$$V_{r2} = \tilde{V}_2;\ V_{\theta 2} = \Omega\,\tilde{r}_2;\ V_z = 0$$

It follows from these relations that the squares of the velocities are related by:

$$\tilde{V}_1^2 = V_1^2 + (\Omega\, r_1)^2;\ V_2^2 = \tilde{V}_2^2 + (\Omega\, r_2)^2 \tag{4.29}$$

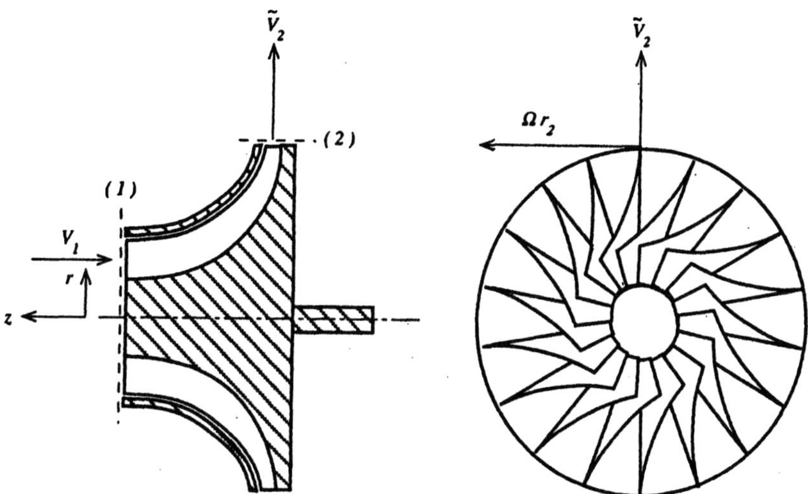

Figure 4.9 A cross-section of a centrifugal pump, and an axial view of the rotor.

Substituting this relation into Bernoulli's equation 4.28, noting that there is steady flow through the rotor and neglecting gravity, we find:

$$p_2 = p_1 + \rho(\Omega\, r_2)^2 + \rho\left(\frac{V_1^2}{2} - \frac{\tilde{V}_2^2}{2}\right) \tag{4.30}$$

Usually, the inlet and outlet flow areas are equal. If so, then by mass conservation $V_1 = \tilde{V}_2$, and therefore the last term of equation 4.30 is zero. To obtain large pressure increases, it is necessary to have a high rotational tip speed $\Omega\, r_2$ for the rotor. This speed is limited by the material stresses in the rotor.

In the inertial reference frame, the unsteady term in Bernoulli's equation 4.13 is significant compared to the other terms and cannot be ignored. To see why this is so, substitute equations 4.29 and 4.30 into 4.13 and neglect gravity to find:

$$\int_1^2 \frac{\partial \mathbf{V}}{\partial t} \cdot d\mathbf{s} = -(\Omega\, r_2)^2$$

By referring to equation 4.30, we can see that the unsteady term is comparable in size to the pressure rise divided by the density.

Some pumps and compressors do not utilize centrifugal acceleration to change the fluid pressure. *Axial flow machines* (*e.g.*, compressors, turbines, fans and propellors) have negligible radial flow ($V_r = \tilde{V}_r = 0$; $r_1 = r_2$) and achieve their pressure change by altering the tangential component of velocity $\tilde{V}_\theta$ relative to the rotor. For such machines the inlet and exit conditions for the rotor can be given as:

$$\tilde{V}_{r1} = 0; \quad \tilde{V}_{\theta 1} = \Omega r_1; \quad \tilde{V}_{z1} = -V_1$$

$$\tilde{V}_{r2} = 0; \quad \tilde{V}_{\theta 2} = \xi \Omega r_1; \quad \tilde{V}_{z2} = -V_1$$

where ξ is a parameter defining the change in tangential speed ($\xi = 1$ means no change) and where, by mass conservation, the axial speed is unchanged through the rotor. The corresponding relations for the square of the velocities are:

$$\tilde{V}_1^2 = V_1^2 + (\Omega r_1)^2; \quad \tilde{V}_2^2 = V_1^2 + \xi^2(\Omega r_1)^2$$

Substituting this in Bernoulli's equation 4.28, neglecting gravity and noting that the flow is steady, we find:

$$p_2 = p_1 + \rho(1 - \xi^2)\frac{(\Omega r_2)^2}{2} \tag{4.31}$$

In compressors, fans and propellors, ξ is only slightly less than unity because of the limited lift that can be generated by the rotating airfoils. Thus the pressure rise available in the rotor is smaller than for a centrifugal machine having the same tip speed Ωr. For jet engines and gas turbine compressors, where large pressure rises are thermodynamically desirable, several stages of compression are needed for axial flow machines. On the other hand, turbines can have large pressure drops (ξ large and negative) because their blades can be highly loaded.

The inviscid flow described in this section neglects the effects of viscosity. Viscous effects degrade the performance of rotating machinery, resulting in less pressure rise in compressors than that given by equations 4.30 and 4.31. These effects will be discussed later in the chapter 8.

4.7 Special Flows

Integrals of Euler's equation that closely resemble Bernoulli's equation can be derived for flows that meet certain restrictions. In some problems it may be more profitable to use such integrals in place of Bernoulli's equation.

4.7.1 Barotropic Flow

In a *barotropic flow* the density ρ is a function of the pressure p alone, *i.e.*, $\rho = \rho\{p\}$. For the fluids with which we ordinarily deal, a flow will be barotropic when the fluid entropy $\hat{s}$ is a constant along a streamline. In chapter 8, we will show that the fluid entropy can remain constant in an inviscid flow. For a barotropic fluid, the third integral of equation 4.10 can be evaluated by noting that the second law of thermodynamics, equation 1.10, requires that:

$$\nabla \hat{h} = \frac{1}{\rho}\nabla p + T\,\nabla \hat{s} \tag{4.32}$$

Thus the third integral of equation 4.10, evaluated along a streamline, becomes:

$$\int_1^2 \left(\frac{1}{\rho}\nabla p\right)\cdot d\mathbf{s} = \int_1^2 \nabla \hat{h}\cdot d\mathbf{s} - \int_1^2 T\nabla \hat{s}\cdot d\mathbf{s}$$
$$= \hat{h}_2 - \hat{h}_1$$

by use of equation 1.46 and by noting that $\nabla \hat{s}\cdot d\mathbf{s} = 0$ since $\hat{s}$ does not vary along a streamline. Thus the version of equation 4.13 for a barotropic flow that corresponds to Bernoulli's equation is:

$$\int_1^2 \frac{\partial \mathbf{V}}{\partial t}\cdot d\mathbf{s} + \left(\frac{V_2^2}{2} + \hat{h}_2 - \mathbf{g}\cdot\mathbf{R_2}\right) - \left(\frac{V_1^2}{2} + \hat{h}_1 - \mathbf{g}\cdot\mathbf{R_1}\right) = 0$$

$$(\textit{if}\ \ \mathbf{V}\cdot(\nabla \hat{s}) = 0) \tag{4.33}$$

This equation is mainly useful for steady, inviscid compressible flow because the first term is difficult to evaluate when the flow is unsteady.

4.7.2 Irrotational Flow

An irrotational flow is one for which $\nabla \times \mathbf{V}$ is zero everywhere in the flow field. In chapter 11, we will show that an incompressible flow can be irrotational. For the present purpose, we will consider a constant-density flow ($\nabla\rho = 0$) that is irrotational ($\nabla \times \mathbf{V} = 0$).

Referring to equation 4.10, the right hand side is zero for an irrotational flow. For a constant-density flow, the integral containing the pressure gradient can be integrated along any curve, not just along a streamline. Thus equation 4.10 assumes the form:

$$\int_1^2 \frac{\partial \mathbf{V}}{\partial t}\cdot d\mathbf{c} + \left(\frac{V_2^2}{2} + \frac{p_2}{\rho} - \mathbf{g}\cdot\mathbf{R_2}\right) - \left(\frac{V_1^2}{2} + \frac{p_1}{\rho} - \mathbf{g}\cdot\mathbf{R_1}\right) = 0$$

$$(\textit{if}\ \ \nabla\rho = 0;\ \nabla \times \mathbf{V} = 0) \tag{4.34}$$

Note that this equivalent of Bernoulli's equation applies along any line that lies within the flow. In steady flow, this means that the sum, $V^2/2 + p/\rho - \mathbf{g}\cdot\mathbf{R}$, has the same value everywhere in the flow field. Thus a knowledge of the velocity field $\mathbf{V}\{\mathbf{R}\}$ enables us to calculate the pressure field $p\,\{\mathbf{R}\}$ but not vice versa.

4.7.3 Flow on a Bernoulli Surface

A *Bernoulli surface* is a surface in the flow that is everywhere tangent to the plane containing the vectors $\mathbf{V}$ and $\nabla \times \mathbf{V}$ at the point of tangency. A streamline would lie in a Bernoulli surface, as would a "streamline" of the vector field, $\nabla \times \mathbf{V}$.[12] The vector product $\mathbf{V} \times (\nabla \times \mathbf{V})$ has the direction of the normal to the Bernoulli surface. If we choose the curve $\mathcal{C}$ of equation 4.10 to lie in a Bernoulli surface, then the scalar product in the integral on the right side of equation 4.10 is zero. Denoting the line element of such a curve by $d\mathbf{b}$, equation 4.10 has the form:

$$\int_1^2 \frac{\partial \mathbf{V}}{\partial t} \cdot d\mathbf{b} + \left(\frac{V_2^2}{2} + \frac{p_2}{\rho} - \mathbf{g} \cdot \mathbf{R_2}\right) - \left(\frac{V_1^2}{2} + \frac{p_1}{\rho} - \mathbf{g} \cdot \mathbf{R_1}\right) = 0$$

$$(\textit{if}\ \ \nabla \rho \cdot d\mathbf{b} = 0;\ \mathbf{V} \times (\nabla \times \mathbf{V}) \cdot d\mathbf{b} = 0) \qquad (4.35)$$

For a steady flow, the sum $V^2/2 + p/\rho - \mathbf{g} \cdot \mathbf{R}$ has the same value everywhere on a Bernoulli surface. Equation 4.35 is valid between points that lie on a Bernoulli surface but not necessarily on the same streamline, even though the flow is not irrotational.

4.7.4 Beltrami Flow

A *Beltrami flow* is one for which $\mathbf{V}$ and $\nabla \times \mathbf{V}$ are everywhere parallel so that $\mathbf{V} \times (\nabla \times \mathbf{V}) = 0$ everywhere, and the right-hand side of equation 4.10 is zero for any curve $\mathcal{C}$. If the density ρ is constant, then equation 4.10 has the form:

$$\int_1^2 \frac{\partial \mathbf{V}}{\partial t} \cdot d\mathbf{c} + \left(\frac{V_2^2}{2} + \frac{p_2}{\rho} - \mathbf{g} \cdot \mathbf{R_2}\right) - \left(\frac{V_1^2}{2} + \frac{p_1}{\rho} - \mathbf{g} \cdot \mathbf{R_1}\right) = 0$$

$$(\textit{if}\ \ \nabla \rho = 0;\ \mathbf{V} \times (\nabla \times \mathbf{V}) = 0) \qquad (4.36)$$

Even though the flow is not irrotational, the sum $V^2/2 + p/\rho - \mathbf{g} \cdot \mathbf{R}$ is constant throughout a steady Beltrami flow. Beltrami flows are not common and exist only in special circumstances.

---

[12]The vector $\nabla \times \mathbf{V}$ is called the *vorticity* and its "streamline" a *vortex line*. Vorticity is discussed in more detail in chapter 11.

4.8 Problems

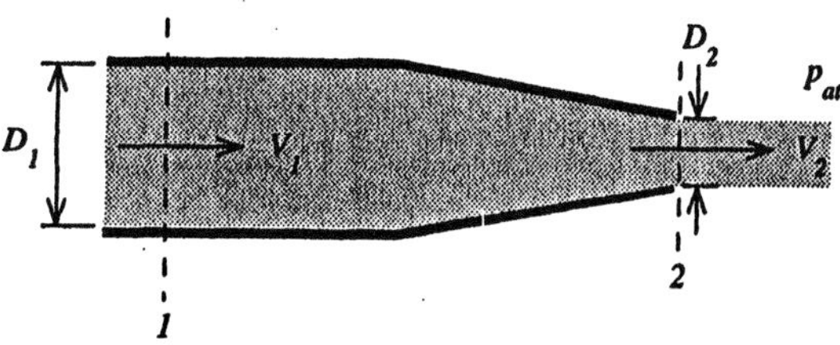

Figure P 4.1

Problem 4.1

A steady stream of water ($\rho = 1.0E(3)\,kg/m^3$) leaves a fire hose nozzle at a speed $V_2 = 40\,m/s$ and at atmospheric pressure $p_{at} = 1.0E(5)\,Pa$, as shown in figure P 4.1. The nozzle diameter $D_2 = 5\,cm$, and the hose diameter $D_1 = 10\,cm$. (a) Calculate the speed V_1 of the water in the fire hose. (b) Calculate the pressure p_1 in the fire hose.

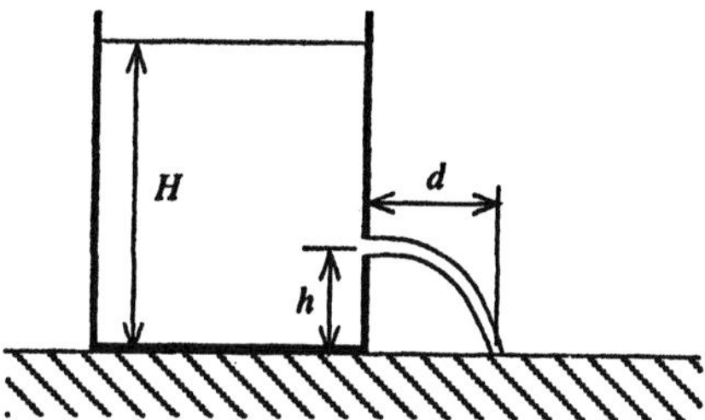

Figure P 4.2

Problem 4.2

A storage tank open to the atmosphere at its top contains water to an unknown depth H above its base, as shown in figure P 4.2. An engineer opens a valve located at a distance h above the base level and observes that the stream of water leaving the valve horizontally curves downward to reach the base level at a distance d from the base of the tank. Derive an expression for the water depth H in terms of the measured distances h and d.

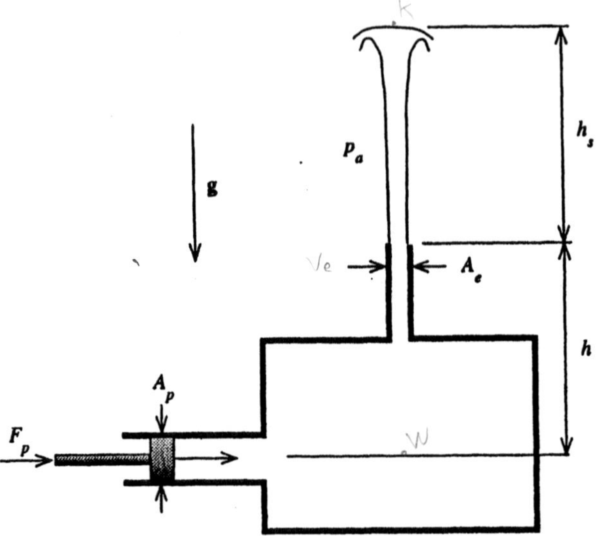

Figure P 4.3

Problem 4.3

In figure P 4.3, a container filled with water has an upward-facing orifice of area A_e through which a steady stream of water, propelled by the inward motion of a piston inside a tube of area A_p that is connected to the container, rises to a height h_s above the orifice. Derive expressions for (a) the constant velocity V_p at which the piston moves and (b) the force F_p that must be exerted on the piston, assuming frictionless motion. (c) Noting that FV_p is the power required to move the piston, calculate the power needed for a fountain rising $10\,m$, supplied through an orifice of diameter $D_e = 1\,in$.

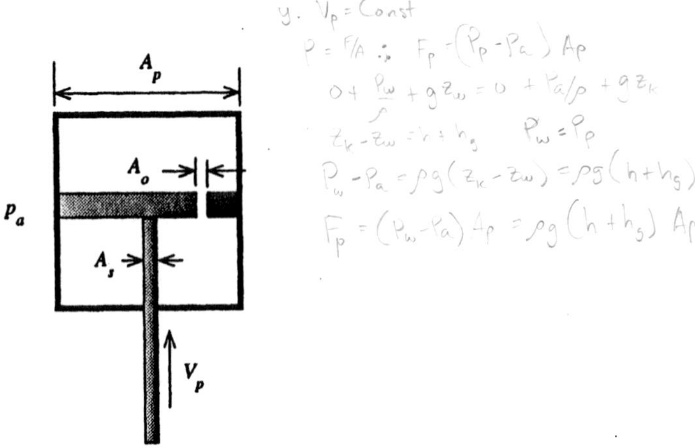

Figure P 4.4

Problem 4.4

An automobile shock absorber consists of a cylinder filled with oil of density $\rho = 9E(2)\,kg/m^3$ and fitted with a piston containing an orifice, as shown in figure P 4.4. (The cylinder, shaft and orifice cross-sectional areas are $A_p = 15\,cm^2$, $A_s = 2\,cm^2$ and $A_o = 1.0\,cm^2$.) Calculate the force F required to push the piston into the cylinder at a speed $V_p = 1.0\,m/s$.

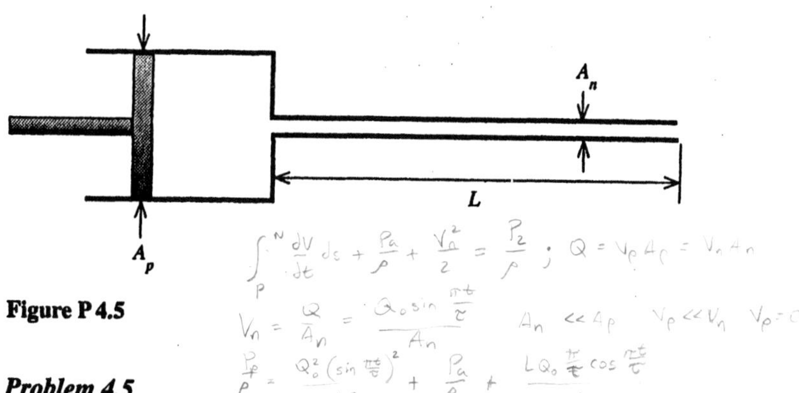

Figure P 4.5

Problem 4.5

In a medical experiment, it is desired to inject quickly a small quantity of radioactive fluid into the vein of an animal. The volumetric flow rate Q from the syringe, shown in figure P 4.5, is to be:

$$Q = Q_0 \sin \frac{\pi t}{\tau} \qquad 0 \leq t \leq \tau$$
$$= 0 \qquad t \geq \tau$$

Assuming the flow is incompressible and that the venous pressure is atmospheric, derive an expression for the pressure $p\{t\}$ in the chamber of the syringe if the needle area A_n is much less than the piston area A_p, and then evaluate this expression for $A_n = 1.0\,mm^2$, $A_p = 1.0\,cm^2$, $L = 10\,cm$, $Q_0 = 10\,cm^3/s$, $\tau = 0.1\,s$, $\rho = 2E(3)\,kg/m^3$.

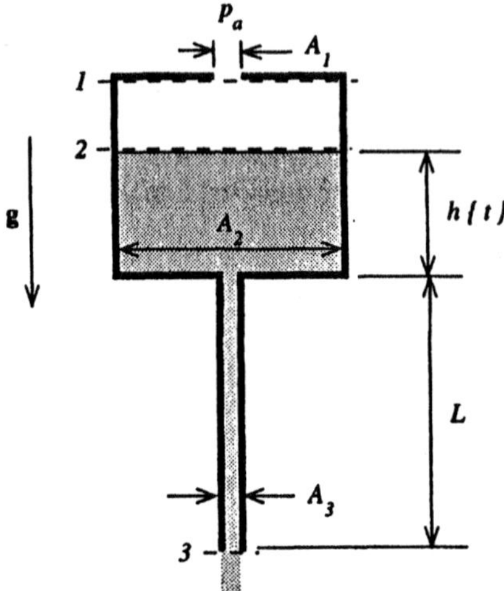

Figure P 4.6

Problem 4.6

A tank of cross-sectional area A_2 is partially filled with water to a depth $h\{t\}$ that decreases with time as the water drains out through a pipe of cross-sectional area A_3 and length L attached to its bottom, as shown in figure P 4.6. The tank is closed at the top, except for a small hole of area A_1 that admits air from the atmosphere to replace the water that drains out. Both A_1 and A_3 are small compared to A_2. (a) Assuming a quasi-steady flow, *i.e.*, that time-dependent acceleration effects are negligible, derive expressions for the velocity V_3 of the water leaving the pipe and the pressure p_2 of the air in the tank. (You may assume that the air density ρ_a, as well as the water density ρ_w, is constant.) (b) Derive an expression of inequality that ensures that the assumption of quasi-steady flow is valid.

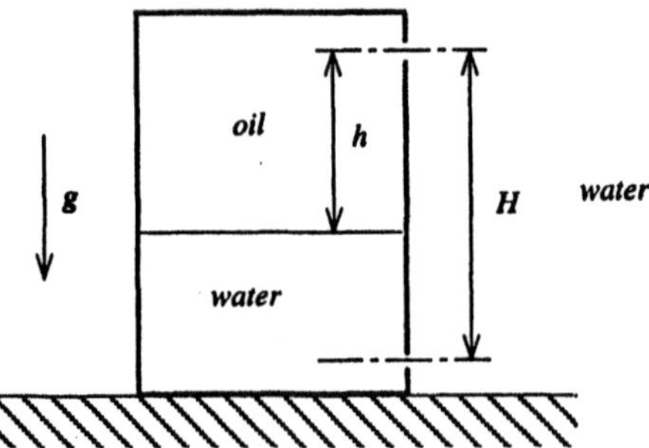

Figure P 4.7

Problem 4.7

A barrel of toxic oil has been thrown illegally into a pond. As shown in figure P 4.7, the barrel sits on the bottom of the pond, surrounded by water. Two holes in the side of the barrel, of equal area and separated by a height H, allow the water to enter the barrel through the bottom hole and oil to leak out through the top hole. This flow is caused by the difference in density between that of the oil, ρ_o, and that of water, $\rho_w > \rho_o$. Derive an expression for the velocity V of the oil out through the top hole when the oil/water interface is a distance h below the top hole.

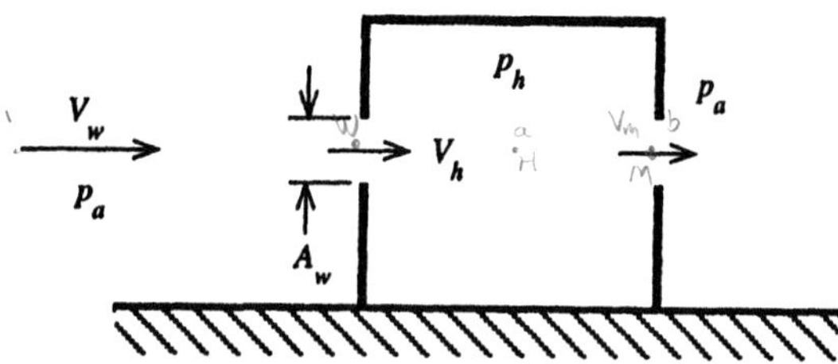

Figure P 4.8

Problem 4.8

A wind of speed $V_w = 10\,m/s$ blows against a house, sketched in Figure P 4.8. On the upwind side of the house a window of area $A_w = 1.0\,m^2$ is open to the wind. Another window of the same size on the downwind side of the house is also open. The pressure outside the house on the downwind side is the same as that far upwind of the house, p_a. (a) Assuming incompressible flow ($\rho = 1.2\,kg/m^3$), calculate the velocity V_h of the wind blowing through the upwind window and the pressure p_h inside the house. (b) If the volume $\mathcal{V}$ of the house is $500\,m^3$, what time τ is required for the volume flow through the window to equal the volume of the house?

a. $V_w = 10\,m/s$ $A_w = 1.0\,m^2$ $\rho = 1.2\,kg/m^3$

Solve: V_h, p_h

$A_m V_m = V_h A_h$

$P_a = 1.2\,kg/m^3$

$$\frac{P_a}{\rho} + \frac{V_w^2}{2} = \frac{P_h}{\rho} + \frac{V_h^2}{2}$$

$$\frac{V_h^2}{2} + \frac{P_a}{\rho} = \frac{P_h}{\rho}$$

$$V_h = \frac{V_w}{\sqrt{2}} = 7.071\ m/s$$

$$P_h = P_a + \frac{\rho V_h^2}{2} = 101\,030\ Pa$$

b. $Q = V_h A_w$

$$\tau = \frac{\mathcal{V}}{Q} = \frac{\mathcal{V}}{V_h A_w} = \frac{500\,m^3}{(7.071)(1.0)} = 70.7\ s.$$

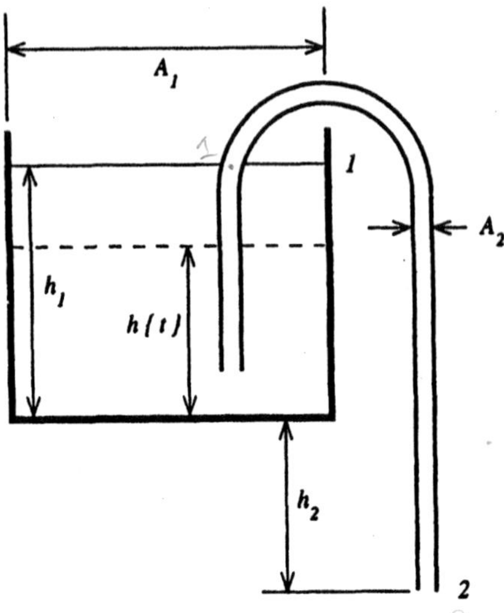

Figure P 4.9

Problem 4.9

A large container filled with water to a depth $h_1 = 20\,cm$ is to be drained by a siphon of total length $L = 1.0\,m$ whose lower end is a distance $h_2 = 40\,cm$ below the bottom of the container, as shown in figure P 4.9. The container cross-sectional area $A_1 = 1.0\,m^2$ is much larger than the area $A_2 = 1.0\,cm^2$ of the siphon tube. At the time $t = 0$ at which the siphon starts, it has been filled with water and the lower end uncovered to let the water start to flow. As time proceeds, the water height $h\{t\}$ in the container decreases. (a) At the start of the flow, calculate the acceleration of a fluid particle in the siphon. (b) After enough time has passed to achieve a quasi-steady flow in the siphon, calculate the velocity V_2 of the fluid leaving the tube when the level $h\{t\} = 10\,cm$. (c) Derive and solve a differential equation for $h\{t\}$.

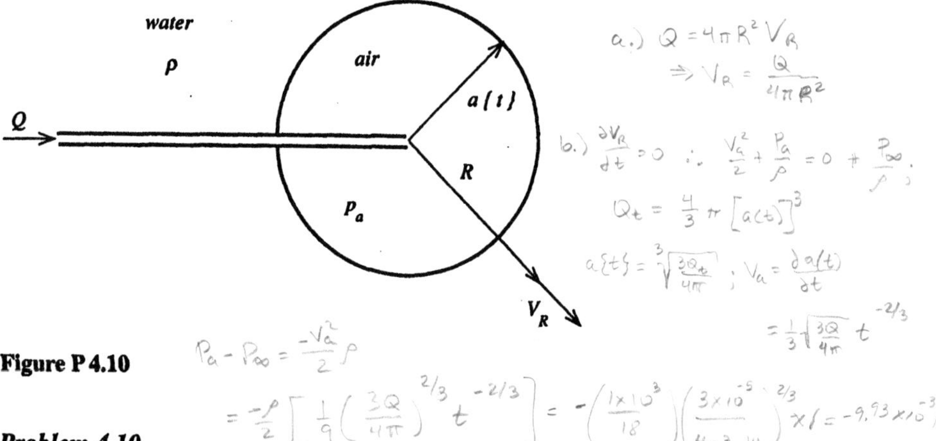

Figure P 4.10

Problem 4.10

A tube of small diameter is inserted into a container of water. Beginning at time $t = 0$, a constant volumetric flow rate $Q = 10\,cm^3/s$ of air is forced through the tube, forming a bubble of radius $a\{t\}$ that increases with time, as shown in figure P 4.10. The water moves out radially because of the expanding bubble, and far from the bubble ($R \gg a$), the water pressure p_∞ is uniform (gravity can be neglected). (a) Utilizing mass conservation, derive an expression for the radial velocity V_R of the water at a distance R from the center of the bubble, assuming that the air density is constant. (b) Assuming the air pressure p_a in the bubble is uniform, derive an expression for the pressure difference $p_a - p_\infty$ at any time t and calculate its numerical value at $t = 1.0\,s$.

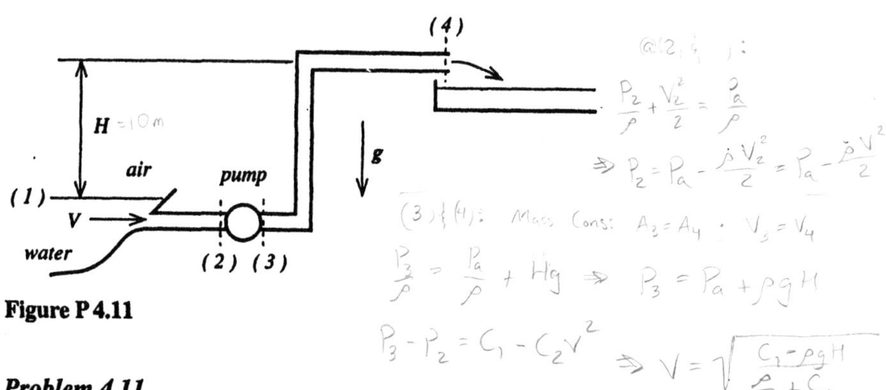

Figure P 4.11

Problem 4.11

A centrifugal pump is used to draw water from a lake and pump it into a storage reservoir, as shown in figure P 4.11. The elevation of the pipe exit (4) is a distance $H = 10\,m$ above the surface of the lake. The pressure rise $p_3 - p_2$ across the pump is related to the flow velocity V in the pipe by:

$$p_3 - p_2 = c_1 - c_2 V^2$$

where $c_1 = 2E(5)\,Pa$ and $c_2 = 5E(2)\,kg/m^3$.

(a) Assuming steady inviscid flow, calculate the pipe flow velocity V. (b) If the total length L of the pipe is $30\,m$, calculate the fluid acceleration dV/dt when the pump is first turned on, the fluid initially being stationary ($V = 0$).

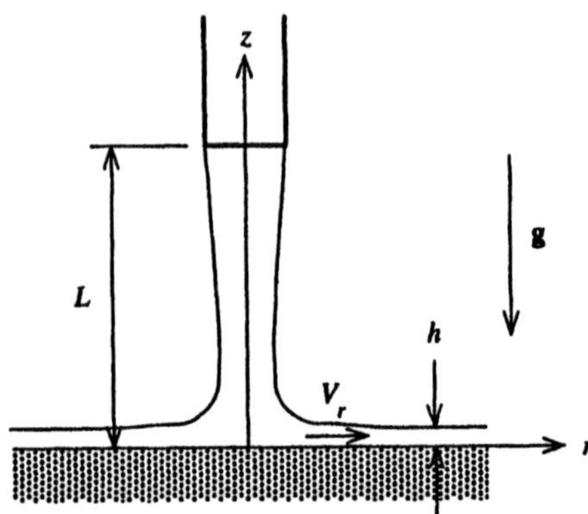

Figure P 4.12

Problem 4.12

Water flows at a uniform speed out of a round pipe vertically downward onto a horizontal flat plate a distance $L = 10\,cm$ below, as shown in figure P 4.12. The volumetric flow rate is $Q = 100\,cm^3/s$, and the pipe area is $A = 1.0\,cm^2$. The water spreads out smoothly over the plate so that at a distance r from the flow axis the radial velocity $V_r\{r\}$ is uniform, *i.e.*, independent of z. The flow is axisymmetric so that the thickness of the water layer $h\{r\}$ depends only upon r. Derive expressions for $V_r\{r\}$ and $h\{r\}$, assuming that $h \ll L$, and calculate their values at $r = 10\,cm$.

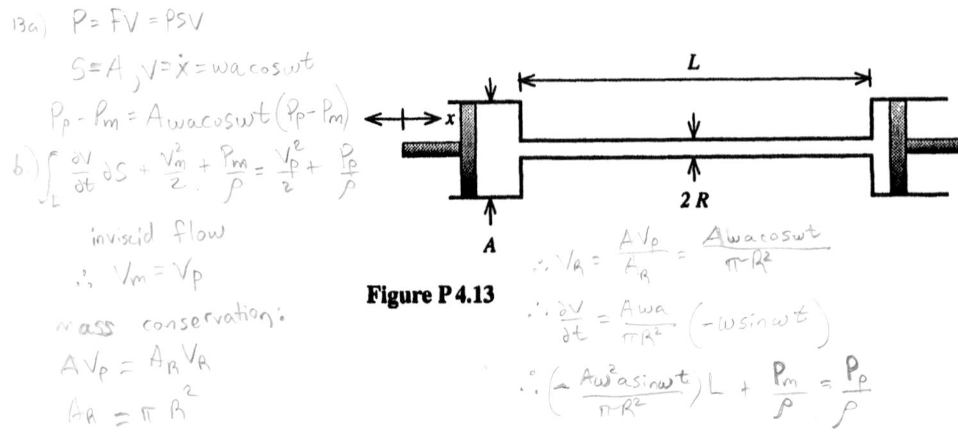

Figure P 4.13

Problem 4.13

An hydraulic system for transmitting power consists of an oscillating pump connected to an identical oscillating motor by a long circular tube of length L and radius R (see figure P4.13). The pump and motor have pistons of equal areas A and undergo simple harmonic displacements $x = a \sin \omega t$, where a is the amplitude and ω the angular frequency of the motion. The fluid in the system is incompressible, having a density ρ.

In answering the following questions, assume that the pressure in either cylinder is identical to the pressure in the pipe at the point of attachment. (a) If there is a difference between the pump cylinder pressure p_p and the motor pressure p_m at any instant, there will be a corresponding difference in the instantaneous power input $\mathcal{P}_p$ of the pump and the power output $\mathcal{P}_m$ of the motor. Derive an expression for the power difference $\mathcal{P}_p - \mathcal{P}_m$ in terms of $p_p - p_m$ and the parameters A, ω and a. (b) Assuming inviscid unsteady flow in the pipe, derive an expression for the pressure difference $p_p - p_m$ in terms of the parameters ρ, L, A, a, ω and R. (c) Prove that the integral of the power difference, $\mathcal{P}_p - \mathcal{P}_m$, over one cycle time is zero.

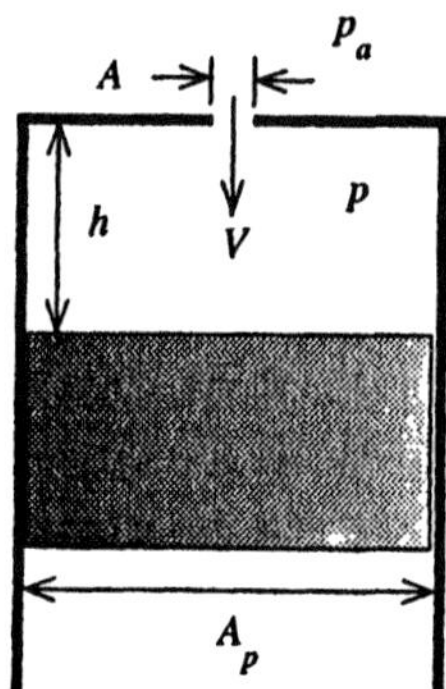

Figure P4.14

Problem 4.14

The spark plug is removed from one cylinder of an automotive engine and the intake and exhaust valves are fixed in their closed positions. As the engine runs, air is alternately ingested and expelled through the spark plug opening whose minimum flow area is $A = 1\,cm^2$, as sketched in figure P4.14. The air volume in the engine cylinder has the form of a circular cylinder of base area (piston area) $A_p = 100\,cm^2$

and height h that changes with time t according to:

$$h = h_0 + \frac{S}{2}(1 - \cos \omega t)$$

where $S = 10\,cm$ is the piston stroke, h_0 the clearance distance and $\omega = 30\,s^{-1}$ the angular speed of the crankshaft. (a) Assuming that the air density ρ is constant, derive an expression for the velocity $V\{t\}$ of the air flowing through the spark plug opening into the cylinder. What is its maximum numerical value? (b) What is the pressure $p\{t\}$ in the cylinder during inflow? During outflow? What is the maximum numerical value of $p - p_a$? (c) As the speed of the engine increases, the power expended in pumping the air into and out of the cylinder increases as ω^n. What is the numerical value of n?

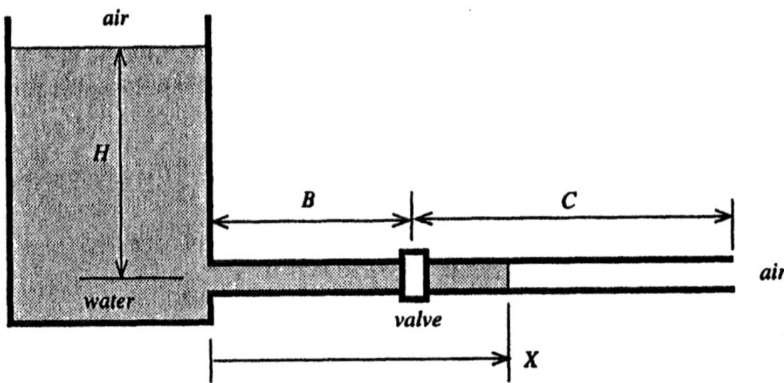

Figure P 4.15

Problem 4.15

A tank of water, open at the top to the atmosphere, supplies a long horizontal pipe attached to the base of the tank at a distance $H = 10\,m$ below the air/water interface (see figure P 4.15). The pipe cross-sectional area is very much smaller than that of the tank. A stop valve is located along the pipe at a distance $B = 20\,m$ from the tank. At a distance $C = 100\,m$ beyond the valve, the pipe is open to the atmosphere. Initially ($t < 0$), the valve is closed and stationary water fills the pipe to the left of the valve while atmospheric air fills the pipe section to the right. At time $t = 0$, the valve is opened suddenly, and water begins to flow along the pipe toward the open end with a speed $V\{t\}$ that changes with time. Assuming that the air pressure in the pipe ahead of the advancing water column is atmospheric pressure, (a) what is the numerical value of the acceleration DV/Dt of the water in the pipe at the instant the valve is opened? (b) It is observed that the leading edge of the water column moving to the right of the valve approaches a steady speed before it reaches the end of the pipe. What is the numerical value of that steady speed? (c) Derive the differential equation or equations which, if integrated, would give the position X of the head of the water column as a function of time t after the opening of the valve. (d) The assumption that the air pressure ahead of the water column is atmospheric pressure would introduce an error in the solution of (c) if the pipe length C is very large. Determine an upper limit to C that will ensure that the solution to (c) is not significantly in error.

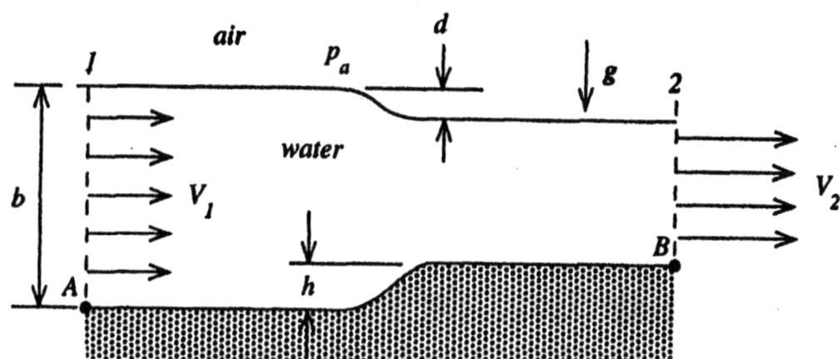

Figure P 4.16

Problem 4.16

In a laboratory water channel, a steady two-dimensional inviscid flow of water moves from left to right, as shown in figure P 4.16. At the upstream and downstream sections 1 and 2, the flow velocities are horizontal and uniform from top to bottom of the water stream. It is observed that, as the fluid flows over the downstream step of height $h = 3\,cm$, the upper surface of the water is depressed a distance $d = 1\,cm$ below its upstream level (see the figure). The upstream depth of water is $b = 10\,cm$. (a) Calculate the gage pressures p_A and p_B at the bottom surface of the channel at both the upstream and downstream locations A and B. (b) What is the numerical value of the ratio V_2/V_1 of the downstream to the upstream velocity? (c) What is the numerical value of the upstream velocity V_1?

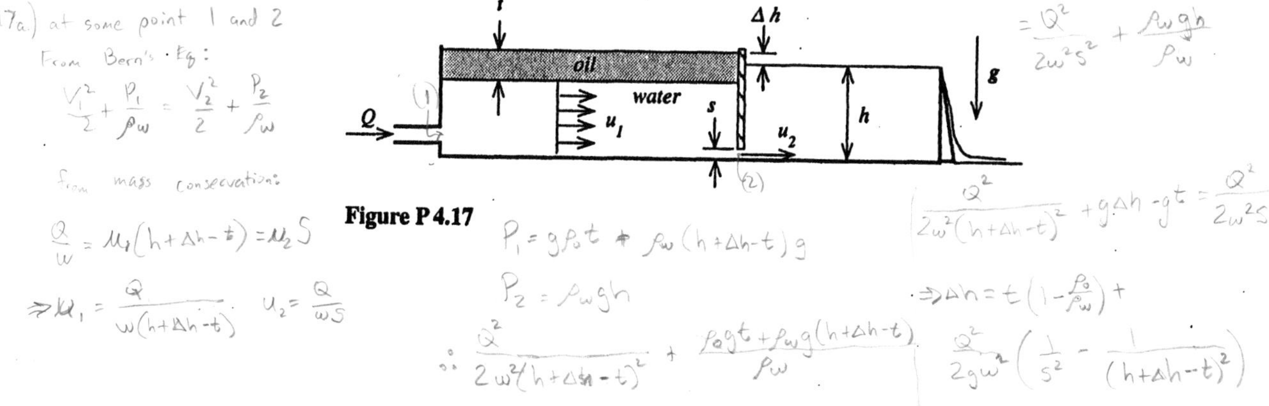

Figure P 4.17

Problem 4.17

A tank is used to separate oil from a flow of water slightly contaminated with oil droplets. As shown in figure P 4.17, the contaminated water flows into the tank at a volume flow rate per width W of tank of Q/W. Oil droplets rise to the surface, where a stationary layer of oil, of uniform thickness t, is retained by a barrier at the midlength of the tank. A thin slot at the bottom of the barrier, of height $s \ll h$, permits oil-free water to flow into a second compartment whose depth of water, h, is maintained by an outflow barrier.

(a) Assuming that the flow of water beneath the oil layer is inviscid, derive an expression for the difference in surface levels, Δh, as a function of the heights t, h and s, the flow rate Q/W and the oil and water densities, ρ_o and ρ_w.

(b) As the flow rate is increased, it is observed that the lower surface of the oil layer close to the barrier moves downward near the slot, until oil begins to flow beneath the barrier. Assuming inviscid flow of water and a motionless layer of oil, derive an expression for the critical volume flow rate Q/W at which the oil begins to leak through the slot.

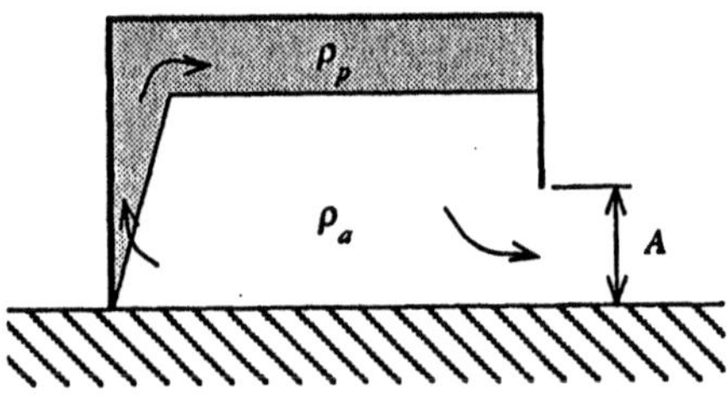

Figure P 4.18

Problem 4.18

A large room has a small doorway of area A open to the outside atmosphere. On the floor of the room opposite the doorway, a fire starts to burn. The fire consumes fuel at a steady mass rate of $\dot{m}$. For each kilogram of fuel burned, f kilograms of air are consumed, forming products of combustion having a density ρ_p that is less than the air density ρ_a.

Soon after the fire starts, a layer of combustion products forms along the ceiling of the room, as illustrated in figure P 4.18. This layer has a uniform density of ρ_p and does not mix with the cooler air below. The growing layer of combustion products displaces air from the room through the doorway.

(a) Assuming inviscid incompressible flow of the room air, derive an expression for the air velocity V_a through the doorway in terms of the parameters $\dot{m}$, A, f, ρ_a and ρ_p. (b) Derive an expression for the pressure difference Δp by which the room air pressure at floor level exceeds the exterior atmospheric pressure at the same level.

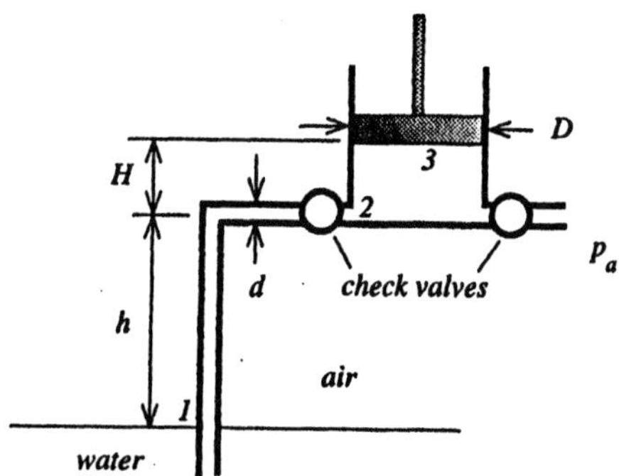

Figure P 4.19

Problem 4.19

A reciprocating pump of diameter $D = 10\,cm$ and stroke $S = 10\,cm$ is used to pump water from a flooded cellar of a home to a height $h = 3\,m$ above the water surface where it is discharged to the ground surface outside the cellar, as shown in figure P 4.19. The pump, connected to suction and discharge lines of equal diameters $d = 2\,cm$, is equipped with check valves that permit flow only from the cellar during the upward stroke of the piston and out through the discharge pipe during the downward stroke. On both upward and downward strokes, the piston speed $V_p = 10\,cm/s$ is constant throughout the stroke distance S. (a) During the upward stroke, what is the speed V of the water in the suction pipe? What is the pressure p_2 at the exit of the suction pipe? What is the pressure p_3 at the piston face when it is a distance H above the exit of the suction pipe? Neglecting the weight of the piston, what force F_u is required to lift the piston? (b) During the downward stroke, what is the pressure p_3 at the piston face? What force F_d is required to push down the piston? (c) What work must be done on the piston to pump each kilogram of water from the cellar? What power P is required to operate the pump?

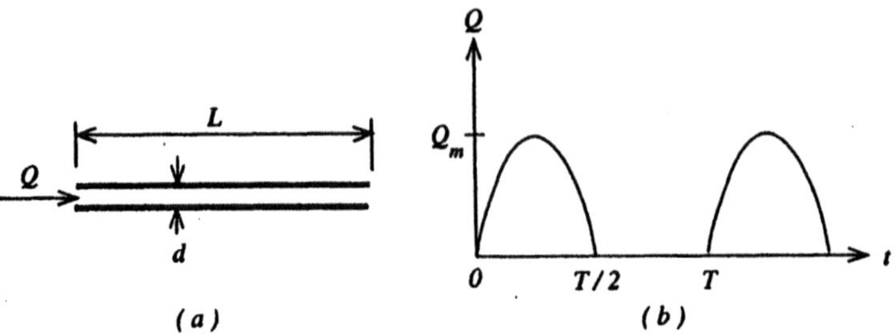

Figure P 4.20

Problem 4.20

A simple model of the human arterial system, shown in (*a*) of figure P 4.20, consists of a rigid circular tube of length L and diameter d. At the left end, blood is delivered from the left ventricle of the heart at a periodic volume flow rate $Q\{t\}$ that varies with time:

$$Q\{t\} = Q_m \sin\left(\frac{2\pi t}{T}\right) \qquad 0 \leq t \leq \frac{T}{2}$$

$$= 0 \qquad \frac{T}{2} \leq t \leq T$$

as shown in sketch (*b*) of figure P 4.20, where Q_m is the maximum flow rate and T is the period of the heart beat. At the right end of the model artery, the pressure p_2 (measured above the constant venous pressure level) is related to the volume flow rate Q_c into the capillary bed by:

$$p_2 = kQ_c$$

where k is a constant characterizing the conductance of the capillary system. Assuming that the arterial flow is incompressible, (a) what is the volume flow rate Q_c into the capillary bed as a function of time? (b) Derive an expression for the pressure p_2 at the right end of the artery, as a function of time. (c) Derive an expression for the pressure p_1 at the left end (entrance) of the artery, as a function of time. (d) Based upon your answer to (c), describe the effect upon the pressure p_1 if the diameter d of the artery is decreased because of fatty deposits.

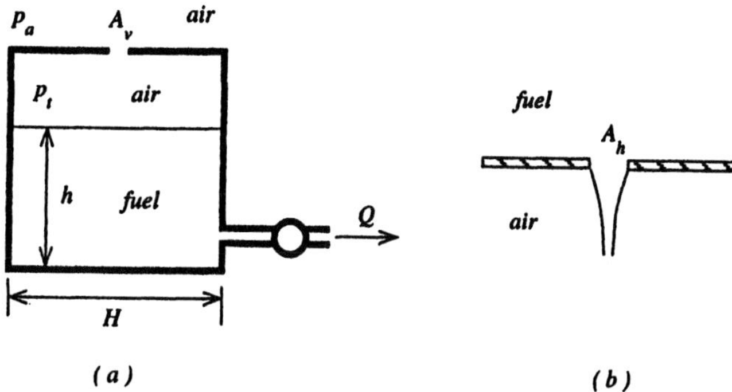

Figure P 4.21

Problem 4.21

A cubical fuel tank of side dimension $H = 1.0\,m$ contains fuel of density $\rho_f = 9.0E(2)\,kg/m^3$, partially filling the tank to a depth h as shown in figure P 4.21(a). The fuel is withdrawn from the tank and supplied to a burner by a pump having a volume flow rate of $Q = 4.0\,l/min$. As the fuel leaves the tank, air enters from the atmosphere through a vent hole of area $A_v = 10\,mm^2$ located in the top of the tank. Assuming that the flow of air into the tank is incompressible, (a) calculate the velocity V_v of the air passing through the vent hole. (b) Calculate the amount by which the air pressure p_t in the tank falls below the atmospheric pressure p_a, assuming that the air density $\rho_a = 1.2\,kg/m^3$.

At another time, when the fuel pump is not operating ($Q = 0$), a leak of area $A_h = 1.0\,mm^2$ develops in the bottom of the tank, as shown in figure P 4.21(b). (c) What is the ratio V_v/V_h of the air velocity through the vent to the fuel velocity through the leak? (d) Derive an expression for the velocity V_h of the leaking fuel as a function of the height h of the fuel in the tank. (e) Calculate the time required for a full tank of fuel to leak away completely.

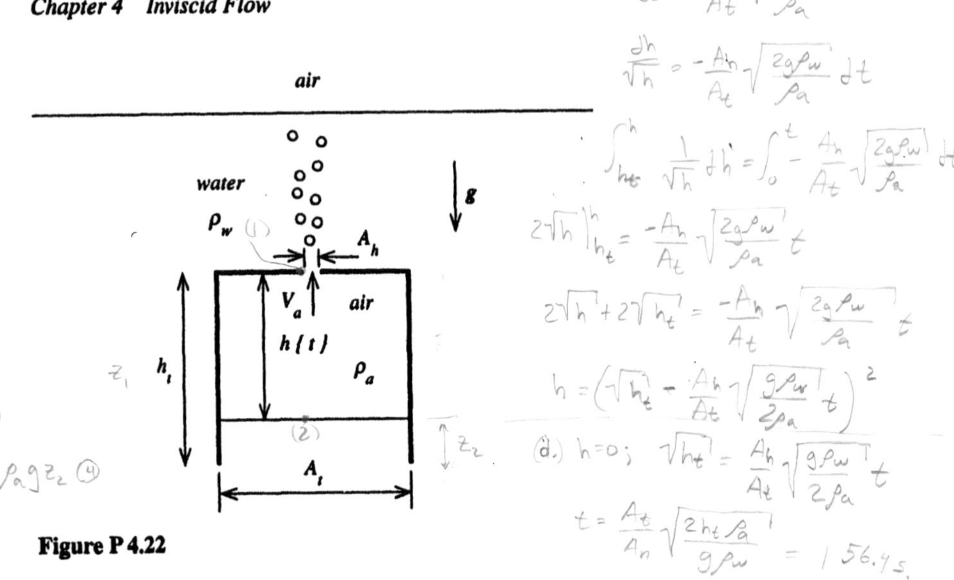

Figure P 4.22

Problem 4.22

A cylindrical tank, open at the bottom, is submerged below the surface of a pond having water of density $\rho_w = 1E(3)\,kg/m^3$, as shown in figure P 4.22. The top of the tank contains a small opening of area $A_h = 1\,cm^2$. The tank, of cross-sectional area $A_t = 1\,m^2$ and height $h_t = 1\,m$, is initially full of air having a density $\rho_a = 1.2\,kg/m^3$ that remains constant as the air leaks out through the opening at the top. The depth of air in the tank, $h\{t\}$, decreases with time from its initial value of h_t as the air leaks out through the hole at a time-varying speed $V_a\{t\}$.

Derive expressions for (a) the air velocity V_a and (b) the derivative dh/dt in terms of the parameters ρ_a, ρ_w, A_t, A_h, g and the variable h. (c) By integration, derive an expression for the depth $h\{t\}$ as a function of time t and the parameters ρ_a, ρ_w, A_t, A_h and g. (d) From your answer to (c), calculate the time for the tank to be emptied of air.

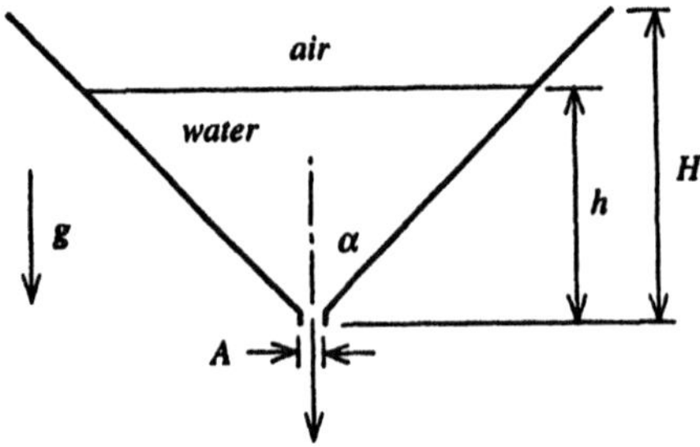

Figure P 4.23

Problem 4.23

As sketched in figure P 4.23, a conical funnel of height H and semi-vertex angle α has a small outlet hole of area $A \ll H^2$. At $t = 0$ the funnel is filled with water, which is then allowed to drain out through the hole. Assuming that, at any time, the flow is quasi-steady and inviscid, derive an expression for the time t required for the funnel to become empty.

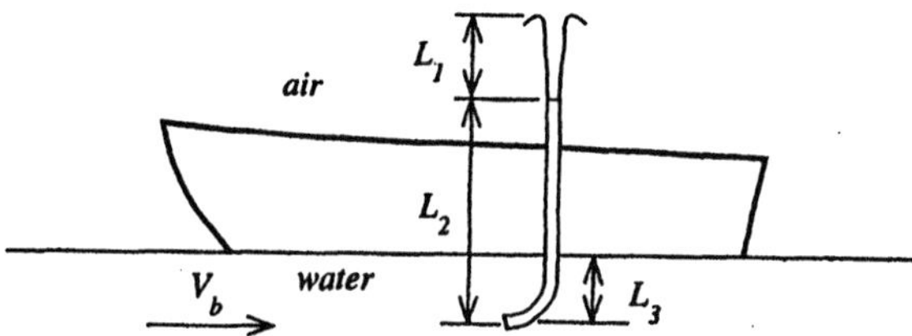

Figure P 4.24

Problem 4.24

An inventor has suggested an inexpensive, simple method for determining the speed at which a small motor boat is travelling. His invention is nothing more than a pipe with a 90° elbow at the lower end. When placed in the water, as shown in figure P 4.24, a fountain of water flows from the upper end of the pipe. To calculate the boat's speed, the maximum height L_1 that the water stream reaches above the end of the pipe is measured to be $1.0\,m$. In addition, the length of the pipe $L_2 = 1.5\,m$ and the depth of the opening below the surface $L_3 = 0.5\,m$ are known. Considering the flow with respect to the boat to be steady, calculate the value of the boat speed V_b.

Problem 4.25

A spherical gas bubble is formed in an inviscid, incompressible, unbounded liquid. The pressure p_∞ far away from the bubble is constant. If the initial pressure in the bubble is greater than p_∞, it is observed that the bubble undergoes radial oscillations, expanding until its pressure falls below p_∞ and then contracting back to a pressure above p_∞, and so on. We are interested in deriving some of the equations needed to describe this motion.

(a) Use the conservation of mass in the liquid to obtain an algebraic expression that relates the radial velocity V of the liquid at a distance r from the center of the bubble to the velocity V_b of the surface of the bubble at the distance r_b. (b) By considering the equation of motion for unsteady, inviscid flow along a radial streamline, derive an ordinary differential equation that relates the bubble radius r_b and its time derivatives, dr_b/dt and d^2r_b/dt^2, to the instantaneous bubble pressure p_b, the pressure p_∞ and the liquid density ρ.

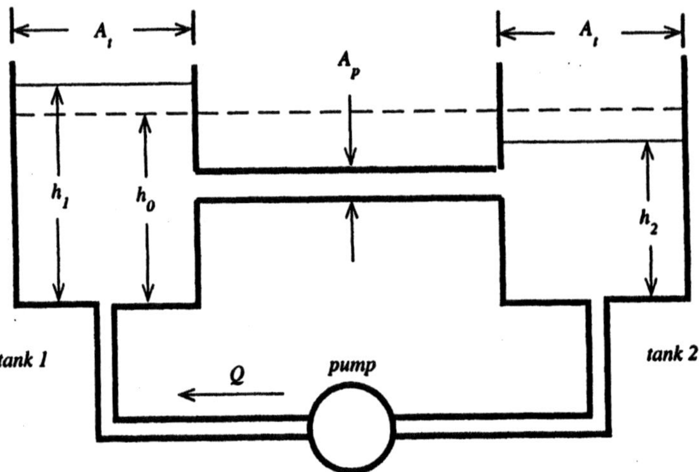

Figure P 4.26

Problem 4.26

Two identical tanks of cross-sectional area A_t are connected by both a pipe of cross-sectional area A_p ($\ll A_t$) and a pump and its connecting piping as shown in figure P 4.26. Initially, the tanks are filled with water to a height h_0 above the tank bottoms. When turned on, the pump provides a constant volume flow rate Q from tank 2 to tank 1, and the water level $h_1\{t\}$ in tank 1 rises and that in tank 2, $h\{t\}$, falls. After a long time, the two water levels stabilize at fixed values, $h_1\{\infty\}$ and $h_2\{\infty\}$.

Considering the flow through the connecting pipe to be inviscid, (a) derive an expression for $h_1\{\infty\}$ in terms of A_p, Q and h_0. (b) Derive an expression for the derivative $dh_1\{t\}$ in terms of the parameters A_p, A_t, Q and h_0. (c) By integrating this expression for the time interval t since the start of the pump, derive an expression for t as a function of A_p, A_t, Q, h_0 and h_t, using the integral relation:

$$\int_0^z \frac{z\,dz}{a - bz} = \frac{1}{b^2}\left(a \ln\left[\frac{a}{a - bz}\right] - bz\right)$$

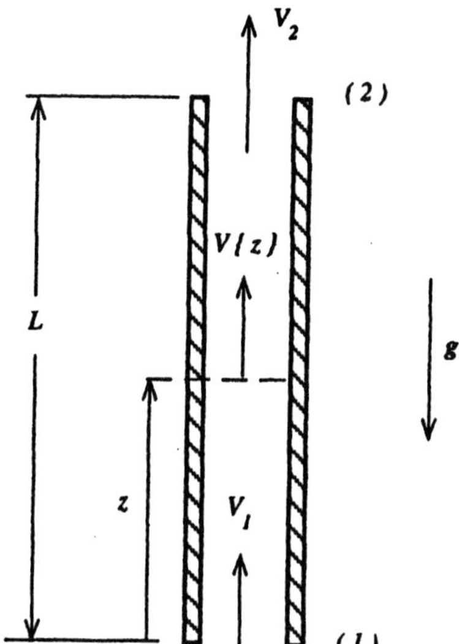

Figure P 4.27

Problem 4.27

A vertical pipe of length L in a boiler is heated from the outside by hot combustion gases, thereby boiling the water inside the pipe, which flows upward. Pure water enters the bottom (1) with a speed V_1 while a mixture of water and steam leaves the top (2) with a speed V_2. The mean density ρ of the mixture decreases from bottom to top as more water is converted from water to steam. The velocity of the steam/water mixture is observed to increase linearly with distance z from the bottom according to the relation:

$$V\{z\} = V_1 + (V_2 - V_1)\frac{z}{L}$$

The flow in the pipe may be regarded as frictionless. In terms of the known quantities V_1, V_2, L and the inlet density ρ_1, derive expressions for (a) the density $\rho\{x\}$ at any point in the column and (b) the pressure difference $p_1 - p_2$ across the column.

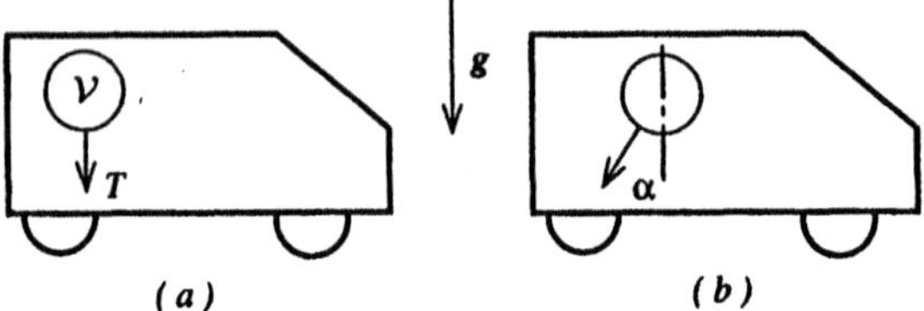

Figure P 4.28

Problem 4.28

In an automobile waiting at a stoplight (figure P 4.28(a)), a child holds a string attached to a balloon. The balloon, of volume $\mathcal{V} = 0.1\,m^3$, is filled with helium having a density $\rho_{He} = 0.14\,kg/m^3$. The air in the automobile has a density $\rho_a = 1.2\,kg/m^3$. (a) Neglecting the mass of the balloon material, calculate the tension T in the string.

When the stoplight changes to green, the automobile accelerates uniformly to 100 km/h in 10 s. The balloon is observed to swing forward, making an angle α with the vertical, as shown in figure P 4.28(b). (b) Calculate the angle α and the tension T in the string.

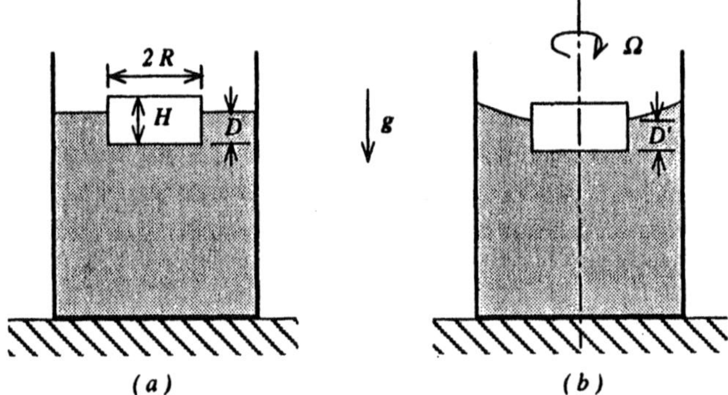

Figure P 4.29

Problem 4.29

A bar of soap, having the shape of a circular cylinder (radius R, height H and specific gravity SG), is allowed to float freely on the surface of water of density ρ in a partially filled beaker, as shown in figure P 4.29 (a). It is observed that the bottom of the bar of soap is at a depth D below the water surface. (a) Derive an expression for the depth D in terms of some or all of the known quantities R, H, SG and ρ.

The beaker is placed on a turntable rotating about a vertical axis at an angular speed

Ω. After a long time, the bar of soap and the water in the beaker rotate as a rigid body and the depth D' of the edge of the bar of soap below the water surface is observed to be different than that for case (a), as shown in figure P4.29 (b). (b) Derive an expression for D' in terms of some or all of the known quantities R, H, SG, Ω and ρ.

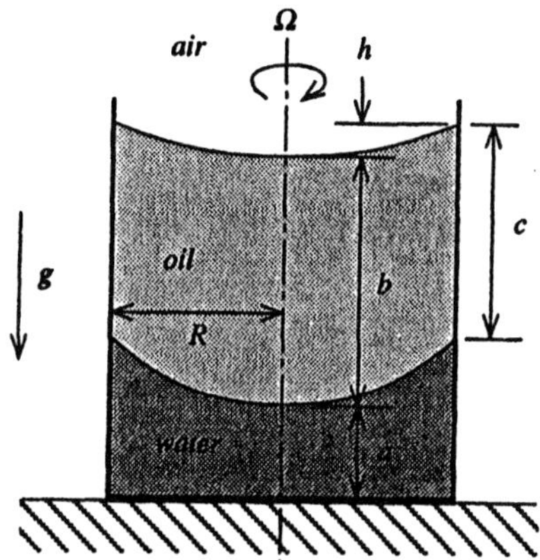

Figure P4.30

Problem 4.30

A cylindrical glass beaker is partially filled with two immiscible liquids, a layer of water below and a layer of oil on top. The water density ρ_w is greater than the oil density ρ_o. The top of the beaker is open to the atmosphere. The beaker and its contents are rotated about the vertical axis at a constant angular speed Ω and the fluids eventually rotate as rigid bodies. The free surface of the oil and the oil-water interface are observed to reach a curved shape as shown in figure P4.30, the respective layers of water and oil having the depths a and b.

(a) Derive an expression for the pressure $p\{z, r\}$ *in the oil* at a height z above the base of the beaker and a radius r from the centerline, using as parameters only the known quantities a, b, g, p_a, ρ_o, ρ_w and Ω. (b) Using the results of (a), derive an expression for the height h by which the oil-air interface at the radius R rises above the interface at the centerline. (c) Derive an expression for the pressure at the oil-water interface on the centerline. (d) As in part (a), derive an expression for the pressure $p\{z, r\}$ *in the water* by using your answer to part (c) and then derive an expression for the distance c shown in figure P4.30.

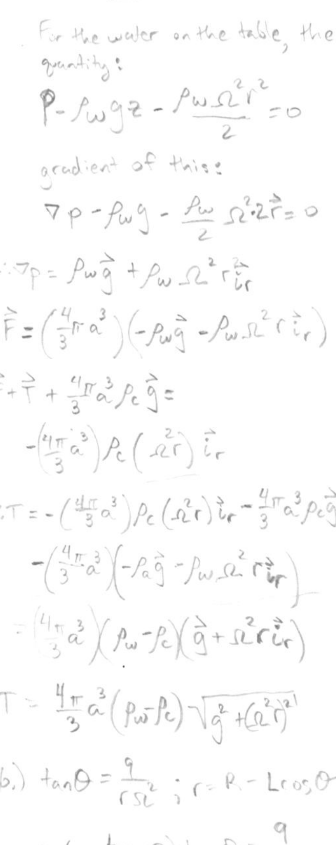

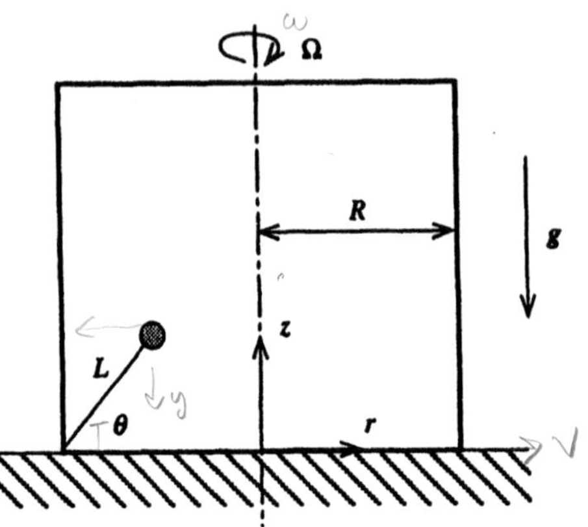

Figure P 4.31

Problem 4.31

A sealed cylindrical container of radius R, filled with water of density ρ_c and shown in cross-section in figure P 4.31, is rotated about its vertical axis at a fixed angular frequency Ω. A very small spherical cork of radius $a \ll R$ and density $\rho_c < \rho_w$ is attached by a thread of length $L < R$ to the outer bottom edge of the container. After a sufficient length of time, the container, water and cork rotate as a rigid body, maintaining the configuration of figure P 4.31.

(a) Derive expressions for the pressure force **F** on the cork and the tension **T** in the string, in terms of the parameters ρ_w, ρ_c, g, Ω and r. (b) Derive an expression from which the angle θ may be found.

Bibliography

Batchelor, G. K. 1967. *An Introduction to Fluid Mechanics*. Cambridge: Cambridge University Press.

Sabersky, Rolf H., Allan J. Acosta, and Edward G. Hauptman. 1989. *Fluid Flow, A First Course in Fluid Mechanics*. 3rd ed. New York: Macmillan Publishing Co.

5 Conservation of Momentum

5.1 Introduction

In treating the dynamics of solid bodies, we sometimes find it useful to employ the momentum of a body as the variable defining its state of motion. Newton's law of motion, which states that the time rate of change of momentum is equal to the net force applied to the body, may then be integrated on time so that the change in momentum is equal to the time integral of the force, or impulse. When two bodies collide, we know that the sum of the momenta is unchanged in the collision because the net impulse applied to the two bodies during the collision is zero. It seems likely that we can clarify our understanding of the motion of a fluid by considering the momentum of a fluid particle as the variable to be used in describing the fluid flow.

In this chapter, we derive and use integral expressions for the motion of a fluid that utilize the fluid momentum as the dependent variable. The momentum per unit volume of fluid is just the product of the density ρ (mass per unit volume) times the velocity $\mathbf{V}$. We do not utilize a distinct symbol for the momentum per unit volume but write the product $\rho\mathbf{V}$ whenever we need to specify the momentum per unit volume in a formula. Depending upon the application, we use either a linear momentum integral or an angular momentum integral to relate flow quantities of interest.

Before we develop a general form of the law of conservation of momentum as applied to a fluid flow, we will find it convenient to derive a general integral relation for any fluid property, called Reynolds' transport theorem, that relates the Langrangian and Eulerian forms of the time rate of change of that property for a finite amount of fluid.

Figure 5.1 A material volume $\mathcal{M}$, which contains the same fluid as it moves and deforms following the motion of a fluid, is shown at three successive times t_1, t_2 and t_3.

5.2 Reynolds' Transport Theorem

Consider a fluid scalar property b which is the amount of this property per unit mass of fluid. For example, b might be a thermodynamic property, such as the internal energy or enthalpy per unit mass, or the electric charge per unit mass of fluid. In general, we would be interested in fluid properties that arise in the expression of the laws of conservation of mass, momentum, energy, electric charge, etc. In this chapter, we consider the momentum of a fluid, so that b could be a scalar component of the momentum per unit mass of fluid, *i.e.* the velocity **V**. Thus we can think of b as being the property u, v or w.

The laws of physics are expressed as applying to a fixed mass of material. In a fluid flow, we identify a fixed mass of material by the *material volume* $\mathcal{M}$. Unlike our control volume $\mathcal{V}$, the material volume $\mathcal{M}$ moves with the fluid, always enclosing the same fluid particles. Thus its position, shape, volume and surface area change with time. A material volume of fluid is illustrated in figure 5.1, showing how it moves and changes shape at succeeding times t_1, t_2, t_3.

The total amount of the property b inside the material volume $\mathcal{M}$, designated by B, may be found by integrating the property per unit volume, ρb, over the material volume $\mathcal{M}$:

$$B \equiv \iiint_{\mathcal{M}} \rho b \, d\mathcal{V} \tag{5.1}$$

The laws of physics may be expressed in terms of the time rate of change of the property B. For example, the first law of thermodynamics would require that the time rate of increase of energy in the material volume $\mathcal{M}$ would equal the rate at which heat is added to the volume minus the rate at which the volume does work on its surroundings. We therefore need to find an expression for dB/dt in terms of fluid properties. Since the material volume moves with the fluid, its time rate of change may be found by integrating the product of the material derivative of b times the density ρ, over the material volume $\mathcal{M}$:

$$\frac{dB}{dt} = \iiint_{\mathcal{M}} \rho \frac{Db}{Dt}\, d\mathcal{V} \tag{5.2}$$

Equation 5.2 is not in a convenient form for evaluating the volume integral because it is not expressed in Eulerian variables. However, the integrand may be rewritten by expanding the material derivative:

$$\begin{aligned} \rho \frac{Db}{Dt} &= \rho \frac{\partial b}{\partial t} + \rho(\mathbf{V} \cdot \nabla) b \\ &= \rho \frac{\partial b}{\partial t} + \nabla \cdot (\rho \mathbf{V} b) - b \nabla \cdot (\rho \mathbf{V}) \\ &= \rho \frac{\partial b}{\partial t} + \nabla \cdot (\rho \mathbf{V} b) + b \frac{\partial \rho}{\partial t} \\ &= \frac{\partial}{\partial t}(\rho b) + \nabla \cdot (\rho b \mathbf{V}) \end{aligned} \tag{5.3}$$

where we have used equation 1.38 in line two and the mass conservation equation 3.13 in line three.

The right side of equation 5.3 is expressed in Eulerian coordinates. We may replace the integration of $\rho\, Db/Dt$ over the material volume $\mathcal{M}$ by an integration of the right side of 5.3 over a fixed control volume $\mathcal{V}$ that coincides with the material volume at any instant of time:

$$\begin{aligned} \iiint_{\mathcal{M}} \rho \frac{Db}{Dt}\, d\mathcal{V} &= \iiint_{\mathcal{V}} \left(\frac{\partial}{\partial t}(\rho b) + \nabla \cdot (\rho b \mathbf{V}) \right) d\mathcal{V} \\ &= \iiint_{\mathcal{V}} \frac{\partial}{\partial t}(\rho b)\, d\mathcal{V} + \iint_{S} \rho b \mathbf{V} \cdot \mathbf{n}\, dS \end{aligned} \tag{5.4}$$

where we have used the divergence theorem 1.44 to convert a volume integral to a surface integral. Combining equation 5.2 with 5.4, we find *Reynolds' transport theorem:*[1]

$$\begin{aligned} \frac{dB}{dt} &= \iiint_{\mathcal{V}} \frac{\partial}{\partial t}(\rho b)\, d\mathcal{V} + \iint_{S} \rho b \mathbf{V} \cdot \mathbf{n}\, dS \\ &= \frac{d}{dt} \iiint_{\mathcal{V}} \rho b\, d\mathcal{V} + \iint_{S} \rho b \mathbf{V} \cdot \mathbf{n}\, dS \end{aligned} \tag{5.5}$$

---

[1]Osborne Reynolds (1842-1912) elucidated the nature of turbulent motion of fluids. His other contributions to fluid mechanics include the theory of lubrication and vortex production. He held many patents on pumps and turbines.

where we may take the time derivative outside the volume integral since the control volume is fixed in space.

Reynolds' transport theorem, equation 5.5, contains two terms that contribute to the rate of change of B within the material volume. The first is the time rate of change of B within the fixed control volume $\mathcal{V}$ that occupies the same location as the material volume, and the second is the net outflow of B across the surface $\mathcal{S}$ of the fixed control volume $\mathcal{V}$.[2] This second quantity is usually called the *flux* of B.

Reynolds' transport theorem expresses the time derivative of a fluid property within a material volume (a Lagrangian derivative) in terms of the Eulerian description of the changes within a fixed control volume. It is a mathematical relationship, not a law of physics. It can be seen that it is simply a formula for differentiating a volume integral whose limits of integration vary with time. For example, suppose we have a flow in which $\mathbf{V} = u\{x, t\}\mathbf{i}_x$, and we choose a material volume of fluid that lies between $x_1\{t\}$ and $x_2\{t\}$ and has a constant cross-sectional area A so that $d\mathcal{V} = A\,dx$. The property B would be:

$$B = \iiint_{\mathcal{M}} \rho b\, d\mathcal{V} = A \int_{x_1\{t\}}^{x_2\{t\}} \rho b\, dx$$

Differentiating with respect to time, we have:

$$\begin{aligned}\frac{dB}{dt} &= A \int_{x_1\{t\}}^{x_2\{t\}} \frac{\partial}{\partial t}(\rho b)\, dx + A(\rho b)_2 \frac{dx_2}{dt} - A(\rho b)_1 \frac{dx_1}{dt} \\ &= A \int_{x_1\{t\}}^{x_2\{t\}} \frac{\partial}{\partial t}(\rho b)\, dx + A(\rho b u)_2 - A(\rho b u)_1\end{aligned}$$

which is the equivalent of the transport theorem for this special case where $(\mathbf{V}\cdot\mathbf{n})_2 = u_2$ and $(\mathbf{V}\cdot\mathbf{n})_1 = -u_1$.

Example 5.1

Assuming that $b = 1$, show that Reynolds' transport theorem reduces to the mass conservation equation 3.9.

---

[2]Note that $\mathbf{V}\cdot\mathbf{n}$ is the volumetric outflow rate per unit surface area, and ρb is the amount of B per unit volume so that $\rho b \mathbf{V}\cdot\mathbf{n}$ is the rate of outflow of B per unit surface area.

Solution
If $b = 1$, equation 5.1 defines B as the mass of material in the material volume:

$$B = \iiint_{\mathcal{M}} \rho \, d\mathcal{V}$$

By definition the material volume always encloses the same amount of material so that $dB/dt = 0$. Substituting in Reynolds' transport theorem 5.5,

$$0 = \frac{d}{dt} \iiint_{\mathcal{V}} \rho \, d\mathcal{V} + \iint_{S} \rho \mathbf{V} \cdot \mathbf{n} \, dS$$

5.3 Linear Momentum

The momentum of the fluid in a material volume, denoted by the symbol $\mathbf{M}$, is the integral of the momentum per unit volume $\rho\mathbf{V}$ over the material volume:

$$\mathbf{M} \equiv \iiint_{\mathcal{M}} \rho \mathbf{V} \, d\mathcal{V} \qquad (5.6)$$

$\mathbf{M}$ is also called the *linear momentum* to distinguish it from angular momentum, which we will treat later in section 5.5.

Let us now apply Reynolds' transport theorem to determine the form of $d\mathbf{M}/dt$. Taking $b = \mathbf{V}$ so that dB/dt becomes $d\mathbf{M}/dt$, equation 5.5 becomes:[3]

$$\frac{d\mathbf{M}}{dt} = \frac{d}{dt} \iiint_{\mathcal{V}} \rho \mathbf{V} \, d\mathcal{V} + \iint_{S} \rho \mathbf{V} (\mathbf{V} \cdot \mathbf{n}) \, dS \qquad (5.7)$$

The first term on the right side of equation 5.7 is the time rate of increase of momentum within the control volume $\mathcal{V}$, while the second is the net outflow of momentum from the control volume.

---

[3]In deriving Reynolds' transport theorem, we assumed that b is a scalar property. Because the theorem applies to each of the three scalar components of a vector property $\mathbf{b}$, by adding the transport theorem for each of these components vectorially, we find that the theorem applies as well for the vector property $\mathbf{b}$.

Figure E 5.2

Example 5.2

A tube of constant cross-sectional area A is filled with a liquid of density ρ. The tube is held in a vertical position, and the upper and lower end are opened to the atmosphere at time $t = 0$, allowing the fluid to drain out of the tube (see figure E 5.2). As the fluid drains, the length $L\{t\}$ of the column of liquid within the tube is measured as a function of time. Derive an expression for the time derivative of the momentum of the fluid inside the tube as a function of ρ, A and $L\{t\}$.

Solution

Select a control volume enclosing the entire interior of the tube, as shown in figure E5.2. Denoting the downward velocity of the fluid by V, apply the conservation of mass to the liquid inside the control volume:

$$\frac{d}{dt}\iiint_V \rho\, dV + \iint_S \rho \mathbf{V} \cdot \mathbf{n}\, dS = 0$$

$$\frac{d}{dt}(\rho AL) + \rho VA = 0$$

$$\frac{dL}{dt} + V = 0$$

$$V = -\frac{dL}{dt}$$

Next apply equation 5.7 to determine the time derivative of the downward momentum, dM/dt:

$$\begin{aligned}\frac{dM}{dt} &= \frac{d}{dt}\iiint_{\mathcal{V}} \rho V \, d\mathcal{V} + \iint_{S} \rho V(\mathbf{V}\cdot\mathbf{n})\, dS \\ &= \frac{d}{dt}(\rho VAL) + \rho V(VA) \\ &= \rho A\left(\frac{d}{dt}(VL) + V^2\right) \\ &= \rho A\left(V\frac{dL}{dt} + L\frac{dV}{dt} + V^2\right)\end{aligned}$$

Replacing V by $-dL/dt$ and simplifying:

$$\begin{aligned}\frac{dM}{dt} &= \rho A\left(-\left(\frac{dL}{dt}\right)^2 + L\left(-\frac{d^2L}{dt^2}\right) + \left(-\frac{dL}{dt}\right)^2\right) \\ &= \rho AL\left(-\frac{d^2L}{dt^2}\right)\end{aligned}$$

5.3.1 Newton's Law of Motion

To apply Newton's law of motion to the material volume that coincides with our control volume at a particular time, we have to determine the forces that act on the fluid within the control volume. First of all, there will be the force caused by the pressure p acting inward on the control surface S:

$$\textbf{pressure force} = \iint_S (-p\mathbf{n})\, dS$$

Secondly, a viscous stress $\boldsymbol{\tau}$ acting on the control surface will give rise to a viscous force:

$$\textbf{viscous force} = \iint_S \boldsymbol{\tau}\, dS$$

Finally, a gravitational force acts on all the fluid inside the control volume:

$$\textbf{gravitational force} = \iiint_{\mathcal{V}} \rho\mathbf{g}\, d\mathcal{V}$$

Newton's law of motion requires that the time rate of change of momentum of the fluid in the material control volume, $d\mathbf{M}/dt$, be equal to the sum of all the forces acting on the fluid in the control volume:[4]

$$\frac{d\mathbf{M}}{dt} = \iint_S (-p\mathbf{n})\, dS + \iint_S \boldsymbol{\tau}\, dS + \iiint_{\mathcal{V}} \rho\mathbf{g}\, d\mathcal{V} \tag{5.8}$$

5.3.2 The Linear Momentum Theorem

Now we may combine Newton's law, equation 5.8, with Reynold's transport theorem for momentum, equation 5.7, to obtain the *linear momentum theorem*:

$$\frac{d}{dt}\iiint_{\mathcal{V}} \rho\mathbf{V}\, d\mathcal{V} + \iint_S \rho\mathbf{V}(\mathbf{V}\cdot\mathbf{n})\, dS$$

$$= \iint_S (-p\mathbf{n})\, dS + \iint_S \boldsymbol{\tau}\, dS + \iiint_{\mathcal{V}} \rho\mathbf{g}\, d\mathcal{V} \tag{5.9}$$

In words, the linear momentum theorem requires that the sum of the time rate of change of the momentum in the control volume plus the net outflow of momentum from the control volume must equal the sum of the surface forces and body forces acting on the fluid in the control volume.

Note that the momentum theorem is a vector equation, which is the equivalent of three scalar equations. In some applications, we may be interested in only one component of the motion and may not need to use all three components of equation 5.9.

Example 5.3

For the flow described in example 5.2, derive an expression for the viscous force $\iint \boldsymbol{\tau}\, dS$ exerted on the fluid inside the tube.

Solution

Having evaluated dM/dt in example 5.2, we may use equation 5.8 to solve for the (downward) viscous force. The pressure force is zero because the pressure equals atmospheric pressure on the upper and lower surfaces of the control surface $\mathcal{S}$. The gravity force, ρgAL, acts downward. Thus equation 5.8 becomes:

$$-\rho AL\frac{d^2L}{dt^2} = 0 + \iint_S \tau\, dS + \rho gAL$$

---

[4]If surface tension, electric or magnetic forces act on the fluid in the control volume, they would have to be included as well.

$$\iint_S \tau \, dS = -\rho AL \left(g + \frac{d^2L}{dt^2} \right)$$

(For an inviscid flow, $\tau = 0$, and thus $d^2L/dt^2 = -dV/dt = -g$, *i.e.*, the fluid is in a state of free fall downward with an acceleration of g. For a very viscous fluid, the downward velocity would be observed to be constant so that $dV/dt = -d^2L/dt^2 = 0$, and the (downward) viscous force would then equal $-\rho gAL$, *i.e.*, the viscous force is equal in magnitude, but opposite in direction to, the gravity force ρgAL acting on the fluid in the tube.)

In many applications, there will be identifiable outflow and inflow streams. In chapter 3, we denoted the mass outflow rate as $\dot{m}_{out}$. We shall therefore denote the momentum flux of the outflow stream by the combination $(\dot{m}\mathbf{V})_{out}$ and that of the inflow stream by $(\dot{m}\mathbf{V})_{in}$. Using this notation, the linear momentum theorem equation 5.9 takes the form:[5]

$$\frac{d}{dt} \iiint_{\mathcal{V}} \rho \mathbf{V} \, d\mathcal{V} + (\dot{m}\mathbf{V})_{out} - (\dot{m}\mathbf{V})_{in}$$
$$= \iint_S (-p\mathbf{n}) \, dS + \iint_S \boldsymbol{\tau} \, dS + \iiint_{\mathcal{V}} \rho \mathbf{g} \, d\mathcal{V} \qquad (5.10)$$

External Forces

In deriving the linear momentum theorem, equation 5.9, we assumed that the control volume contained only the fluid of interest. In many applications, however, it may simplify the analysis to select a fixed control volume that encloses solid objects, such as the walls of a container or even a moving piston. Any solid object in contact with a fluid within the control volume will exert a force on that fluid which equals the integral of the fluid stress $(-p)\mathbf{n} + \boldsymbol{\tau}$ over the surface of the fluid in contact with the solid object. This force on the fluid, which we call an external force $\mathbf{F}_{ex}$, must be sustained by some agent outside the control volume, such as a table top that supports a glass filled with water or the hand that moves a teaspoon mixing cream into a cup of coffee. To account for such forces, we may add to the right-hand side of the linear

---

[5]Note that the direction of the vector $\dot{m}\mathbf{V}$ is the same as that of the velocity vector $\mathbf{V}$ but that the vector difference between the outflow and inflow quantities enters into the momentum theorem, equation 5.10.

momentum theorem, equation 5.10, a term $\Sigma \mathbf{F}_{ex}$ which is the vector sum of all such forces:[6]

$$\frac{d}{dt}\iiint_{\mathcal{V}} \rho \mathbf{V}\, d\mathcal{V} + (\dot{m}\mathbf{V})_{out} - (\dot{m}\mathbf{V})_{in}$$
$$= \iint_{S} (-p\mathbf{n})\, dS + \iint_{S} \boldsymbol{\tau}\, dS + \iiint_{\mathcal{V}} \rho \mathbf{g}\, d\mathcal{V} + \Sigma \mathbf{F}_{ex} \qquad (5.11)$$

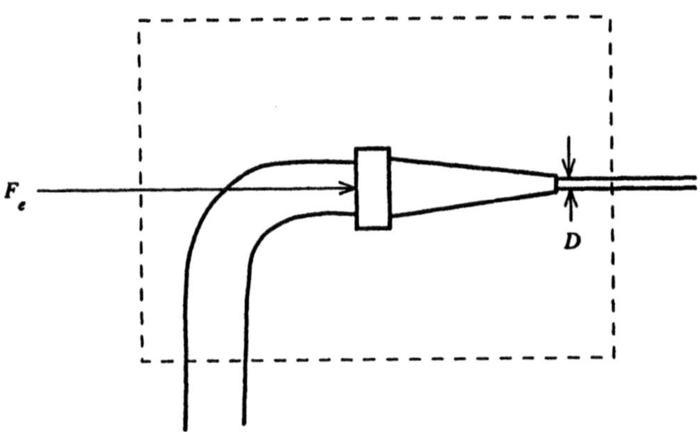

Figure E 5.4

Example 5.4

A fireman is directing a stream of water from a fire hose nozzle. The water stream volume flow rate Q is 150 gal/min, and the nozzle exit diameter $D = 1\ in$. Calculate the force F_e that the fireman exerts on the hose nozzle to hold it in place.

Solution

Choose a control volume that surrounds the hose nozzle, as sketched in figure E 5.4. Now apply the horizontal component of the linear momentum theorem, equation 5.11, noting that the unsteady term is zero because the flow is steady, that the pressure force term is zero because the pressure is constant (atmospheric pressure) on the vertical control surface, that the viscous force term is zero because the viscous stress is zero

[6]In the rare case where the solid object undergoes a substantial acceleration, the force $\mathbf{F}_{ex}$ acting on the fluid will differ from that exerted by the external agent by the product of the mass of the solid object times its acceleration.

on the control surface and that the inflow momentum has no horizontal component, leaving:

$$\dot{m}V_{out} = F_e$$

$$F_e = \rho Q V_{out}$$

Evaluating these terms:

$$Q = \frac{150\,gal}{min} \times \frac{1\,min}{60\,sec} \times \frac{3.785E(-3)\,m^3}{1\,gal} = 9.463E(-3)\,m^3/s$$

$$V_{out} = \frac{4Q}{\pi D^2} = \frac{4 \times 9.463E(-3)}{\pi(2.54E(-2)\,m)^2} = 18.68\,m/s$$

$$F_e = 1E(3)kg/m^3 \times 9.463E(-3)\,m^3 \times 18.68\,m/s = 176.8\,N$$

$$= 176.8\,N \times \frac{1\,lbf}{4.448\,N} = 39.74\,lbf$$

Note the considerable force required to direct this moderate flow of water.

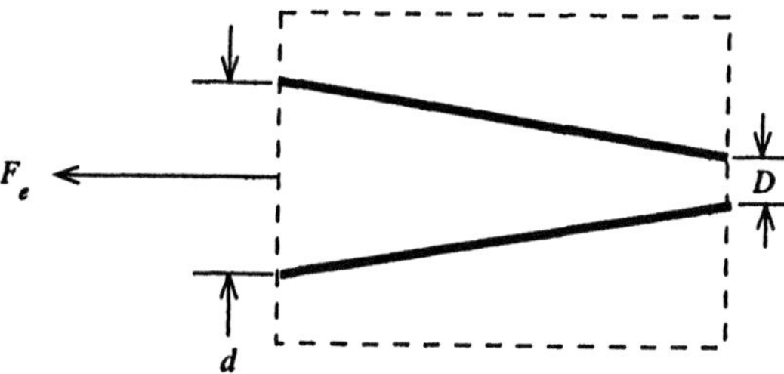

Figure E 5.5

Example 5.5

The fire hose nozzle of example 5.4 tapers from a diameter $D = 1\,in$ at the open end to a diameter $d = 2\,in$ at the end attached to the hose, as shown in figure E 5.5. Calculate the force F_c exerted on the nozzle by the coupling that attaches it to the fire hose. (Assume that the flow in the nozzle is inviscid.)

Solution

Choosing the control volume shown in figure E 5.5, the horizontal component of the linear momentum integral equation 5.11 becomes:

$$(\dot{m}V)_{out} - (\dot{m}V)_{in} = (p_{in} - p_{out})A_{in} - F_c$$

$$F_c = (p_{in} - p_{out})A_{in} - \dot{m}(V_{out} - V_{in})$$

In evaluating the pressure force, only that portion of the control surface (A_{in}) where the pressure exceeds atmospheric pressure by the amount $p_{in} - p_{atm} = p_{in} - p_{out}$ contributes to the net pressure force.

To determine the pressure difference $p_{in} - p_{out}$, we need to use Bernoulli's equation for steady flow:

$$\frac{p_{in}}{\rho} + \frac{V_{in}^2}{2} = \frac{p_{out}}{\rho} + \frac{V_{out}^2}{2}$$

$$p_{in} - p_{out} = \frac{\rho}{2}(V_{out}^2 - V_{in}^2)$$

so that the momentum equation becomes:

$$F_c = \frac{\rho}{2}(V_{out}^2 - V_{in}^2)A_{in} - \dot{m}(V_{out} - V_{in})$$

But by mass conservation,

$$\rho V_{in}A_{in} = \rho V_{out}A_{out}$$

$$V_{in} = V_{out}\frac{A_{out}}{A_{in}}$$

Replacing V_{in} by this value in the momentum equation,

$$F_c = \frac{\rho}{2}V_{out}^2\left(1 - \frac{A_{out}^2}{A_{in}^2}\right)A_{in} - \rho A_{out}V_{out}^2\left(1 - \frac{A_{out}}{A_{in}}\right)$$

$$= \rho Q V_{out}\left[\frac{1}{2}\left(\frac{A_{in}}{A_{out}} + \frac{A_{out}}{A_{in}}\right) - 1\right]$$

Since $\rho Q V_{out} = 176.8\,N$ from example 5.4 and $A_{in} = (d/D)^2 A_{out} = 4A_{out}$,

$$F_c = 176.8\,N\left[\frac{1}{2}\left(4 + \frac{1}{4}\right) - 1\right] = 176.8\,N \times \frac{9}{8} = 198.9\,N$$

5.4 Applications of the Linear Momentum Theorem

The linear momentum theorem applies to all flows, whether viscous or inviscid, compressible or incompressible, subsonic or supersonic. Because it is an integral theorem, we must know something about the flow streams entering and leaving the control volume in order to apply the theorem to a particular flow. Nevertheless, it is a powerful tool for analyzing many important flows. In this section, we consider several useful examples.

5.4.1 Propulsion

Rockets, aircraft and ships are propelled by devices that increase the momentum of fluid passing through the propulsion system. In the case of rockets, the propelling fluid is the rocket's fuel, while for aircraft and ships the momentum of the ambient fluid is increased by a jet engine or propeller. The propulsive force exerted by such an engine is used to accelerate the vehicle and/or overcome the resistance to motion through the ambient fluid.

Rocket Motors

A rocket motor is supplied with a combustible mixture of fuel and oxidant, which upon burning, reaches a high pressure and temperature. The combustion products flow out of the rocket motor through a nozzle, accelerating to a high velocity as the fluid pressure falls to the much lower ambient value. A propulsive force is exerted on the rocket motor as a consequence of the pressure force acting on the internal surface of the motor, including the nozzle. By the linear momentum theorem, this force equals the increase of momentum of the combustion products.

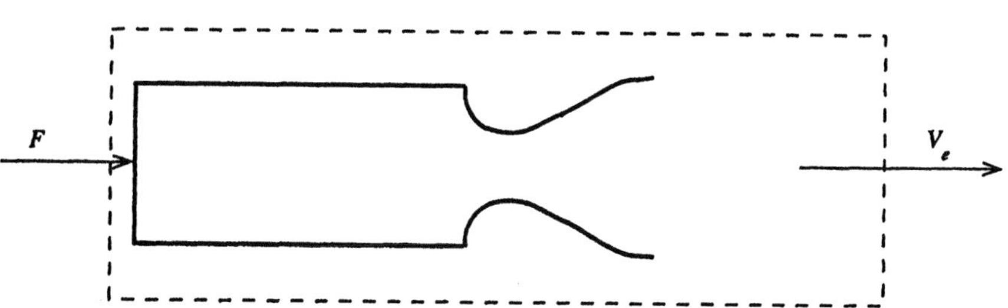

Figure E 5.6

Example 5.6

A solid fuel rocket engine is fired horizontally on a test stand, as shown in figure E 5.6. The fuel burns at a mass rate $\dot{m} = 2\,kg/s$, and the rocket exhaust velocity V_e is $200\,m/s$. Calculate the restraining force F required to hold the rocket in place.

Solution

Choose a control volume surrounding the rocket and intercepting the exhaust plume far enough downstream that the plume pressure equals atmospheric pressure (see figure E 5.6). Apply the horizontal component of the linear momentum theorem, equation 5.11,

$$\frac{d}{dt}\iiint_{\mathcal{V}} \rho \mathbf{V}\, d\mathcal{V} + (\dot{m}\mathbf{V})_{out} - (\dot{m}\mathbf{V})_{in}$$

$$= \iint_{S} (-p\mathbf{n})\, dS + \iint_{S} \tau\, dS + \iiint_{\mathcal{V}} \rho \mathbf{g}\, d\mathcal{V} + \Sigma \mathbf{F}_{ex}$$

$$0 + \dot{m}V_e - 0 = 0 + 0 + 0 + F$$

Note that the unsteady term, the inflow momentum and the gravity force are zero, as well as the pressure force and viscous force because the pressure is constant and because the viscous stress is zero on the control surface. Evaluating the force F,

$$F = \dot{m}V_e$$

$$= 2\,kg/s \times 200\,m/s = 400\,N$$

The flow of combustion gases from a rocket engine is a compressible flow. The maximum velocity of the fluid in the exhaust plume, V_e, is determined by the thermodynamic properties of the propellent gas. The greater this value, the the higher the speed that the rocket will attain when its fuel has been consumed.

Jet Engines

The familiar aircraft jet engine ingests, through its intake, ambient air, which is subsequently compressed, mixed with fuel, burned and expanded through the turbine stage and exhaust nozzle. The exiting gas has a speed higher than flight speed so that the momentum of the fluid passing through the engine is increased, giving rise to the thrust that the engine exerts on the airframe. In the ducted fan jet engine, some of the inflow air passes through a compressor stage and then is accelerated in a nozzle without passing through the combustion chamber. In either case, the function of the jet engine is to increase the momentum of the fluid entering the engine so as to produce thrust on the airframe.

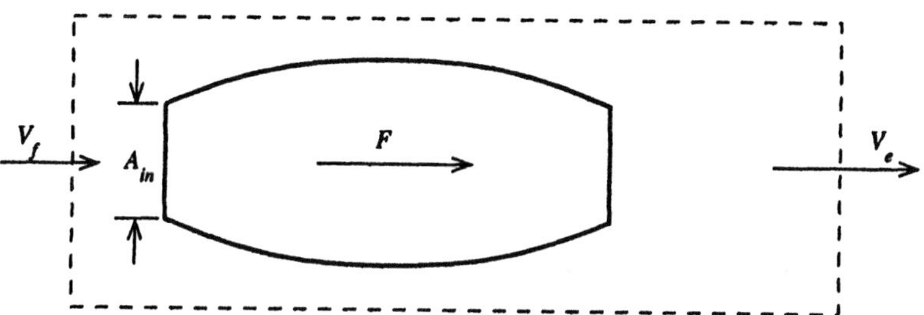

Figure E 5.7

Example 5.7

A jet engine, sketched in figure E 5.7, attached to an airplane in steady flight at a speed $V_f = 250\, m/s$, ingests air at a density $\rho_a = 0.4\, kg/m^3$ through an inlet area $A_{in} = 1.0\, m^2$. Fuel is fed to the engine at a mass flow rate $\dot{m}_f = 2\, kg/s$. The exhaust jet has a speed $V_e = 500\, m/s$ when it reaches ambient atmospheric pressure. Calculate the value of the force F exerted on the engine by the airframe.

Solution

Apply the linear momentum theorem, equation 5.11, in the horizontal flow direction while noting that the unsteady term and the pressure, viscous and gravity forces are zero:

$$\frac{d}{dt}\iiint_{\mathcal{V}} \rho \mathbf{V}\, d\mathcal{V} + (\dot{m}\mathbf{V})_{out} - (\dot{m}\mathbf{V})_{in}$$

$$= \iint_S (-p\mathbf{n})\, dS + \iint_S \boldsymbol{\tau}\, dS + \iiint_{\mathcal{V}} \rho \mathbf{g}\, d\mathcal{V} + \Sigma \mathbf{F}_{ex}$$

$$0 + (\dot{m}V)_{out} - (\dot{m}V)_{in} = 0 + 0 + 0 + F$$

The net momentum flux $(\dot{m}V)_{out} - (\dot{m}V)_{in}$ includes only the flow passing through the jet engine. Any flow through the control volume that bypasses the engine has equal inflow and outflow velocity and no net momentum flux. Next calculate $\dot{m}_{in}$:

$$\dot{m}_{in} = \rho_a V_f A_{in} = 0.4\, kg/m^3 \times 250\, m/s \times 1.0\, m^2 = 100\, kg/s$$

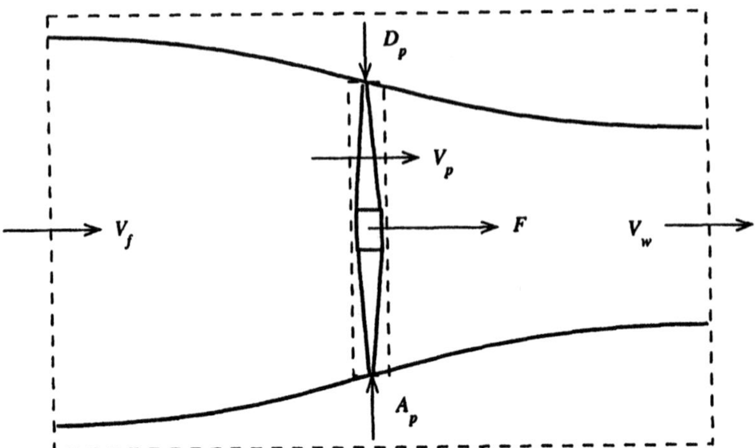

Figure 5.2 In the flow through an airplane propeller, the streamlines that pass by the propeller tips enclose the fluid that passes through the propeller. The approaching fluid, having the flight speed V_f, is accelerated by the propeller to a speed V_p at the propeller and to an even greater speed V_w in the propeller wake far downstream, resulting in a thrust force F.

Noting that $\dot{m}_{out} = \dot{m}_{in} + \dot{m}_f = 102\, kg/s$ by mass conservation, F is found to be:

$$F = (\dot{m}V)_{out} - (\dot{m}V)_{in} = 102\, kg/s \times 500\, m/s - 100\, kg/s \times 250\, m/s = 2.6E(4)\, N$$

Propellers

Although invented by Leonardo daVinci in the fifteenth century, the screw propeller was not utilized as a practical method of ship and aircraft propulsion until the nineteenth and twentieth centuries, respectively, when engines of sufficient power became available. For both ships and aircraft, the principle of propeller propulsion is the same. Rapidly rotating blades develop a lift force having a component in the direction of the axis of rotation that sums to a thrust force in the axial direction. The rotating blades cause the fluid to accelerate in the axial direction, so that the fluid that has passed through the propeller moves at a higher speed than that at which it approached the vessel or airplane. This increase in momentum of the propelled fluid equals the propeller thrust.

The flow through a propeller is sketched in figure 5.2. Upstream of the propeller, the fluid is accelerated from the flight speed V_f to a higher speed V_p as it passes through the propeller, an acceleration caused by a pressure p_{in} in front of the propeller that is lower than atmospheric pressure. The fluid acted upon by the propeller experiences a pressure rise to a value p_{out} that is higher than atmospheric pressure. In the wake region behind the propeller, the fluid accelerates to an even higher speed V_w as the pressure falls to atmospheric. The propeller acts like a pump to increase the pressure of the fluid flowing through it by an amount $p_{out} - p_{in}$, thereby increasing the flow speed from V_f to V_w. As the flow accelerates, the streamlines converge to preserve the constant mass flow rate of fluid passing through the propeller.

To analyze this propeller flow, first select a control volume surrounding the propeller alone, as shown by the inner dashed line in figure 5.2. Assuming incompressible flow, mass conservation requires that the inflow and outflow velocities be equal:

$$(\rho V A_p)_{in} = (\rho V A_p)_{out} = \rho V_p A_P = \dot{m} \tag{5.12}$$

since both the inflow and outflow areas equal the propeller flow area $A_p = \pi D_p^2/4$. Applying the linear momentum equation 5.11 to this control volume, we can find the thrust force $\underset{\cdot}{F}$:

$$\frac{d}{dt}\iiint_{\mathcal{V}} \rho \mathbf{V}\, d\mathcal{V} + (\dot{m}\mathbf{V})_{out} - (\dot{m}\mathbf{V})_{in}$$

$$= \iint_S (-p\mathbf{n})\, dS + \iint_S \boldsymbol{\tau}\, dS + \iiint_{\mathcal{V}} \rho \mathbf{g}\, d\mathcal{V} + \Sigma \mathbf{F}_{ex}$$

$$0 + \dot{m}V_p - \dot{m}V_p = p_{in}A_p - p_{out}A_p + 0 + 0 + F$$

$$F = (p_{out} - p_{in})A_p \tag{5.13}$$

We may also express the thrust force F in terms of the flight and wake speeds, the equivalent of the inflow and outflow speeds of the jet engine. To do so, assume that the flow upstream and downstream of the propeller is a steady inviscid flow, and apply Bernoulli's equation to each region:[7]

$$\frac{p_a}{\rho} + \frac{V_f^2}{2} = \frac{p_{in}}{\rho} + \frac{V_p^2}{2}$$

$$\frac{p_{out}}{\rho} + \frac{V_p^2}{2} = \frac{p_a}{\rho} + \frac{V_w^2}{2}$$

where the fluid pressure far upstream and downstream is the atmospheric pressure p_a.

[7] In writing the second of these equations, we assume that the flow behind the propeller is purely axial, having no significant tangential component. The rotation of the propeller will induce some tangential component downstream, but this will be small if the propeller speed is high enough compared to V_f, as discussed in section 5.6.3 below.

Adding these equations we find the pressure increase across the propeller:

$$\frac{p_{out} - p_{in}}{\rho} = \frac{V_w^2}{2} - \frac{V_f^2}{2} \tag{5.14}$$

Substituting this in equation 5.13, the thrust force F may also be expressed as:

$$F = \rho A_p \left(\frac{V_w^2}{2} - \frac{V_f^2}{2} \right) \tag{5.15}$$

Thus the propeller thrust force increases with an increase in wake speed V_w in excess of the flight speed V_f, although not simply in proportion to that increase.

To find the flow speed V_p through the propeller, we apply the linear momentum equation 5.11 to the outer control volume shown in figure 5.2:

$$\frac{d}{dt} \iiint_{\mathcal{V}} \rho \mathbf{V} \, d\mathcal{V} + (\dot{m}\mathbf{V})_{out} - (\dot{m}\mathbf{V})_{in}$$

$$= \iint_{S} (-p\mathbf{n}) \, dS + \iint_{S} \boldsymbol{\tau} \, dS + \iiint_{\mathcal{V}} \rho \mathbf{g} \, d\mathcal{V} + \Sigma \mathbf{F}_{ex}$$

$$0 + \dot{m} V_w - \dot{m} V_f = 0 + 0 + 0 + F$$

$$F = \dot{m}(V_w - V_f)$$

where only the fluid inside the streamlines passing through the propeller tip contributes to the net change in momentum flux, $(\dot{m}V)_{in} - (\dot{m}V)_{out}$. Substituting this in equation 5.15 and using equation 5.12, we find:

$$V_p = \frac{V_w + V_f}{2} \tag{5.16}$$

The propeller flow speed V_p is the arithmetic mean of the flight speed V_f and the wake speed V_w. As the propeller develops more thrust, both the wake speed increment and the mass flow through the propeller increase, and the thrust increases as the product of these quantities, as can be seen by combining equations 5.15 and 5.16:

$$F = \rho A_p V_p \left(V_w - V_f \right) \tag{5.17}$$

In a reference frame fixed in the fluid through which the propeller moves, the passage of the propeller leaves a wake of fluid moving with a speed $V_w - V_f$ in the direction opposite to the motion of the vehicle. This moving stream of fluid contains kinetic energy that is created by the expenditure of some of the power supplied by the propulsion motor driving the propeller. It is relatively simple to calculate the power involved in propelling the vehicle if we note that the power P_p supplied to the propeller is the product of the volumetric flow rate through the propeller $V_p A_p$ times the pressure rise $p_{out} - p_{in}$:

$$P_p = V_p A_P (p_{out} - p_{in})$$
$$= V_p F \tag{5.18}$$

where we have used equation 5.13 to eliminate $p_{out} - p_{in}$. However, the power P_v delivered to the vehicle is the product of the thrust force F times the vehicle speed V_f, which is less than the propeller power P_p. The ratio of these two quantities is called the *propulsive efficiency* η_{prop}:

$$\eta_{prop} \equiv \frac{P_v}{P_p} = \frac{FV_f}{FV_p} = \frac{2V_f}{V_w + V_f} \tag{5.19}$$

where equation 5.16 was used to eliminate V_p. Since V_w must exceed V_f in order to produce some thrust, the propulsive efficiency is always less than unity. The lost power is expended in the kinetic energy of the wake. To reduce this power loss and to increase the efficiency of propulsion, we should use the largest possible area A_p so as to minimize the velocity difference $V_w - V_f$ while still providing the needed thrust. The size of ship and aircraft propellers are limited by practical considerations that yield propulsion efficiencies that are generally near 90%.

The blade area of an airplane propeller is only several percent of the area A_p because it is possible to develop the required thrust force even with such a relatively small blade area. However, for ship propellers, the blade area is a large fraction of A_p in order to avoid *cavitation*, *i.e.*, the formation of vapor bubbles on the blade surface.

Example 5.8

An airplane flying at a speed $V_f = 200\,km/h$ through air at a density $\rho = 1.2\,kg/m^3$ requires a propulsive force F of $3E(3)\,N$. The airplane is fitted with a propeller of diameter $D_p = 2\,m$.

Calculate (a) the wake speed V_w, (b) the propulsive efficiency η_{prop} and (c) the engine power P_p for this aircraft.

Solution

(a) The flight speed $V_f = 200E(3)\,m/3600\,s = 55.56\,m/s$. The wake speed may be found by solving equation 5.15 for V_w:

$$V_w^2 = V_f^2 + \frac{2F}{\rho A_p}$$

$$= (55.56\,m/s)^2 + \frac{2 \times 3E(3)\,N}{1.2\,kg/m^3 \times \pi \times 1\,m^2}$$

$$V_w = 68.40\,m/s$$

(b) The propulsive efficiency may be calculated from equation 5.19:

$$\eta_{prop} = \frac{2V_f}{V_w + V_f} = \frac{2 \times 55.56}{68.40 + 55.56} = 89.64\%$$

(c) The engine power $P_p = FV_p$ may be found using equation 5.16:

$$P_p = FV_p = \frac{F(V_w + V_f)}{2} = \frac{3E(3)\,N \times (68.40 + 55.56)\,m/s}{2} = 185.9\,kW$$

5.4.2 Wind Turbines

A wind turbine is a device for producing useful mechanical power from the motion of the wind. Although windmills have been used in past centuries for grinding grain and pumping water, in recent years, efficient wind turbines have been constructed that generate electrical power for use far from the site of the turbine. These turbines are mounted on towers that permit the turbine axis, which is usually horizontal, to be oriented in the direction of the wind. The turbine blades (usually three) look much like airplane propeller blades, although commonly they are much greater in diameter in order to generate economical amounts of power.

The flow of the wind past the rotating wind turbine blades is identical to that through the propeller illustrated in figure 5.2, except that the flow directions are reversed, *i.e.*, the flow is from right to left in figure 5.2. In other words, the wind speed, V_w, is higher than the wake speed, V_f, the wind decelerating as it passes through the wind turbine. The air pressure decreases across the wind turbine ($p_{in} > p_{out}$), which extracts power P_{wt} from the wind flow at a rate found by combining equations 5.18, 5.15 and 5.16:

$$P_{wt} = \rho A_p \left(\frac{V_w + V_f}{2}\right)^2 (V_w - V_f) \tag{5.20}$$

One might think that maximum power would be extracted from the wind turbine when its wake speed $V_f = 0$, *i.e.*, the entire kinetic energy of the wind would be drawn off by the turbine. In fact, the maximum power is extracted when the wake speed $V_f = V_w/3$,[8] at which value of V_f the power is:

---

[8]To show that this is so, set the derivative $\partial P_{wt}/\partial V_f$ of equation 5.20 equal to zero, and solve for V_f.

Figure 5.3 Wind turbines arrayed in a row on a wind farm. These turbines, rated at 400 *kW* each, have diameters of 33 *m* and their axes are 30 *m* above ground. (Photograph courtesy of U. S. Windpower.)

$$max\, P_{wt} = \frac{16}{27}\left(\frac{1}{2}\rho A_p V_w^3\right) \tag{5.21}$$

where the term in parentheses is the flux of wind kinetic energy passing through an area of A_p, *i.e.*, the product of a mass flow rate $\rho V_w A_p$ times the wind kinetic energy per unit mass of air, $V_w^2/2$. Because the maximum power that can be delivered by a wind turbine varies as the cube of the wind speed, the most economical sites for wind turbines are those that have high sustained wind speeds.

Figure 5.3 shows an array of wind turbines on a "wind farm." They are usually spaced 2–3 diameters apart in the cross-wind direction and 5–10 diameters apart in the downwind direction, with respect to the direction of the prevailing wind. These

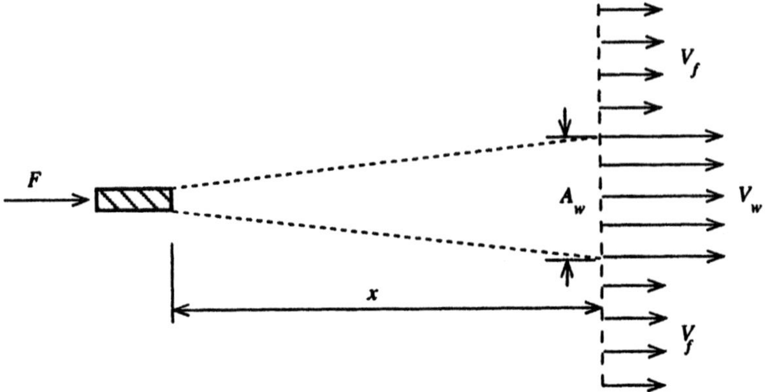

Figure 5.4 The wake behind a jet engine expands in cross-sectional area A_w as the distance x from the jet engine increases while the wake speed V_w decreases toward the flight speed V_f.

wind turbines have an active control system to orient them in the direction of the wind so that the turbine blades are upstream of the support tower.

Example 5.9

Calculate the maximum power that can be generated by a wind turbine of diameter $D = 6\,m$ when the wind speed is 20 miles per hour and the air density is $1.2\,kg/m^3$.

Solution
Using equation 5.21, the maximum power is:

$$max\,P_{wt} = \frac{8}{27}(1.2\,kg/m^3)\left(\frac{\pi}{4}(6)^2\,m^2\right)\left(20\,mph \times \frac{0.447\,m/s}{mph}\right)^3 = 7.183\,kW$$

5.4.3 Wakes and Jets

In section 5.4.1, we noted that a propeller or jet engine leaves behind a trail of fluid moving with respect to the medium through which the vehicle travels. This "river" of fluid is called a *wake*. In time, the wake fluid mixes with the surrounding medium and slows down. The change in velocity of the wake flow is such that it conserves momentum.

The flow in the wake of a jet engine attached to an airplane is illustrated in figure 5.4, where the reference frame is attached to the aircraft so that the flow is steady. As the distance x behind the aircraft increases, the wake speed V_w approaches a value equal to the flight speed V_f while the wake cross-sectional area A_w increases without limit. The manner in which V_w changes with increasing A_w may be determined by applying the linear momentum equation 5.11 to a control volume that surrounds the jet engine and the wake at a sufficient distance from the aircraft that the pressure on the control surface equals atmospheric pressure everywhere, giving:

$$\frac{d}{dt}\iiint_{\mathcal{V}} \rho \mathbf{V}\, d\mathcal{V} + (\dot{m}\mathbf{V})_{out} - (\dot{m}\mathbf{V})_{in}$$

$$= \iint_S (-p\mathbf{n})\, dS + \iint_S \boldsymbol{\tau}\, dS + \iiint_{\mathcal{V}} \rho \mathbf{g}\, d\mathcal{V} + \Sigma \mathbf{F}_{ex}$$

$$0 + \dot{m}(V_w - V_f) = 0 + 0 + 0 + F$$

where $\dot{m}$ is the mass flow rate of the fluid passing through the wake at the distance x:

$$\dot{m} = \rho_w A_w V_w$$

Combining these equations, we find:

$$V_w(V_w - V_f) = \frac{F}{\rho_w A_w}$$

$$V_w = \frac{V_f + \sqrt{V_f^2 + 4F/\rho_w A_w}}{2} \rightarrow V_f + \frac{F}{\rho_w A_w V_f} \tag{5.22}$$

where the asymptotic value of V_w holds for $F/\rho A_w V_f^2 \ll 1$. At large distances from the aircraft, observations show that the wake area A_w increases as x^2 so that the velocity excess in the wake, $V_w - V_f$, varies as x^{-2}.

Example 5.10

For the jet engine of example 5.7, calculate the wake speed V_w if the wake area $A_w = 100\, m^2$ at $x = 100\, m$.

Solution

Assuming that the asymptotic form of equation 5.22 is applicable,

$$V_w = V_f + \frac{F}{\rho_w A_w V_f}$$

$$= 250\, m/s + \frac{2.6E(4)\, N}{0.4\, kg/m^3 \times 250\, m/s \times 100\, m^2} = 252.6\, m/s$$

Notice that this satisfies the condition for the asymptotic form.

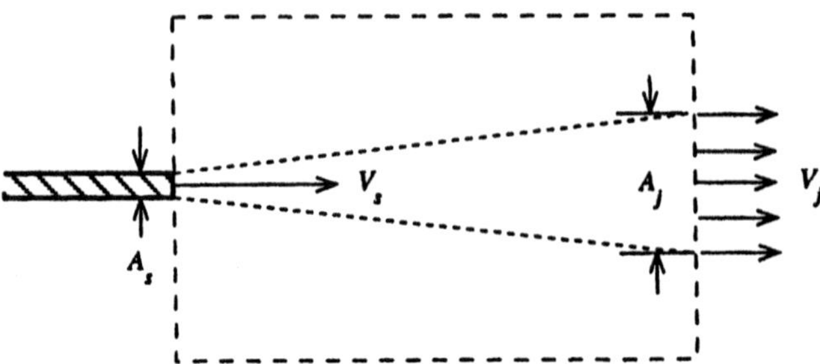

Figure 5.5 A jet directed into a stationary medium spreads with distance x from the jet source as it entrains surrounding fluid. The jet fluid slows as the jet area A_j increases, but the jet momentum remains constant.

The flow of fluid from a directed source into stationary surroundings forms a *jet* flow. Like a wake, the stream of fluid from the source grows in cross-sectional area as it mixes with the surrounding ambient fluid, as illustrated in figure 5.5. The jet speed V_j decreases with distance x from the source in a manner that conserves the momentum in the jet. If we apply the linear momentum equation 5.11 to the control volume shown in figure 5.5, where the pressure is constant on the control surface, we find the jet speed to be:

$$\frac{d}{dt}\iiint_{\mathcal{V}} \rho\mathbf{V}\,d\mathcal{V} + (\dot{m}\mathbf{V})_{out} - (\dot{m}\mathbf{V})_{in}$$

$$= \iint_S (-p\mathbf{n})\,dS + \iint_S \tau\,dS + \iiint_{\mathcal{V}} \rho\mathbf{g}\,d\mathcal{V} + \Sigma\mathbf{F}_{ex}$$

$$0 + \rho_j A_j V_j^2 - \rho_s A_s V_s^2 = 0 + 0 + 0 + 0$$

$$V_j = V_s\sqrt{\frac{\rho_s A_s}{\rho j A j}} \tag{5.23}$$

where the subscript s identifies the conditions of the flow at the jet source. Like a wake, the jet area A_j is observed to grow as x^2 at large distances from the source.

Example 5.11

A jet of water of speed $V_j = 1\ m/s$ issues through a hole of diameter $D_s = 3\ cm$ into a

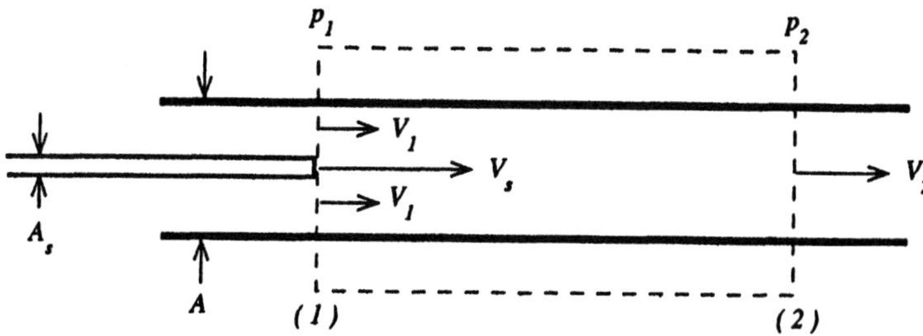

Figure 5.6 A jet pump consists of a coaxial jet of high speed fluid injected into a pipe of lower speed fluid. The mixing of the two streams produces a rise in pressure downstream.

swimming pool. If the jet diameter $D_j = 10\,cm$ at a distance $x = 1\,m$ from the source, what is the value of the jet speed V_j at that point?

Solution

Using equation 5.23 and noting that $\rho_j = \rho_s$ for water,

$$V_j = V_s\sqrt{\frac{\rho_s A_s}{\rho j A j}} = V_s\frac{D_s}{D_j} = 1\,m/s \times \frac{3\,cm}{10\,cm} = 0.3\,m/s$$

The jet velocity decreases inversely with the jet diameter.

The Jet Pump

One practical use of the fluid jet is to pump fluids to a higher pressure by use of a jet of fluid injected into a pipe of moving fluid. Such a *jet pump* is illustrated in figure 5.6 which shows a jet of velocity V_s and area A_s aligned with the axis of a pipe of area A at a point where the pipe flow velocity is V_1 and the pressure is p_1. At a distance downstream, where the two streams have completely mixed, the velocity is V_2, and the pressure p_2 is greater than p_1. The amount of the pressure rise $p_2 - p_1$ depends upon the velocities V_1 and V_s and the area ratio A_s/A in a manner that may be found by applying mass and momentum conservation to the fluid in the control volume shown in figure 5.6.

We will consider the case for which the jet fluid and the pumped fluid are both incompressible and have the same density ρ. Applying mass conservation to the steady flow of fluid across the control surface of figure 5.6,

$$\rho A V_2 = \rho(A - A_s)V_1 + \rho A_s V_s$$

Next use the linear momentum equation 5.11 for the same control volume, but assume that the viscous force on the pipe walls is negligible ($\tau = 0$):

$$\frac{d}{dt}\iiint_{\mathcal{V}} \rho \mathbf{V}\, d\mathcal{V} + (\dot{m}\mathbf{V})_{out} - (\dot{m}\mathbf{V})_{in}$$

$$= \iint_{S} (-p\mathbf{n})\, dS + \iint_{S} \tau\, dS + \iiint_{\mathcal{V}} \rho \mathbf{g}\, d\mathcal{V} + \Sigma \mathbf{F}_{ex}$$

$$0 + \rho A V_2^2 - \rho (A - A_s) V_1^2 - \rho A_s V_s^2$$

$$= (p_1 - p_2) A + 0 + 0 + 0$$

Eliminating V_2 between these two equations and solving for the pressure rise,

$$p_2 - p_1 = \frac{A_s}{A}\left(1 - \frac{A_s}{A}\right) \rho (V_s - V_1)^2 \tag{5.24}$$

The maximum pressure rise that we could expect would be that for inviscid flow of the jet decelerating from the speed V_s to $V_2 = V_s A_s / A$, or a pressure rise of $(\rho V_s^2/2)(1 - A_s^2/A^2)$. Dividing equation 5.24 by this pressure rise, we have a dimensionless form of the jet pump equation:

$$\frac{p_1 - p_2}{\frac{1}{2}\rho V_s^2 \left[1 - (A_s/A)^2\right]} = 2\left(\frac{A_s/A}{1 + A_s/A}\right)\left(1 - \frac{V_1}{V_s}\right)^2 \tag{5.25}$$

Since $A_s/A \leq 1$ and $V_1/V_s \leq 1$, the right side of equation 5.25 is always less than one.

The jet pump allows us to pump a greater volume flow rate ($V_1[A - A_s]$) than that needed to supply the jet ($V_s A_s$), albeit with a lower pressure rise than that needed for the jet supply.

Example 5.12

A jet pump consists of a jet of diameter $D_s = 1\, in$ inside a pipe of diameter $D = 3\, in$. The jet volumetric flow rate Q_s is $100\, GPM$ (gallons per minute). Calculate the pressure rise in the jet pump when the volumetric flow rate Q_1 is $500\, GPM$.

Solution

In SI units, the flow areas and flow rates are:

$$A_s = \frac{\pi}{4}(2.54E(-2)\, m)^2 = 5.067E(-4)\, m^2; \qquad A = 9A_s = 4.560E(-3)\, m^2$$

$$Q_s = \frac{100\,gal}{min} \times \frac{3.785E(-3)\,m^3}{gal} \times \frac{min}{60\,s} = 6.308E(-3)\,m^3/s;$$

$$Q_1 = 5\,Q_s = 3.154E(-2)\,m^3/s$$

and the velocities V_1 and V_s are:

$$V_1 = \frac{Q_1}{A - A_s} = \frac{3.154E(-2)\,m^3/s}{4.560E(-3)\,m^2 - 5.067E(-4)\,m^2} = 7.781\,m/s$$

$$V_s = \frac{Q_s}{A_s} = \frac{6.308E(-3)\,m^3/s}{5.067E(-4)\,m^2} = 12.45\,m/s$$

Substituting these values in equation 5.24,

$$p_2 - p_1 = \frac{1}{9}\left(1 - \frac{1}{9}\right)(1E(3)kg/m^3)(12.45\,m/s - 7.781\,m/s)^2 = 2.153E(3)\,Pa$$

5.4.4 Turbine Blades and Blade Rows

Fluid flowing through a turbine produces power by exerting a force on a moving turbine blade. The blade is shaped so as to produce a change in direction of the fluid flowing past it, thereby producing a force caused by the change in momentum of the flowing fluid.

An example of such a flow is illustrated in figure 5.7, which shows a jet of water of speed V_n issuing from a nozzle of flow area A_n and impinging on a turbine blade moving at a constant speed V_b that is less than the jet speed. The blade is shaped to nearly reverse the flow of water, redirecting it to either side of the nozzle stream in equal amounts. The force F_b exerted on the blade balances the momentum change of the redirected fluid.

To analyze this flow, choose a control volume surrounding the blade, as sketched in figure 5.7, that is fixed with respect to the blade, which is an inertial reference frame.[9] The speed V_r of the water stream relative to the blade, *i.e.*, the speed measured relative to the control volume, is:

$$V_r = V_n - V_b$$

Applying the linear momentum theorem equation 5.11 to the flow through the control volume while noting that the mass flow rate into (and out of) the control volume is

---

[9]We are ignoring here the circular motion of the blade about the turbine axis, which produces an acceleration normal to the plane of figure 5.7 and does not affect the momentum balance in the plane of the figure.

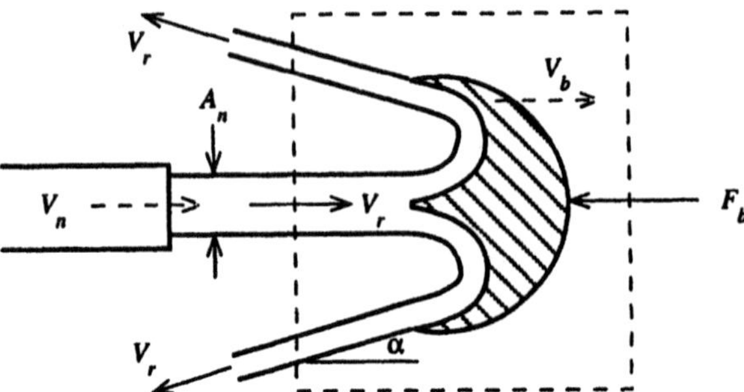

Figure 5.7 In the flow of water impinging on the blade of a Pelton turbine, a nozzle directs a stream of high velocity water tangential to the turbine disc to which are affixed radial blades (whose cross-sectional shape is sketched above). The blades move in the direction of the water jet, with at least one blade always intercepting the stream. The velocities of the fluid relative to the blade are indicated by solid arrows while the velocities of the fluid and blade in the laboratory reference frame are shown by dashed arrows.

$\rho A_n V_r$, we find:

$$\frac{d}{dt}\iiint_{\mathcal{V}} \rho \mathbf{V}\, d\mathcal{V} + (\dot{m}\mathbf{V})_{out} - (\dot{m}\mathbf{V})_{in}$$

$$= \iint_S (-p\mathbf{n})\, dS + \iint_S \boldsymbol{\tau}\, dS + \iiint_{\mathcal{V}} \rho \mathbf{g}\, d\mathcal{V} + \Sigma \mathbf{F}_{ex}$$

$$0 + \rho A_n V_r(-V_r \cos\alpha - V_r) = 0 + 0 + 0 - F_b$$

$$F_b = \rho A_n V_r^2(1 + \cos\alpha) = \rho A_n (V_n - V_b)^2 (1 + \cos\alpha) \tag{5.26}$$

The power P_b produced by the moving blade is the product of the force F_b times the blade velocity V_b:

$$P_b = F_b V_b = \rho A_n V_b (V_n - V_b)^2 (1 + \cos\alpha) \tag{5.27}$$

The blade power is zero both when the blade speed is zero ($V_b = 0$) and when the blade speed equals the nozzle flow speed ($V_b = V_n$).

The turbine power P_t is greater than the blade power P_b because there is a time during which more than one blade is developing power. To determine the turbine power, we begin by noting that, if the time between blades intercepting the jet is Δt, then the fluid mass processed by each blade is $\rho A_n V_n \Delta t$. However, the mass flow rate into each blade is $\rho A_n V_r$, and so the blade is producing power for the period $(\rho A_n V_n \Delta t)/(\rho A_n V_r) = V_n \Delta t / V_r$. Thus the time-averaged power developed by the turbine is V_n/V_r times the blade power P_b given by equation 5.27:

$$P_t = \frac{V_n}{V_r} P_b = \rho A_n V_n V_b (V_n - V_b)(1 + \cos\alpha) \tag{5.28}$$

The turbine power P_t is a maximum when $V_b = V_n/2$:[10]

$$max\ P_b = \rho A_n \frac{V_n^2}{2}\left(V_n - \frac{V_n}{2}\right)(1 + \cos\alpha) = \frac{(1 + \cos\alpha)}{2}\left(\rho A_n \frac{V_n^3}{2}\right) \tag{5.29}$$

The power available in the nozzle is the mass flow rate $\rho A_n V_n$ times the kinetic energy per unit mass of the water, $V_n^2/2$, which equals $\rho A_n V_n^3/2$. The maximum power that can be developed by the blade is $(1 + \cos\alpha)/2 \leq 1$ times the available power.

Example 5.13

A small Pelton turbine is supplied with water from a reservoir whose surface is a height $h = 100\,m$ above the turbine elevation. The area A_n of the turbine jet stream is $1.0\,in^2$, and the blade angle $\alpha = 20°$.

Calculate (a) the nozzle velocity V_n, (b) the maximum power P_t of the turbine and (c) the blade speed V_b and force F_b when maximum power is being produced.

Solution

(a) Applying Bernoulli's equation for steady inviscid flow from the reservoir surface to the nozzle exit:

$$0 + \frac{p_a}{\rho} + gh = \frac{V_n^2}{2} + \frac{p_a}{\rho} + 0$$

$$V_n = \sqrt{2gh} = \sqrt{2 \times 9.807\,m/s^2 \times 100\,m} = 44.29\,m/s$$

(b) Maximum power may be found from equation 5.29:

$$max\ P_b = \frac{(1 + \cos 20°)}{2}\left((1E(3)\,kg/m^3)(2.54E(-2)\,m)^2 \frac{(44.29\,m/s)^3}{2}\right)$$

$$= 27.18\,kW$$

(c) At maximum power, using equation 5.26,

$$V_b = \frac{V_n}{2} = 22.15\,m/s$$

$$F_b = (1E(3)\,kg/m^3)(2.54E(-2)\,m)^2([44.29 - 22.15]\,m/s)^2(1 + \cos 20°) = 613.4\,N$$

---

[10] To prove this, set $\partial P_t/\partial V_b = 0$, and solve for V_b.

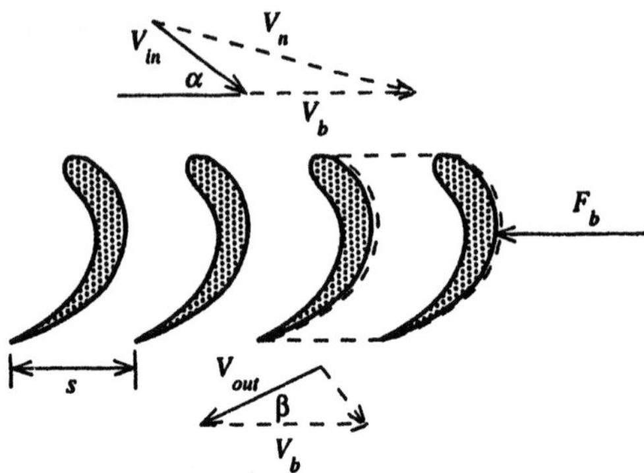

Figure 5.8 The flow across a blade row of a turbine, showing the inflow and outflow velocities in the laboratory reference frame (dashed arrows) and the blade row reference frame (solid arrows), as viewed in the inward radial direction. The airfoil-shaped blades turn the flow from the relative inflow angle α to the relative outflow angle β.

Most gas and steam turbines utilize rows of blades that are supplied with fluid around the entire circumference of the turbine wheel. The fluid velocity has tangential and axial components but no radial component. Looking radially downward on the row of blades, we would see a flow as sketched in figure 5.8, where the blades are spaced a distance s apart. The incoming flow has an absolute speed V_n, and a speed relative to the blade of V_{in}, the vector difference being the blade speed V_b, as shown in the vector diagram at the top of figure 5.8. The blades are shaped to turn this flow through a large angle, from the inlet angle α to the exit angle β. The outflow speed relative to the blade, V_{out}, will equal the inflow speed if there is no pressure drop from inflow to outflow, called an *impulse stage*, or it may increase because the pressure falls, known as a *reaction stage*.

To find the force F_b acting on a single blade of such a turbine row, select a control surface that encloses one blade (as shown by the dashed line in figure 5.8), which has surfaces that lie along the right-hand surface of adjacent blades. With this choice, the pressure force on the portion of the control surface in contact with the blades will add to zero because the differential force on similarly located area elements of the adjacent blades touching the control volume are equal and opposite in direction. In the tangential direction, there will be no contribution to the momentum integral from the pressure force.

If we denote the height of the blade by h, applying mass conservation to this control volume results in:

$$\rho_{in}(V_{in} \sin \alpha)(sh) = \rho_{out}(V_{out} \sin \beta)(sh) \tag{5.30}$$

Now apply the linear momentum theorem in the tangential direction:

$$\frac{d}{dt}\iiint_{\mathcal{V}} \rho \mathbf{V}\, d\mathcal{V} + (\dot{m}\mathbf{V})_{out} - (\dot{m}\mathbf{V})_{in}$$

$$= \iint_S (-p\mathbf{n})\, dS + \iint_S \boldsymbol{\tau}\, dS + \iiint_{\mathcal{V}} \rho \mathbf{g}\, d\mathcal{V} + \Sigma \mathbf{F}_{ex}$$

$$0 + \rho_{in}(V_{in} \sin \alpha)(sh)(-V_{out} \cos \beta) - \rho_{out}(V_{out} \sin \beta)(sh)(V_{in} \cos \alpha) = 0 + 0 + 0 - F_b$$

$$F_b = \rho_{in} V_{in}(sh)(\sin \alpha)(V_{in} \cos \alpha + V_{out} \cos \beta)$$

$$= \rho_{in} V_{in}^2(sh)(\sin \alpha \cos \alpha)\left(1 + \frac{\rho_{in} \tan \alpha}{\rho_{out} \tan \beta}\right) \tag{5.31}$$

where we have used the mass conservation equation 5.30 to replace V_{out}.

Example 5.14

Derive an expression for the force F_b on the blade of an impulse stage for which the blade speed is chosen so as to minimize the kinetic energy of the downstream flow.

Solution

For an impulse stage, equation 5.31 is simplified by noting that $\rho_{in} = \rho_{out}$ and $V_{in} = V_{out}$ because there is no pressure change in the stage and therefore that $\beta = \alpha$ by mass conservation equation 5.30. (The turbine blade shape is thus symmetric about the blade centerline.) To minimize the kinetic energy of the flow leaving the stage, we choose the blade speed $V_{out} \cos \beta = V_b$ so that there will be no tangential component of the flow leaving the stage (see the lower vector diagram in figure 5.8). With these restrictions, equation 5.31 becomes:

$$F_b = 2\rho_{in} V_b^2(sh) \tan \alpha$$

5.4.5 Horizontal Flows with a Free Surface

The flow of water in channels, over spillways and through sluice gates may be accompanied by changes in velocity brought about by obstructions to the flow. The forces acting on these obstructions can be determined by applying the linear momentum integral to these flows. In this section, we consider two examples of such flows.

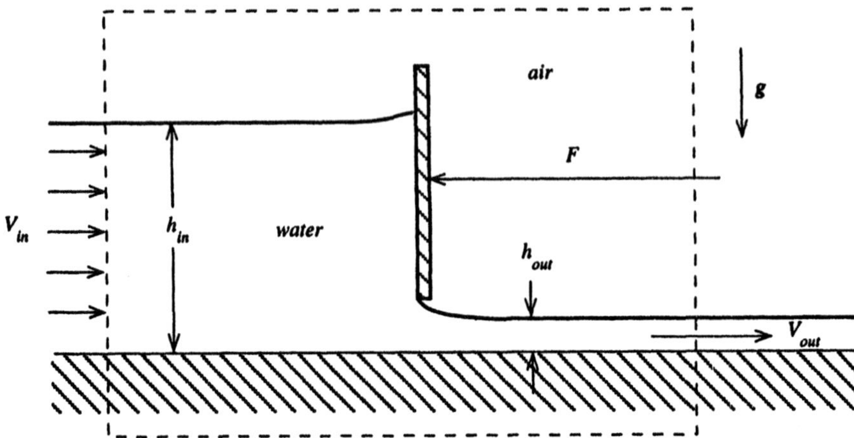

Figure 5.9 The two-dimensional flow of water underneath a sluice gate is a uniform, horizontal flow both far upstream and downstream from the gate. The restraining force holding the gate in place results from a linear momentum balance.

Flow through a Sluice Gate

Figure 5.9 depicts the two-dimensional flow beneath a sluice gate inserted in a wide channel of flowing water. Far upstream and downstream, the flow velocities V_{in} and V_{out}, respectively, are horizontal and uniform from the bottom to the air/water interface at the top, and the depth of the fluid layers, h_{in} and h_{out}, are also constant. Close to the gate, the velocity and layer depth vary, as the incoming fluid accelerates to pass under the opening in the gate. We therefore select a control volume enclosing the gate which extends far enough upstream and downstream that the flow conditions are known.

In applying the linear momentum theorem to the control volume shown by a dashed line in figure 5.9, we will need to evaluate the integral of the pressure on the control surface. However, we need only to consider the portion of the flow where the pressure differs from atmospheric by the amount $p - p_a$, which occurs in the upstream and downstream water layers since a uniform pressure of p_a integrates to zero. In these water layers, the streamlines are horizontal and uncurved and the pressure distribution is hydrostatic:

$$p + \rho g z = p_a + \rho g h_{in}$$

so that the net pressure integral on the upstream flow, for example, becomes:

$$\int_0^{h_{in}} (p - p_a)\, dz = \rho g \int_0^{h_{in}} (h_{in} - z) dz = \rho g \left| h_{in} z - \frac{z^2}{2} \right|_0^{h_{in}} = \frac{1}{2} \rho g h_{in}^2 \tag{5.32}$$

Now we may utilize the linear momentum theorem 5.11 in the horizontal direction to find the horizontal force per unit width of gate, F/W, needed to restrain the sluice gate:

$$\frac{d}{dt} \iiint_{\mathcal{V}} \rho \mathbf{V}\, d\mathcal{V} + (\dot{m}\mathbf{V})_{out} - (\dot{m}\mathbf{V})_{in}$$

$$= \iint_S (-p\mathbf{n})\, dS + \iint_S \boldsymbol{\tau}\, dS + \iiint_{\mathcal{V}} \rho \mathbf{g}\, d\mathcal{V} + \Sigma \mathbf{F}_{ex}$$

$$0 + \rho V_{out}^2 (W h_{out}) - \rho V_{in}^2 (W h_{in})$$

$$= W \int_0^{h_{in}} (p - p_a)\, dz - W \int_0^{h_{out}} (p - p_a)\, dz + 0 + 0 - F$$

$$\frac{F}{W} = \frac{1}{2} \rho g (h_{in}^2 - h_{out}^2) + \rho h_{in} V_{in}^2 - \rho h_{out} V_{out}^2 \tag{5.33}$$

where we have considered the viscous force acting on the stream bed as negligible. We may now eliminate V_{in} from this expression by applying mass conservation to the control volume:

$$\rho V_{in} (W h_{in}) = \rho V_{out} (W h_{out})$$

$$V_{in} = \left(\frac{h_{out}}{h_{in}} \right) V_{out} \tag{5.34}$$

Substituting in equation 5.33 and simplifying,

$$\frac{F}{W} = \frac{1}{2} \rho g (h_{in}^2 - h_{out}^2) - \rho g h_{out} V_{out}^2 \left(1 - \frac{h_{out}}{h_{in}} \right) \tag{5.35}$$

The first term of this expression is simply the force that would exist if the sluice gate were closed so as to separate two static layers of depths h_{in} and h_{out}. The second term reduces this amount in proportion to the flow through the sluice gate.

To simplify this expression further, we assume that the flow through the gate is an inviscid flow so that Bernoulli's equation for steady flow may be applied to a streamline along the air/water interface:

$$\frac{p_a}{\rho} + \frac{V_{in}^2}{2} + g h_{in} = \frac{p_a}{\rho} + \frac{V_{out}^2}{2} + g h_{out}$$

$$\left(1 - \frac{h_{out}^2}{h_{in}^2} \right) \frac{V_{out}^2}{2} = g (h_{in} - h_{out})$$

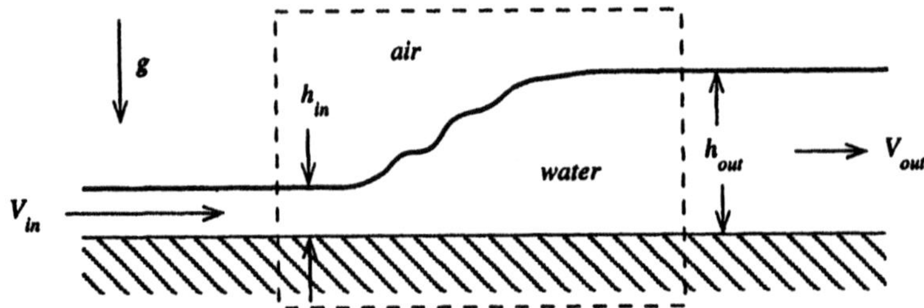

Figure 5.10 An hydraulic jump in a channel flow is always accompanied by an increase in the stream depth downstream.

$$V_{out}^2 = \frac{2gh_{in}^2}{h_{in} + h_{out}} \tag{5.36}$$

where we have used mass conservation equation 5.34 in the second line to eliminate V_{in}. If we now substitute this expression for V_{out} into equation 5.35 for the force on the sluice gate and simplify the ensuing expression, we will find:

$$\frac{F}{W} = \frac{1}{2}\rho g \left(\frac{(h_{in} - h_{out})^3}{h_{in} + h_{out}} \right) \tag{5.37}$$

The Hydraulic Jump

It is possible for a water flow in a channel without an obstruction to undergo a spontaneous increase in level, as seen in figure 5.10. Such a flow is called an *hydraulic jump*. The region of transition from the uniform flow upstream to another uniform flow downstream is irregular and random, being a turbulent flow with small wavelets on the surface that appear to break. While we cannot describe the flow in this transition region, we can relate the upstream and downstream conditions in the hydraulic jump by utilizing the linear momentum theorem.

If we apply mass and momentum conservation to the control volume in figure 5.10, we will obtain equations identical to 5.33 - 5.35, except with $F = 0$. Solving equation 5.35 (with $F = 0$) for V_{out} and then V_{in} (using 5.34), we find:

$$\begin{aligned} \frac{V_{out}^2}{gh_{out}} &= \frac{1}{2}\left(\frac{h_{in}}{h_{out}}\right)\left(\frac{h_{in}}{h_{out}} + 1\right) \leq 1 \\ \frac{V_{in}^2}{gh_{in}} &= \frac{1}{2}\left(\frac{h_{out}}{h_{in}}\right)\left(\frac{h_{out}}{h_{in}} + 1\right) \geq 1 \end{aligned} \tag{5.38}$$

where the inequality follows from the observation that the water level always increases ($h_{out} \geq h_{in}$) across an hydraulic jump.

The conclusion to be drawn from equation 5.38 is that the upstream flow must have a sufficiently high speed and small depth that $V_{in} \geq \sqrt{gh_{in}}$. Such a flow is called a *supercritical flow* because it is moving faster than small gravity waves can propagate on the air/water interface. On the other hand, the downstream flow ($V_{out} \leq \sqrt{gh_{out}}$) is a *subcritical flow*.

For the flow beneath the sluice gate illustrated in figure 5.9, the outflow velocity V_{out} of equation 5.36 observes the relation:

$$\frac{V_{out}^2}{gh_{out}} = \frac{2(h_{in}/h_{out})^2}{(h_{in}/h_{out})+1} \geq 1 \tag{5.39}$$

where the inequality holds when $h_{in} \geq h_{out}$, which is always the case for flow beneath the sluice gate. Thus it is possible for an hydraulic jump to exist downstream of a sluice gate. However, the final stream depth downstream of such an hydraulic jump is always less than the level upstream of the sluice gate.

5.5 Angular Momentum

The vector product $\mathbf{R} \times m\mathbf{V}$ of the position vector $\mathbf{R}$ of a moving particle times its momentum $m\mathbf{V}$ is called the moment of momentum, or more commonly, the *angular momentum* of the particle. In both classical and quantum mechanics, angular momentum is a significant descriptor of motion because in many circumstances it is a time invariant.[11] Are there fluid flows for which the angular momentum would be a useful descriptor? In this section, we consider some examples of flows for which the use of angular momentum as a flow variable helps to simplify the analysis.

5.5.1 Newton's Law of Angular Momentum

Newton's law of motion may be expressed in terms of angular momentum by multiplying the law of linear momentum by $\mathbf{R}$:

$$\mathbf{R} \times \frac{d}{dt}(m\mathbf{V}) = \mathbf{R} \times \mathbf{F}$$

$$\frac{d}{dt}(\mathbf{R} \times m\mathbf{V}) - \frac{d\mathbf{R}}{dt} \times m\mathbf{V} = \mathbf{R} \times \mathbf{F}$$

$$\frac{d}{dt}(\mathbf{R} \times m\mathbf{V}) = \mathbf{R} \times \mathbf{F} \tag{5.40}$$

[11]The constancy of the angular momentum of a planet moving about the sun was embodied in Kepler's laws of planetary motion, the predecessor of Newton's law of motion.

where the second term on the second line of equation 5.40 is zero because $d\mathbf{R}/dt \times \mathbf{V} = \mathbf{V} \times \mathbf{V} = 0$. Expressed in words, Newton's law states that *the time rate of change of angular momentum of a mass equals the moment of the force acting on the mass.* Of course, this is not a second law in addition to that of linear momentum but merely another manner of stating the same law of motion.

We now wish to apply this law to the fluid contained within a material control volume. We begin by denoting the angular momentum of the fluid in the control volume by $\mathbf{H}$:

$$\mathbf{H} \equiv \iiint_{\mathcal{M}} (\mathbf{R} \times \rho\mathbf{V})\, d\mathcal{V} \tag{5.41}$$

so that $d\mathbf{H}/dt$ is the time rate of change of angular momentum. We next need to evaluate the moment of the forces acting on the fluid in the control volume. These forces consist of the pressure force $p(-\mathbf{n})$ and the viscous force $\boldsymbol{\tau}$ acting on the surface $\mathcal{S}$ and of the gravity force $\rho\mathbf{g}$ acting on the volume $\mathcal{V}$. Multiplying each of these forces by $\mathbf{R}$, integrating over the surface $\mathcal{S}$ or volume $\mathcal{V}$ as appropriate, summing and equating to $d\mathbf{H}/dt$, we have *Newton's law of angular momentum* for the fluid in a material control volume:

$$\frac{d\mathbf{H}}{dt} = \iint_{\mathcal{S}} (\mathbf{R} \times [-p\mathbf{n}])\, d\mathcal{S} + \iint_{\mathcal{S}} (\mathbf{R} \times \boldsymbol{\tau})\, d\mathcal{S} + \iiint_{\mathcal{V}} (\mathbf{R} \times \rho\mathbf{g})\, d\mathcal{V} \tag{5.42}$$

This is the angular momentum equivalent of the linear momentum equation 5.8.

5.5.2 The Angular Momentum Theorem

The time rate of change of angular momentum, $d\mathbf{H}/dt$, in Newton's law of angular momentum, equation 5.42, refers to a material volume. To express this in terms of a fixed control volume that occupies the same position as the material volume at a particular instant of time we apply Reynolds' transport theorem 5.5 with $b = \mathbf{R} \times \rho\mathbf{V}$:

$$\frac{d\mathbf{H}}{dt} = \frac{d}{dt}\iiint_{\mathcal{V}} (\mathbf{R} \times \rho\mathbf{V})\, d\mathcal{V} + \iint_{\mathcal{S}} (\mathbf{R} \times \rho\mathbf{V})(\mathbf{V} \cdot \mathbf{n})\, d\mathcal{S} \tag{5.43}$$

Substituting this equation 5.43 in Newton's law of angular momentum, equation 5.42, we find the *angular momentum theorem* for a fluid flow:

$$\begin{aligned} \frac{d}{dt}\iiint_{\mathcal{V}} (\mathbf{R} \times \rho\mathbf{V})\, d\mathcal{V} &+ \iint_{\mathcal{S}} (\mathbf{R} \times \rho\mathbf{V})(\mathbf{V} \cdot \mathbf{n})\, d\mathcal{S} \\ &= \iint_{\mathcal{S}} (\mathbf{R} \times [-p\mathbf{n}])\, d\mathcal{S} + \iint_{\mathcal{S}} (\mathbf{R} \times \boldsymbol{\tau})\, d\mathcal{S} + \iiint_{\mathcal{V}} (\mathbf{R} \times \rho\mathbf{g})\, d\mathcal{V} \end{aligned} \tag{5.44}$$

Like the linear momentum theorem, this vector equation has three scalar components. A product such as $\mathbf{R} \times \mathbf{V}$ has a direction that is perpendicular to both $\mathbf{R}$ and $\mathbf{V}$. For flows with axial symmetry, such as is encountered in turbomachines, the only nonzero component of angular momentum has the direction of the axis of symmetry.

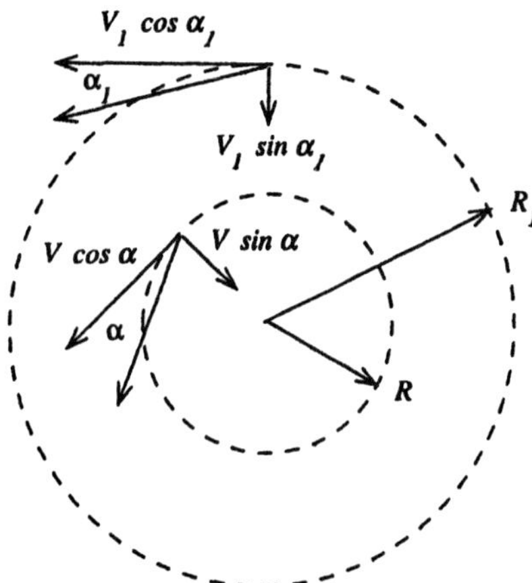

Figure E 5.15

Example 5.15

A steady, plane axisymmetric flow of incompressible, inviscid fluid consists of a swirling, inward flow having a velocity $\mathbf{V}$ that lies at an angle α with respect to the tangent to a circle of radius R, as shown in figure E 5.15. Using cylindrical coordinates, the velocity $\mathbf{V}$ may be expressed as:

$$\mathbf{V} = (V \cos\alpha)\mathbf{i}_\theta - (V \sin\alpha)\mathbf{i}_r$$

where, because of axial symmetry, V and α depend only upon R.

The flow speed V_1 and angle α_1 are known at the inflow radius R_1. (a) Neglecting gravity, derive an expression for the speed V and flow angle α at any radius $R \leq R_1$. (b) Assuming steady inviscid flow, derive an expression for the pressure difference $p_1 - p$ along a streamline between R_1 and R.

Solution

(a) In applying the angular momentum theorem, equation 5.44, to the control volume contained between the two concentric circles of figure E 5.15, the moment of the pressure force is zero since the pressure is constant on the circumference of a circle (because of axial symmetry) and the moments of the viscous and gravity forces are zero by assumption. Thus the angular momentum of the inflow and outflow terms of 5.44 must add to zero:

$$0 + (R_1 \rho V_1 \cos \alpha_1)\mathbf{i}_z(-V_1 \sin \alpha_1) 2\pi R_1 + (R\rho V \cos \alpha)\mathbf{i}_z(V \sin \alpha) 2\pi R = 0 + 0 + 0$$

$$R^2 V^2 \cos \alpha \sin \alpha = R_1^2 V_1^2 \cos \alpha_1 \sin \alpha_1$$

where we have used the product $\mathbf{R} \times \mathbf{V} = (RV \cos \alpha)\mathbf{i}_z$ and have assumed a unit width of control volume in the z direction. It is also necessary that mass be conserved:

$$\rho V \sin \alpha (2\pi R) = \rho V_1 \sin \alpha_1 (2\pi R_1)$$

$$RV \sin \alpha = R_1 V_1 \sin \alpha_1$$

By squaring the second of these equations and dividing by the first, we find:

$$\tan \alpha = \tan \alpha_1$$

$$\alpha = \alpha_1$$

Thus the angle α is constant throughout this flow field, and a streamline is a constant angle spiral. Since α is constant, the mass conservation equation can be solved for V:

$$V = V_1 \left(\frac{R_1}{R} \right)$$

(b) Using Bernoulli's equation for inviscid, incompressible flow along a streamline,

$$\frac{p_1}{\rho} + \frac{V_1^2}{2} = \frac{p}{\rho} + \frac{V^2}{2}$$

$$p_1 - p = \frac{\rho}{2}(V^2 - V_1^2) = \frac{\rho}{2} V_1^2 \left[\left(\frac{R_1}{R} \right)^2 - 1 \right]$$

This relation is inapplicable at small R because it would require a negative pressure p, which is physically impossible.

The flow in this example illustrates the wind flow in a tornado, where the speed increases and the pressure decreases toward the center of the tornado.

In applying the linear momentum theorem to practical flows, we noted that there are usually easily identifiable inflow and outflow streams for which the flux of momentum could be written as, for example, $(\dot{m}\mathbf{V})_{in}$. We can adopt the same notation for the flux of angular momentum by writing it as $(\mathbf{R} \times \dot{m}\mathbf{V})_{in}$. With this change, the angular momentum theorem, equation 5.44, takes the form:

$$\frac{d}{dt}\iiint_{\mathcal{V}}(\mathbf{R} \times \rho\mathbf{V})\,d\mathcal{V} + (\mathbf{R} \times \dot{m}\mathbf{V})_{out} - (\mathbf{R} \times \dot{m}\mathbf{V})_{in}$$

$$= \iint_{S}(\mathbf{R} \times [-p\mathbf{n}])\,dS + \iint_{S}(\mathbf{R} \times \boldsymbol{\tau})\,dS + \iiint_{\mathcal{V}}(\mathbf{R} \times \rho\mathbf{g})\,d\mathcal{V} \quad (5.45)$$

Moment of External Forces

When the control volume encloses solid structures, it is possible to exert both a force and a couple, or torque, on the fluid inside the control volume. For rotating compressors, pumps and turbines, the torque applied to the rotating shaft is the means whereby the pressure and velocity of the fluid are changed between inflow and outflow. It is therefore necessary to take into account the sum of the moments of these external forces, $\Sigma\mathbf{T}_{ex}$, by adding this term to the right side of equation 5.45:

$$\frac{d}{dt}\iiint_{\mathcal{V}}(\mathbf{R} \times \rho\mathbf{V})\,d\mathcal{V} + (\mathbf{R} \times \dot{m}\mathbf{V})_{out} - (\mathbf{R} \times \dot{m}\mathbf{V})_{in}$$

$$= \iint_{S}(\mathbf{R} \times [-p\mathbf{n}])\,dS + \iint_{S}(\mathbf{R} \times \boldsymbol{\tau})\,dS + \iiint_{\mathcal{V}}(\mathbf{R} \times \rho\mathbf{g})\,d\mathcal{V} + \Sigma\mathbf{T}_{ex} \quad (5.46)$$

For axisymmetric flows in turbomachines, the moment of the external forces $\Sigma\mathbf{T}_{ex}$ is a couple vector in the direction of the axis of rotation.

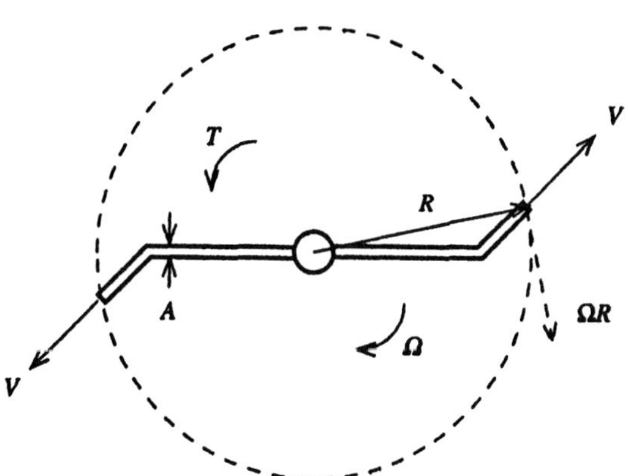

Figure E 5.16

Example 5.16

A lawn sprinkler consists of a rotating tube of internal area $A = 10\,mm^2$, inside which water flows with a speed $V = 5\,m/s$, being supplied by a hose attached at the center of rotation. Viewed from above in figure E 5.16, the sprinkler rotates in a clockwise direction, the tip of the tube making an angle $\alpha = 30°$ with respect to the tangential direction. The tip radial dimension is $R = 10\,cm$. Because of friction in the bearing of the rotor, the rotor is retarded by a counterclockwise torque $T = 2E(-2)\,Nm$, as indicated in figure E 5.16.

Calculate (a) the angular speed Ω of the sprinkler rotor and (b) the velocity **V** of the fluid stream relative to the ground.

Solution

(a) Selecting cylindrical coordinates with the z-axis pointing out of the plane of the figure and a control volume of radius R, the velocity **V** of the fluid measured in the inertial reference frame of the ground is:

$$\mathbf{V} = (V \sin\alpha)\,\mathbf{i}_r + (V\cos\alpha - \Omega R)\,\mathbf{i}_\theta$$

and the product $\mathbf{R} \times \mathbf{V}$ is:

$$\mathbf{R} \times \mathbf{V} = R\,\mathbf{i}_r \times \mathbf{V} = R(V\cos\alpha - \Omega R)\,\mathbf{i}_z$$

Noting that the mass flow rate in each arm is ρAV and substituting in the angular momentum theorem equation 5.46, we find:

$$\frac{d}{dt}\iiint_{\mathcal{V}}(\mathbf{R} \times \rho\mathbf{V})\,d\mathcal{V} + (\mathbf{R} \times \dot{m}\mathbf{V})_{out} - (\mathbf{R} \times \dot{m}\mathbf{V})_{in}$$

$$= \iint_S (\mathbf{R} \times [-p\mathbf{n}])\,dS + \iint_S (\mathbf{R} \times \boldsymbol{\tau})\,dS + \iiint_{\mathcal{V}}(\mathbf{R} \times \rho\mathbf{g})\,d\mathcal{V} + \Sigma\mathbf{T}_{ex}$$

$$0 + 2R(V\cos\alpha - \Omega R)(\rho AV)\mathbf{i}_z + 0 = 0 + 0 + 0 + T\,\mathbf{i}_z$$

$$\Omega = \left(\frac{V}{R}\right)\cos\alpha - \frac{T}{2\rho AR^2V}$$

$$= \left(\frac{5\,m/s}{0.1\,m}\right)\cos 30° - \frac{2E(-2)\,Nm}{(2)(1E(3)\,kg/m^3)(1E(-5)\,m^2)(0.1\,m)^2(5\,m/s)}$$

$$= 23.30\,s^{-1}$$

(b) The velocity **V** in the ground reference frame is:

$$\mathbf{V} = V\sin\alpha\,\mathbf{i}_r + (V\cos\alpha - \Omega R)\mathbf{i}_\theta$$

$$= (5\,m/s)\sin 30^\circ\,\mathbf{i}_r + ((5\,m/s)\cos 30^\circ - (23.30\,s^{-1})(0.1\,m))\mathbf{i}_\theta$$

$$= (2.5\,m/s)\mathbf{i}_r + (2.0\,m/s)\mathbf{i}_\theta$$

Note that if there were no friction ($T = 0$), then $V\cos\alpha - \Omega R = 0$, and the velocity $\mathbf{V}$ is purely radial, *i.e.*, the fluid flow has no angular momentum. Although the fluid stream leaving the sprinkler has the form of a rotating spiral, the fluid droplets move in the radial direction.

5.6 Applications of the Angular Momentum Theorem

The most common applications of the angular momentum theorem involve rotating machinery such as pumps, fans, compressors, turbines and propellers. Such applications enable us to determine the torque and mechanical power absorbed (or delivered) by these machines, an integral part of the design of turbomachinery. The relationship between torque and speed of a turbomachine must match that of the electric motor, generator or engine to which it is attached in order for power to be efficiently transmitted between them. Thus the angular momentum theorem is essential to the understanding of how turbomachines operate.

5.6.1 Centrifugal (Radial Flow) Compressors and Pumps

Centrifugal turbomachines increase the pressure of the fluid passing through them by virtue of the high angular speed Ω of the rotor, which subjects the fluid to a centrifugal acceleration that must be maintained by a strong radial pressure gradient. In section 4.6.3, we found that the pressure rise in an inviscid incompressible flow through a centrifugal pump equaled $\rho\Omega^2 R_{out}^2$, where R_{out} is the rotor tip radius, assuming equal inflow and outflow areas.

For any centrifugal device having an axial inflow (*i.e.*, no inflow angular momentum) and a radial outflow, such as that illustrated in figure 4.9, we may determine the rotor torque T by applying the angular momentum theorem 5.46 to a control volume surrounding the rotor:

$$\frac{d}{dt}\iiint_{\mathcal{V}} (\mathbf{R} \times \rho\mathbf{V})\,d\mathcal{V} + (\mathbf{R} \times \dot{m}\mathbf{V})_{out} - (\mathbf{R} \times \dot{m}\mathbf{V})_{in}$$

$$= \iint_S (\mathbf{R} \times [-p\mathbf{n}])\,dS + \iint_S (\mathbf{R} \times \boldsymbol{\tau})\,dS + \iiint_{\mathcal{V}} (\mathbf{R} \times \rho\mathbf{g})\,d\mathcal{V} + \Sigma\mathbf{T}_{ex}$$

$$0 + \dot{m}(RV_\theta)_{out}\,\mathbf{i}_z - 0 = 0 + 0 + 0 + \mathbf{T}$$

$$\mathbf{T} = \dot{m}(RV_\theta)_{out}\,\mathbf{i}_z \tag{5.47}$$

where $(V_\theta)_{out}$ is the tangential component of the outflow velocity (measured in the inertial reference frame of the stationary stator) and where $\dot{m}$ is the mass flow rate through the machine. If the flow relative to the rotor at its exit is purely radial, then $(V_\theta)_{out} = \Omega R_{out}$, and the torque becomes:

$$\mathbf{T} = \dot{m}\Omega^2 R_{out}^2\,\mathbf{i}_z \qquad \textit{if} \;\; (V_\theta)_{out} = \Omega R_{out} \tag{5.48}$$

The power P_{in} required to operate the machine is the product $\mathbf{\Omega}\cdot\mathbf{T}$ of the angular speed $\mathbf{\Omega} = \Omega\,\mathbf{i}_z$ and the torque $\mathbf{T}$:

$$\begin{aligned} P_{in} &= \mathbf{\Omega}\cdot\mathbf{T} = \dot{m}\Omega(RV_\theta)_{out} \\ &= \dot{m}\Omega^2 R_{out}^2 \qquad \textit{if} \;\; (V_\theta)_{out} = \Omega R_{out} \end{aligned} \tag{5.49}$$

Example 5.17

The lawn sprinkler device of example 5.16 may be operated as a reaction turbine by attaching an electric generator that applies a restraining torque $T\mathbf{i}_z$ to the rotor and absorbs an amount of power $-\mathbf{\Omega}\cdot\mathbf{T} = -(-\Omega\mathbf{i}_z)\cdot(T\mathbf{i}_z) = \Omega T$.

(a) For any angular speed Ω, derive an expression for the power P_{out} of the turbine. (b) Derive an expression for the angular speed that maximizes the power, and for the maximum power. (c) Assuming inviscid flow through the device, supplied from a pressure vessel, derive an expression for the pressure p_s of the fluid in the pressure vessel.

Solution

(a) From example 5.16, the torque T and power P_{out} are:

$$T = (2\rho AV)R(V\cos\alpha - \Omega R) = \dot{m}R(V\cos\alpha - \Omega R)$$

$$P_{out} = \Omega T = \dot{m}\Omega R(V\cos\alpha - \Omega R)$$

(b) The maximum value of P_{out} occurs when $dP_{out}/d\Omega = 0$:

$$\dot{m}R(V\cos\alpha - 2\Omega R) = 0$$

$$\Omega R = \frac{1}{2}V\cos\alpha$$

For this value of ΩR, the maximum power is:

$$\textit{max}\;\; P_{out} = \dot{m}(\frac{1}{2}V\cos\alpha)(\frac{1}{2}V\cos\alpha) = \dot{m}\frac{1}{4}(V\cos\alpha)^2$$

(c) For steady inviscid flow through the rotor, we may apply the form of Bernoulli's equation for a steadily rotating reference frame, equation 4.28, between the inflow on the axis and the outflow at R (neglecting gravity):

$$\frac{V_{in}^2}{2} + \frac{p_{in}}{\rho} = \frac{V^2}{2} + \frac{p_a}{\rho} - \frac{(\Omega R)^2}{2}$$

However, the inflow is related to the conditions in the pressure vessel by Bernoulli's equation for the laboratory inertial reference frame (again neglecting gravity):

$$0 + \frac{p_s}{\rho} = \frac{V_{in}^2}{2} + \frac{p_{in}}{\rho}$$

Combining these equations,

$$p_s = p_a + \frac{\rho}{2}[V^2 - (\Omega R)^2]$$

These simple relations exemplify the operating characteristics of turbines. For example, the torque exerted by the rotor decreases linearly from a value of $\dot{m}RV\cos\alpha$ when the rotor is stationary ($\Omega = 0$) to zero when the rotor is free-wheeling at a speed $\Omega = (V\cos\alpha)/R$. At these extremes, the power P_{out} is zero. When the rotor speed is zero, the pressure difference $p_s - p_a$ is $\rho V^2/2$, which is to be expected since the flow is steady in the laboratory inertial reference frame. Conversely, when the rotor is free-wheeling, $p_s - p_a = (\rho/2)(V^2 - (V\cos\alpha)^2) = (\rho/2)(V\sin\alpha)^2$, where $V\sin\alpha$ is the fluid velocity in the inertial reference frame. At the condition of maximum power, the pressure difference $p_s - p_a$ equals $(\rho/2)V^2(1 - [\cos\alpha]^2/4)$.

5.6.2 Axial Flow Turbines and Compressors

In jet engines and gas or steam turbine power plants, the flow of fluid has very little radial component, the velocity having mainly axial and tangential components. These are called axial flow machines to distinguish them from centrifugal machines. The change in pressure across a rotor stage is related to the change in relative tangential velocity, as explained in section 4.6.3. In contrast to the centrifugal machine, the inflow stream may have some angular momentum, especially in the case of a turbine rotor, that must be taken into account in applying the angular momentum theorem.

A particular example of a turbine stage, consisting of a stator blade row and a rotor blade row, is illustrated in figure 5.11(a). In this case, the stator and rotor blades are designed to accept purely axial inflow and to turn the flow through a large angle while the flow accelerates because of a pressure drop. The increase in speed and the angle of deflection are such that the tangential component of the exit velocity is equal to the

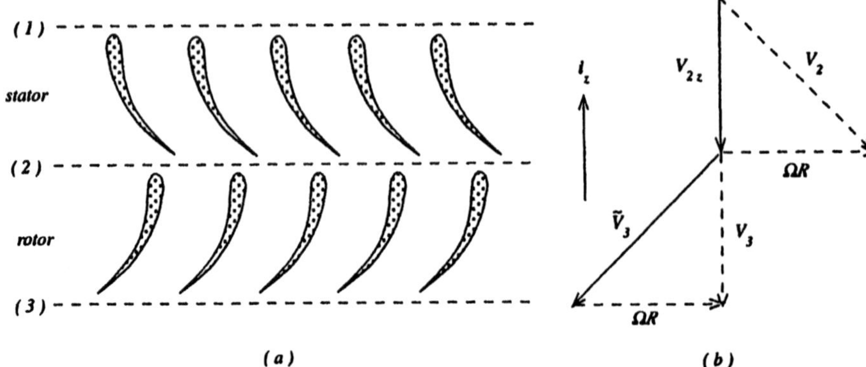

Figure 5.11 As shown in (*a*), an axial flow turbine stage consists of a set of stationary stator blades and a row of moving rotor blades having a tangential speed ΩR to the right. As shown in the velocity diagram (*b*), each blade row turns the incoming flow from a purely axial (relative) flow to a flow with a relative tangential component of ΩR. (Dashed vectors are velocities relative to the stationary stator while solid vectors are velocities relative to the moving rotor.)

rotor tangential speed ΩR. The vector diagram of the absolute and relative velocities across the stator and rotor are shown in figure 5.11(*b*).

We may apply the angular momentum theorem to the rotor in order to calculate the torque $\mathbf{T}$ applied to the rotor:

$$\frac{d}{dt}\iiint_{\mathcal{V}}(\mathbf{R}\times\rho\mathbf{V})\,d\mathcal{V}+(\mathbf{R}\times\dot{m}\mathbf{V})_{out}-(\mathbf{R}\times\dot{m}\mathbf{V})_{in}$$

$$=\iint_{S}(\mathbf{R}\times[-p\mathbf{n}])\,dS+\iint_{S}(\mathbf{R}\times\boldsymbol{\tau})\,dS+\iiint_{\mathcal{V}}(\mathbf{R}\times\rho\mathbf{g})\,d\mathcal{V}+\Sigma\mathbf{T}_{ex}$$

$$0+0-(R\dot{m}\Omega R)\mathbf{i}_z=0+0+0+\mathbf{T}$$

$$\mathbf{T}=-\dot{m}\Omega R^2\mathbf{i}_z$$

and then calculate the power $P=-\boldsymbol{\Omega}\cdot\mathbf{T}$ delivered by the rotor:

$$P=-(\Omega\mathbf{i}_z)\cdot(-\dot{m}\Omega R^2\mathbf{i}_z)=\dot{m}(\Omega R)^2$$

If we had applied the angular momentum theorem to *both* the stator and the rotor, we would have found that the net torque would be zero because both the stator inflow and the rotor outflow have no tangential component of velocity. Thus the torque applied to the stator is equal in magnitude, but opposite in direction, to that applied to the rotor. This stator torque is transmitted to the turbine foundation and would be counterbalanced by the torque transmitted to the foundation of the electric generator being driven by the gas turbine.[12]

If this flow were incompressible and inviscid, then we could calculate the pressure drop across the stage. For the stator, applying Bernoulli's equation between 1 and 2,

$$\frac{p_1}{\rho} + \frac{V_1^2}{2} = \frac{p_2}{\rho} + \frac{V_2^2}{2}$$

$$\frac{p_1}{\rho} - \frac{p_2}{\rho} = \frac{V_2^2}{2} - \frac{V_1^2}{2}$$

For the rotor, Bernoulli's equation 4.28 in the rotating reference frame is:

$$\frac{p_2}{\rho} + \frac{V_{2z}^2}{2} - \frac{(\Omega R)^2}{2} = \frac{p_3}{\rho} + \frac{\tilde{V}_3^2}{2} - \frac{(\Omega R)^2}{2}$$

$$\frac{p_2}{\rho} - \frac{p_3}{\rho} = \frac{\tilde{V}_3^2}{2} - \frac{V_{2z}^2}{2} = \frac{V_3^2}{2} + \frac{(\Omega R)^2}{2} - \frac{V_{2z}^2}{2}$$

where we have used the relation $\tilde{V}_3^2 = V_3^2 + (\Omega R)^2$ of figure 5.11(*b*). However, incompressible mass conservation requires that:

$$V_1 = V_{2z} = V_3$$

assuming no change in the flow area through the stage. Adding the two equations for pressure change and utilizing the constancy of the axial component of velocity required by mass conservation, as well as the relation $V_2^2 = V_{2z}^2 + (\Omega R)^2$ of figure 5.11(*b*), we find:

$$p_1 - p_3 = \rho(\Omega R)^2$$

Note that in this special case the power P equals $(\dot{m}/\rho)(p_1 - p_3)$, or the product of the volumetric flow rate $\dot{m}/\rho$ times the pressure drop across the stage.[13]

---

[12] In the case of an aircraft jet engine, where the inflow and outflow streams have no angular momentum, there is no net torque applied to the airframe, the compressor plus turbine rotor and stator torques each summing to zero.

[13] In thermodynamic cycles employing trubomachines to produce mechanical power, such as the steam turbine or gas turbine plant, the power produced by the turbine exceeds that absorbed by the feed water pump or the compressor because the volumetric flow rate $\dot{m}/\rho$ through the tubine is increased (compared with that through the pump or compressor) by heating the fluid in the boiler or combustion chamber at constant pressure, thereby decreasing ρ and increasing $\dot{m}/\rho$, since $\dot{m}$ is the same for turbine and pump or compressor. Of course, in these examples, the flow is not incompressible so that the power is not exactly the product of volumetric flow rate times pressure change, but the effect of decreasing the density of the working fluid by heating is to cause the turbine power to exceed the pump or compressor power and thereby make net power available for useful work.

Figure 5.12 A 226 *Megawatt* axial-flow gas turbine unit, shown with the upper half of the casing removed. Compressor stages in the upper left and gas turbine stages in the lower right sections of the picture. (Photograph courtesy of GE Industrial and Power Systems)

Figure 5.12 shows a view from above a large stationary gas turbine unit having the upper half of the casing removed. The fluid flows axially from the upper left to the lower right, first through the 18-stage compressor, then through the combustion chambers (not shown) and finally through the 3-stage turbine. Note the gradual shortening of the compressor blades through successive stages as the air density rises during compression. In the turbine stages, the blades become longer when the density falls.

5.6.3 Propellers

The fluid flowing into a propeller on an airplane or a ship has no angular momentum, but due to the rotation of the propeller, some angular momentum is added to the fluid in the direction of rotation of the propeller. As we shall see below, the amount of this angular momentum depends upon the angular velocity Ω of the propeller and its thrust F.

The thrust force F exerted by the propeller is the result of a lift force generated by the moving propeller blade. Figure 5.13 shows a vector diagram of the fluid velocity $\tilde{\mathbf{V}}_b$ relative to the blade, which is the vector sum of the axial speed $\mathbf{V}_p$ of the fluid past the propeller minus the tangential speed of rotation of the propeller blade, $\mathbf{\Omega}\times\mathbf{R} = \Omega\,\mathbf{i}_z \times R\,\mathbf{i}_r = \Omega R\,\mathbf{i}_\theta$. The propeller blade has a helical shape so as to

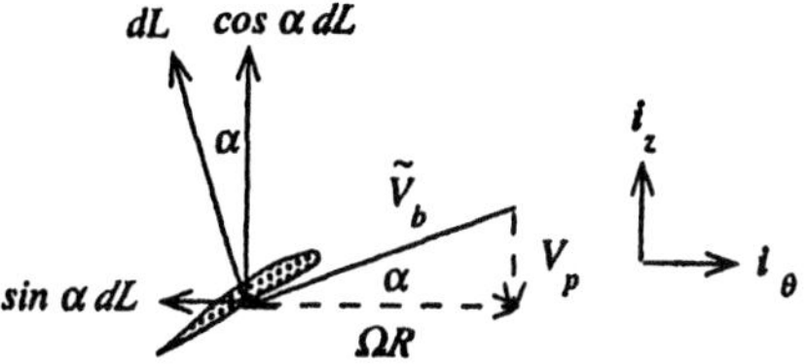

Figure 5.13 The fluid velocity $\tilde{\mathbf{V}}_b$ relative to a propeller blade is a combination of the axial flow velocity $\mathbf{V}_p$ through the propeller and the tangential velocity $\Omega r\, \mathbf{i}_\theta$ of the propeller blade. This relative flow produces a lift force increment dL that has components in the axial and tangential directions.

produce a small angle of attack of the blade cross-section with respect to the relative flow, as illustrated in figure 5.13. An incremental lift force dL acting on an increment of propeller span ds is generated by the propeller airfoil. The lift force dL has a direction perpendicular to $\tilde{\mathbf{V}}_b$, making an angle α with respect to the axis of rotation of the propeller.[14] The blade lift force increment dL has a component $(\cos\alpha)dL$ in the axial direction and a tangential component of $(\sin\alpha)dL$, that respectively contribute increments dF and dT to the thrust F and torque T:

$$dF = (\cos\alpha)dL$$

$$dT = R(\sin\alpha)dL \tag{5.50}$$

Eliminating dL and noting that $\tan\alpha = V_p/\Omega R$,

$$dT = R(\sin\alpha)\left(\frac{dF}{\cos\alpha}\right) = R(\tan\alpha)dF = \frac{V_p}{\Omega}dF$$

which can be integrated along the span of the propeller to give the magnitude of the torque T in relation to the thrust F:

$$T = \frac{V_p F}{\Omega} \tag{5.51}$$

Note that this relation is consistent with the requirement that the power P_p delivered to the propeller, which equals ΩT, is also equal to $V_p F$ by equation 5.18.

Most propellers operate under conditions where the propeller tip speed ΩR_p (where R_p is the radius of the propeller) is much greater than V_p so that $V_p/\Omega R_p$ and α are small. It therefore follows from equation 5.51 that $T \ll R_p F$.

The conservation of angular momentum, equation 5.46, tells us that, for a propeller, the angular momentum of the propeller wake, $(\mathbf{R} \times \dot{m}\mathbf{V})_{out}$, equals the torque T and

---

[14] We are neglecting here any drag force of the airfoil, which would act in the direction of $\tilde{\mathbf{V}}_b$.

is thus proportional to the thrust F by equation 5.51:

$$0 + (\mathbf{R} \times \dot{m}\mathbf{V})_{out} - 0 = 0 + 0 + 0 + \mathbf{T}$$

$$(\mathbf{R} \times \dot{m}\mathbf{V})_{out} = \frac{V_p F}{\Omega} \mathbf{i}_\theta \tag{5.52}$$

In single engine propeller-driven aircraft, the propeller torque is transmitted to the engine mount and airframe. In order to prevent the airplane from rolling, the aelerons are trimmed to counteract this torque. In twin engine aircraft, it is customary to have engines with opposite rotation so that there is no net torque applied to the airframe.

5.7 Problems

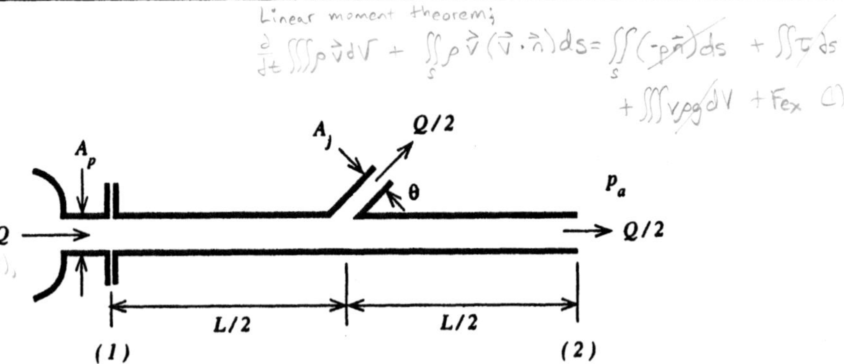

Figure P 5.1

Problem 5.1

A liquid of density ρ flows from a reservoir into a straight pipe of constant area A_p and total length L. The pipe is connected to the reservoir at a flange held by bolts at section 1, as shown in figure P 5.1. The volume flow rate $Q\{t\}$ into the pipe is a function of time t:

$$Q\{t\} = Kt$$

where K is a constant. At a point one half way along the pipe, one half of the flow ($Q\{t\}/2$) is expelled through a short side jet of area A_j, at an angle θ to the pipe axis. The rest of the liquid exits through the open end at section 2 where the external ambient pressure is p_a.

Assuming inviscid flow, derive expressions for (a) the time-dependent liquid pressure $p_1\{t\}$ at the pipe inlet section 1 and (b) the net tensile force F in the bolts holding the pipe to the reservoir in terms of the parameters ρ, K, A_p, L, θ and p_a.

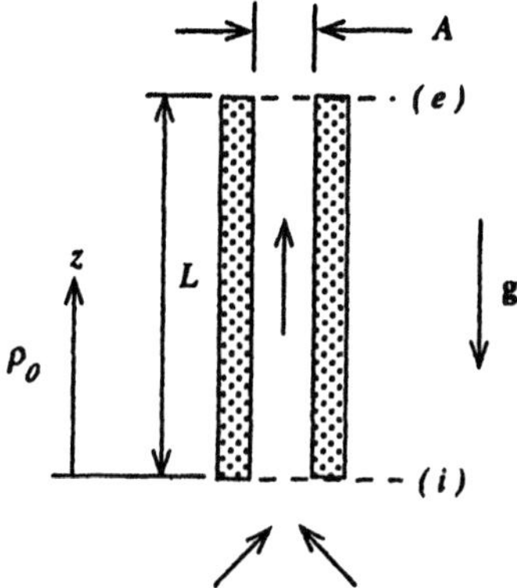

Figure P 5.2

Problem 5.2

In figure P 5.2, a natural convection heat exchanger consists of two heated plates between which ambient air moves vertically upward. Heat is supplied through the heat exchanger walls to an air channel of uniform cross-sectional area A. There is sufficient mixing so that the flow speed and air temperature are nearly uniform over any cross-section, giving rise to a vertical density distribution in the channel:

$$\rho\{z\} = \rho_0\left(1 - \frac{z}{4L}\right)$$

where ρ_0 is the uniform density of the unheated air outside the exchanger.

Assuming that wall friction is negligible in the heat exchanger channel, that the inflow air from the surrounding region to the inlet section i, at $z = 0$, is inviscid and incompressible, that the flow exits from the top of the exchanger, at section e, where $z = L$, at a pressure equal to the surrounding ambient pressure at that level, that the flow is steady and that the surrounding atmosphere has a static pressure distribution with constant density ρ_0, derive an expression for the inlet velocity V_i in terms of the parameters L, g, A, and ρ_0, as necessary.

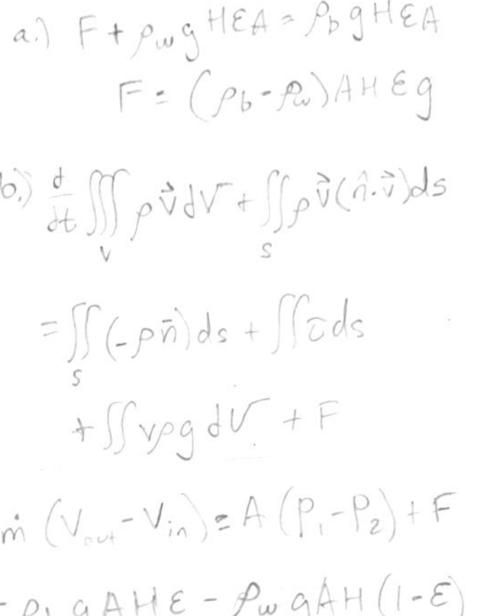

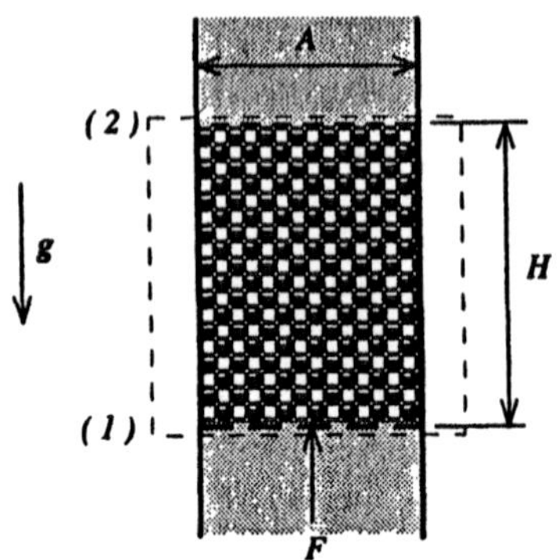

Figure P 5.3

Problem 5.3

A vertical pipe filled with water of density $\rho_w = 1E(3)\,kg/m^3$ and having a cross-sectional area $A = 1E(2)\,cm^2$ contains a layer of glass beads of height $H = 20\,cm$ that is held in place by a screen placed underneath the bead layer, as shown in figure P 5.3. The beads have a density $\rho_b = 2.4E(3)\,kg/m^3$ and the fraction of the layer volume AH occupied by the beads is $\epsilon = 0.52$.

(a) When there is no flow of water, an upward force F is exerted on the bed of beads by the screen underneath it. Derive an expression for the force F in terms of the parameters ρ_w, ρ_b, ϵ, A, H and g by considering a force balance on the beads in the control volume shown in figure P 5.3. (You may neglect any vertical force exerted on the beads by the pipe walls.) (b) Water is pumped upward at a flow rate such that the screen exerts no force on the bed of beads ($F = 0$). By applying the momentum theorem to the control volume, calculate the pressure drop $p_1 - p_2$ between the bottom and top of the bead layer.

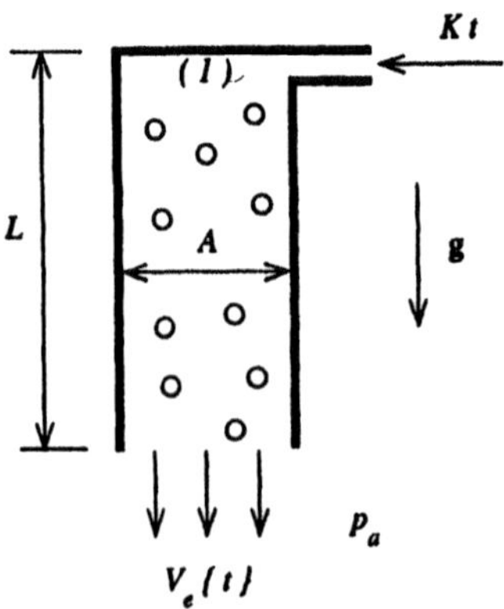

Figure P 5.4

Problem 5.4

Water flows into a chamber of cross-sectional area A and Length L through an inlet port at the top of the chamber. The chamber is situated in a gravitational field $\mathbf{g}$ as shown in figure P 5.4. The mass inflow rate $\dot{m}$ changes with time according to:

$$\dot{m} = Kt$$

where K is a dimensional constant. As water enters the chamber, an electric current flows through it at the top of the chamber, causing partial electrolysis and the formation of bubbles of hydrogen and oxygen. The resulting mixture of water and bubbles has an average density ρ_0 that is uniform in the chamber and does not change with time. The mixture flows rapidly downwards through the chamber (all components have the same velocity) in a turbulent, viscous manner and is ejected into the atmosphere, where the pressure is p_a.

(a) Derive an expression for the time-dependent exit velocity $V_e\{t\}$ at the chamber outlet. (b) Assuming that the shear stresses on the walls of the chamber may be neglected, derive an expression for the time-dependent pressure $p_1\{t\}$ at the top of the chamber.

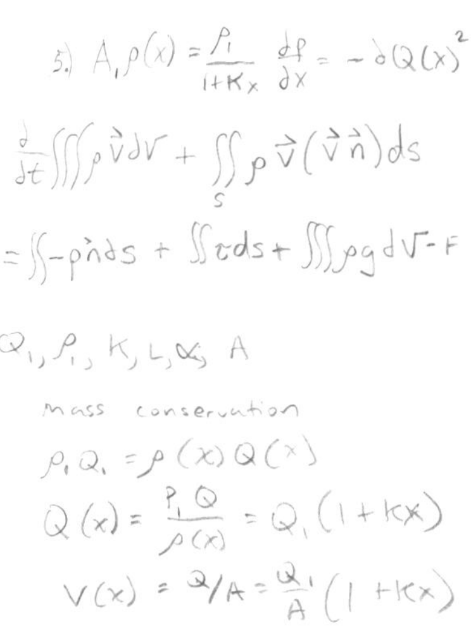

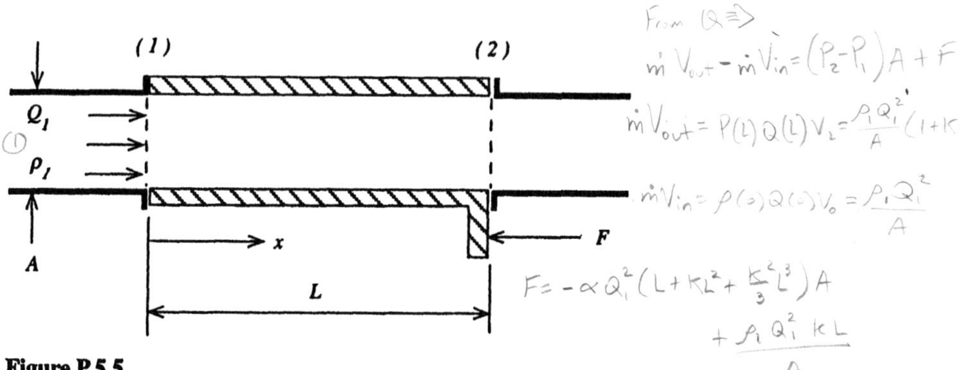

Figure P 5.5

Problem 5.5

An experiment is being conducted in a straight test section of length L and cross-sectional area A, as seen in figure P 5.5. The flow is steady and approximately uniform across the test section at every cross-section. The fluid density varies with axial position x along the axis of the test section according to:

$$\rho\{x\} = \frac{\rho_1}{1 + Kx}$$

where ρ_1 is the density at the entrance to the test section and where K is a dimensional constant. Because of viscous, turbulent flow in the test section, there is a pressure gradient in the x-direction that depends upon the local volume flow rate $Q\{x\}$ at each location x along the test section, according to:

$$\frac{dp}{dx} = -\alpha(Q\{x\})^2$$

where α is a dimensional constant. Derive an expression for the force F needed to restrain the test section from moving in the flow direction. Express this answer in terms of the known parameters Q_1, ρ_1, K, L, α and A.

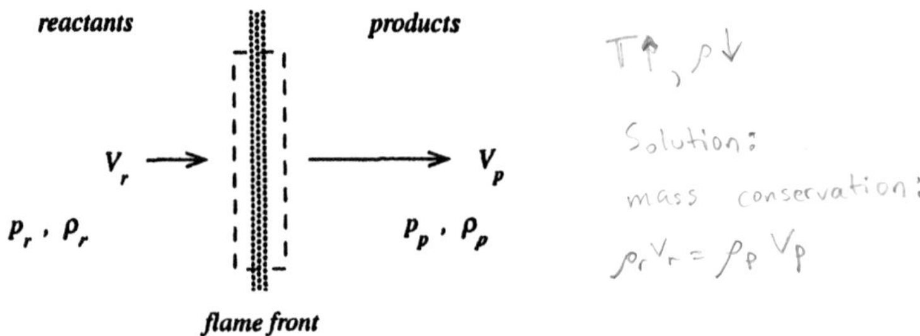

Figure P 5.6

Problem 5.6

A plane flame front is a very thin zone within which reactants (fuel and air) are transformed into products (water vapor, carbon dioxide, etc.). In this process, the fluid temperature increases and the density decreases.

In figure P 5.6, a plane flame front is observed to remain stationary and normal to a uniform steady flow of reactants. The upstream flow properties are measured to be $p_r = 1E(5)\,Pa$, $\rho_r = 1\,kg/m^3$ and $V_r = 0.1 m/s$. By thermodynamic calculation, the downstream product density $\rho_p = 0.2\,kg/m^3$. By applying the mass and momentum integrals to the control volume shown in the figure, calculate the downstream velocity V_p and pressure difference $p_r - p_p$.

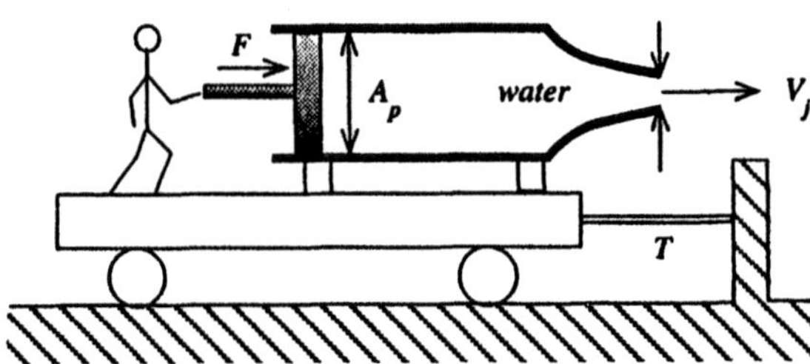

Figure P 5.7

Problem 5.7

A cylinder equipped with a frictionless piston of area A_p and filled with water of density ρ is mounted on a wheeled cart, as shown in figure P 5.7. A person standing on the cart exerts a force F on the piston, causing a water jet to be ejected into the surrounding air through the nozzle on the right of the cylinder. The cart is restrained from moving to the left by a rope tied to a stantion. Derive expressions for the velocity V_j of the water jet and the tension T in the rope in terms of ρ, F, A_j and A_p.

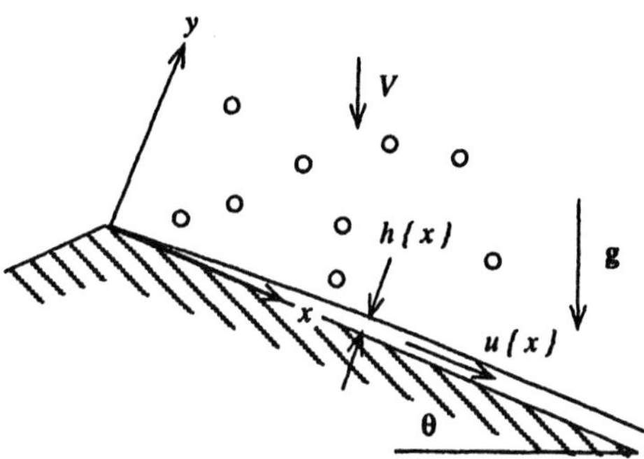

Figure P 5.8

Problem 5.8

A violent rainstorm hits a roof inclined at an angle θ from the horizontal in figure P 5.8. The rain pours down at a mass flow rate $\dot{m}$ per unit horizontal area, and each rain drop falls with a purely vertical speed of V. A steady state is established in which a water layer flows down the inclined roof while the raindrops splatter violently on the top part of the water layer. In analyzing the flow, you may make the following assumptions: (*i*) friction between the roof and the water layer is negligible, (*ii*) the angle of the water surface relative to the roof is very small ($dh/dx \ll 1$), but the roof inclination θ is not necessarily small, (*iii*) the rain drops lose their momentum instantly upon hitting the upper edge of the water layer and (*iv*) the water velocity $u\{x\}$ in the direction x is independent of y at any section x.

Derive expressions for (a) the pressure p_l just below the water surface and (b) the pressure $p\{x, y\}$ at any location x as a function of y in terms of the known parameters $\dot{m}$, V, θ, g, ρ and p_a and the unknown thickness $h\{x\}$. (c) By applying the mass conservation and linear momentum theorems to a control volume that encloses the liquid layer between $x = 0$ and a location x, derive an integral equation from which $h\{x\}$ could be calculated in terms of the known parameters.

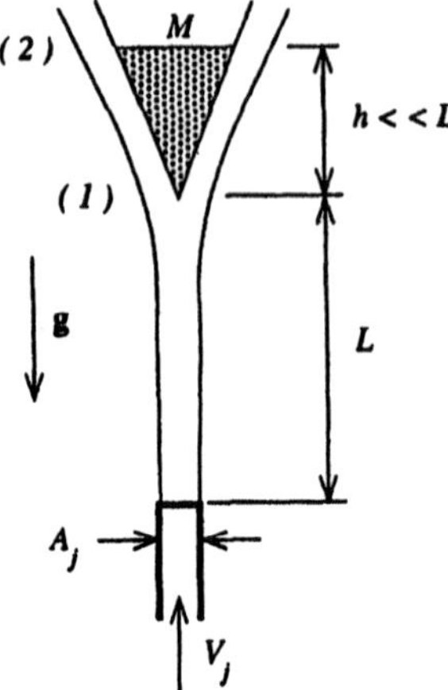

Figure P 5.9

Problem 5.9

A circular cone of mass M is suspended at a distance L above a jet of water, as shown in figure P 5.9. The side of the cone makes an angle θ with the horizontal, and its altitude $h \ll L$. Derive expressions for the velocity of the jet fluid at 1 and 2, as well as for the distance L.

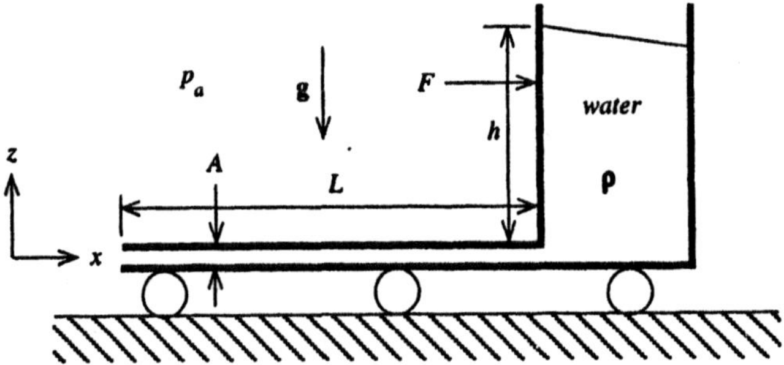

Figure P 5.10

Problem 5.10

A horizontal force F is applied to a cart on frictionless rollers so as to make it move with a constant acceleration a. The cart contains a large reservoir containing water at the bottom of which is a pipe of cross-sectional area A and length L that is open to the atmosphere at the left end, as shown in figure P 5.10. Water flows inviscidly from the reservoir, whose height is h, through the pipe to the atmosphere.

Derive an expressions for (a) the exit velocity $\tilde{V}$ ***relative*** to the cart and (b) the applied force F required to maintain the cart acceleration a in terms of the known variables ρ, g, a, h, L, A, and the mass M of the cart and its fluid.

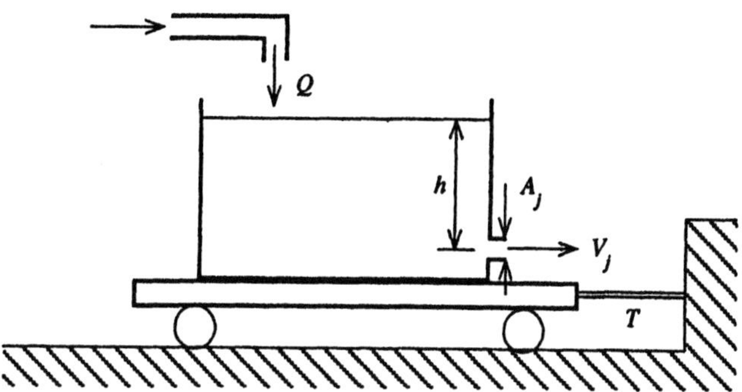

Figure P 5.11

Problem 5.11

In figure P 5.11, a water tank of uniform cross-sectional area A is placed on a cart, mounted on frictionless rollers. Water in the tank discharges to the atmosphere through a very short nozzle of area $A_j \ll A$. The water inside the tank is maintained at a constant height h above the nozzle by an external supply at a volumetric flow rate Q. The cart is restrained from moving by the cable seen in figure P 5.11. The mass of the tank, cart and water is M.

In terms of the parameters h, A_j and the water density ρ, derive expressions for (a) the volumetric flow rate Q and (b) the tensile force T in the cable. (c) At time $t = 0$, the cable is cut, and the cart begins to move to the left at a velocity $V_c\{t\}$ that increases with time. Derive an expression for $V_c\{t\}$.

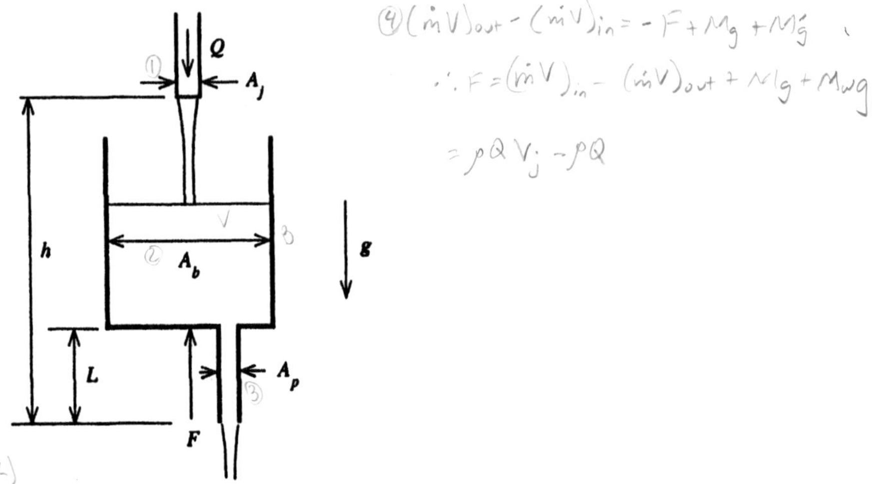

Figure P 5.12

Problem 5.12

A stream of water from a vertical jet of volume flow rate Q and area A_j falls into a bucket having a cross-sectional area A_b, as shown in figure P 5.12. The water in the bucket drains through a pipe of area $A_p \ll A_b$ and length L. The bottom of the pipe is a distance h below the exit of the water jet. Derive an expression for the force F required to hold the bucket stationary in terms of the parameters given in figure P 5.12 and the mass M of the bucket.

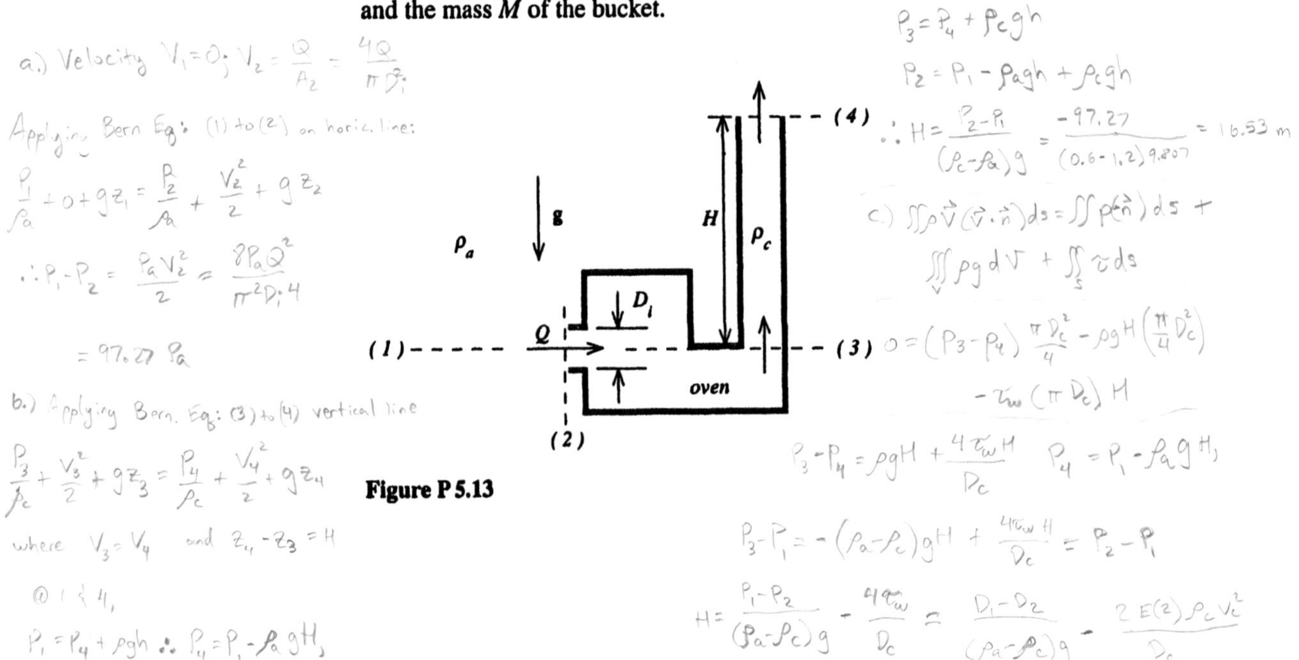

Figure P 5.13

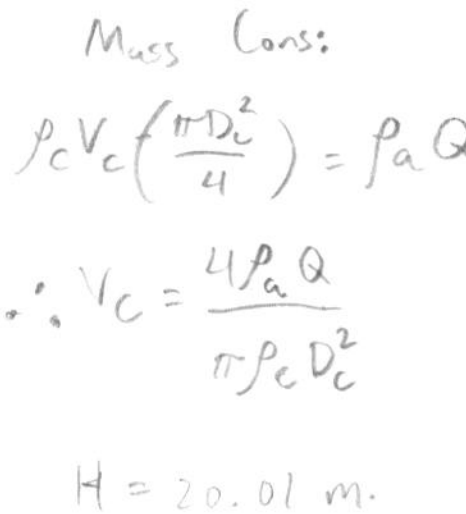

Problem 5.13

An industrial drying oven, heated by infrared lamps, requires a supply of cooling air at a rate $Q = 10\,m^3/s$. To induce this airflow, a vertical chimney of height H will be added downstream of the oven (see figure P 5.13), the buoyant flow up the chimney supplying enough draft to suck atmospheric air into the oven. You are asked to specify the height H of the chimney.

The diameter of the oven inlet is $D_i = 1\,m$ while the chimney diameter is to be $D_c = 1.5\,m$. The air is heated in the oven so that its density in the chimney is $\rho_c = 0.6\,kg/m^3$. There is no pressure change $p_2 - p_3$ between the oven inlet 2 and the chimney inlet 3. The atmospheric density is $\rho_a = 1.2\,kg/m^3$.

(a) Assuming incompressible inviscid flow from the atmosphere to the oven inlet, calculate the numerical value of the pressure change $p_1 - p_2$, where 1 is far from the inlet but at the same elevation. (b) Assuming the flow up the chimney is incompressible and inviscid, calculate the height H of the chimney needed to induce the flow $Q = 10\,m^3/s$. (c) Assuming that the flow in the chimney is incompressible but viscous and that a wall shear stress of $\tau_w = 5E(-3)\rho_c V_c^2$ exists on the chimney wall, calculate the height H of the chimney.

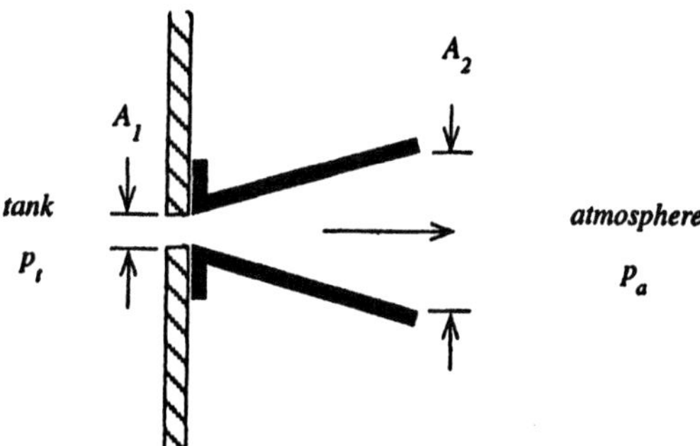

Figure P 5.14

Problem 5.14

A liquid flows steadily from a tank at a pressure p_t to the atmosphere at a pressure p_a through a conical nozzle attached to the tank, of inlet area A_1 and outlet area A_2, as shown in figure P 5.14. The flow may be considered inviscid.

(a) Derive an expression for the force F required to hold the nozzle in place against the tank, in terms of the parameters A_1, A_2, p_t and p_a. (b) Express the condition on p_t that ensures that the pressure p_1 at the nozzle entrance should be positive in terms of A_1, A_2 and p_a.

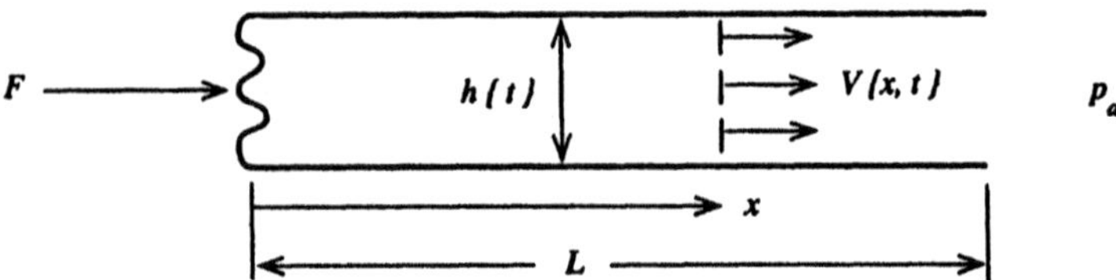

Figure P 5.15

Problem 5.15

Two thin rectangular plates, of dimensions L by W, are connected at one end by a closed-end flexible bellows arrangement, as sketched in figure P 5.15. The parallel plates are separated by a distance $h\{t\}$ that is decreasing at a constant rate w, *i.e.*, $dh/dt = -w$. The fluid between the plates is inviscid and incompressible, having a density ρ. The plate width W is much greater than the length L so that we can assume that the flow is quasi-unidirectional and that the velocity $V\{x, t\}$ is uniform at any cross-section x. The pressure at the outlet $x = L$ is atmospheric pressure, p_a.

(a) Derive an expression for the fluid velocity $V\{x, t\}$ in terms of the parameters x, h and w. (b) Express $p\{x, t\}$ as a function of ρ, x, L and $h\{t\}$. (c) Using the momentum theorem, derive an expression for the force F required to keep the plates from moving horizontally.

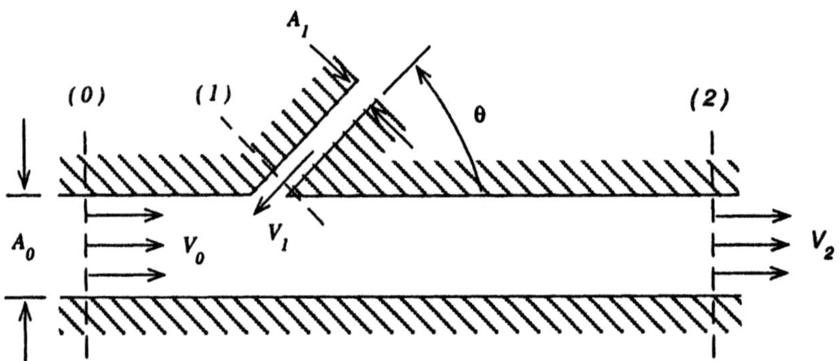

Figure P 5.16

Problem 5.16

An incompressible fluid of density ρ flows steadily along a duct of constant cross-sectional area A_0 with an upstream speed V_0. It is diluted by the injection of a different fluid of equal density which enters at an angle θ with respect to the duct axis through a port of normal area A_1 and with a speed V_1 (see figure P 5.16). The two streams mix in a viscous and turbulent manner. The resulting mixture, of density equal to that upstream, acquires a uniform speed V_2 downstream of the mixing region.

(a) Derive an expression for the pressure drop $p_0 - p_2$ between stations 0 and 2 in terms of the given inflow parameters ρ, A_0, A_1, θ, V_0 and V_1. (b) Derive expressions for the vertical and horizontal components of the force that the fluid exerts on the channel walls.

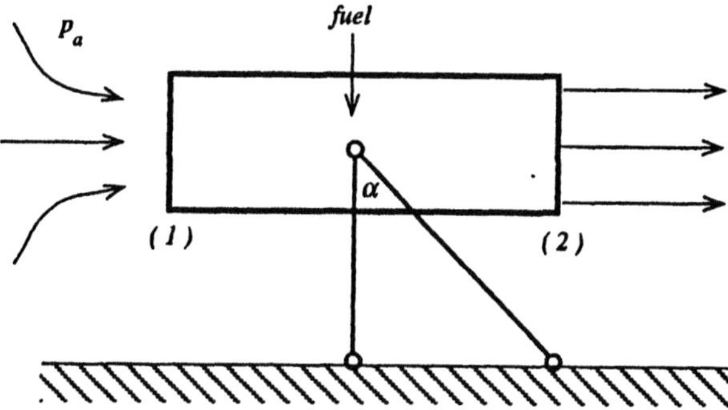

Figure P 5.17

Problem 5.17

A gas turbine engine is being tested under steady flow conditions in the rig shown in figure P 5.17. The engine thrust is balanced by the tension force F in the diagonal strut, which makes an angle $\alpha = 45°$ with the vertical. The mass flow rate of air from the ambient atmosphere into the engine is $\dot{m}_a = 1E(2)\,kg/s$ at a density $\rho_1 = 1.2\,kg/m^3$. The mass flow rate of fuel into the engine is $\dot{m}_f = 2\,kg/s$. The exhaust gas density is $\rho_2 = 0.4\,kg/m^3$. The intake and exhaust flow areas are both equal, $A_1 = A_2 = 1\,m^2$. The atmospheric pressure $p_a = 1E(5)\,Pa$.

Calculate (a) the inlet and outlet velocities, V_1 and V_2, (b) the inlet pressure p_1 and (c) the tension F in the diagonal strut.

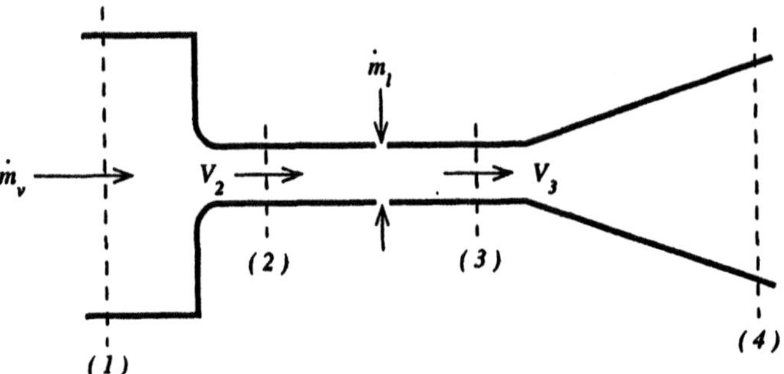

Figure P 5.18

Problem 5.18

In the apparatus in figure P 5.18, steam at the steady mass flow rate $\dot{m}_v$ and density ρ_v accelerates in an inviscid flow from a very large low speed reservoir (section 1) through a nozzle to the speed V_2 in a duct of constant cross-sectional area A. Spray nozzles surrounding the duct inject liquid water radially at a sufficient mass flow rate $\dot{m}_l$ to condense all the steam so that at section 3 only liquid flows at the speed V_3 and density ρ_l. The liquid stream is decelerated in an inviscid flow through a diffuser to a very low speed at section 4.

The flow is steady, and the pressure changes between 1 and 2 and between 3 and 4 are small enough so that the flows through the nozzle and diffuser may be treated as incompressible, having densities ρ_v and ρ_l, respectively. (a) Derive expressions for the dimensionless velocity ratio, V_3/V_2, and the dimensionless pressure change, $(p_2 - p_3)/(\rho_v/2)V_2^2$, in terms of the known parameters ρ_v, ρ_l, $\dot{m}_v$ and $\dot{m}_l$. (b) Show that the downstream pressure p_4 is greater than the upstream pressure p_1 if $(\dot{m}_l/\dot{m}_v) < \sqrt{\rho_l/\rho_v} - 1$.

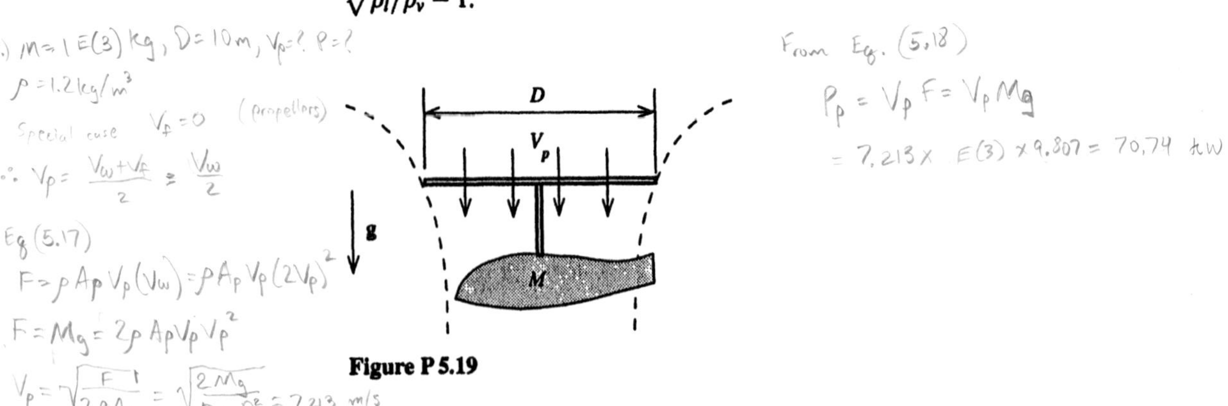

Figure P 5.19

Problem 5.19

A helicopter, having a mass $M = 1E(3)\,kg$ and rotor diameter $D = 10\,m$, is hovering in stationary flight in figure P5.19. The air drawn into the rotor passes downward through it at an unknown axial speed V_p. Calculate the speed V_p and the power required to sustain the hovering flight if the air density is $\rho = 1.2\,kg/m^3$.

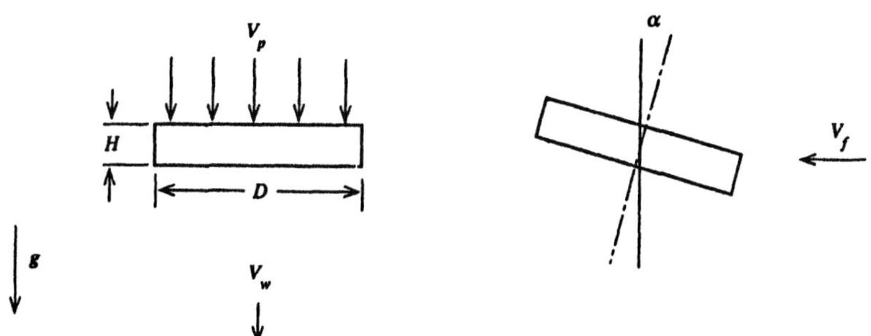

Figure P 5.20

Problem 5.20

A "flying saucer" consists of a pair of counter-rotating propellers (and engine) inside a cylindrical shroud of diameter D and axial height H. The propellers induce an air flow of uniform velocity V_p through the device, whose mass is M.

The flying saucer hovers motionless in a stationary atmosphere of density ρ, as shown on the left in figure P5.20. Derive expressions for (a) the axial velocity V_p through the device and (b) the power P_p required to maintain this flow in terms of the parameters M, D and ρ.

When the axis of the device is tilted by a small angle α from the vertical, it moves horizontally at a constant speed V_f, as shown on the right in figure P5.20. The horizontal drag force $\mathcal{D}$ exerted by the air on the moving device is

$$\mathcal{D} = (\rho/2)V_f^2 C_{\mathcal{D}}(HD)$$

where the drag coefficient $C_{\mathcal{D}}$ is a constant. For this motion, derive expressions for (c) the flight velocity V_f and (d) the power P_p required to maintain it in terms of the parameters M, D, H, α, $C_{\mathcal{D}}$ and ρ.

Problem 5.21

An inventor proposes to mount a horizontal axis wind turbine on a ship. The power generated by the wind turbine will be used to drive a ship's propeller and thus to propel the ship through the water. Assuming (*i*) that it is an ideal wind turbine with

100% conversion of wind turbine power P_{wt} to ship propulsion power, (*ii*) that the drag force on the ship's hull resisting its movement through the water is much smaller than the thrust force F acting on the wind turbine and (*iii*) that the wind turbine is operated so as to generate the maximum power possible, calculate the ratio V_s/V_a of the ship speed V_s to the wind speed V_a (relative to the water) when the ship is proceeding directly into the wind.

Problem 5.22

It is proposed to build a tidal current turbine at a location in a tidal estuary where the tidal current velocity V_t varies sinusoidally with time over a tidal period of 12.5 h. The maximum tidal current velocity $(V_t)_m = 2.5\,m/s$. A surplus ship's propeller, having a diameter $D = 1\,m$, will be used as the underwater equivalent of a wind turbine to drive an electric generator. The device will be submerged well below the surface of the water but clear of the bed of the estuary.

(a) If the propeller acted as an ideal turbine, calculate the maximum power that could be generated and the time-averaged power. (b) The characteristics of the propeller, acting as a power producing device, are not known. It is decided to operate the device conservatively, with the axial flow speed V_p at the propeller equal to $0.9V_t$. For such light loading, the ratio of the propeller tip speed ΩR to the current speed V_t, as calculated from the propeller's pitch/diameter ratio, is determined to be 3. Assuming ideal flow through the propeller/turbine, calculate the power P, the angular speed Ω, the torque T and the thrust force F at the time of maximum current when $V_t = 2.5\,m/s$.

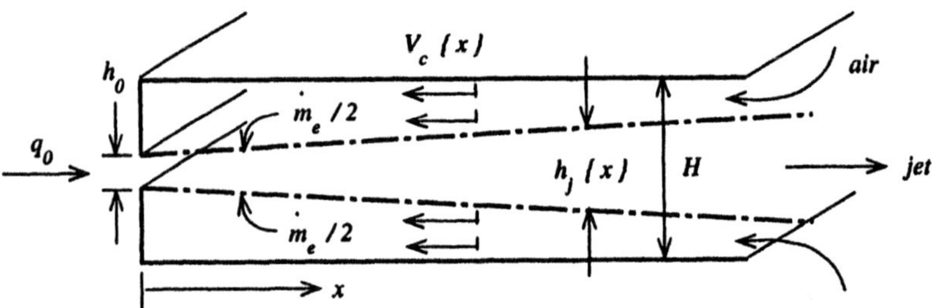

Figure P 5.23

Problem 5.23

A long, two-dimensional duct of constant height H is open to the atmosphere (pressure p_a) at one end. It is closed at the other end, except at the centerline, where a high-speed, turbulent air jet is injected axially with a volume flow rate per unit width of q_0, as shown in figure P 5.23. The thickness of the jet at the point of injection is h_0. The jet traverses the duct along the centerline and exits at the open end without touching the walls.

The turbulent jet "entrains" air into itself: surrounding air at its periphery is drawn into the jet and mixes with the high-momentum air in the jet. This entrainment gives rise to an air counterflow above and below the turbulent jet (see figure P 5.23). The jet velocity and thickness $h_j\{x\}$ therefore change slightly with position x in the channel.

The density ρ of the air in the jet and the counterflow is constant and uniform. The counterflowing air may be modeled as flowing inviscidly, but the jet flow is not inviscid. The mass rate of air entrainment $\dot{m}_e$ on the top plus the bottom of the jet, per unit width times length of the jet, varies with position x according to:

$$\dot{m}_e = 2\rho K x$$

In terms of the known parameters q_0, ρ, K, H, and h_0, derive expressions for (a) the volume flow rate per unit width of the jet, $q\{x\}$, as a function of position x along the jet axis, (b) the inflow velocity $V_c\{x\}$ of the counterflow air as a function of x, and (c) the pressure $p\{x\}$ of the air in the outer, counterflowing region as a function of x. (d) Show how you would use the linear momentum integral to derive an expression from which the jet thickness $h_j\{x\}$ could be found in terms of the parameters listed above.

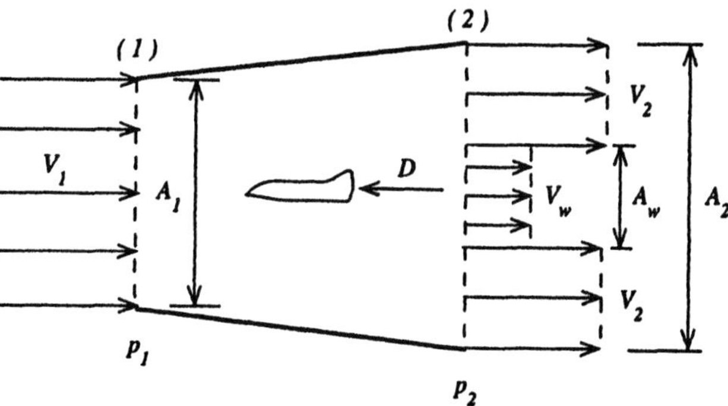

Figure P 5.24

Problem 5.24

A variable-area wind tunnel is used to determine the drag force D on a model of an airplane. The upstream flow in the tunnel has a uniform speed V_1 and pressure p_1 all across the inlet area A_1 of the tunnel test section. Although the downstream pressure p_2 is also uniform across the outlet flow area A_2, there is a core of the flow, of area A_w, within which the uniform speed V_w is less than the uniform speed V_2 in the remainder of the exiting flow (see figure P 5.24).

(a) The test engineer adjusts the exit area A_2 until the pressure p_2 and velocity V_2 match the inlet values, p_1 and V_1. This will simulate free flight in the atmosphere because the streamlines passing far from the aircraft model are little disturbed. Assuming incompressible flow and negligible wall friction in the tunnel, derive expressions for the downstream area A_2 and the model drag force D in terms of A_1, V_1, A_w, V_w, p_1 and ρ, the quantities measured in the test.

(b) In a second test, the engineer maintains the tunnel walls parallel to each other so that $A_2 = A_1$ but provides for sucking some air out of the tunnel through porous walls at a volume flow rate Q, the suction flow leaving the tunnel with a tangential velocity component equal to V_1. The rate Q is adjusted until $V_2 = V_1$ and $p_2 = p_1$. Derive expressions for the volume flow rate Q and drag force D in terms of the measured quantities A_1, V_1, A_w, V_w, p_1 and ρ.

(c) In a third test, the engineer keeps the walls parallel, as in (b), turns off the suction ($Q = 0$) and measures all the flow quantities upstream and downstream (p_1, A_1, V_1, p_2, $A_2 = A_1$, $V_2 A_w$ and V_w). In terms of these quantities, derive expressions for the drag force D, as well as a relationship among them that expresses the conservation of mass.

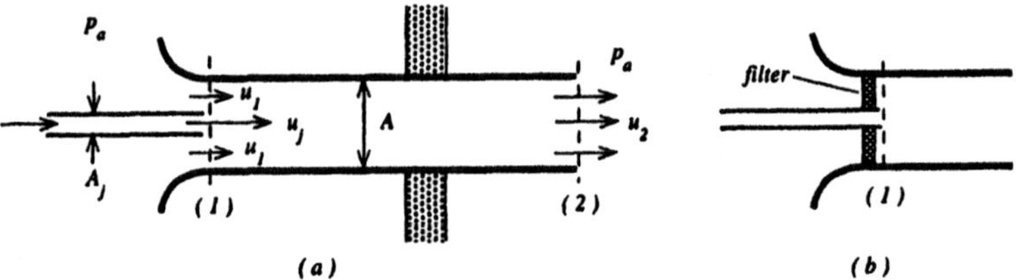

Figure P 5.25

Problem 5.25

In figure P 5.25(a), a ventilation duct of cross-sectional area A is used to exhaust air from a work space to the exterior of a building. Instead of using a blower, the duct is equipped with a high-speed air jet (nozzle area A_j and speed u_j), which induces an air flow through the duct by turbulent mixing with the air in the duct. Although the flow of room air into the duct inlet can be considered as inviscid, the duct flow involves viscous mixing but with negligible wall friction. Both the duct and the jet flows have the same constant density ρ, and the entrance and exit velocities (u_j, u_1 and u_2) are uniform across their respective flow areas (A_j, $A - A_j$ and A). The ambient pressure inside and outside the building is the same and equal to p_a.

(a) Derive a set of algebraic equations from which you can find the inlet velocity u_1 in terms of the known quantities A, A_j, u_j, p_a and ρ, but do not solve for u_1. Show that you have sufficient equations to find u_1 explicitly. (b) In a test of the system, it is found that the volume flow through the duct is proportional to the volume flow rate of the jet. Show that this experimental result is consistent with the equations you derived in (a). (c) It is decided to place a filter at the entrance to the duct to catch dust particles, as shown in figure P 5.25(b). The pressure drop Δp across the filter is related to the flow by:

$$\Delta p = C\left(\rho\frac{u_1^2}{2}\right)$$

where C is a constant. Derive the modification to the equations of part (a) needed to permit determining the flow speed u_1 with the filter in place.

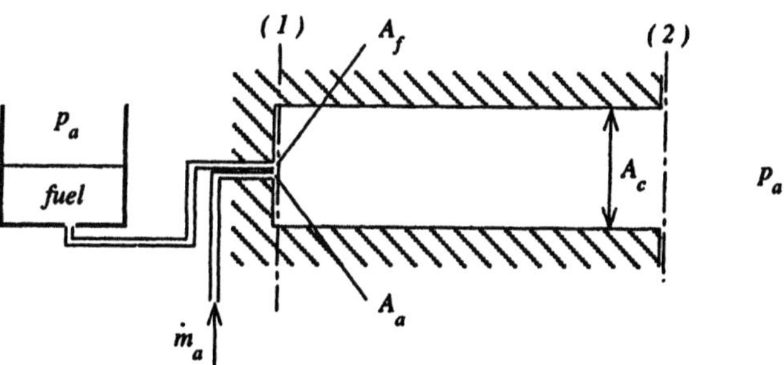

Figure P 5.26

Problem 5.26

In a simple fuel/air mixing chamber sketched in figure P 5.26, fuel is drawn from a fuel tank by a subatmospheric pressure p_1 produced by an adjacent air jet having a known mass flow rate of air, $\dot{m}_a$. Under steady flow operation, fuel enters the mixing chamber as a steady, uniform stream that breaks into droplets due to the combined effects of surface tension and air-flow turbulence. These droplets then mix with the air (but do not vaporize) and exit with the same speed as the air in a stream of uniform velocity at section (2). The effects of wall friction, both in the liquid fuel supply line and in the mixing chamber, are small and may be neglected. The air density ρ_a is constant and much smaller than the fuel density ρ_f. At (1), the fuel and air jet areas, A_f and A_a respectively, are much smaller than the mixing chamber flow area A_c.

(a) Assuming inviscid flow in the fuel line, derive an expression for the mass flow rate of fuel $\dot{m}_f$ in terms of the (unknown) pressure p_1 and other parameters of the problem listed above. (b) By applying the linear momentum principle to the flow in the mixing chamber, derive another expression that relates the unknowns, $\dot{m}_f$ and p_1, and, by combining with the results of (a), derive an expression for the mass flow ratio $\dot{m}_f/\dot{m}_a$ in terms of the known flow areas and fluid densities.

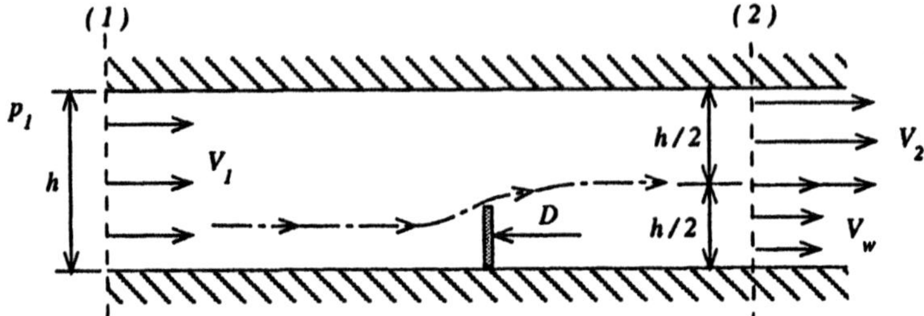

Figure P 5.27

Problem 5.27

A model of a snow fence is tested in a wind tunnel to determine the drag force D exerted on the fence by the wind. (A snow fence is an open meshwork fence designed to collect wind driven snow.) Figure P 5.27 shows the fence in the wind tunnel, whose width normal to the figure is W. Upstream of the fence, at section 1, the pressure p_1 and the velocity V_1 are measured and found to be uniform across the tunnel cross-section. Downstream of the fence, at section 2, the velocity is not uniform. In the lower half of the tunnel, a velocity V_w is measured, which is smaller than V_1. By injecting smoke in the tunnel to locate streamlines, it is concluded that the lower speed fluid in the lower half of the tunnel at section 2 is fluid that has passed through the snow fence while fluid in the upper half of the tunnel at 2 has passed above it. (These flows are separated by the dividing streamline shown in figure P 5.27.)

Assume that the flow is incompressible of density ρ. In terms of the known quantities p_1, V_1, V_w, ρ, W and h, derive expressions for (a) the velocity V_2 in the upper half of the tunnel at 2, (b) the pressure p_2 at 2, assuming that the flow in the upper half of the tunnel, which does not pass through the snow fence, is inviscid, and (c) the force D needed to restrain the snow fence (assuming the frictional forces on the wind tunnel walls are negligible). (d) By examining the streamlines that pass through the fence, it is concluded that there is no change in the velocity V_f of the air as it passes through the fence but that there is a pressure drop Δp across the fence because of viscous effects. Assuming that the flow upstream and downstream of the fence (but not through the fence) is inviscid, derive an expression for Δp. (Note that V_f is not measured.)

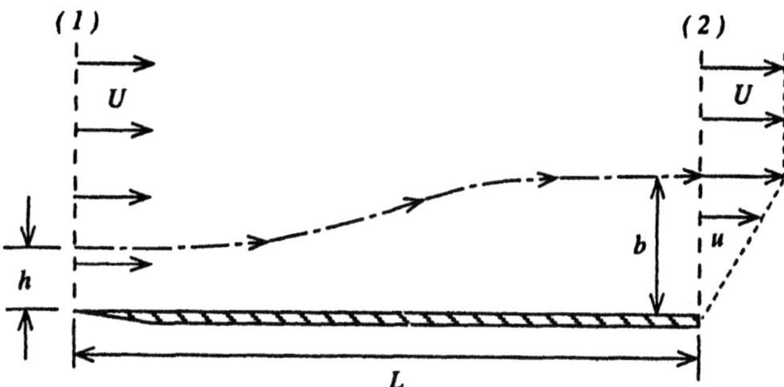

Figure P 5.28

Problem 5.28

A flat plate of length L is placed in a wind tunnel, aligned with the flow, as illustrated in figure P 5.28. Measurements of the flow above the plate reveal that the horizontal velocity profile at the leading edge, section 1, is uniform (see figure P 5.28) and has a speed U. At the trailing edge, section 2, the velocity profile is linear out to a distance b, but uniform beyond that point:

$$u = U\left(\frac{y}{b}\right), \quad 0 \le y \le b; \qquad u = U, \quad y \ge b$$

Figure P 5.28 shows the streamline which ends at a distance b from the trailing edge; its distance h from the leading edge is not measured.

In terms of the measured values b, L, U and the density ρ, derive expressions for (a) the distance h and (b) the drag force D (per unit distance normal to the plane of figure P 5.28) that the air flow exerts on the upper surface of the plate, assuming the pressure is uniform throughout the flow field. (c) The plate is porous. Air is blown upward through the plate, emerging with an upward speed $v = U/10$ all along the upper surface of the plate. There is no horizontal component of this air at the plate surface. Derive expressions for h and D in terms of b, L, U and ρ.

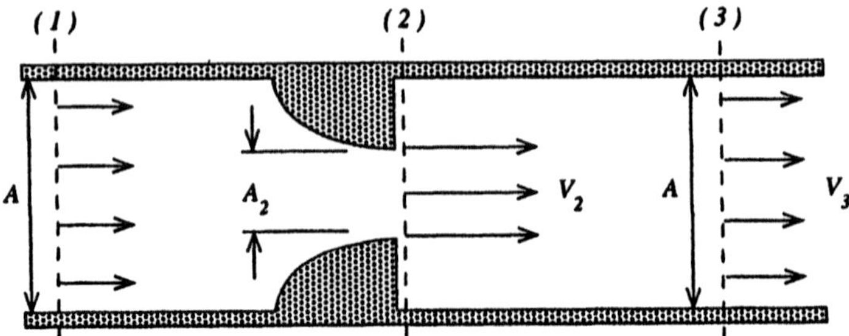

Figure P 5.29

Problem 5.29

An incompressible fluid of density ρ flows steadily in a circular pipe of cross-sectional area A that encloses a nozzle, as shown in figure P 5.29. At sections 1, 2 and 3, the pressures p_1, p_2 and p_3 are uniform across the cross-sectional area A. The velocities V_1 and V_3 are also uniform across the area A, while at the nozzle exit the velocity V_2 is uniform across the area A_2 but is zero across the remaining area, $A - A_2$.

(a) Given the known quantities A, A_2 and V_1, derive expressions for the velocities V_2 and V_3. (b) Due to viscous effects downstream of the nozzle, Bernoulli's equation does not apply to the flow from 2 to 3. Using the momentum integral, derive an expression for $p_3 - p_2$ in terms of ρ, V_1, A_2 and A. (You may neglect the viscous shear forces along the pipe wall between 2 and 3.) (c) Assuming the flow between 1 and 2 is inviscid, prove that $p_3 < p_1$. (d) Derive an expression for the force F exerted by the fluid on the nozzle.

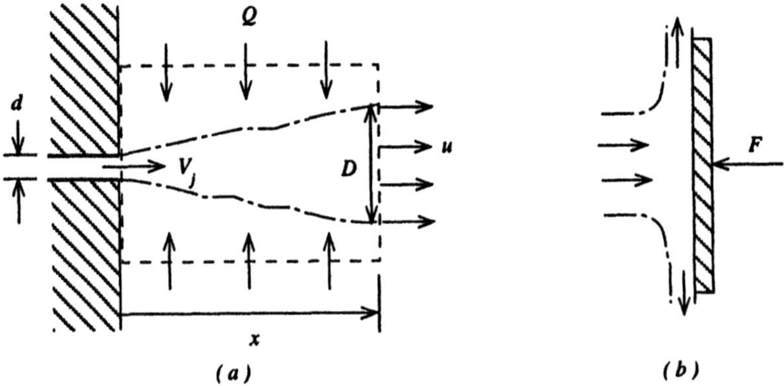

Figure P 5.30

Problem 5.30

An axisymmetric jet of air (exit velocity V_j) is ejected into a stationary atmosphere from a pipe of diameter d embedded in a wall, as shown in figure P 5.30(a). In the jet, the fluid velocity u is uniform throughout the cross-section of diameter D. The jet diameter D is observed to grow linearly with the distance x from the wall:

$$D = d + Kx$$

where K is a dimensionless proportionality constant. The jet entrains some of the surrounding air, which flows inward radially toward the jet, as sketched in figure P 5.30(a).

(a) Assuming that the pressure inside the jet is the same as atmospheric pressure and that the density of the jet fluid is the same as the density ρ of the atmosphere, derive an expression for the jet velocity u as a function of x and the parameters V_j, d and K by applying the linear momentum theorem to the control volume in figure P 5.30(a). (b) At a large distance from the jet orifice the jet impinges on a solid wall, as shown in figure P 5.30(b). Derive an expression for the force F required to hold this second wall stationary. (c) Derive an expression for the volumetric flow rate $Q\{x\}$ of atmospheric air entrained into the jet between $x = 0$ and a point x in terms of x, V_j, d and K.

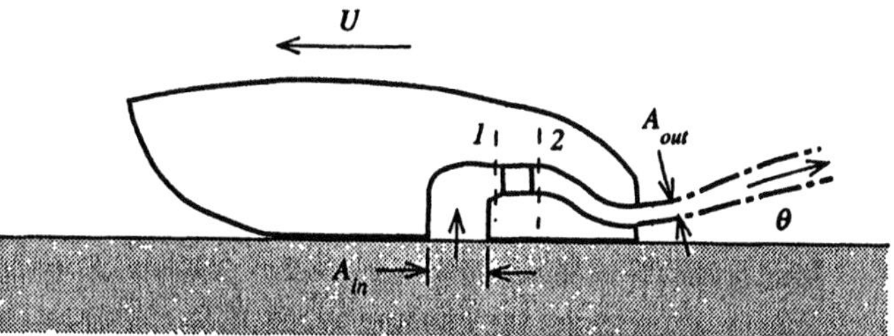

Figure P 5.31

Problem 5.31

A powerboat is propelled forward by a backward pointing water jet that is inclined at an angle θ above the horizontal (see figure P 5.31). The jet is produced by a pump that sucks water into an inlet of area A_{in} and discharges through an outlet of much smaller area A_{out} at a mass flow rate of $\dot{m}$. The water inlet is flush with the bottom of the hull and is horizontal. The mass density of the water is ρ.

(a) Derive an expression for the horizontal thrust F created by the propulsion system when the boat is at rest. Give your answer in terms of $\dot{m}$, A_{out}, ρ and θ. (b) When moving through the water at a speed U, the hull experiences a horizontal force F

resisting the motion, caused by friction and wave motion of the water on the surface of the hull:

$$F = KU^2$$

where K is a constant. Derive an expression for the cruising speed U in terms of $\dot{m}$, A_{out}, ρ, θ and K. (c) Assuming that the pressure p_{in} at the intake is atmospheric and that the flow is inviscid from the intake to section 1 (just upstream from the pump) and from section 2 (just downstream of the pump) to the exit, derive an expression for $p_2 - p_1$ if $A_1 = A_2$. (You may neglect gravity.) Give your answer in terms of $\dot{m}$, A_{in}, A_{out} and ρ.

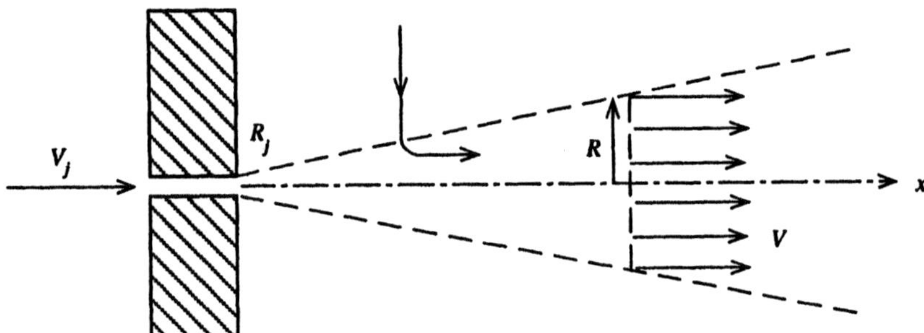

Figure P 5.32

Problem 5.32

A steady flow air stream emerges from a wall through a hole of radius $R_j = 2E(-3)\,m$ with a speed $V_j = 10\,m/s$ into a stationary atmosphere of constant pressure. As the stream mixes with the surrounding atmosphere, it forms a conical jet whose radius R increases linearly with distance x from the wall according to:

$$R = \alpha x$$

where the proportionality constant has the value $\alpha = 0.15$. As shown in figure P 5.32, the velocity V of the fluid in the jet is constant throughout the jet cross-sectional area πR^2. The fluid leaving the orifice in the wall mixes with the surrounding air, drawing it inward in a radial direction and accelerating it in the axial direction (see figure P 5.32).

(a) Assuming that the air leaving the orifice has the same density ρ_a as the surrounding atmosphere, calculate the numerical value of the velocity V of the jet fluid at a distance $x = 10\,m$ from the wall.

In a second experiment, the fluid leaving the orifice is argon gas, having a density

ρ_j that is greater than that of the surrounding air, ρ_a. Because of mixing with the surrounding air, the volume fraction of argon, $\phi\{x\}$, which is uniform within the jet cross-section at any distance x, decreases with increasing x. At a location in the jet where the argon volume fraction is ϕ, the jet density ρ is the sum of the argon density $\phi\rho_j$ and the air density $(1-\phi)\rho_a$.

(b) By considering also the mass flow of argon, derive (but do not solve) two equations whose solutions would give V and ϕ as functions of x.

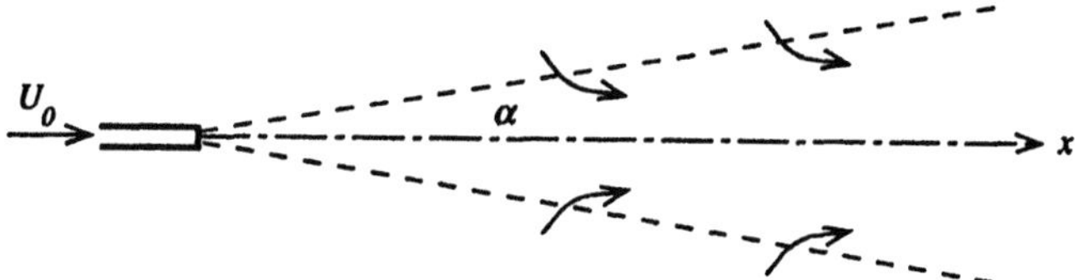

Figure P 5.33

Problem 5.33

A liquid of density ρ_s is sprayed from a nozzle into still air of density ρ_a, as shown in figure P 5.33. The liquid stream emerges from the nozzle at a speed U_0 and breaks up into small droplets, entraining air and accelerating it in the direction of the spray axis x. The mass flow rate of the liquid leaving the nozzle is $\dot{m}$. It is observed that the liquid droplets are uniformly distributed across a cross-section of a solid cone of half angle α and have the same speed U within that cross-section.

By using photography, the speed $U\{x\}$ of the spray droplets is measured as a function of distance x from the tip of the nozzle. (a) Derive an expression for the mean speed $V\{x\}$ of the air within the spray cone, as a function of x and the known flow variables, assuming that V is uniform across the jet cross-section. (b) The spray is observed to consist of droplets of uniform diameter d. Derive an expression for the number of droplets per unit volume, $n\{x\}$, as a function of x and the flow parameters.

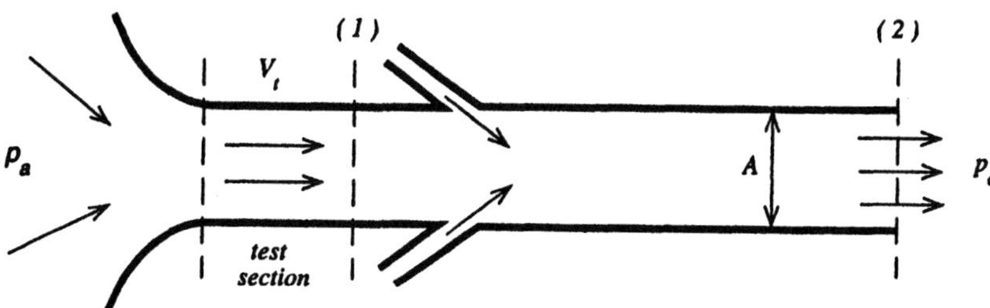

Figure P 5.34

Problem 5.34

It is proposed to build a wind tunnel in which flow is induced by jet pumps, as depicted in figure P 5.34. The tunnel consists of a duct of constant area A, with a well-rounded inlet leading to a test section. Downstream of the test section, air jets with a horizontal velocity component V_j are injected into the tunnel at a total mass flow rate $\dot{m}_j$. The high-speed air of the jets mixes in a turbulent, viscous fashion with the air coming through the test section, and by the time the mixture reaches the exit section (2), all the air has the same uniform velocity.

In analyzing this flow, you may assume that the air density ρ is constant, that between (1) and (2) the wall shear force is negligible and that the flow of air from the atmosphere to 1 is inviscid.

We are interested in determining the relation between the tunnel speed V_t and the tunnel design variables V_j, $\dot{m}_j$, A and ρ. (a) By examining the flow between (1) and (2), derive an expression for the pressure p_1 at station (1) in terms of the tunnel speed V_t and the design variables. (b) By considering the flow upstream of (1), derive another relation between p_1 and V_t. (c) From the results of (a) and (b), derive a relation between V_t and the design variables.

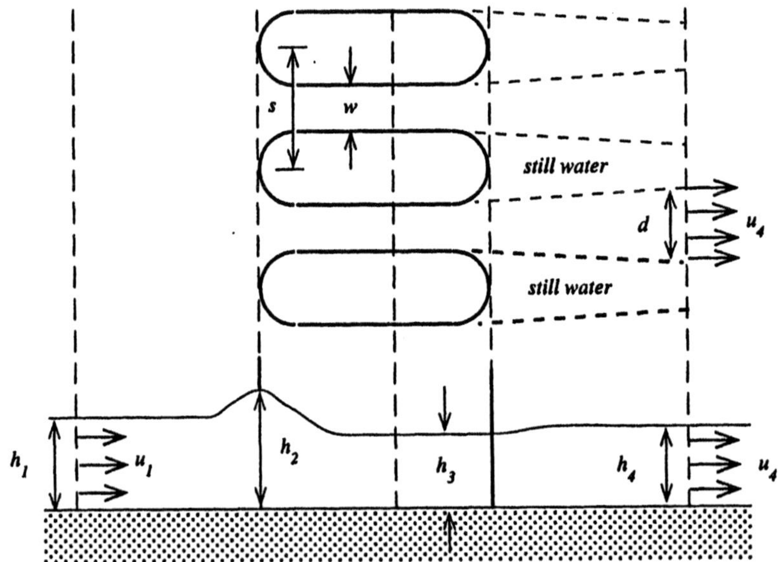

Figure P 5.35

Problem 5.35

A bridge over a river is supported by a series of caissons having a centerline separation distance s but only a channel width w between the parallel sides of adjacent caissons, as shown in the upper half of figure P 5.35 (plan view). The river has a level bed. Upstream of the bridge the river depth is h_1, and the flow speed u_1 is uniform from top to bottom and side to side.

When viewed from the side (lower half of figure P 5.35, elevation) the surface of the river is observed to rise to a height h_2 at the upstream end of the caisson and then fall to a lower height h_3 in the channel between the caissons. It is further observed that downstream of the bridge the flow speed u_4 is uniform within a channel of width d but essentially zero within a wake region of width $s - d$. Also, the water surface has a uniform height h_4.

(a) Assuming that the flow is inviscid between 1 and 3, derive expressions for the heights h_2 and h_3 in terms of the flow parameters h_1, u_1, s and w. (b) Derive an expression for the downstream speed u_4 in terms of the parameters u_1, h_1, s, d and h_4. (c) Neglecting friction on the river bed, derive an expression for the horizontal force exerted on a single caisson by the river flow.

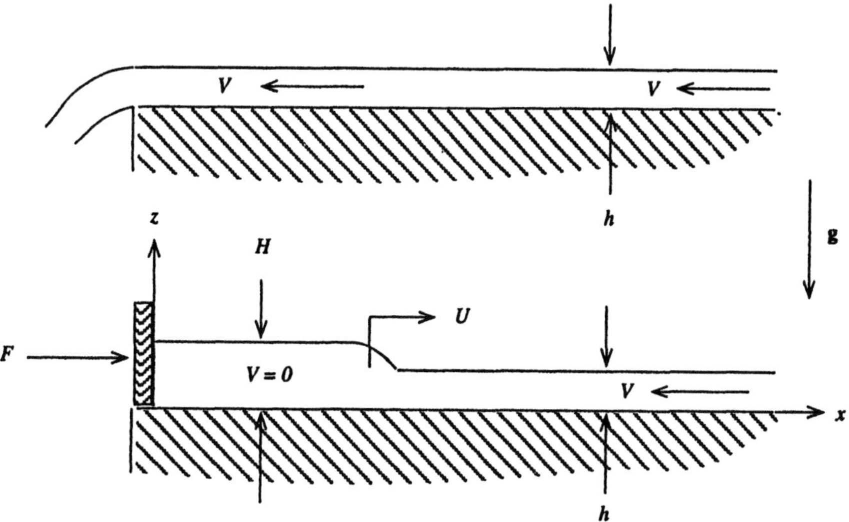

Figure P 5.36

Problem 5.36

A horizontal channel of width W carries a steady flow of water having a depth h

and speed V that is uniform throughout the channel, as shown in the upper half of figure P 5.36. At time $t = 0$, a gate is suddenly inserted into the flowing stream at the downstream end ($x = 0$), and it is observed that a thick layer of stationary water is formed adjacent to the gate. The front edge of this layer moves upstream at a constant speed U, into the oncoming flow (see the lower half of figure P 5.36).

(a) By considering the conservation of momentum in the x-direction, derive an expression for the horizontal force F required to support the gate in a stationary position, in terms of the observed flow quantities h, V and U. (b) By invoking the conservation of mass, derive an expression for the thickness H of the stationary layer in terms of the same quantities. (c) If the front edge of the stationary layer is an hydraulic jump characterized by the relations of equation 5.38, derive an expression for the front speed U in terms of the observed flow quantities V and h.

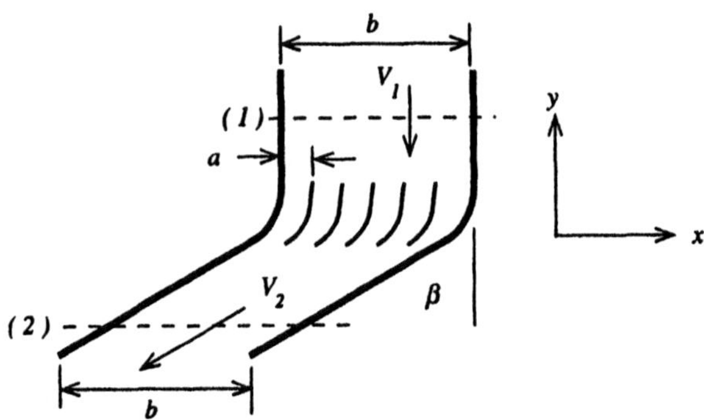

Figure P 5.37

Problem 5.37

Water flows in a horizontal rectangular duct of depth $W = 1\,m$ and width $b = 0.6\,m$, which undergoes a turn of angle $\beta = 60°$ (see figure P 5.37). There are guide vanes in the turn at equally spaced intervals of $a = 0.1\,m$. Assuming inviscid flow, calculate (a) the pressure change $p_1 - p_2$ between the flow upstream and downstream of the turn and (b) the x and y components of the force $\mathbf{F}$ exerted by the water on *one* guide vane.

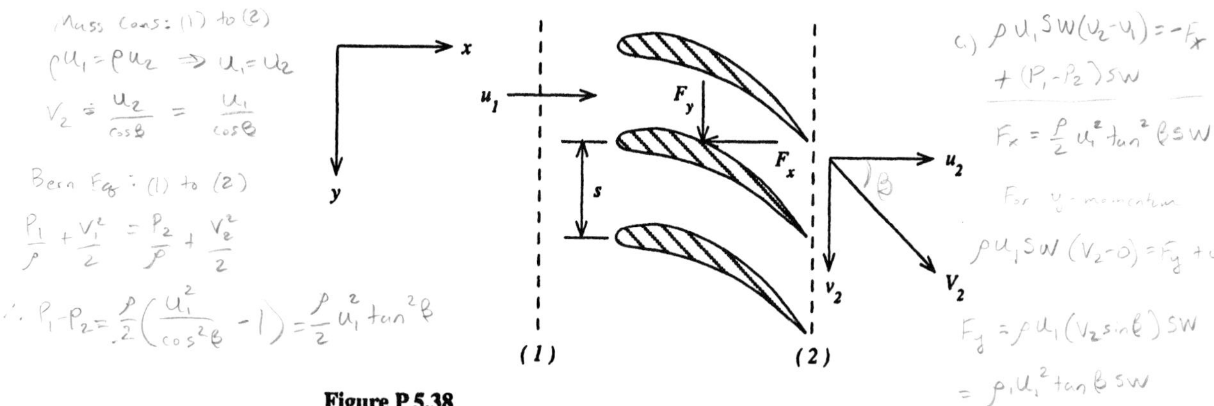

Figure P 5.38

Problem 5.38

An axial flow hydroturbine inlet stator consists of a series of blades that turn the inlet axial flow through an angle β, as shown in figure P 5.38. The blade spacing is s, and the height of the blade (dimension normal to the plane of figure P 5.38) is W. In terms of the flow parameters S, W, u_1, β and the density ρ, derive expressions for (a) the exit flow speed V_2, (b) the pressure drop $p_1 - p_2$ and (c) the force components F_x and F_y that restrain a single blade.

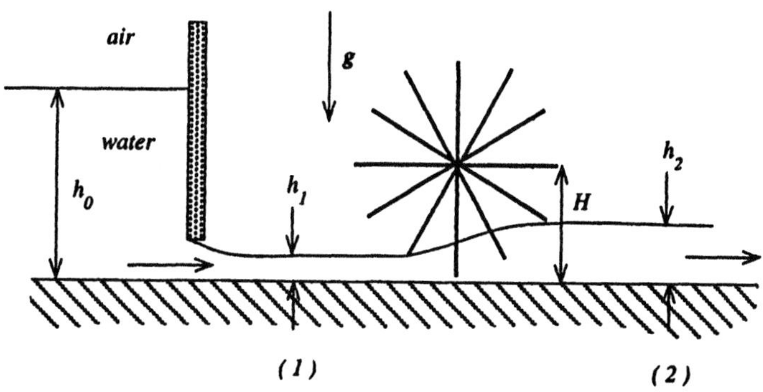

Figure P 5.39

Problem 5.39

The paddle wheel of an old sawmill is located in a water stream of width $W = 2\,m$. The water driving the mill is supplied by a reservoir whose water surface is a height $h_0 = 5\,m$ above the stream bed. The water is accelerated by passage under a sluice gate, approaching the paddle wheel at a depth of $h_1 = 1\,m$ (see figure P 5.39). The flow downstream of the paddle wheel has a depth $h_2 = 2\,m$.

The flow from the reservoir to station 1 may be considered inviscid, but between 1 and 2, where the paddle wheel churns in the water, viscous dissipation occurs, and the flow cannot be considered inviscid. However, the viscous shear force on the stream bed is negligibly small. Calculate (a) the velocities V_1 and V_2 at stations 1 and 2, and (b) the horizontal force F_x exerted by the flowing stream on the paddle wheel.

Problem 5.40

The water wheel of problem 5.39 is mounted with its axis a distance $H = 4\,m$ above the stream bed, as shown in figure P 5.39. Calculate the clockwise torque T exerted on the wheel by the stream of water when $V_1 = 8.858\,m/s$.

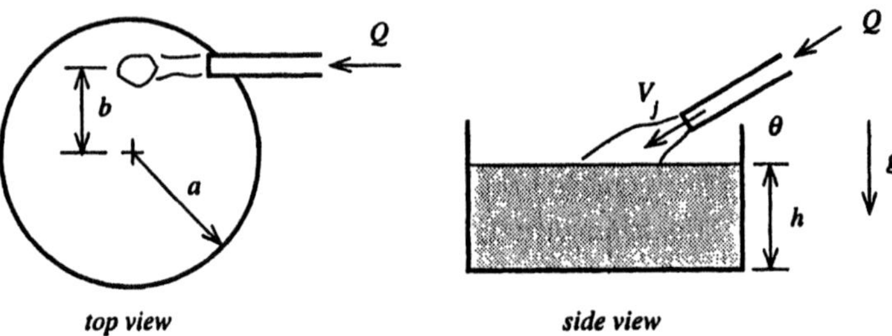

Figure P 5.41

Problem 5.41

At $t = 0$, a circular tank of radius $a = 0.5\,m$ contains water at rest having a depth $h = 0.2\,m$. Between $0 \leq t \leq \tau$, a water hose is played onto the surface of the water in the tank in figure P 5.41, the hose having a volume flow rate $Q = 2E(-3)\,m^3/s$ and an exit velocity $V_j = 1\,m/s$. The jet hits tangentially on the water surface at a radius $b = 0.3\,m$, with an angle $\theta = 30°$ above the horizontal. After a time $\tau = 100\,s$, the hose is turned off. Eventually, because of turbulent mixing within the water, all the water in the tank ends up rotating like a rigid body at an angular velocity Ω. Neglecting any frictional forces between the water and the tank walls, calculate the value of Ω.

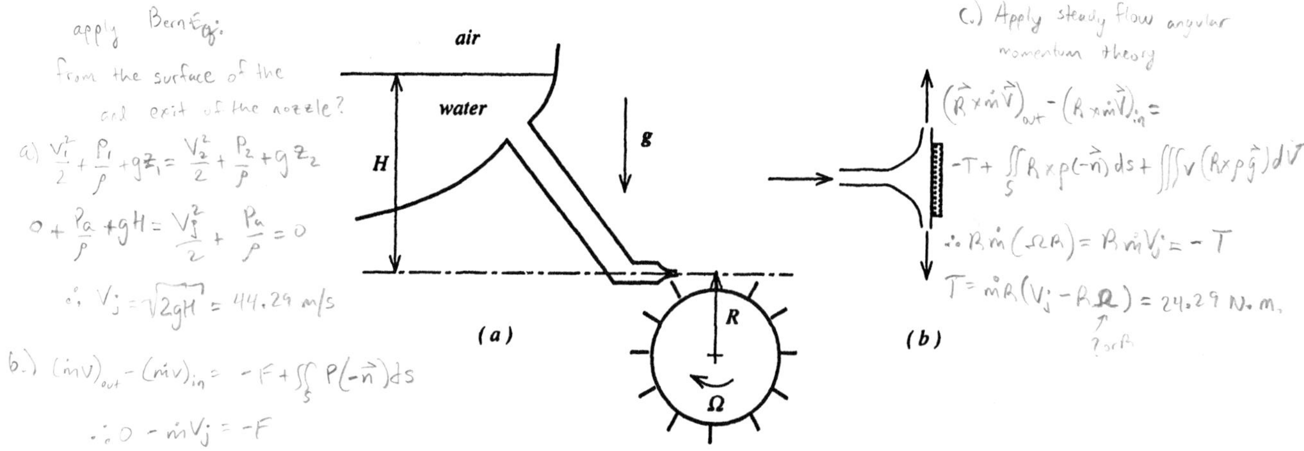

Figure P 5.42

Problem 5.42

A rudimentary water turbine is driven by a jet of water emerging from a nozzle into the surrounding atmosphere. The fluid is supplied from a large reservoir whose surface is a height $H = 100\,m$ above the nozzle, as sketched in figure P 5.42(a). The turbine blades are flat plates extending radially outward from the turbine wheel which intercept the jet stream. The radial distance from the turbine axis to the jet centerline, $R = 1\,m$, is much larger than the height of the blades. Relative to the moving blades, the jet stream is deflected sideways, perpendicular to the axis of the jet, as shown in figure P 5.42(b).

(a) Neglecting friction in the supply duct, calculate the speed V_j of the jet fluid. (b) If the turbine shaft is locked against rotation, calculate the force F exerted by the jet on the blade which intercepts it when the mass flow rate of the jet $\dot{m} = 1\,kg/s$. (c) If the turbine shaft rotates at an angular speed of $\Omega = 20$ *radians/s*, calculate the torque T exerted on the turbine shaft by the fluid stream.

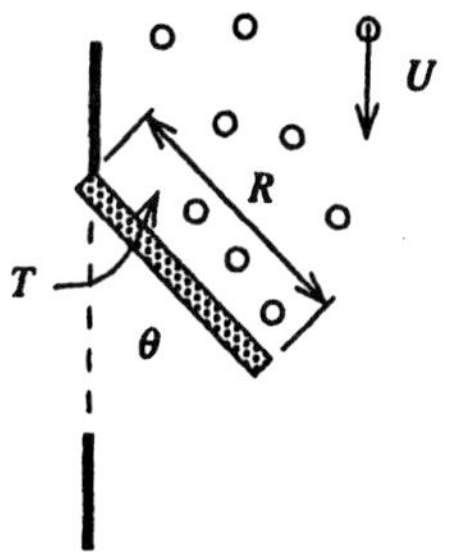

Figure P 5.43

Problem 5.43

On a rainy day, a window of height R and width W, hinged at the top, is opened against a falling rain, as shown in figure P 5.43. The rain drops have a droplet velocity of U and the rain rate (volume of water collected per unit horizontal area per unit time) is $\dot{h}$. All of the rain drops that fall on the window lose all their momentum to the window.

(a) Derive an expression for the mass of rain per unit volume of air, ρ_r. (b) Neglecting the gravitational force, derive an expression for the counterclockwise torque T required to hold the window fixed at $\theta = 45°$. (c) If the window is opened at a constant angular velocity Ω from $\theta = 0$ to $\theta = 90°$, derive an expression for the mass of rain that strikes the window. (In parts (a)–(c), express your answer in terms of ρ, U, $\dot{h}$, R, W and Ω only.)

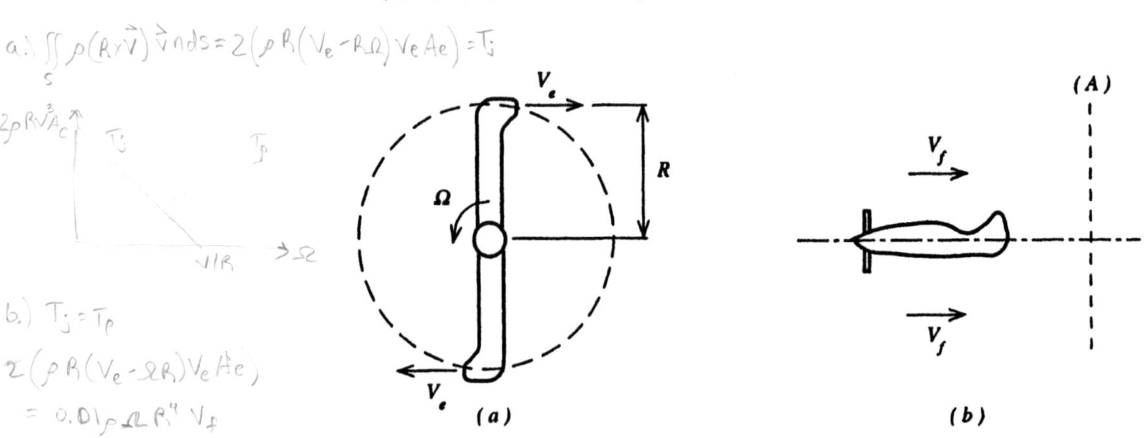

Figure P 5.44

Problem 5.44

An inventor is designing a new method for propelling a light aircraft. The method, sketched in figure P 5.44(a), consists of pumping air through the interior of a conventional propeller, from which it emerges at the propeller tip in a jet of velocity V_e (relative to the propeller) through an orifice of area A_e, the jet density being the same as atmospheric density ρ. The jet action rotates the propeller, supplying a torque T_j. The aircraft engine drives a compressor which supplies the air jets. The torque T_p required to turn the propeller at an angular velocity Ω is:

$$T_p = 0.01\rho\Omega R^4 V_f$$

where R is the propeller radius and V_f is the flight speed of the airplane. (a) Derive an expression for the torque T_j supplied by the jets to the propeller as a function of the parameters ρ, V_e, A_e, Ω and R. For fixed values of ρ, V_e, A_e and R, sketch how T_j varies as a function of Ω, finding the intercepts on the T and Ω axes. On the same sketch, indicate how the required propeller torque T_p varies with Ω. (b) The inventor chooses $V_e = 3V_f$ and $A_e = 0.01R^2$. Calculate the numerical value of $\Omega R/V_f$ resulting from this choice. (c) Downstream of the aircraft, what is the value of the integral $\iint \rho(\mathbf{R} \times \mathbf{V})\mathbf{V} \cdot \mathbf{n}\, dS$ evaluated across a plane (A) transverse to the propeller axis of rotation, as shown in figure P 5.44(b)? Justify your answer.

Bibliography

Landau, L. D., and E. M. Lifshitz. 1959. *Fluid Mechanics*. Vol. 6, *Course of Theoretical Physics*. Translated by J. B. Sykes and W. H. Reid. London: Pergamon Press.

Sabersky, Rolf H., Allan J. Acosta, and Edward G. Hauptman. 1989. *Fluid Flow, A First Course in Fluid Mechanics*. 3rd ed. New York: Macmillan Publishing Co.

6 Laminar Viscous Flow

6.1 Introduction

Having already noted that viscous effects can be neglected only under some circumstances, such as a static fluid or a flow at large Reynolds number[1] without boundary effects, we need to consider how to include viscous effects when they are important. While we will be able to formulate a differential equation of motion of a viscous fluid (called the Navier-Stokes equation), we will find it very difficult to apply it to any but the simplest flows. No such simple relationship as Bernoulli's equation exists for a viscous flow. Nevertheless, by considering a few simple viscous flows we can learn quite a bit about viscous flows in general that will be helpful in understanding the complex flows that arise in engineering applications.

A major complication that arises in understanding viscous flows is that some of these flows misbehave terribly—such a miscreant is called a *turbulent flow*. Turbulent flows are characterized by unsteady, chaotic velocity fields even when the controlling conditions of the flow are time independent. We depend very much on empirical descriptions of turbulent flows because our analytical and numerical skills cannot describe adequately the immensely detailed nature of such flows. Nevertheless, by studying the viscous flows that are not turbulent, which we call *laminar flows*, we can develop enough physical insight to help us understand the principal features of turbulent flows.

In this chapter, we first consider the formulation of the basic equation of motion of a viscous fluid, the Navier-Stokes equation. We then move on to some simple yet practical solutions of the Navier-Stokes equation for laminar flows, primarily steady, through channels or around solid bodies. In most of these examples, the fluid variables

[1] As a reminder, the Reynolds number of a flow is VL/ν, where V and L are a typical velocity and length of the flow field and ν is the kinematic viscosity of the fluid.

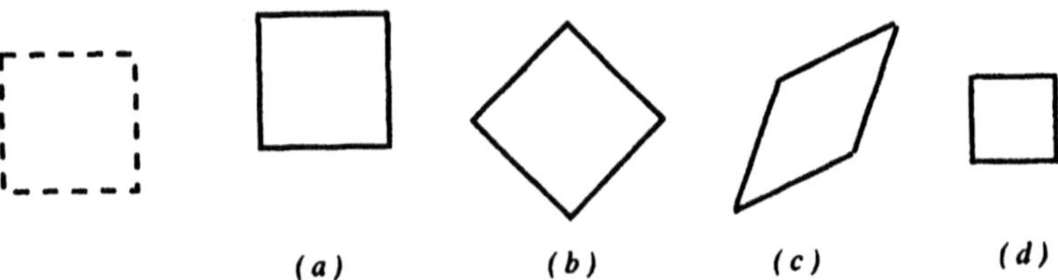

Figure 6.1 The motion of a fluid element, initially defined by the dashed figure, can be described as a combination of (*a*) translation, (*b*) rigid body rotation, (*c*) shear and (*d*) pure compression. Only the deformations of (*c*) and (*d*) give rise to viscous stresses.

depend upon only one spatial coordinate, but in a few, the dependence is upon two coordinates because the flow is more complex. Although the Reynolds number of these flows may be high, they are not inviscid because of the pervasive influence of nearby solid boundaries.

In the following chapter 7, we shall consider turbulent (viscous) flow, primarily from an empirical point of view. We will describe the conditions that make a viscous flow turbulent rather than laminar. For the time being, we will consider only laminar flows that, even if unsteady, are not random or chaotic.

6.2 The Viscous Stress

In hydrostatics, the only stress on an element of fluid surface is a normal stress, called the pressure, which is the same regardless of the orientation of the surface element (Pascal's Law).[2] However, in a viscous flow the stress $\boldsymbol{\sigma}$ will in general not be normal to the surface—in fact, it will usually have a tangential component proportional to the viscosity. We will find it convenient to consider the fluid stress $\boldsymbol{\sigma}$ to be the sum of a pressure stress $(-p)\mathbf{n}$ in the normal direction and a *viscous stress* $\boldsymbol{\tau}$ that is proportional to the viscosity:

$$\boldsymbol{\sigma} \equiv (-p)\mathbf{n} + \boldsymbol{\tau} \tag{6.1}$$

We have previously used this decomposition of the stress into a pressure and viscous component when deriving the linear and angular momentum theorems, equations 5.11 and 5.46.

A viscous stress arises when the shape of a fluid element is changed in a flow. The motion of a fluid element may be considered as the sum of four components, as illustrated in figure 6.1: (*a*) simple translation, (*b*) simple rotation as a rigid body,

---

[2]See section 2.3.1.

(*c*) shearing motion and (*d*) pure compression that is equal in all directions. Only the shearing motion (*c*) and compression (*d*) will give rise to a viscous stress that is proportional to the *rate* of these motions since there is no change in shape of the fluid element in the translation (*a*) and rotation (*b*).

Let us now determine the viscous stress that arises from the shearing motion. Consider first the simplest possible shear flow, that between two parallel planes as shown in figure 6.2(*a*). A fluid element is distorted when the upper surface slides relative to the lower surface. For a Newtonian fluid, a shear stress on the lower and upper surfaces of a fluid element is generated in proportion to the velocity gradient and the fluid viscosity:

$$\tau_{xy} = \mu \frac{\partial u}{\partial y} \tag{6.2}$$

in accordance with equation 1.5. (The symbol τ_{xy} denotes that the shear stress acts in the x direction on a surface element whose normal is in the y direction.) But there must be a complementary stress τ_{yx} on the vertical faces of the element in figure 6.2(*b*) that is equal to τ_{xy} because otherwise the element would accelerate in rotation at an infinite angular velocity as its size became infinitesimal.[3] Such a relation must hold for any direction of the axes:

$$\tau_{xy} = \tau_{yx}; \qquad \tau_{yz} = \tau_{zy}; \qquad \tau_{xz} = \tau_{zx} \tag{6.3}$$

Now consider a shear flow in which the boundary plates are oriented in the y direction, as shown in figure 6.2(*c*). The shear stress τ_{yx} on the surface of this element is:

$$\tau_{yx} = \mu \frac{\partial v}{\partial x} = \tau_{xy}$$

Any combination of the flows in figure 6.2(*b*) and (*c*) results in viscous stresses that are simply the sum of these separate contributions:

$$\tau_{xy} = \tau_{yx} = \mu \left(\frac{\partial u}{\partial y} + \frac{\partial v}{\partial x} \right)$$

This result holds for shearing motion in the x, y plane of figure 6.2. Similar relations must also hold for shear in the y, z and z, x planes. In summary, the viscous shear stresses are:

$$\tau_{xy} = \tau_{yx} = \mu \left(\frac{\partial u}{\partial y} + \frac{\partial v}{\partial x} \right)$$

---

[3]The moment of the shear forces acting on the element of figure 6.2(*b*) varies as the square of the element length whereas its moment of inertia varies as the fourth power of that length so that the angular acceleration would vary as the minus two power of the element length.

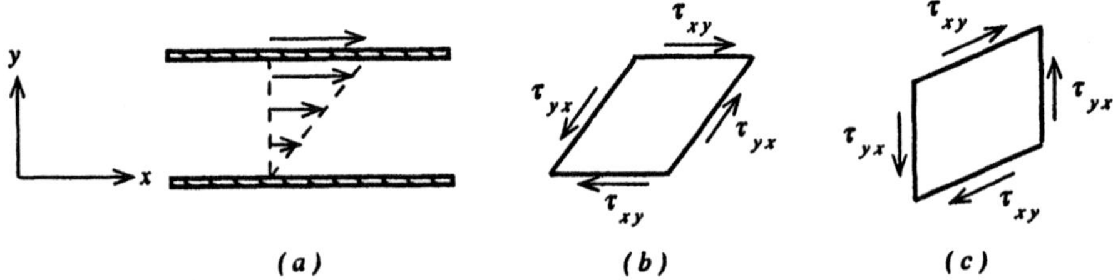

Figure 6.2 The simple shearing motion of (*a*) gives rise to the viscous shear stresses τ_{xy} and τ_{yx} shown in (*b*). If the shearing motion is in the *y* direction, the viscous shear stresses are those of (*c*).

$$\tau_{yz} = \tau_{zy} = \mu \left(\frac{\partial v}{\partial z} + \frac{\partial w}{\partial y} \right)$$

$$\tau_{zx} = \tau_{xz} = \mu \left(\frac{\partial w}{\partial x} + \frac{\partial u}{\partial z} \right) \tag{6.4}$$

Is there a viscous stress component normal to a surface in a flow? We might guess by analogy with the viscous shear stresses that the normal viscous stress τ_{xx} acting on a plane normal to the x axis might be:

$$\tau_{xx} = \mu \left(\frac{\partial u}{\partial x} + \frac{\partial u}{\partial x} \right) = 2\mu \left(\frac{\partial u}{\partial x} \right)$$

However, this is only partly correct. There is another contribution to τ_{xx}, whose correct form is:

$$\tau_{xx} = 2\mu \left(\frac{\partial u}{\partial x} \right) + \left(\mu_B - \frac{2}{3}\mu \right) \nabla \cdot \mathbf{V} \tag{6.5}$$

where μ_B is the *bulk viscosity* of the fluid.[4] For an incompressible fluid, $\nabla \cdot \mathbf{V} = 0$, the second term on the right of equation 6.5 is zero, and our initial guess was correct. Thus the normal viscous stresses in an incompressible fluid become:

$$\tau_{xx} = 2\mu \left(\frac{\partial u}{\partial x} \right)$$

$$\tau_{yy} = 2\mu \left(\frac{\partial v}{\partial y} \right)$$

[4]See, for example, Robert S. Brodkey, 1967, *The Phenomena of Fluid Motions* (Reading:Addison-Wesley Publishing Co.). The bulk viscosity of a monatomic perfect gas is zero. For a compressible fluid undergoing pure compression of figure 6.1(*d*), where $\partial u/\partial x = \partial v/\partial y = \partial w/\partial z = \nabla \cdot \mathbf{V}/3$, $\tau_{xx} = \tau_{yy} = \tau_{zz} = \mu_B \nabla \cdot \mathbf{V}$. Only in extreme cases of compressible flow (*e.g.*, shock waves or high frequency sound waves) might stresses proportional to the bulk viscosity be significant.

$$\tau_{zz} = 2\mu\left(\frac{\partial w}{\partial z}\right) \qquad (\textit{if}\ \nabla\cdot\mathbf{V} = 0) \tag{6.6}$$

Equations 6.4 and 6.6 together define the viscous stress $\boldsymbol{\tau}$ for an incompressible fluid.

Example 6.1

The velocity **V** of a steady air flow above a horizontal surface ($z = 0$) is:

$$\mathbf{V} = (az - bz^2)\,\mathbf{i}_x + cz\,\mathbf{i}_y$$

where the constants a, b and c have the numerical values of:

$$a = 1.0\,s^{-1},\ b = 0.1\,m^{-1}\,s^{-1},\ c = 2.0\,s^{-1}$$

Calculate the numerical values of all the viscous stress components (equations 6.4 and 6.6) at $z = 1.0\,m$ if $\mu = 1.82E(-5)\,Pa\,s$.

Solution
Since **V** depends only upon z, the only non-zero derivatives of **V** are:

$$\frac{\partial u}{\partial z} = a - 2bz; \qquad \frac{\partial v}{\partial z} = c$$

and the only non-zero viscous stress components are:

$$\tau_{xz} = \tau_{zx} = \mu(a - 2bz) = [1.82E(-5)\,Pa\,s](1.0\,s^{-1} - 2[0.1\,m^{-1}\,s^{-1}][1.0\,m])$$

$$= 1.456E(-5)\,Pa$$

$$\tau_{yz} = \tau_{zy} = \mu c = [1.82E(-5)\,Pa\,s](2\,s^{-1}) = 3.64E(-5)\,Pa$$

6.3 The Viscous Force

Just as the pressure stress $(-p)\mathbf{n}$ gives rise to a pressure force per unit volume of $-\nabla p$, the viscous stress $\boldsymbol{\tau}$ will give rise to a viscous force per unit volume, which we denote by **f**. If this force is divided by the density ρ, then the ratio, $\mathbf{f}/\rho$, is the viscous force per unit mass. Adding this to the pressure force per unit mass, $-\nabla p/\rho$, and the gravity force per unit mass, **g**, we would have the total force acting on a unit mass of viscous fluid. Setting this equal to the acceleration of a fluid particle, $D\mathbf{V}/Dt$, would give us an equation of motion for a viscous fluid. In effect, we would be adding an additional term, $\mathbf{f}/\rho$, to the right side of Euler's equation 4.5 to account for the

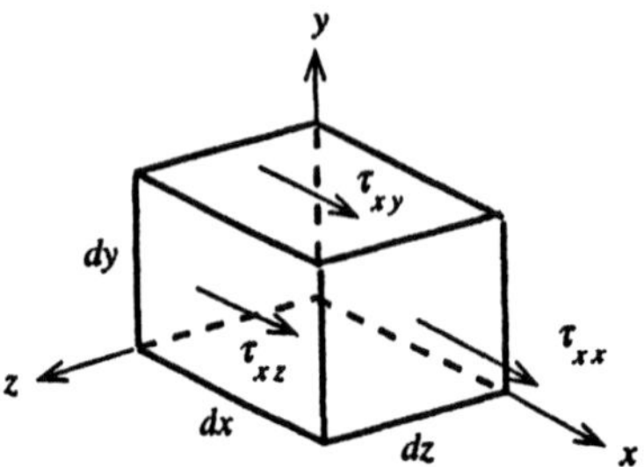

Figure 6.3 In finding the x component of the viscous force acting on a fluid element of volume $dx\,dy\,dz$ we need only to account for the viscous shear stresses τ_{xy} and τ_{xz} acting on the surfaces normal to the y- and z-axes and the viscous normal stress τ_{xx} acting on the surfaces normal to the x-axis.

viscous force which was neglected in deriving Euler's equation. This new equation is the called the Navier-Stokes equation.

To derive an expression for the viscous force per unit volume, **f**, consider the fluid element of volume $dx\,dy\,dz$ shown in figure 6.3. For the time being, consider only the viscous stresses acting in the x direction, τ_{xx}, τ_{xy} and τ_{xz}. The normal viscous stress τ_{xx} increases with distance x from a value of τ_{xx} at one face of the element to a value of $\tau_{xx} + (\partial\tau_{xx}/\partial x)\,dx$ at $x + dx$. The net difference between these stresses, $(\partial\tau_{xx}/\partial x)\,dx$, acts over an area $dy\,dz$ and contributes an amount $(\partial\tau_{xx}/\partial x)\,dx\,dy\,dz$ to the x component of the force acting on the fluid element or an amount $(\partial\tau_{xx}/\partial x)$ to the x component of the force per unit volume, $\mathbf{f}\cdot\mathbf{i}_x$. There will be similar contributions from the stresses τ_{xy} and τ_{xz}, due to their increases in the y and z directions, respectively, giving a total x component of the viscous force per unit volume of:

$$\mathbf{f}\cdot\mathbf{i}_x = \frac{\partial\tau_{xx}}{\partial x} + \frac{\partial\tau_{xy}}{\partial y} + \frac{\partial\tau_{xz}}{\partial z} \tag{6.7}$$

For an incompressible fluid, we may replace the viscous stresses τ_{xx}, τ_{xy} and τ_{xz} by their values in equations 6.4 and 6.6:

$$\mathbf{f}\cdot\mathbf{i}_x = \frac{\partial}{\partial x}\left(2\mu\frac{\partial u}{\partial x}\right) + \frac{\partial}{\partial y}\left(\mu\left[\frac{\partial u}{\partial y} + \frac{\partial v}{\partial x}\right]\right) + \frac{\partial}{\partial z}\left(\mu\left[\frac{\partial w}{\partial x} + \frac{\partial u}{\partial z}\right]\right) \tag{6.8}$$

This formidable expression can be simplified if we assume that μ is constant everywhere in the flow field and can be taken outside the partial derivatives to give:[5]

[5]In viscous flows with heat transfer, the viscosity μ may vary considerably because it is a function of temperature. The simplification of constant μ may not be appropriate in some heat transfer applications. In such cases, the variation of μ requires the addition of a term, $(\partial\mathbf{V}/\partial x + \nabla u)\cdot\nabla\mu$, to the right side of equation 6.9.

$$\begin{aligned}\mathbf{f}\cdot\mathbf{i}_x &= \mu\left(\frac{\partial^2 u}{\partial x^2}+\frac{\partial^2 u}{\partial y^2}+\frac{\partial^2 u}{\partial z^2}\right)+\mu\frac{\partial}{\partial x}\left(\frac{\partial u}{\partial x}+\frac{\partial v}{\partial y}+\frac{\partial w}{\partial z}\right)\\ &= \mu\nabla^2 u+\mu\frac{\partial}{\partial x}(\nabla\cdot\mathbf{V})\\ &= \mu\nabla^2 u \end{aligned} \tag{6.9}$$

where the condition of incompressibility eliminates the term containing $\nabla\cdot\mathbf{V}$. Since the x component of $\mathbf{f}$ is $\mu\nabla^2 u$, the y and z components must be $\mu\nabla^2 v$ and $\mu\nabla^2 w$, respectively. Having found the three components of $\mathbf{f}$, we may write the viscous force term in vector form as:

$$\begin{aligned}\mathbf{f} &= (\mu\nabla^2 u)\,\mathbf{i}_x+(\mu\nabla^2 v)\,\mathbf{i}_y+(\mu\nabla^2 w)\,\mathbf{i}_z\\ &= \mu\nabla^2\mathbf{V} \qquad \textit{if}\ \nabla\cdot\mathbf{V}=0; \qquad \mu = \textit{const.}\end{aligned} \tag{6.10}$$

6.4 The Navier-Stokes Equation of Motion

Having found the viscous force per unit volume, $\mathbf{f}$ in equation 6.10, we may divide it by ρ to obtain the viscous force per unit mass and add to the pressure and gravity forces per unit mass in Euler's equation 4.5 to obtain the form of the *Navier-Stokes equation* [6] for an incompressible fluid with constant viscosity:

$$\frac{D\mathbf{V}}{Dt} = -\frac{1}{\rho}\nabla p+\mathbf{g}+\nu\nabla^2\mathbf{V} \qquad \textit{if}\ \nabla\cdot\mathbf{V}=0; \qquad \mu=\textit{const.} \tag{6.11}$$

where we have replaced μ/ρ by ν. When the fluid density is constant throughout the flow field, it is convenient to introduce the variable $p^* \equiv p-\rho\mathbf{g}\cdot\mathbf{R}$, defined in equation 4.7, so that the gravity term is combined implicitly with the pressure term, as in equation 4.8:

$$\frac{D\mathbf{V}}{Dt} = -\frac{1}{\rho}\nabla p^* +\nu\nabla^2\mathbf{V} \qquad \textit{if}\ \rho,\ \mu=\textit{const.} \tag{6.12}$$

[6] This equation was developed independently by Navier and Stokes. Claude-Louis-Marie-Henri Navier (1785-1836) contributed to engineering and mechanics, both solid and fluid. While he first formulated the Navier-Stokes equations, he did not have the concept of shear. George Gabriel Stokes (1842-1912) made many contributions to viscous flows, including the complete formulation of the Navier-Stokes equation and low speed (Stokes) flow. He also analyzed water and light waves and made contributions to the theory of sound waves. The unit of kinematic viscosity in the *cgs* system of units, the *stoke* (see table 1.5), is named in his honor.

The x, y and z components of the Navier-Stokes equation, expressed in cartesian coordinates, are:

$$\frac{\partial u}{\partial t}+u\frac{\partial u}{\partial x}+v\frac{\partial u}{\partial y}+w\frac{\partial u}{\partial z}=-\frac{1}{\rho}\frac{\partial p^*}{\partial x}+\nu\left(\frac{\partial^2 u}{\partial x^2}+\frac{\partial^2 u}{\partial y^2}+\frac{\partial^2 u}{\partial z^2}\right) \tag{6.13}$$

$$\frac{\partial v}{\partial t}+u\frac{\partial v}{\partial x}+v\frac{\partial v}{\partial y}+w\frac{\partial v}{\partial z}=-\frac{1}{\rho}\frac{\partial p^*}{\partial y}+\nu\left(\frac{\partial^2 v}{\partial x^2}+\frac{\partial^2 v}{\partial y^2}+\frac{\partial^2 v}{\partial z^2}\right) \tag{6.14}$$

$$\frac{\partial w}{\partial t}+u\frac{\partial w}{\partial x}+v\frac{\partial w}{\partial y}+w\frac{\partial w}{\partial z}=-\frac{1}{\rho}\frac{\partial p^*}{\partial z}+\nu\left(\frac{\partial^2 w}{\partial x^2}+\frac{\partial^2 w}{\partial y^2}+\frac{\partial^2 w}{\partial z^2}\right) \tag{6.15}$$

Example 6.2

For the flow of example 6.1, calculate the value of ∇p^*.

Solution

Solving the Navier-Stokes equation 6.12 for ∇p^*:

$$\nabla p^* = -\rho\frac{D\mathbf{V}}{Dt}+\mu\nabla^2\mathbf{V}$$

All the components of $D\mathbf{V}/Dt$ are zero because all the derivatives of $\mathbf{V}$ are zero except $\partial\mathbf{V}/\partial z$, which is multiplied by $w = 0$. The only nonzero term of $\nabla^2\mathbf{V}$ is the z derivative, so that:

$$\begin{aligned}\nabla p^* &= \mu\frac{\partial^2\mathbf{V}}{\partial z^2} = \mu(2b\,\mathbf{i}_x)\\ &= [1.82E(-5)\,Pa\,s](2)(0.1\,m^{-1}\,s^{-1})\,\mathbf{i}_x = [3.64E(-6)\,Pa/m]\,\mathbf{i}_x\end{aligned}$$

In cylindrical coordinates, the form of the Navier-Stokes equation is more complicated because there are non-differential terms in both $D\mathbf{V}/Dt$ (see equation 4.4) and $\nabla^2\mathbf{V}$. The components in the radial ($\mathbf{i}_r$), tangential ($\mathbf{i}_\theta$) and axial ($\mathbf{i}_z$) directions are, respectively:

$$\begin{aligned}&\left(\frac{\partial V_r}{\partial t}+V_r\frac{\partial V_r}{\partial r}+\frac{V_\theta}{r}\frac{\partial V_r}{\partial \theta}+V_z\frac{\partial V_r}{\partial z}-\frac{V_\theta^2}{r}\right)\\ &=-\frac{1}{\rho}\frac{\partial p^*}{\partial r}+\nu\left[\frac{\partial^2 V_r}{\partial r^2}+\frac{1}{r}\left(\frac{\partial V_r}{\partial r}\right)+\frac{1}{r^2}\left(\frac{\partial^2 V_r}{\partial \theta^2}\right)+\frac{\partial^2 V_r}{\partial z^2}-\frac{V_r}{r^2}-\frac{2}{r^2}\left(\frac{\partial V_\theta}{\partial \theta}\right)\right]\end{aligned} \tag{6.16}$$

$$\left(\frac{\partial V_\theta}{\partial t} + V_r\frac{\partial V_\theta}{\partial r} + \frac{V_\theta}{r}\frac{\partial V_\theta}{\partial \theta} + V_z\frac{\partial V_\theta}{\partial z} + \frac{V_\theta V_r}{r}\right)$$
$$= -\frac{1}{\rho r}\frac{\partial p^*}{\partial \theta} + \nu\left[\frac{\partial^2 V_\theta}{\partial r^2} + \frac{1}{r}\left(\frac{\partial V_\theta}{\partial r}\right) + \frac{1}{r^2}\left(\frac{\partial^2 V_r}{\partial \theta^2}\right) + \frac{\partial^2 V_\theta}{\partial z^2} - \frac{V_\theta}{r^2} + \frac{2}{r^2}\left(\frac{\partial V_r}{\partial \theta}\right)\right] \tag{6.17}$$

$$\left(\frac{\partial V_z}{\partial t} + V_r\frac{\partial V_z}{\partial r} + \frac{V_\theta}{r}\frac{\partial V_z}{\partial \theta} + V_z\frac{\partial V_z}{\partial z}\right)$$
$$= -\frac{1}{\rho}\frac{\partial p^*}{\partial z} + \nu\left[\frac{\partial^2 V_z}{\partial r^2} + \frac{1}{r}\left(\frac{\partial V_z}{\partial r}\right) + \frac{1}{r^2}\left(\frac{\partial^2 V_z}{\partial \theta^2}\right) + \frac{\partial^2 V_z}{\partial z^2}\right] \tag{6.18}$$

Example 6.3

A viscous liquid flows inside a very long straight tube of circular cross-section. The streamlines of the flow are all parallel to the tube axis so that the fluid velocity components in the radial and tangential directions are zero (V_r, $V_\theta = 0$). Since the flow has axial symmetry, the axial velocity component, V_z, depends only upon the radial coordinate r. Using cylindrical coordinates, write the simplified form of the Navier-Stokes equation for this flow.

Solution

Dropping all terms in equations 6.16-6.18 that contain V_r, V_θ or the z or θ derivatives of V_z, since these terms are all zero, there remains:

$$0 = -\frac{1}{\rho}\frac{\partial p^*}{\partial r}$$

$$0 = -\frac{1}{\rho r}\frac{\partial p^*}{\partial \theta}$$

$$0 = -\frac{1}{\rho}\frac{\partial p^*}{\partial z} + \nu\left[\frac{\partial^2 V_z}{\partial r^2} + \frac{1}{r}\left(\frac{\partial V_z}{\partial r}\right)\right]$$

Rearranging,

$$\frac{\partial p^*}{\partial r} = 0; \qquad \frac{\partial p^*}{\partial \theta} = 0$$

$$\left[\frac{\partial^2 V_z}{\partial r^2} + \frac{1}{r}\left(\frac{\partial V_z}{\partial r}\right)\right] - \frac{1}{\mu}\frac{\partial p^*}{\partial z} = 0$$

We can conclude from the first two equations that p^* does not depend upon r or θ, and thus at most depends upon z. However, the third equation requires that $\partial p^*/\partial z$ be at most a function of r only. The only way to satisfy these requirements is for $\partial p^*/\partial z$ to be a constant, which means that the bracketed term in the third equation is also a constant. The solution to this differential equation is treated below in section 6.5.4 on circular Poiseuille flow.

6.4.1 Boundary and Initial Conditions

The Navier-Stokes equation is a partial differential equation with four independent variables (*e.g.*, x, y, z, t) and four dependent variables (*e.g.*, u, v, w, p^*).[7] Together with the mass conservation equation, equation 3.13, there are four scalar equations (three components of the Navier-Stokes equation and the mass conservation equation) that may be solved for the four unknown dependent variables. Needless to say, there is no general solution to this formidable set of equations.

Before we can solve viscous flow problems by integrating the equations of motion and mass conservation, we must specify the physical conditions that constrain the flow at its boundaries, called *boundary conditions*, and the state of the flow at an initial time (if the flow is time dependent), called the *initial condition*. These conditions are generally recognizable from the nature of the flow.

Where a viscous fluid meets a solid boundary, there is no relative motion of the fluid and the solid, *i.e.*, the fluid and the solid at a boundary point have the same velocity:

$$\mathbf{V} = \mathbf{V}_{solid} \tag{6.19}$$

This is often termed the *non-slip* condition, meaning that the fluid does not slip along the solid surface the way a different solid might. The physical reason for the non-slip condition is that the fluid molecules hitting the solid wall collide so frequently with the solid wall molecules that they have no average motion that is different from the wall molecules.[8] The non-slip boundary condition applies to the three components of the fluid velocity $\mathbf{V}$ everywhere along the surface of a solid boundary.

At the interface between two immiscible fluids, such as air and water, the fluid stress acting on one fluid (say σ_a) is equal in magnitude, but opposite in direction, to

---

[7] Having assumed that ρ and μ are constant in deriving the form 6.12 of the Navier-Stokes equation, they are not dependent variables but parameters of the flow.

[8] In gas flows at extremely low pressures, there can be slip between the gas and a solid wall when the gas molecules move distances between collisions that are not negligibly small compared with the wall length.

the fluid stress acting on the other fluid (σ_b):[9]

$$\sigma_a = -\sigma_b$$

or

$$[(-p)\mathbf{n}]_a + \tau_a = -([(-p)\mathbf{n}]_b + \tau_b) \tag{6.20}$$

where we have used equation 6.1 to decompose the fluid stress into a pressure stress $(-p)\mathbf{n}$ and a viscous stress τ. In the case of inviscid flow, the only fluid stress is the pressure stress $(-p)\mathbf{n}$, a normal stress, and equation 6.20 requires only that the pressure in each fluid be the same at the interface. However, for viscous flows the sum of the pressure and the viscous stresses must balance at the fluid interface. In almost all practical cases, the viscous stress is tangential to the interface so that the pressure and viscous stresses are separately equal (but opposite) at the interface:

$$p_a = p_b; \qquad \tau_a = -\tau_b \tag{6.21}$$

where we have used the fact that $\mathbf{n}_a = -\mathbf{n}_b$.

In the case of flows around bodies, the boundary conditions far from the body ($\mathbf{R} \to \infty$) are usually those of a uniform flow, *i.e.*, the pressure p^* and velocity $\mathbf{V}$ are constant.[10] For such uniform flows, the viscous stress is zero, and the flow satisfies the Navier-Stokes equation and mass conservation far from the body.

Initial conditions for unsteady flows usually correspond to a fluid at rest ($p^* = constant$ and $\mathbf{V} = 0$). We shall consider one such viscous flow in section 6.5.9.

6.5 Applications of the Navier-Stokes Equation

In this section, we consider laminar viscous flows for which the the velocity field depends upon only one or two spatial dimensions or upon one spatial dimension and time. Because we restrict our analysis to only one or two independent variables and because the boundary conditions are appropriately simple, we are usually able to obtain analytical solutions to the velocity and pressure fields. Despite this simplicity, these flows are practically useful because they exhibit the physical effects of viscosity that are present in similar but more complicated flows. In some cases, these flows are directly applicable to engineering problems.

[9]This assumes that the effects of interfacial tension are negligible.

[10]They are usually denoted by p^*_∞ and $\mathbf{V}_\infty$.

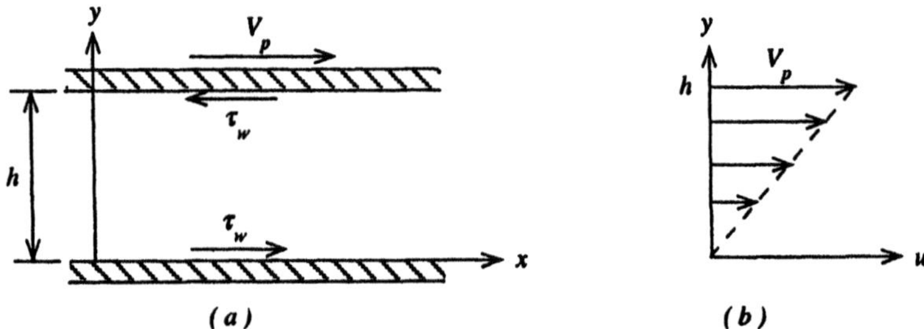

Figure 6.4 Above, (*a*) shows a plane Couette flow of a viscous fluid between two parallel plates separated by a distance h. The upper plate moves with a velocity $\mathbf{V} = V_p\, \mathbf{i}_x$ while the lower one is held stationary. A shear stress τ_w acts on the upper and lower plates. In (b), the variation in the fluid velocity u is seen as a function of y.

6.5.1 Plane Couette Flow

The steady viscous flow between two parallel plates, one of which moves parallel to the other at a fixed distance apart, is called a *plane Couette flow*.[11] Figure 6.4(*a*) shows a fluid flow between two parallel plates spaced a distance h apart, the lower plate being stationary and the upper plate moving with a constant velocity $\mathbf{V} = V_p\, \mathbf{i}_x$ in the x direction. The fluid extends to infinity in the x and z directions, the only change in the velocity occurring in the y direction. The fluid moves only in the x direction with a velocity $\mathbf{V} = u\{y\}\, \mathbf{i}_x$ that depends upon y alone, and there is no pressure change imposed on the flow:

$$\mathbf{V} = u\{y\}\, \mathbf{i}_x; \qquad p^* = const. \tag{6.22}$$

Along a streamline of the flow, which is a horizontal line in figure 6.4(*a*), there is no change in $\mathbf{V}$ so that the material derivative $D\mathbf{V}/Dt = 0$. Also, $\nabla p^* = 0$. Substituting these relations in the Navier-Stokes equation 6.12 and noting that only y derivatives are non-zero,

$$0 = 0 + \nu\left(\frac{\partial^2 u}{\partial y^2}\right)$$

$$\frac{d^2 u}{dy^2} = 0$$

[11]M. Maurice Couette (1858-1943) invented a viscometer that measured the wall stress in what is now called Couette flow.

This is a total differential equation because u depends only upon y. Integrating twice on y:

$$u = c_1 y + c_2$$

The constants of integration (c_1, c_2) may be determined by applying the boundary conditions of equation 6.19 that the velocity $u\{0\} = 0$ at the fixed plate ($y = 0$) and $u\{h\} = V_p$ at the moving plate ($y = h$), giving:

$$u = V_p \left(\frac{y}{h}\right) \tag{6.23}$$

This variation of velocity, which is linear with vertical distance y, is plotted in figure 6.4(b) with u as abscissa and y as ordinate; this plot of $u\{y\}$ is called the *velocity profile.*[12]

The shear stress $\boldsymbol{\tau}_w$ acting on the upper plate, as shown in figure 6.4(a), is equal and opposite to the shear stress $\tau_{xy}\,\mathbf{i}_x$ acting on the fluid at $y = h$:

$$\boldsymbol{\tau}_w = -\tau_{xy}\,\mathbf{i}_x = -\mu\left(\frac{\partial v}{\partial x} + \frac{\partial u}{\partial y}\right)_{y=h} \mathbf{i}_x = -\mu\left(\frac{V_p}{h}\right)\mathbf{i}_x \tag{6.24}$$

An equal but opposite shear stress acts on the lower plate. Note that the shear stress $\tau_{xy} = \tau_{yx}$ is constant in the flow field since $\partial u/\partial y = V_p/h = constant$.

Of practical interest is the fluid volumetric flow rate in the x direction induced by the moving plate. Denoting this rate by Q and the width of the channel in the z direction (normal to the plane of figure 6.4[a]) by W, the volumetric flow rate is calculated by integrating u over the flow area Wh:

$$Q = \iint \mathbf{V}\cdot\mathbf{n}\,dS = W\int_0^h u\,dy = \frac{WV_p}{h}\left.\frac{y^2}{2}\right|_0^h = \frac{1}{2}WV_p h$$

$$\frac{Q}{W} = \frac{1}{2}V_p h \tag{6.25}$$

Often it is important to determine the volume flow averaged velocity, denoted by $\bar{V}$:

$$\bar{V} \equiv \frac{Q}{Wh} = \frac{1}{2}V_p \tag{6.26}$$

---

[12]Note that the velocity field satisfies the mass conservation law for a compressible fluid, $\nabla \cdot \mathbf{V} = \partial u/\partial x + \partial v/\partial y + \partial w/\partial z = 0$.

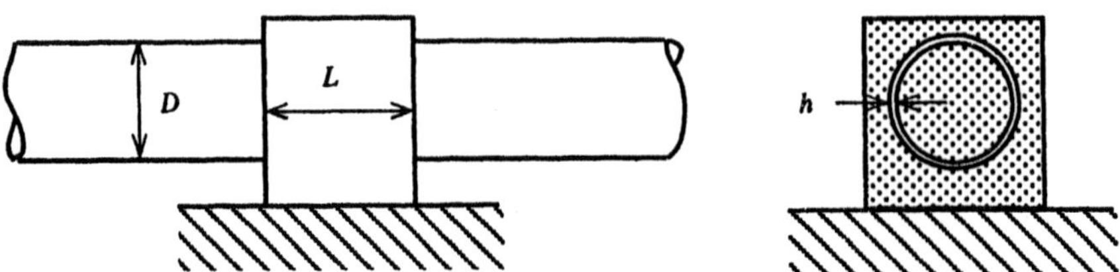

Figure E 6.4

Example 6.4

A journal bearing supports a circular shaft of diameter $D = 10\,cm$ turning at 3600 revolutions per minute. The bearing length $L = 10\,cm$. The gap betwen the shaft and the bearing, $h = 0.1\,mm$, is filled with a lubricant whose viscosity is $\mu = 6.7E(-5)\,Pa\,s$. Calculate the torque T applied to the shaft to overcome the friction in the bearing and the power P consumed in the bearing by friction.

Solution

The flow of lubricating oil in the bearing gap is a very nearly a plane Couette flow because the gap width h is very much less than the radius of curvature $D/2$ of the shaft surface. Using equation 6.24, the shear stress τ_w exerted on the surface of the shaft is:

$$\tau_w = \mu\frac{V_p}{h} = \mu\frac{\Omega D}{2h}$$

where Ω is the angular velocity of the shaft. The torque T is the product of the shear stress τ_w times the radius $D/2$ times the surface area πDL of the bearing:

$$T = \tau_w\frac{D}{2}(\pi DL) = \frac{\pi\mu\Omega LD^3}{4h}$$

$$= \frac{\pi(6.7E(-5)\,Pa\,s)(2\pi \times 60\,s^{-1})(0.1\,m)(0.1\,m)^3}{4 \times 1.0E(-4)\,m} = 1.984E(-2)\,Nm$$

The power P consumed is the product of the torque T times the angular speed Ω:

$$P = T\Omega = 1.984E(-2)\,Nm \times (2\pi \times 60\,s^{-1}) = 7.479\,W$$

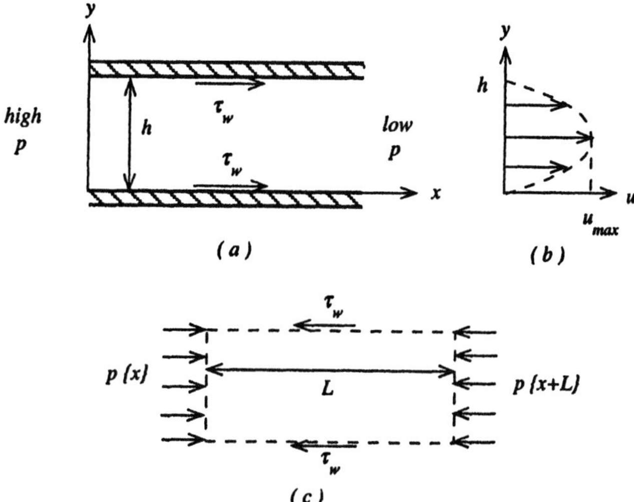

Figure 6.5 Above, (*a*) shows that in plane Poiseuille flow between parallel walls, the fluid flows from high to low pressure, exerting a shear stress on the walls in the direction of flow. In (*b*), the velocity profile is parabolic, and in (*c*) the shear stress acting on the fluid in a control volume is balanced by the pressure gradient.

6.5.2 Plane Poiseuille Flow

The steady laminar viscous flow in a channel or tube from a region of higher pressure to one of lower pressure is called *Poiseuille flow.*[13] The flow between parallel plates separated by a fixed distance h, illustrated in figure 6.5(*a*), is called *plane Poiseuille flow*. Assuming the higher pressure region is on the left and the lower pressure on the right, the pressure gradient $\partial p^*/\partial x$ would be negative while the fluid velocity is to the right, *i.e.*, the fluid is moving from a region of higher pressure ($-x$) to one of lower pressure ($+x$). Like plane Couette flow, the velocity $\mathbf{V}$ lies in the x direction and is a function of y alone, but the pressure gradient, which also lies in the x direction, is not zero but a constant:

$$\mathbf{V} = u\{y\}\,\mathbf{i}_x; \qquad \nabla p^* = \frac{\partial p^*}{\partial x}\,\mathbf{i}_x = constant$$

[13]Jean Marie Louis Poiseuille (1799-1869) was a French physician who studied the flow of blood in capillaries. The unit of viscosity in the *cgs* system of units, called the *poise* (see table 1.5), was named after him.

Substituting this form of the velocity and pressure fields in the Navier-Stokes equation 6.12 and multiplying by ρ, we find:

$$0 = -\frac{\partial p^*}{\partial x} + \mu\left(\frac{\partial^2 u}{\partial y^2}\right)$$

$$\mu\frac{d^2 u}{dy^2} = \frac{dp^*}{dx}$$

where we have used total derivatives to indicate that u and p^* depend only upon y and x, respectively. Integrating twice upon y,

$$\mu u = \left(\frac{dp^*}{dx}\right)\frac{y^2}{2} + c_1 y + c_2$$

By applying the boundary conditions of zero velocity at the wall, ($u\{0\} = 0$; $u\{h\} = 0$), the constants of integration (c_1, c_2) can be determined and the velocity $u\{y\}$ found to be:

$$u = \frac{1}{2\mu}\left(-\frac{dp^*}{dx}\right)y(h-y) \tag{6.27}$$

which is plotted in figure 6.5(*b*).[14] (We keep the negative sign inside the parenthesis enclosing dp^*/dx to remind us that the fluid velocity is positive when dp^*/dx is negative.) The velocity distribution of figure 6.5(*b*) is parabolic in y, the maximum velocity u_{max} occurring at the center of the channel ($y = h/2$):

$$u_{max} = \frac{h^2}{8\mu}\left(-\frac{dp^*}{dx}\right) \tag{6.28}$$

The volumetric flow rate Q through a channel cross-section of area Wh is found by integrating equation 6.27:

$$Q = W\int_0^h u\,dy = \frac{W}{2\mu}\left(-\frac{dp^*}{dx}\right)\int_0^h y(h-y)\,dy = \frac{W}{2\mu}\left(-\frac{dp^*}{dx}\right)\left|\frac{y^2 h}{2} - \frac{y^3}{3}\right|_0^h$$

$$\frac{Q}{W} = \frac{h^3}{12\mu}\left(-\frac{dp^*}{dx}\right) \tag{6.29}$$

The mean flow velocity $\bar{V}$ is:

$$\bar{V} = \frac{Q}{Wh} = \frac{h^2}{12\mu}\left(-\frac{dp^*}{dx}\right) = \frac{2}{3}u_{max} \tag{6.30}$$

The shear stress τ_w acting on the lower wall (and the upper wall as well) may be

[14]This velocity distribution satisfies incompressible mass conservation, $\nabla \cdot \mathbf{V} = 0$.

found by evaluating the velocity gradient $\partial u/\partial y$ at the wall ($y = 0$):

$$\tau_w = \mu \left(\frac{\partial u}{\partial y}\right)_{y=0} \mathbf{i}_x = \mu \frac{h}{2\mu}\left(-\frac{dp^*}{dx}\right)\mathbf{i}_x = \frac{h}{2}\left(-\frac{dp^*}{dx}\right)\mathbf{i}_x \tag{6.31}$$

Notice that the wall shear stress τ_w does not depend explicitly upon the fluid viscosity (equation 6.31) but that, in contrast to plane Couette flow, the fluid velocity does (equation 6.27). In plane Poiseuille flow, the flow speed adjusts until the wall shear stress reaches a value that balances the imposed pressure gradient, as in equation 6.31.

The relation 6.31 between wall shear stress and pressure gradient expresses the force balance on a control volume that encloses the fluid within the channel, as shown in figure 6.5(c). Assuming the length of this volume is L and applying the linear momentum theorem 5.11, noting that $(\dot{m}\mathbf{V})_{out} = (\dot{m}\mathbf{V})_{in}$ and $\mathbf{F}_e = 0$:

$$\frac{d}{dt}\iiint_{\mathcal{V}} \rho\mathbf{V}\,d\mathcal{V} + (\dot{m}\mathbf{V})_{out} - (\dot{m}\mathbf{V})_{in}$$

$$= \iint_S (-p\mathbf{n})\,dS + \iint_S \boldsymbol{\tau}\,dS + \iiint_{\mathcal{V}} \rho\mathbf{g}\,d\mathcal{V} + \Sigma\mathbf{F}_{ex}$$

$$0 = Wh\left[p^*\{x\} - p^*\{x+L\}\right]\mathbf{i}_x - 2(\tau_w\,\mathbf{i}_x)LW$$

$$\tau_w = \frac{h}{2L}\left[p^*\{x\} - p^*\{x+L\}\right] = \frac{h}{2}\left(-\frac{dp^*}{dx}\right)$$

where the pressure and gravity terms of 5.11 have been combined by use of p^* in place of p.

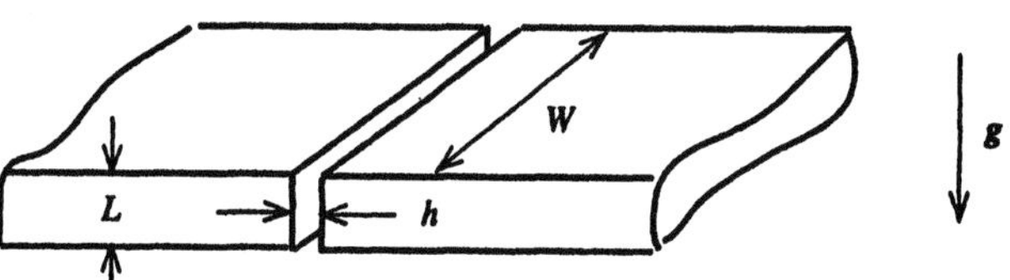

Figure E 6.5

Example 6.5

A flat roof of a building is constructed of precast concrete slabs of width $W = 1.0\,m$ and depth $L = 0.1\,m$, as shown in figure E 6.5. Through an oversight, an end joint between two slabs was not sealed, leaving a crack of width $h = 1.0\,mm$. When it rains, the crack fills with water which leaks into the interior of the building. Calculate the volume flow rate of rainwater through the crack assuming steady laminar viscous (plane Poiseuille) flow with $\mu = 1.13E(-3)\,Pa\,s$.

Solution
The volumetric flow rate through the crack is that of equation 6.29:

$$Q = \frac{h^3 W}{12\mu}\left(-\frac{dp^*}{dx}\right)$$

Measuring x vertically downward from the upper surface, p^* is:

$$p^* = p - \rho \mathbf{g} \cdot \mathbf{R} = p - \rho(g\,\mathbf{i}_x) \cdot (x\,\mathbf{i}_x) = p - \rho g x$$

and, assuming atmospheric pressure above and below the roof, $-dp^*/dx$ becomes:

$$-\frac{dp^*}{dx} = -\frac{(p_a - \rho g L) - p_a}{L} = \rho g$$

Consequently, the volume flow rate Q is:

$$\begin{aligned} Q &= \frac{\rho g h^3 W}{12\mu} \\ &= \frac{(1.0E(3)\,kg/m^3)(9.807\,m/s^2)(1.0E(-3)\,m)^3)(1.0\,m)}{12 \times 1.13E(-3)\,Pa\,s} = 7.232E(-4)\,m^3/s \\ &= 0.7232\,l/s \end{aligned}$$

Flow Down an Inclined Plane

In some applications, a viscous liquid can flow while in contact with only one solid surface, the fluid motion being caused by a component of the gravity force parallel to the solid surface. Such a plane Poiseuille flow is illustrated in figure 6.6(a), showing a plane surface inclined above the horizontal by an angle ϕ and covered with a liquid layer of constant thickness h that flows parallel to the plate in the downhill direction. The upper surface of the fluid ($y = h$) is in contact with the air, in which the pressure is constant ($p = p_a$ at $y = h$) and which exerts a negligible shear stress on the liquid surface ($\tau_{xy} = 0$ at $y = h$).

To describe this motion, select the x component of the Navier-Stokes equation 6.11, noting that $D\mathbf{V}/Dt = 0$, $\partial p/\partial x = 0$ because the air pressure is constant and the

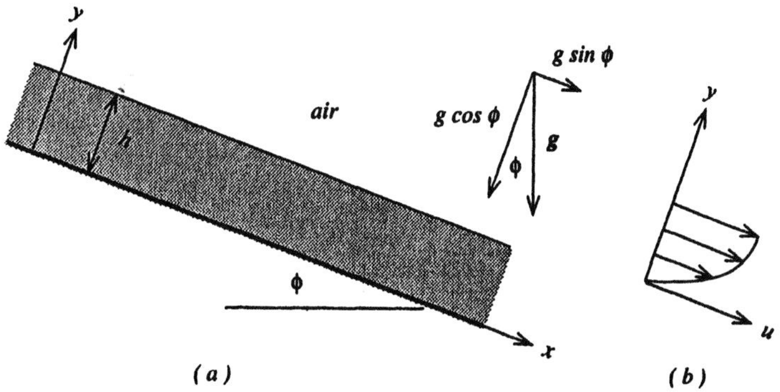

Figure 6.6 In the flow of a viscous liquid in contact with air down an inclined plane, the flow parallel to the plane is caused by the component of gravity in the direction of the plane surface.

x component of **g** is $g \sin\phi$:

$$0 = 0 + g\sin\phi + \nu\frac{\partial^2 u}{\partial y^2}$$

$$\frac{d^2u}{dy^2} = -\frac{g\sin\phi}{\nu}$$

Integrating twice on y:

$$u = -\frac{g(\sin\phi)y^2}{2\nu} + c_1 y + c_2$$

and applying the boundary conditions that $u = 0$ at $y = 0$ and $\tau_{xy} = \mu du/dy = 0$ at $y = h$ enables us to find the velocity distribution $u\{y\}$:

$$u = \frac{g\sin\phi}{\nu}\left(hy - \frac{y^2}{2}\right) \tag{6.32}$$

This velocity profile, sketched in figure 6.6(b), is parabolic with a maximum value u_{max} at the upper surface ($y = h$):

$$u_{max} = \frac{gh^2\sin\phi}{2\nu} \tag{6.33}$$

The volumetric flow rate per unit distance normal to the plane of the flow, Q/W, is found by integrating equation 6.32 on y:

$$\frac{Q}{W} = \int_0^h u\,dy = \frac{g\sin\phi}{\nu}\left|\frac{hy^2}{2} - \frac{y^3}{6}\right|_0^h = \frac{gh^3\sin\phi}{3\nu} \tag{6.34}$$

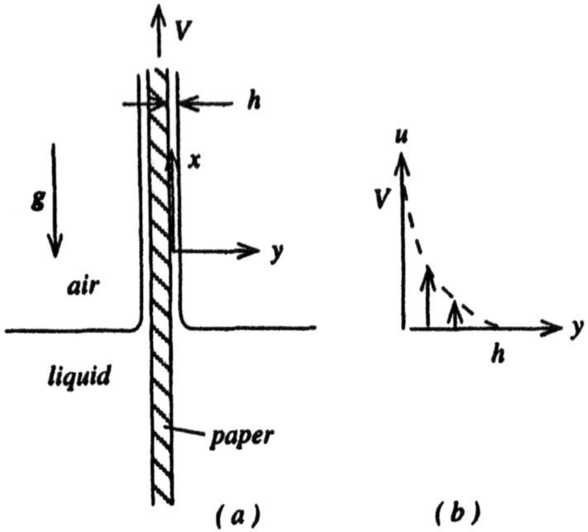

Figure E 6.6

Example 6.6

A paper coating process consists of pulling a continuous sheet of paper upward through the surface of a pool of coating liquid at a speed $V = 0.1\ m/s$. A film of coating liquid of constant thickness h adheres to the paper, as shown in figure E 6.6(*a*). The liquid speed $u\{h\}$ at the outside edge of the liquid layer ($y = h$) is zero (see figure E 6.6(*b*)). Calculate the film thickness h if the liquid kinematic viscosity $\nu = 1.0E(-6)\ m/s$.

Solution

Taking the x axis vertically upward and **g** downward, the Navier-Stokes equation 6.11 in the vertical direction is:

$$0 = -g + \nu \frac{\partial^2 u}{\partial y^2}$$

$$\frac{d^2 u}{dy^2} = \frac{g}{\nu}$$

Integrating twice results in:

$$u = \frac{g y^2}{2\nu} + c_1 y + c_2$$

Applying the boundary condition that $du/dy = 0$ at $y = h$,

$$0 = \frac{gh}{\nu} + c_1$$

$$u = \frac{g}{\nu}\left(\frac{y^2}{2} - hy\right) + c_2$$

But $u = V$ at $y = 0$ so that $c_2 = V$ and:

$$u = \frac{g}{\nu}\left(\frac{y^2}{2} - hy\right) + V$$

If $u = 0$ at $y = h$ then:

$$0 = -\frac{gh^2}{2\nu} + V$$

$$h = \sqrt{\frac{2\nu V}{g}} = \sqrt{\frac{2(1.0E(-6)\,m^2/s)(0.1\,m/s)}{9.807\,m/s^2}} = 1.428E(-4)\,m$$

6.5.3 Combined Plane Couette and Poiseuille Flow

Because the form of the Navier-Stokes equation for both Couette flow and Poiseuille flow is *linear, i.e.*, it contains $u\{y\}$ to the first power, we can add together these two solutions, equations 6.23 and 6.27, to find a flow that satisfies the boundary conditions of Couette flow yet has an impressed pressure gradient:

$$u = V_p\left(\frac{y}{h}\right) + \frac{1}{2\mu}\left(-\frac{dp^*}{dx}\right) y(h - y) \tag{6.35}$$

In figure 6.7 is sketched a combined Couette and Poiseuille flow that is the sum of the flows shown in figures 6.4 and 6.5, the latter being the case of a falling pressure

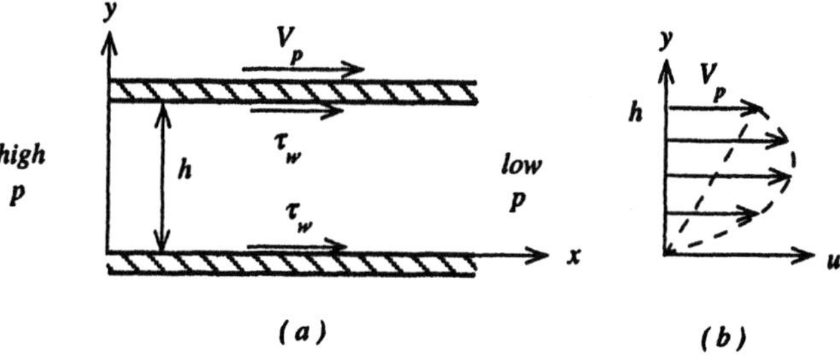

Figure 6.7 Above, (*a*) shows the linear combination of Couette and Poiseuille plane flows between parallel plates with relative motion; (*b*) an impressed pressure gradient results in a velocity profile that is the sum of the two flows.

(negative pressure gradient).[15] The volumetric flow rate Q is simply the sum of the rates for the respective flows, equations 6.25 and 6.29:

$$\frac{Q}{W} = \frac{V_p h}{2} + \frac{h^3}{12\mu}\left(-\frac{dp^*}{dx}\right) \tag{6.36}$$

Likewise, we may find an expression for the shear stress τ_w on the upper (moving) wall by adding together equations 6.24 and 6.31:

$$\tau_w = -\mu\left(\frac{V_p}{h}\right)\mathbf{i}_x + \frac{h}{2}\left(-\frac{dp^*}{dx}\right)\mathbf{i}_x \tag{6.37}$$

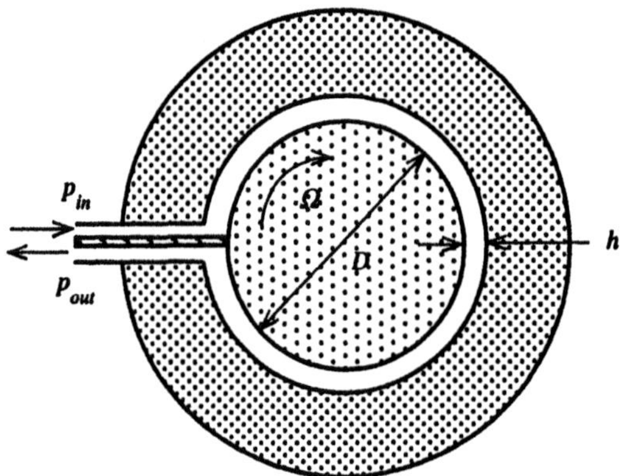

Figure E 6.7

Example 6.7

A friction pump consists of a solid cylinder of diameter D and length W that rotates clockwise at an angular speed Ω inside a hollow coaxial cylinder of inside diameter $D + 2h$, as shown in figure E 6.7. The fluid flow into the pump is pulled clockwise through a complete circle by the friction with the moving inner cylinder surface. The inflow and outflow passages to the pump are separated by a septum that prevents leakage from the higher outflow pressure p_{out} to the lower inflow pressure p_{in}, the pressure difference $\Delta p \equiv p_{out} - p_{in}$ being maintained by the clockwise flow through the pump.

---

[15]If the pressure gradient were positive, the velocity $u\{y\}$ would be less than the value for the Couette flow alone rather than greater than that value as sketched in figure 6.7(b).

Derive expressions for (*a*) the volumetric flow rate Q through the pump and (*b*) the clockwise torque T that must be applied to the rotor for steady operation as functions of the pressure rise, Δp. (*c*) If the power P_{out} produced by the pumped fluid is $Q\Delta p$, derive an expression for the value of Δp that maximizes P_{out} and find the numerical value of the pump efficiency $\eta_p \equiv P_{out}/P_{in}$ for this value of Δp, where the input power is $P_{in} = \Omega T$.

Solution

(*a*) The pump flow is a combined plane Couette and Poiseuille flow in which the wall velocity V_p and pressure gradient dp^*/dx are:

$$V_p = \frac{\Omega D}{2}; \qquad \frac{dp^*}{dx} = \frac{p_{out} - p_{in}}{\pi D} = \frac{\Delta p}{\pi D}$$

Inserting these values in equation 6.36 for the volumetric flow rate Q,

$$Q = W\left[\frac{\Omega Dh}{4} + \frac{h^3}{12\mu}\left(-\frac{\Delta p}{\pi D}\right)\right] = \frac{W\Omega Dh}{4}\left(1 - \frac{h^2\Delta p}{3\pi\mu\Omega D^2}\right)$$

(*b*) If τ_w is the magnitude of the shear stress acting on the wall of the rotating cylinder in opposition to its motion, then the torque T that must be applied to the cylinder in the clockwise direction is:

$$T = \tau_w\frac{D}{2}(\pi DW) = \frac{\pi D^2 W}{2}\left(\frac{\mu\Omega D}{2h} + \frac{h\Delta p}{2\pi D}\right) = \frac{\pi\mu W\Omega D^3}{4h}\left(1 + \frac{h^2\Delta p}{\pi\mu\Omega D^2}\right)$$

where we have used equation 6.37 to evaluate τ_w.

(*c*) The power P_{out} is:

$$P_{out} = Q\Delta p = \frac{W\Omega Dh}{4}\left(1 - \frac{h^2\Delta p}{3\pi\mu\Omega D^2}\right)\Delta p$$

To maximize P_{out}, differentiate with respect to Δp and set equal to 0:

$$1 - \frac{2h^2\Delta p}{3\pi\mu\Omega D^2} = 0; \qquad \Delta p = \frac{3\pi\mu\Omega D^2}{2h^2}$$

which results in a maximum P_{out} of:

$$max\ P_{out} = \frac{W\Omega Dh}{4}\left(\frac{1}{2}\right)\left(\frac{3\pi\mu\Omega D^2}{2h^2}\right) = \frac{3\pi\mu W\Omega^2 D^3}{16h}$$

The corresponding value of P_{in} is:

$$P_{in} = \Omega T = \frac{\pi\mu W\Omega^2 D^3}{4h}\left(\frac{5}{2}\right) = \frac{5\pi\mu W\Omega^2 D^3}{8h}$$

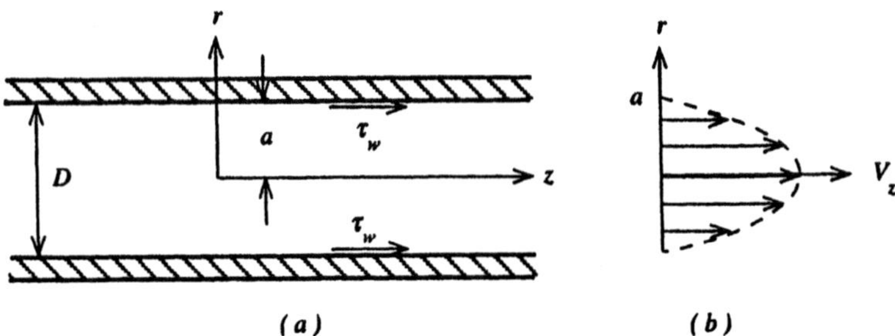

Figure 6.8 Above, (*a*) shows an axisymmetric Poiseuille flow in a circular tube using cylindrical coordinates, and (*b*) shows the axial velocity profile, which is parabolic.

The ratio of these powers is the pump efficiency η_p:

$$\eta_p = \frac{P_{out}}{P_{in}} = \frac{3\pi}{16} \times \frac{8}{5\pi} = 30\%$$

6.5.4 Circular Poiseuille Flow

Laminar flow through a circular tube is of great practical interest because a circular cross-section is the most common form of tube or pipe. In fact, it was the research on blood flow in animal capillaries that first revealed the relationship between volumetric flow rate and pressure change in pipes of circular cross-section. In mammals, both blood flow in capillaries and veins and air flow in lung alveoli are examples of circular Poiseuille flow. Other common examples are the flow through a soda straw or through a hypodermic needle.

Because of the circular symmetry of the container, circular Poiseuille flow is an axially-symmetric flow best described by using cylindrical coordinates, as shown in figure 6.8(*a*). In Example 6.3, we found that the Navier-Stokes equation for this flow simplifies to the form:

$$\frac{d^2V_z}{dr^2} + \frac{1}{r}\left(\frac{dV_z}{dr}\right) = \frac{1}{r}\frac{d}{dr}\left(r\frac{dV_z}{dr}\right) = \frac{1}{\mu}\frac{dp^*}{dz} \tag{6.38}$$

where we have used total derivatives because V_z is a function of the radial distance r alone and p^* is a function of axial distance z alone. Multiplying 6.38 by r and integrating once,

$$r\frac{dV_z}{dr} = \frac{r^2}{2\mu}\frac{dp^*}{dz} + c_1 \tag{6.39}$$

The shear stress at the tube centerline $r = 0$ must be zero because of axial symmetry and thus $dV_z/dr = 0$ at $r = 0$, thereby requiring that $c_1 = 0$. Dividing 6.39 (with $c_1 = 0$) by r and integrating,

$$V_z = \frac{r^2}{4\mu}\frac{dp^*}{dz} + c_2$$

By choosing $c_2 = -a^2(dp^*/dz)(4\mu)$, the velocity V_z at the tube wall $r = a$ becomes zero, as it should be at a stationary wall, and the velocity distribution for circular Poiseuille flow is:

$$V_z = \frac{a^2 - r^2}{4\mu}\left(-\frac{dp^*}{dz}\right) \tag{6.40}$$

where we include the minus sign multiplying dp^*/dz because the latter is negative for positive V_z, *i.e.*, for flow in the z direction.[16] This parabolic velocity distribution is illustrated in figure 6.8(*b*).

The volumetric flow rate Q may be found by integrating the axial velocity V_z across the tube cross-section:

$$Q = \int_0^a V_z(2\pi r)\,dr = \frac{\pi}{2\mu}\left(-\frac{dp^*}{dz}\right)\int_0^a r(a^2 - r^2)\,dr = \frac{\pi}{2\mu}\left(-\frac{dp^*}{dz}\right)\left|\frac{r^2a^2}{2} - \frac{r^4}{4}\right|_0^a$$

$$= \frac{\pi a^4}{8\mu}\left(-\frac{dp^*}{dz}\right) = \frac{\pi D^4}{128\mu}\left(-\frac{dp^*}{dz}\right) \tag{6.41}$$

The volumetric flow rate through a circular tube is very sensitive to the tube diameter D, varying as the fourth power of the diameter. For a fixed pressure difference across a coronary artery, for example, a reduction of the flow area $\pi D^2/4$ by a factor of two would decrease the blood volume flow rate by a factor of four.

The average flow velocity $\bar{V}$ is obtained by dividing the volumetric flow rate Q by the tube area $\pi D^2/4$:

$$\bar{V} = \frac{4Q}{\pi D^2} = \frac{D^2}{32\mu}\left(-\frac{dp^*}{dz}\right) \tag{6.42}$$

An expression for the wall shear stress τ_w may be derived by applying the linear momentum theorem to a cylindrical sample of the fluid inside the circular tube of length L and diameter D. The pressure difference $(-dp^*/dz)L$ acting on the cylinder produces a force $(-dp^*/dz)L(\pi D^2/4)$ in the z direction that must be balanced by the shear stress force $\tau_w(\pi DL)$ acting in the opposite direction:

$$\tau_w(\pi DL) = \frac{\pi D^2 L}{4}\left(-\frac{dp^*}{dz}\right)$$

[16]Note that this satisfies equation 3.15 for incompressible mass conservation.

$$\tau_w = \frac{D}{4}\left(-\frac{dp^*}{dz}\right) = 8\mu\left(\frac{\bar{V}}{D}\right) \tag{6.43}$$

where the last term is found by utilizing equation 6.42.

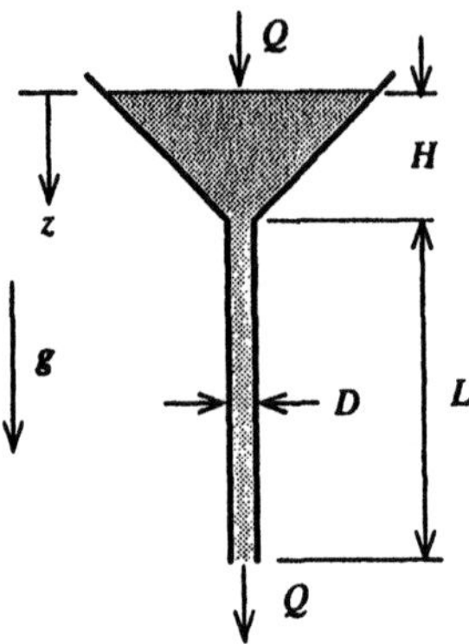

Figure E 6.8

Example 6.8

The kinematic viscosity of a mixture of waste oils is to be measured by use of a laboratory funnel. The oil is poured into the funnel at a steady rate Q, as shown in figure E 6.8, maintaining the level of oil in the funnel at a distance $H = 3\,cm$ above the entrance to the funnel tube, which has a length $L = 30\,cm$ and diameter $D = 3\,mm$. The time required for $100\,cm^3$ to pass through the funnel is measured to be $152\,s$. Calculate the kinematic viscosity ν of the oil mixture.

Solution
Solving equation 6.41 for $\nu = \mu/\rho$:

$$\nu = \frac{\mu}{\rho} = \frac{\pi D^4}{128 Q\rho}\left(-\frac{dp^*}{dz}\right)$$

Assuming that the flow in the top part of the funnel is inviscid, the value of p^* at the entrance to the funnel tube is the same as that at the liquid surface, *i.e.*, p_a. At the funnel exit, $p^* = p_a - \rho g(H + L)$. Thus the pressure gradient is:

$$\frac{dp^*}{dz} = \frac{(p_a - \rho g(H + L)) - p_a}{L} = -\rho g\left(1 + \frac{H}{L}\right)$$

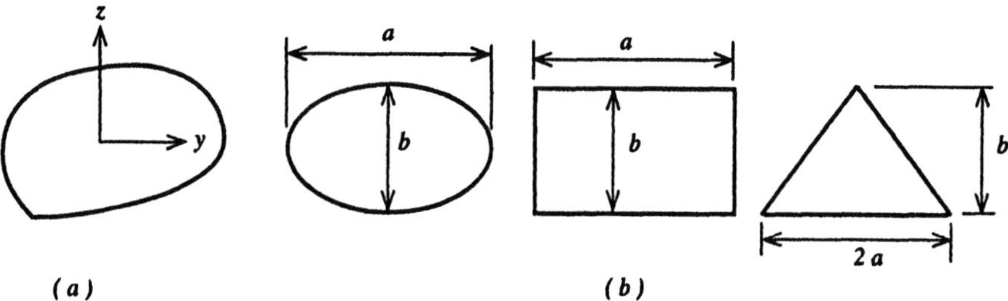

Figure 6.9 Above, (*a*) shows a tube of arbitrary cross section, for which the Navier-Stokes equation for Poiseuille flow may be solved, and (*b*) shows dimensions of elliptical, rectangular and triangular (isosceles) tubes for which the Darcy friction factor f and hydraulic diameter $\mathcal{D}_h$ may be found analytically.

Substituting in the equation for ν:

$$\nu = \frac{\pi D^4 g}{128 Q}\left(1 + \frac{H}{L}\right)$$

$$= \frac{\pi(3.0E(-3)\,m)^4(9.807\,m/s^2)}{128(1.0E(-3)\,m^3 \div 152\,s)}(1 + \frac{3}{30}) = 2.682E(-4)\,m^2/s$$

(If the funnel length L were doubled in this experiment, the flow rate Q would decrease only slightly, in proportion to $1 + H/L$.)

6.5.5 Poiseuille Flow in Noncircular Tubes

Although a circular tube is the most common form, other tube shapes, such as rectangular ventilation ducts, may be utilized in engineering systems. The properties of Poiseuille flow in such tubes may be described by solving the Navier-Stokes equation with boundary conditions suitable to the particular tube shape.

Using cartesian coordinates with the flow in the direction of the x-axis, the Navier-Stokes equation for flow in a duct of arbitrary cross section, as shown in figure 6.9(*a*), is:

$$0 = -\frac{dp^*}{dx} + \mu\left(\frac{\partial^2 u}{\partial z^2} + \frac{\partial^2 u}{\partial y^2}\right) \qquad (6.44)$$

This partial differential equation is a form of *Poisson's equation* and may be solved analytically for some simple shapes, such as an ellipse, a rectangle and an isosceles triangle (see figure 6.9[*b*]), or numerically for any arbitrary shape with the boundary condition that $u = 0$ at the tube wall. The solution to this equation, $u\{y, z\}$, can then be used to determine the mean flow velocity $\bar{V}$ by integrating u over the tube cross-section. We can express this value of $\bar{V}$ approximately by evaluating the magnitude of the terms in equation 6.44:

$$0 = -\frac{dp^*}{dx} + \mu\frac{\bar{V}}{\ell^2}$$

$$\bar{V} \sim \frac{\ell^2}{\mu}\left(-\frac{dp^*}{dx}\right) \tag{6.45}$$

where ℓ is a characteristic dimension of the tube shape that weighs the shorter dimension b more heavily than the longer dimension a of figure 6.9(b). For circular Poiseuille flow, we have the exact relation of equation 6.42, in which the characteristic dimension is the tube diameter D and the proportionality constant is 32. For other shapes, the characteristic dimension is chosen to be the *hydraulic diameter* $\mathcal{D}_h$, defined as:

$$\mathcal{D}_h \equiv \frac{4 \times Area}{Perimeter} \tag{6.46}$$

For a circular tube, the hydraulic diameter is $\mathcal{D}_h = 4(\pi D^2/4)/2\pi D = D$, the tube diameter, while $\mathcal{D}_h = 2h$ for plane Poiseuille flow.

Darcy Friction Factor

The linear relationship 6.45 between the mean flow velocity and the pressure gradient is conventionally expressed in a dimensionless form by defining the *Darcy friction factor* f as:

$$f \equiv \frac{(-dp^*/dx)\,\mathcal{D}_h}{\frac{1}{2}\rho\bar{V}^2} \tag{6.47}$$

where both the numerator and denominator of the right side of equation 6.47 have the dimensions of pressure. Note that if we know the value of f, it is relatively easy to find the pressure drop Δp^* in a tube of length L by inverting equation 6.47:[17]

$$\Delta p^* = \left(-\frac{dp^*}{dx}\right)L = f\left(\frac{1}{2}\rho\bar{V}^2\right)\frac{L}{\mathcal{D}_h} \tag{6.48}$$

For a circular tube, the Darcy friction factor f may be found by substituting equation 6.42 into 6.47, giving:

---

[17]Equation 6.48 is known as the *Darcy-Weisbach equation.* Henry Philibert Gaspard Darcy (1803–1858) conducted extensive experiments on the pressure drop of water flowing in pipes. Julius Weisbach (1806–1871) proposed the form of this equation in his influential treatise on engineering fluid mechanics.

Table 6.1 Friction factor and hydraulic diameter for noncircular tubes

	Ellipse		Rectangle		Isosceles triangle	
b/a	$(Re_{\mathcal{D}_h})f$	$\mathcal{D}_h/b$	$(Re_{\mathcal{D}_h})f$	$\mathcal{D}_h/b$	$(Re_{\mathcal{D}_h})f$	$\mathcal{D}_h/b$
1.0	64.00	1.0	56.91	1.0	52.61	0.828
0.5	67.29	1.298	62.19	1.333	50.49	0.944
0.2	74.41	1.4965	76.28	1.667	48.63	0.990
0.1	77.26	1.548	84.68	1.818	48.31	0.998

$$f = \frac{64}{Re_{\mathcal{D}}}; \quad Re_{\mathcal{D}} = \frac{\rho \bar{V} \mathcal{D}}{\mu} \tag{6.49}$$

In chapter 7 on turbulent flow, we will find that f is as useful a quantity for turbulent flow in channels as it is for laminar viscous (Poiseuille) flow.

If we multiply equation 6.47 by the Reynolds number $Re_{\mathcal{D}_h}$ based on the hydraulic diameter $\mathcal{D}_h$,

$$Re_{\mathcal{D}_h} \equiv \frac{\rho \bar{V} \mathcal{D}_h}{\mu} \tag{6.50}$$

we obtain:

$$(Re_{\mathcal{D}_h})f = \frac{2\mathcal{D}_h^2}{\mu \bar{V}} \left(-\frac{dp^*}{dx} \right) \tag{6.51}$$

The dimensionless group on the right side of equation 6.51 is a function only of the shape of the tube cross-section. For a circular tube, it has the value 64, while for plane Poiseuille flow (a rectangle with $a/b = \infty$), it has the value 48. Table 6.1 lists the values of $(Re_{\mathcal{D}_h})f$ and $\mathcal{D}_h/b$ for channels of elliptical, rectangular and triangular (isosceles) cross-section having several different values of the aspect ratio b/a as defined in figure 6.9. Note that for all of the shapes considered, the value of $(Re_{\mathcal{D}_h})f$ does not differ by more than 25% from the circular tube value of 64.

By combining equations 6.48 and 6.50, we may express the pressure drop Δp^* in a tube of length L in the form:

$$\Delta p^* = \frac{(Re_{\mathcal{D}_h})f}{(\mathcal{D}_h/b)^2} \left(\frac{\mu \bar{V} L}{2b^2} \right) \tag{6.52}$$

where the numerical values of the numerator and denominator of the first factor on the right side of equation 6.52 can be obtained from table 6.1.

6.5.6 Lubrication Flows

When one solid surface slides over another, the frictional resistance can be greatly reduced by separating the surfaces with a thin film of lubricating oil. For example,

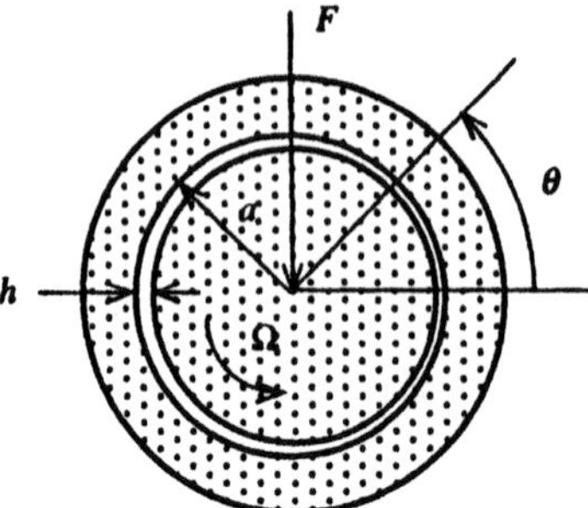

Figure 6.10 A shaft turning in a journal bearing moves laterally to an eccentric position when loaded with a force F.

in automobile engines, the sliding motion of the pistons in the cylinders and the rotational motion of the crankshaft and connecting rods in their bearings are copiously lubricated with engine oil to reduce friction. Equally important, the presence of the oil prevents the moving surfaces from coming into solid contact even though there are large forces pushing them together, thereby preventing scoring and damage to the solid surfaces. The description of the motion of thin films of lubricating fluid undergoing shearing motion and subject to normal forces is called *lubrication theory*. In this section, we will examine a simple case of lubrication flow, that in a journal bearing.

A journal bearing is a hollow cylinder enclosing a solid shaft that rotates about its axis at a speed Ω. As shown in figure 6.10, the journal bearing radius a is slightly larger than that of the shaft, by an amount $\bar{h}$ that is very small compared to the value of a.[18] When the shaft is subject to a vertical force F, it moves to the right, as shown in figure 6.10, becoming eccentric with respect to the journal bearing, and the clearance h between the shaft and the bearing varies with angular position θ:

$$h = \bar{h}(1 - \eta \cos \theta) \tag{6.53}$$

where the eccentricity η is the ratio of the lateral displacement of the shaft to $\bar{h}$.

Why does the shaft move to the right rather than in the direction of the applied force F? To understand this displacement, consider the flow of lubricating oil in the lower semicircle of the clearance space shown in figure 6.10. Due to the Couette flow induced by the rotating shaft, there is an inflow on the left ($\theta = \pi$) of $V_p h/2 = (\Omega a/2)\bar{h}(1+\eta)$ and an outflow on the right ($\theta = 0$) of $(\Omega a/2)\bar{h}(1-\eta)$ for a net inflow of $\Omega a \bar{h} \eta$. Since this is a steady incompressible flow, this net inflow must be counteracted

[18] Typically, $\bar{h}/a = 1E(-3)$.

by a pressure gradient sufficient to induce an equal Poiseuille outflow at $\theta = 0$ and π:

$$\frac{\bar{h}^3}{12\mu}\left(-\frac{dp}{a\,d\theta}\right) = \left(\frac{\Omega a\bar{h}}{2}\right)\eta$$

$$\left(-\frac{dp}{d\theta}\right) = 6\mu\Omega\left(\frac{a}{\bar{h}}\right)^2\eta$$

Thus the pressure in the bottom half of the bearing will exceed that in the top half by an average amount:

$$\Delta p \simeq 6\mu\Omega\left(\frac{a}{\bar{h}}\right)^2\eta$$

which will support a force per unit length of bearing F/W:

$$\frac{F}{W} \simeq a\Delta p \simeq 6\mu\Omega\left(\frac{a^3}{\bar{h}^2}\right)\eta$$

Note that the eccentricity η of the shaft in the bearing is proportional to the applied load F.

To solve this problem exactly, we begin with equation 6.36 for the volume flow rate Q/W in combined Couette and Poiseuille flow, which we solve for the pressure gradient:

$$\frac{dp}{d\theta} = \frac{12\mu a}{h^3}\left(\frac{\Omega a h}{2} - \frac{Q}{W}\right) \tag{6.54}$$

in which h is the function of θ given in equation 6.53. If we integrate equation 6.54 from 0 to 2π, the left side integrates to zero because the pressure is the same at 0 and 2π, and the integral of the right side may be solved for Q/W:

$$\frac{Q}{W} = \frac{\bar{h}\Omega a(1-\eta^2)}{2+\eta^2} \tag{6.55}$$

As the bearing is loaded and η increases, the flow rate Q of lubricating oil decreases.

Using this value of Q/W in equation 6.54 and integrating from 0 to θ, we find the pressure $p\{\theta\}$:

$$p\{\theta\} = p\{0\} - 6\mu\Omega\left(\frac{a}{\bar{h}}\right)^2\frac{\eta\sin\theta(2-\eta\cos\theta)}{(2+\eta^2)(1-\eta\cos\theta)^2} \tag{6.56}$$

The pressure in the upper half of the journal bearing ($0 \le \theta \le \pi$) is always less than $p\{0\}$ while that in the lower half is greater by an equal amount. The upward force of the pressure can be integrated to give the load per unit width:

$$\frac{F}{W} = 12\pi\mu\Omega\left(\frac{a^3}{\bar{h}^2}\right)\frac{\eta}{(\sqrt{1-\eta^2})(2+\eta^2)} \tag{6.57}$$

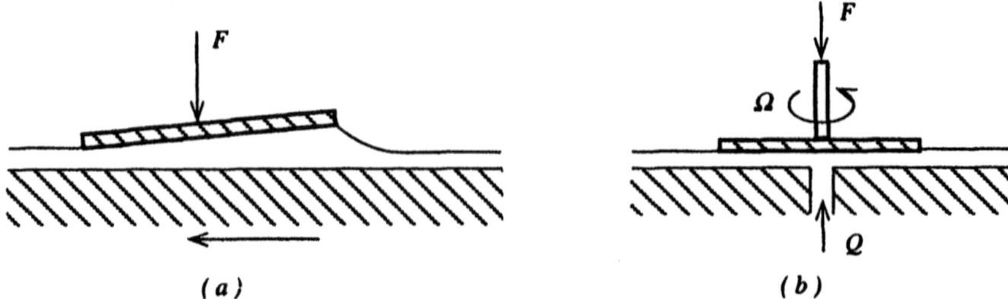

Figure 6.11 In (*a*) above, a slider bearing supports a load F by squeezing lubricant between itself and a moving plane surface, while in (*b*), a forced lubrication thrust bearing sustains a thrust load on a rotating shaft when it is supplied with pressurized lubricant flow Q.

It is also possible to calculate the torque per unit width T/W by integrating the wall shear stress τ_w for the Couette flow component ($\tau_w = \mu\Omega a/h$) and the Poiseuille flow component ($\tau_w = (h/2)(dp/a\,d\theta)$) over the shaft circumference:

$$\frac{T}{W} = 4\pi\mu\Omega\left(\frac{a^3}{\bar{h}}\right)\frac{1+2\eta^2}{(\sqrt{1-\eta^2})(2+\eta^2)} \tag{6.58}$$

The load-carrying capacity of a journal bearing increases with decreasing clearance $\bar{h}$ and increasing eccentricity η. If the eccentricity $\eta = 1$, the journal and shaft would be in contact at $\theta = 0$, which is undesirable. The infinite values for F and T predicted by equations 6.57 and 6.58 when $\eta = 1$ would not be reached in a practical journal bearing because there would be some outflow in the axial direction that would reduce the pressure and load to a finite value. Heavily loaded journal bearings may be supplied with a flow of oil to counteract leakage along the axis and to carry away heat.

In the preceding derivation, we made use of the combined Couette/Poiseuille flow, equation 6.36, to describe the flow in a journal bearing that has a variable thickness h and thereby a variable mean speed $\bar{V}$. But $\bar{V}$ was considered a constant, independent of x, when we derived the solution to these flows, which were simplified because $D\mathbf{V}/Dt = 0$. Have we made an error in forgetting about this restriction?

Not at all. While the acceleration $D\mathbf{V}/Dt$ is not zero in a lubrication flow, it may be small enough compared to the other terms in the Navier-Stokes equation that we can ignore it, in effect considering it to be zero. For the steady flow in a journal bearing, we may estimate the magnitude of $D\mathbf{V}/Dt = \mathbf{V}\cdot\nabla\mathbf{V}$ as approximately $\bar{V}^2/a$ and the viscous force per unit volume $\nu\nabla^2\mathbf{V}$ as about $\nu\bar{V}/\bar{h}^2$ so that the condition that we may neglect the acceleration term is:

$$\frac{\bar{V}^2}{a} \ll \nu\frac{\bar{V}}{\bar{h}^2}$$

$$\frac{\bar{V}\bar{h}}{\nu} \equiv Re_h \ll \frac{a}{\bar{h}}$$

that is, the Reynolds number Re_h, based upon the film thickness $\bar{h}$ and mean speed $\bar{V}$, should be much less than the ratio of the journal radius a to the film thickness. Since $a/\bar{h}$ is generally of the order of $10^2 - 10^3$ in a journal bearing, the Reynolds number can be larger than unity and yet allow us to disregard the acceleration term compared to the viscous term in the Navier-Stokes equation.

Other types of lubricated bearings are shown in figure 6.11. The slider bearing (figure 6.11[*a*]) moves laterally with respect to a plane surface but at a small angle that causes the lubricant pressure to build as the gap between the slider and the flat surface decreases, enabling the slider to sustain a normal load. In hard disk drives of computers, the magnetic pickup slides over the disk surface on a film of air that prevents physical contact. An example of a forced lubrication bearing is shown in figure 6.11(*b*), in which lubricant is introduced from a pressurized supply to the axis of a rotating thrust bearing so as to sustain a thrust F acting on the bearing. In this case, the rotary motion of the shaft does not induce a pressure rise to sustain the applied force.

Example 6.9

A bearing on the crankshaft of an automobile engine has a radius $a = 1.5\,cm$, a length $W = 3\,cm$ and a clearance $\bar{h} = 5E(-5)\,m$. It is filled with lubricating oil of viscosity $\mu = 2E(-2)\,Pa\,s$ and density $\rho = 9E(2)\,kg/m^3$. The crankshaft rotates at 3600 revolutions per minute.

(a) If the eccentricity η is not to exceed 0.5, calculate the maximum load F that the bearing can sustain. (b) At this condition, calculate the torque T and frictional power $P = \Omega T$ of the bearing. (c) Calculate the Reynolds number $Re_h \equiv \bar{V}\bar{h}/\nu$ under this condition.

Solution

(a) The load force F is found from equation 6.57 for $\eta = 0.5$:

$$L = 12\pi\mu\Omega W\left(\frac{a^3}{\bar{h}^2}\right)\frac{\eta}{(\sqrt{1-\eta^2})(2+\eta^2)}$$

$$= 12\pi(2E(-2)Pa\,s)(2\pi \times 60\,s^{-1})(3E(-2)\,m)\left(\frac{(1.5E(-2)\,m)^3}{(5E(-5)\,m)^2}\right)\frac{0.5}{\sqrt{0.75}(2.25)}$$

$$= 2.954E(3)\,N$$

(b) Use equation 6.58 to calculate the torque T and power P:

$$T = 4\pi\mu\Omega W\left(\frac{a^3}{\bar{h}}\right)\frac{1+2\eta^2}{(\sqrt{1-\eta^2})(2+\eta^2)}$$

$$= 4\pi(2E(-2)Pa\,s)(2\pi \times 60\,s^{-1})(3E(-2)\,m)\left(\frac{(1.5E(-2)\,m)^3}{5E(-5)\,m}\right)\frac{1.5}{\sqrt{0.75}(2.25)}$$

$$= 0.1477\,Nm$$

$$P = \Omega T = (2\pi \times 60\,s^{-1})(0.1477\,Nm) = 55.68\,W$$

(c) To calculate the Reynolds number, use equation 6.55 to find $\bar{V} = Q/W\bar{h}$:

$$\frac{\bar{V}\bar{h}}{\nu} = \frac{\Omega a(1-\eta^2)\bar{h}}{\nu(2+\eta^2)} = \frac{\rho\Omega a\bar{h}(1-\eta^2)}{\mu(2+\eta^2)}$$

$$= \frac{(9E(3)\,kg/m^3)(2\pi \times 60\,s^{-1})(1.5E(-2)\,m)(5E(-5)\,m)(0.75)}{(2E(-2)\,Pa\,s)(2.25)} = 42.41$$

Note that $Re_h \ll a/h = 1.5E(-2) \div 5E(-5) = 300$.

6.5.7 Creeping (Stokes) Flow

In a viscous flow where the speed V and length scale L are small, the Reynolds number $Re = VL/\nu$ may be much less than unity. A flow having a very small Reynolds number is called a *creeping flow*, or a *Stokes flow*. In such flows the acceleration $D\mathbf{V}/Dt$ is so small compared to the viscous force per unit mass, $\nu\nabla^2\mathbf{V}$, that we may neglect it in the Navier-Stokes equation 6.12 and obtain the simpler equation of motion of a creeping flow:

$$0 = -\nabla p^* + \mu\nabla^2\mathbf{V} \tag{6.59}$$

after having multiplied through by ρ. Together with the equation of mass conservation, $\nabla \cdot \mathbf{V} = 0$, equation 6.59 can be used to find three-dimensional velocity fields of creeping flow.

The most famous example of creeping flow is that of a viscous fluid flowing over a stationary solid sphere of radius a at a speed V_∞. Such a flow is axisymmetric about the flow direction. Using the coordinates R and ϕ defined in figure 6.12, the radial and tangential velocity components V_R and V_ϕ that satisfy the conditions of $\mathbf{V} = 0$ at the sphere surface $R = a$ and $\mathbf{V} = \mathbf{V}_\infty$ at $R = \infty$ are found to be:[19]

---

[19]For a complete derivation, see F. S. Sherman, 1990, *Viscous Flow* (New York: McGraw-Hill Publishing Co.)

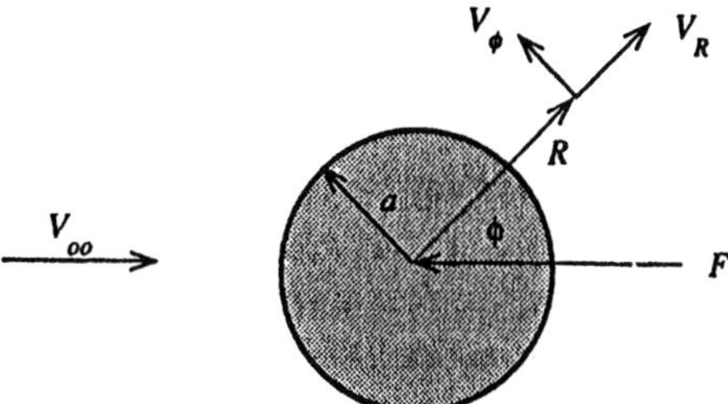

Figure 6.12 Coordinate system for viscous creeping flow of speed V_∞ over a solid sphere of radius a.

$$V_R = V_\infty \cos\phi \left(1 - \frac{3a}{2R} + \frac{a^3}{2R^3}\right)$$

$$V_\phi = V_\infty \sin\phi \left(-1 + \frac{3a}{4R} + \frac{5a^3}{4R^3}\right) \qquad (6.60)$$

where the axial component of velocity, $V_R \cos\phi - V_\phi \sin\phi$, clearly equals V_∞ at $R = \infty$. The corresponding values of pressure p^* and shear stress $\tau_{\phi R}$ are:

$$p^* = p^*_\infty - \frac{3\mu a V_\infty \cos\phi}{2R^2}$$

$$\tau_{\phi R} = -\frac{\mu V_\infty \sin\phi}{R}\left(1 - \frac{3a}{4R} + \frac{5a^3}{4R^3}\right) \qquad (6.61)$$

The pressure p^* is higher in the front of the sphere than in the back, contributing to the drag on the sphere. The wall shear stress, $\tau_{\phi R}$ at $R = a$, acts in the direction to increase the drag. By summing the pressure force and the viscous shear force acting on each element of the surface of the sphere, we can find the total drag force F:[20]

$$F = 6\pi\mu V_\infty a \qquad (6.62)$$

Of the total drag, two-thirds is attributable to the wall shear stress and one-third to the pressure.

The drag on bodies in creeping flow is not sensitive to the body shape. For example, a flat disc of radius a oriented normal or parallel to the flow direction has a drag of

[20]This relationship, first derived by Stokes, figured prominently in the experimental determination of two physical constants, Boltzman's constant and the electric charge of an electron. In the former case, incorporation of Einstein's theory of Brownian motion in the analysis of observations of Brownian motion led to a calculation of Boltzman's constant. The electric charge was determined in the famous Millikan oil drop experiment where charged oil droplets were dragged through a fluid by an electric field.

$16\mu V_\infty a$ or $(32/3)\mu V_\infty a$, respectively, not greatly different than that of the sphere, equation 6.62, especially when one considers the great differences in shape and flow orientation.

We may apply this expression for the drag on a sphere to the calculation of the steady free-fall velocity V_f of a spherical particle of density ρ_p moving downward through a fluid of density ρ_f under the influence of gravity. The gravity force acting on the particle is $\rho_p g(4\pi a^3/3)$, but the upward buoyant force of the surrounding fluid is $\rho_f g(4\pi a^3/3)$ so that the net downward force is $(\rho_p - \rho_f)g(4\pi a^3/3)$. Setting this equal to the drag force F of equation 6.62 and solving for V_f,

$$(\rho_p - \rho_f)g\left(\frac{4\pi a^3}{3}\right) = 6\pi\mu V_f a$$

$$V_f = \frac{2(\rho_p - \rho_f)ga^2}{9\mu} \tag{6.63}$$

The condition that the Reynolds number be small in free fall is:

$$\frac{\rho_f V_f a}{\mu} = \frac{2}{9}\left(\frac{\rho_p - \rho_f}{\rho_f}\right)\left(\frac{\rho_f g^{1/2} a^{3/2}}{\mu}\right)^2 \ll 1 \tag{6.64}$$

which can be satisfied if the term $\rho_f g^{1/2} a^{3/2}/\mu$ is smaller than one.

Example 6.10

A spherical solid particle of diameter $D = 1\,micrometer = 1.0E(-6)\,m$ and density $\rho_p = 2E(3)\,kg/m^3$ falls slowly downward through stationary air of density $\rho_f = 1.206\,kg/m^3$ and viscosity $\mu = 1.80E(-5)\,Pa\,s$. Calculate (a) the free fall velocity V_f and (b) the Reynolds number $Re_D \equiv V_f D/\nu$.

Solution

(a) Using equation 6.63,

$$V_f = \frac{2(\rho_p - \rho_f)ga^2}{9\mu} = \frac{2([2E(3) - 1.206]\,kg/m^3)(9.807\,m/s^2)(1E(-6\,m)^2}{9(1.8E(-5)\,Pa\,s)}$$

$$= 2.420E(-4)\,m/s$$

(b)

$$Re_D = \frac{\rho_f V_f D}{\mu} = \frac{(1.206\,kg/m^3)(2.420E(-4)\,m/s)(2E(-6)\,m)}{1.8E(-6)\,Pa\,s)} = 3.243E(-5)$$

The Reynolds number is well below unity, ensuring the conditions for creeping flow that was assumed in using equation 6.63.

6.5.8 Flow through Porous Media

Just as it does on the surface of the earth, water also flows underground, but only through pores in the soil and underlying geologic structure. Although these pores are very small and account for only a small portion of the underground volume, it is possible for water to move large distances underground, albeit very slowly.[21] Other fluids, such as natural gas and oil, also may be moved out of underground locations by pumping from wells. The movement of contaminated fluid from a solid waste landfill into a potable water aquifer located beneath it is an example of unwanted underground flow. Thus the motion of fluids through porous rock, induced by pressure and gravity forces, can be of great practical importance.

It is possible to filter harmful particles from a fluid stream by passing it through a porous solid whose pores are too small to permit the passage of of the particles. In other cases, the large surface area of the microscopic pores may provide sites for chemical catalysis or adsorbtion of components of the fluid. Flow through filters and catalyst beds is an engineering application of flow through porous media.

Although the fluid passages through a porous solid are irregular in size and length, we might expect the average flow speed through them to be comparable to the flow through a circular tube of diameter D, as given by equation 6.42:

$$\bar{V} = \frac{D^2}{32\,\mu}\left(-\frac{dp^*}{dz}\right)$$

To generalize this relationship to a porous solid, where the pore size and length is variable, we introduce a proportionality constant k, called the *permeability*, that includes an averaging of the pore size, shape and length, to determine the volume flow rate per unit area of solid, V_d, as:[22]

$$\mathbf{V}_d = -\frac{k}{\mu}\nabla p^* \tag{6.65}$$

Equation 6.65 is called *Darcy's Law*. The permeability k, which has the dimensions

---

[21]It is a surprising fact that a larger portion of all the fresh water of the continental land mass is stored in underground aquifers than in the lakes and rivers on its surface.

[22]This is a generalization of equation 6.42 for laminar viscous flow under the assumption that the Reynolds number of the pore flow, $V_d\sqrt{k}/\nu$, is much smaller than one. It also assumes that the pore size is much smaller than the length scale over which p^* and $\mathbf{V}_d$ change.

of length squared, is an approximate measure of the average of the square of the pore diameters. The unit of permeability, called the *darcy*, is $1\ cm^2 = 1E(-4)\ m^2$.[23]

The velocity of the fluid in a pore will be much higher than $\mathbf{V}_d$ because the latter is based upon the volume of fluid passing through a unit area, only a small portion of which is composed of pores. Thus a fluid particle will move a distance L in a time that is much smaller than L/V_d, but it is the quantity V_d that is observable in porous flows and that is of most practical interest.

For incompressible flows, the velocity field $\mathbf{V}_d$ must satisfy the condition of mass conservation, equation 3.17:

$$\nabla \cdot \mathbf{V}_d = 0 \tag{6.66}$$

Equations 6.65 and 6.66 together enable us to find the field of flow in a porous medium. The methods of solution for a Darcy flow are similar to those used in irrotational flows (see chapter 11).

The fraction of the volume of a solid that is occupied by the fluid is called the *porosity*. The porosity of a porous solid determines its capacity to store fluids, but does not explicitly affect the flow field $\mathbf{V}_d$.

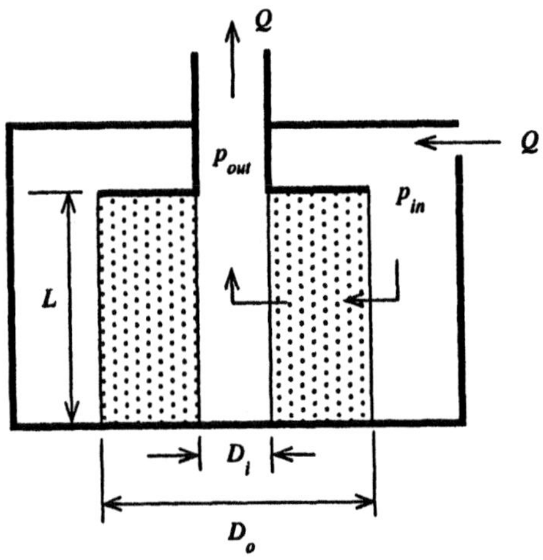

Figure E 6.11

[23]Permeabilities are quite variable. Some typical ranges (in *darcys*) are for sand, 1-100; soil, 0.1-10; sandstone, 1E(-7)-10; limestone, 1E(-6)-1; and brick, 0.01-0.1.

Example 6.11

A diesel fuel filter, shown in figure E 6.11, consists of a porous material in the form of a hollow cylinder of inner diameter $D_i = 3\,cm$, outer diameter $D_o = 10\,cm$ and length $L = 20\,cm$ that is contained within a chamber that provides for a radial inflow of the fuel to be filtered. If the fuel flow rate is $Q = 1\,l/min$, the fuel viscosity $\mu = 2E(-6)\,Pa\,s$ and the fuel filter permeability $k = 0.01\ darcys = 1E(-6)\,m^2$, calculate the pressure drop $p_{in} - p_{out}$ required to force the fuel through the filter.

Solution

If r is the radial distance from the center of the filter, the radial inflow velocity $V_r \equiv -V_d$ is determined by equation 6.65:

$$V_r = \frac{k}{\mu}\frac{dp}{dr}$$

However, the conservation of mass requires that the volume flow rate of the diesel fluid through a cylinder of radius r and length L be equal to Q:

$$2\pi r L V_r = Q$$

Eliminating V_r between these equations,

$$\frac{Q}{2\pi r L} = \frac{k}{\mu}\frac{dp}{dr}$$

$$dp = \frac{\mu Q}{2\pi k L}\frac{dr}{r}$$

Integrating from the inner radius $D_i/2$ to the outer radius $D_o/2$:

$$\begin{aligned} p_{in} - p_{out} &= \frac{\mu Q}{2\pi k L}\ln\left(\frac{D_o}{D_i}\right) \\ &= \frac{(2E(-6)\,Pa\,s)(1\,l/60\,s)}{2\pi(1E(-6)\,m^2)(0.2\,m)}\ln\left(\frac{10}{3}\right) \\ &= 3.194E(-2)Pa \end{aligned}$$

Darcy's Law, equation 6.65, has the same form as Ohm's law for the electric current density $\mathbf{j}$ in an electrical conductor having an electrical conductivity σ:

$$\mathbf{j} = -\sigma \nabla \phi \qquad (6.67)$$

where ϕ is the electric potential, or voltage. The free electrons move through the

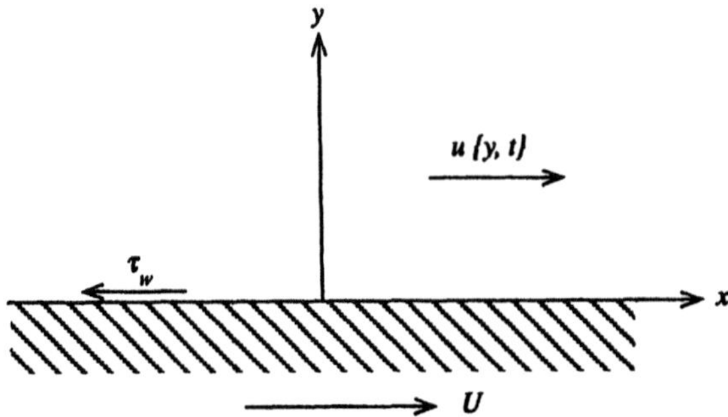

Figure 6.13 An initially stationary fluid ($y \geq 0$) is set into unsteady motion by the horizontal movement of the solid surface ($y = 0$) at a fixed speed U beginning at $t = 0$. The fluid velocity $u\{y, t\}\,\mathbf{i}_x$ is a function of y and t only.

conductor just like the viscous fluid in a porous solid, encountering resistance from the atoms of the electrical conductor. In electrical superconductors, where $\sigma = \infty$, a finite current can flow with no voltage drop ($\nabla\phi = 0$). In superfluid helium, ^4He below 4.2 K, the superfluid phase has zero viscosity and can flow through the normal phase at a finite velocity without a pressure drop, as in Darcy's law with $\mu = 0$. Both superconductivity and superfluidity are examples of quantum behavior exhibited on a macroscopic scale.

6.5.9 Unsteady Laminar Flow

The applications we have considered up to this point have been ones in which the acceleration of a fluid particle, $D\mathbf{V}/Dt$, either has been zero or so much smaller than the viscous force per unit mass, $\nu\nabla^2\mathbf{V}$, that it was negligible. Flows of these types are called *inertia-free*. But, when a flow is unsteady, it may be essential to include the acceleration term in the solution of the Navier-Stokes equation. In this section, we consider two examples of unsteady laminar viscous flow that are sufficiently simple to permit us to obtain analytical solutions to the Navier-Stokes equation. More importantly, they reveal physical features that are common to many complex flows encountered in engineering applications.

Impulsive Shearing Motion of a Plane Wall

A very simple unsteady shearing motion is illustrated in figure 6.13. An initially motionless viscous fluid, located above a horizontal solid plane surface at $y = 0$, extends to infinity in the x, y and z directions. Beginning at time $t = 0$, the solid wall is impulsively set into motion with a velocity $U\,\mathbf{i}_x$ in the x direction. The viscous shear force exerted on the fluid by the moving wall causes the fluid to move in the x direction, more fluid being coaxed into moving as time progresses. The motion is entirely caused by viscous shear forces, there being no imposed pressure gradient in the direction of motion. The velocity $\mathbf{V}$ has only one component, $u\,\mathbf{i}_x$, which depends upon y and t only. Thus the only components of $D\mathbf{V}/Dt$ and $\nu\nabla^2\mathbf{V}$ are $(\partial u/\partial t)\,\mathbf{i}_x$ and $\nu(\partial^2 u/\partial y^2)\,\mathbf{i}_x$, respectively, and the Navier-Stokes equation 6.12 becomes:

$$\frac{\partial u}{\partial t} = \nu\frac{\partial^2 u}{\partial y^2} \tag{6.68}$$

The velocity component $u\{y,t\}$ must satisfy the initial condition that $u\{y,0\} = 0$ and the boundary conditions that $u\{0,t\}) = U$ and $u\{\infty,t\} = 0$.

Equation 6.68 is an example of a well-known partial differential equation of mathematical physics called the *diffusion equation.* The coefficient ν is called a *diffusivity,* whose value determines the rate at which the quantity, u, which is the x-momentum per unit mass, diffuses in the y direction as time proceeds. In this example, ν is then the *momentum diffusivity,* having the SI units of m^2/s.

The partial differential equation 6.68 has an elegant solution that is compatible with the initial and boundary conditions of this flow. This solution may be found by assuming that $u\{y,t\}$ is a function of a particular combination of y and t, denoted by η:

$$\eta \equiv \frac{y}{\sqrt{\nu t}} \tag{6.69}$$

$$u = Uf\{\eta\} \tag{6.70}$$

where $f\{\eta\}$ is a function only of its argument, η. The boundary conditions on f are thus $f\{0\} = 1$ and $f\{\infty\} = 0$.

The substitution of equations 6.69 and 6.70 into the Navier-Stokes equation 6.68 gives rise to a total differential equation for f. To see how this comes about, first evaluate the partial derivatives,

$$\frac{\partial \eta}{\partial t} = \frac{\partial}{\partial t}\left(\frac{y}{\sqrt{\nu t}}\right) = -\frac{1}{2t}\left(\frac{y}{\sqrt{\nu t}}\right) = -\frac{\eta}{2t}; \qquad \frac{\partial \eta}{\partial y} = \frac{\partial}{\partial y}\left(\frac{y}{\sqrt{\nu t}}\right) = \frac{1}{\sqrt{\nu t}};$$

$$\frac{1}{U}\frac{\partial u}{\partial t} = \frac{\partial \eta}{\partial t}\frac{df}{d\eta} = -\frac{\eta}{2t}\frac{df}{d\eta}; \qquad \frac{1}{U}\frac{\partial u}{\partial y} = \frac{\partial \eta}{\partial y}\left(\frac{df}{d\eta}\right) = \frac{1}{\sqrt{\nu t}}\left(\frac{df}{d\eta}\right);$$

$$\frac{1}{U}\frac{\partial^2 u}{\partial y^2} = \frac{1}{\nu t}\frac{d}{d\eta}\left(\frac{df}{d\eta}\right)$$

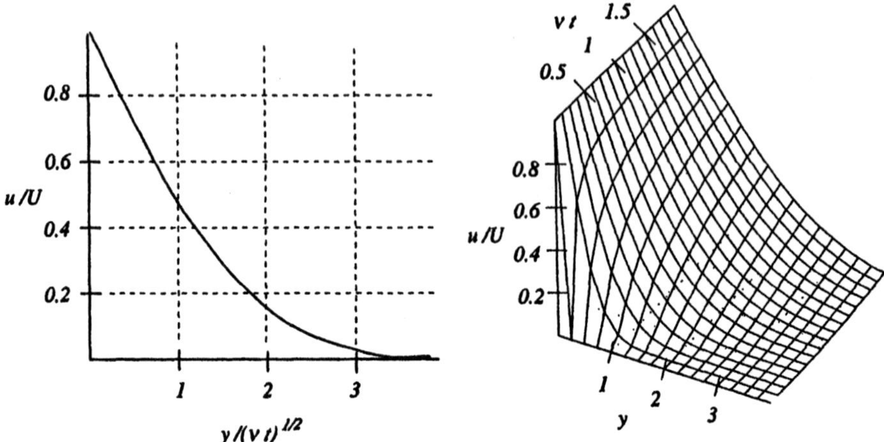

Figure 6.14 *Above left*, the velocity profile $u/U \equiv f$ as a function of $y/(\nu t)^{1/2} \equiv \eta$ for the impulsive motion of a plane wall. *Above right*, the velocity profile u/U as a function of y and the modified time νt.

and then substitute them in equation 6.68 to obtain:

$$-\frac{\eta}{2}\left(\frac{df}{d\eta}\right) = \frac{d}{d\eta}\left(\frac{df}{d\eta}\right)$$

which may be integrated twice:

$$\frac{df}{d\eta} = c_1 e^{-(\eta/2)^2}$$

$$f = c_1 \int_0^{\eta/2} e^{-\xi^2/4}\, d\xi + c_2$$

If we choose $c_1 = -1/\sqrt{\pi}$ and $c_2 = 1$, we will have satisfied the boundary conditions on f and hence u, obtaining:[24]

$$\frac{u\{y,t\}}{U} = 1 - \frac{2}{\sqrt{\pi}} \int_0^{y/2\sqrt{\nu t}} e^{-\xi^2/4}\, d\xi \tag{6.71}$$

The second term of equation 6.71 is called the *error function* of argument $y/2\sqrt{\nu t}$.

[24]The definite integral $\int_0^{\infty} \exp(-\xi^2/4)\, d\xi = \sqrt{\pi}/2$

It is a transcendental function arising in probability theory for which tabulated values are available.[25] Using these values, figure 6.14 shows the velocity profile $u/U \equiv f$ as a function of $y/\sqrt{\nu t} \equiv \eta$. Since this does not show directly the increasing penetration of the moving fluid into the stationary flow, figure 6.14 includes a plot of the velocity profile u/U as a function of y and the modified time νt.

It is useful to have a precise method for defining the thickness of the region of the flow that is affected by the moving wall. One method is to calculate a thickness δ^* such that the volume flow rate in a layer of this thickness having a uniform velocity U is the same as that of the moving fluid:

$$U\delta^* \equiv \int_0^\infty u\,dy = U\sqrt{\nu t}\int_0^\infty f\,d\eta = 2U\left(\frac{\nu t}{\pi}\right)^{1/2}; \qquad \delta^* = 2\left(\frac{\nu t}{\pi}\right)^{1/2} \tag{6.72}$$

where the numerical value of the definite integral is determined by the properties of the error function. As this thickness δ^* is not much different than $\sqrt{\nu t}$, we may say that an approximate measure of the layer thickness is $\sqrt{\nu t}$.

The shear stress τ_w acting on the wall (see figure 6.13) opposes the motion of the wall. The power required to move the wall, per unit of wall area, is the product $\tau_w U$. To find the value of $\tau_w\{t\}$, we determine the velocity gradient at the wall using equation 6.71:

$$\tau_w = -\mu\left(\frac{\partial u}{\partial y}\right)_{y=0} = \mu\left(\frac{U}{\sqrt{\pi\nu t}}\right) \tag{6.73}$$

The shear stress is the same as that on the moving wall of Couette flow in a channel of height $\sqrt{\pi\nu t}$.

Example 6.12

The shaft of the journal bearing in example E 6.4 is impulsively started from rest to the speed Ω. (a) Derive an expression for the torque T as a function of time when the time is small enough that $\sqrt{\nu t} \ll h$. (b) Calculate the numerical value of t that would make $\sqrt{\nu t} = h$ if $\rho = 8.0E(2)\,kg/m^3$.

Solution

(a) Since equation 6.73 shows that the wall shear stress τ_w in the unsteady startup

---

[25] See H. S. Carslaw and J. C. Jaeger, 1947, *Conduction of Heat in Solids* (Oxford: Oxford University Press), or M. Abramowitz and I. Stegun, 1964, *Handbook of Mathematical Functions* (Washington, D.C.: U.S. Department of Commerce).

flow is the same as that for Couette flow when $h = \sqrt{\pi\nu t}$, simply replace h by $\sqrt{\pi\nu t}$ in the expression for T in the solution to example E 6.4:

$$T = \frac{\pi\mu\Omega LD}{4\sqrt{\pi\nu t}} = \frac{\sqrt{\pi\rho\mu}\,\Omega LD}{4\sqrt{t}}$$

(b)

$$t = \frac{h^2}{\nu} = \frac{\rho h^2}{\mu} = \frac{(8E(2)\,kg/m^3)(1E(-4)\,m)^2}{6.7E(-5)\,Pa\,s} = 0.1194\,s$$

Oscillatory Shearing Motion of a Plane Wall

An important alternate form of shearing motion of a plane wall to that illustrated in figure 6.13 is one for which the wall velocity is sinusoidal in time:

$$u\{0, t\} = U \cos \omega t \tag{6.74}$$

where ω is the angular frequency of the sinusoidal motion. The disturbance to the fluid caused by the moving wall is propagated outward into the flow, just as in the case of impulsive flow, but after a time $t = \pi/\omega$, the wall has reversed its motion and sends a cancelling disturbance. Thus the wall motion only penetrates a distance about $\sqrt{\pi\nu/\omega}$ into the fluid, no matter how long the wall continues to oscillate.

The velocity $u\{y, t\}$ of the fluid caused by the oscillatory motion of the plane wall may be found by solving equation 6.68 using the method of separation of variables:[26]

$$\frac{u}{U} = \exp\left[-\left(\frac{\omega}{2\nu}\right)^{1/2} y\right] \cos\left[\omega t - \left(\frac{\omega}{2\nu}\right)^{1/2} y\right] \tag{6.75}$$

which reduces to equation 6.74 when $y = 0$. This velocity profile is a cosine wave whose amplitude dies off exponentially in the y direction by a factor of e in each incremental distance of $\sqrt{2\nu/\omega}$. The motion at a distance y is also a sinusoidal function of time but lags behind that at the wall by the phase angle $y\sqrt{\omega/2\nu}$. Thus, at the distance $y = \pi\sqrt{2\nu/\omega}$, the fluid moves in a direction opposite to that of the wall. A plot of the velocity profile of equation 6.75 is shown in figure 6.15, where it can be seen that the wall disturbance is confined to a region near the wall.

---

[26]See F. S. Sherman, 1990, *Viscous Flow* (New York: McGraw-Hill Publishing Co.).

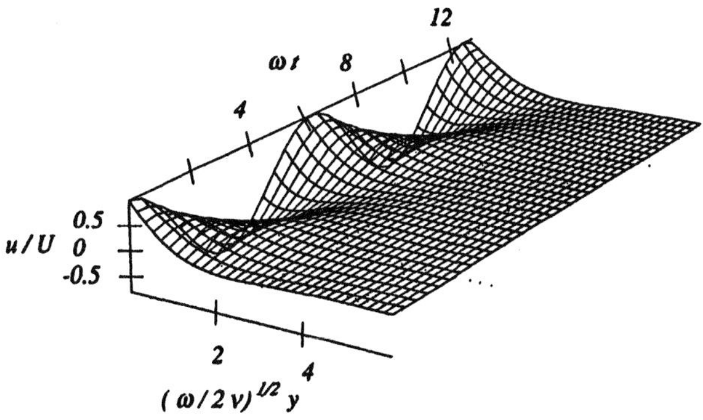

Figure 6.15 The velocity profile $u/U \equiv f$ of the fluid adjacent to an oscillating plane wall as a function of the dimensionless distance $(\omega/2\nu)^{1/2}y$ and the dimensionless time ωt.

To find the wall shear stress τ_w acting on the wall, we multiply the wall velocity gradient by the viscosity:

$$\tau_w = \mu \left(\frac{\partial u}{\partial y}\right)_{y=0} = \mu U \left(\frac{\omega}{2\nu}\right)^{1/2} [\sin \omega t - \cos \omega t] \tag{6.76}$$

The power per unit area of wall, P/A, that must be supplied to maintain the oscillating wall is the product of the external force per unit area applied to the wall, $-\tau_w$, times the wall velocity $u\{0, t\} = U \cos \omega t$:

$$\frac{P}{A} = -\tau_w U \cos \omega t = \mu U^2 \left(\frac{\omega}{2\nu}\right)^{1/2} \left[\cos^2 \omega t - \cos \omega t \sin \omega t\right] \tag{6.77}$$

If we time-average this expression over one cycle, to obtain the average power $\overline{P}$, the second term in brackets averages to zero and the first to $1/2$, giving:

$$\frac{\overline{P}}{A} = \frac{\mu U^2}{2} \left(\frac{\omega}{2\nu}\right)^{1/2} \tag{6.78}$$

This expression may be used to calculate the power lost in oscillatory flows, such as a sound wave propagating inside a tube or the vibration of an elastic solid surrounded by stationary fluid.

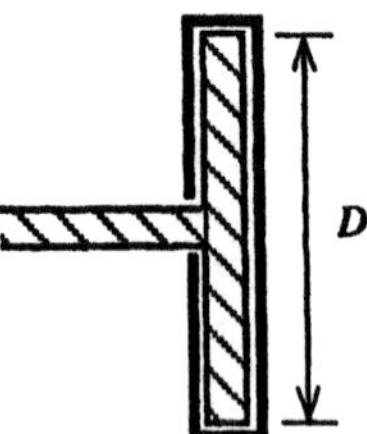

Figure E 6.13

Example 6.13

A torsional vibration damper for a diesel engine, sketched in figure E 6.13, consists of a circular disk of diameter $D = 0.5\,m$ rigidly attached to the end of the engine crankshaft. It is surrounded by a housing that is free to rotate with respect to the crankshaft, the space between the housing and the disk being filled with oil of viscosity $\mu = 1.0\,Pa\,s$ and density $\rho = 9.0E(2)\,kg/m^3$. When the engine runs at a fixed speed, both the housing and the disc rotate at the same speed and there is no angular relative motion between them. However, when a torsional vibration of angular frequency $\omega = 1E(3)\,s^{-1}$ and angular amplitude $\phi = 1E(-3)$ develops, the relative angular displacement of the disc with respect to the housing is $\phi \sin \omega t$, and the relative angular velocity is $\omega\phi \cos \omega t$. Thus at any distance r from the axis, the disc surface is moving tangentially at a speed $r\omega\phi \cos \omega t$ with respect to the oil inside the housing. Calculate the power $\overline{P}$ absorbed by the vibration damper.

Solution

The velocity amplitude U of the disk surface is equal to $r\omega\phi$. Substituting this value into equation 6.78 and integrating over both sides of the disk area, we find the power $\overline{P}$:

$$\overline{P} = \frac{\mu}{2}\left(\frac{\omega}{2\nu}\right)^{1/2} 2\int_0^{D/2} (r\omega\phi)^2(2\pi r)\,dr = 2\pi\mu(\omega\phi)^2\left(\frac{\omega}{2\nu}\right)^{1/2}\int_0^{D/2} r^3\,dr$$

$$= 2\pi\mu(\omega\phi)^2\left(\frac{\omega}{2\nu}\right)^{1/2}\left|\frac{r^4}{4}\right|_0^{D/2} = \frac{\pi}{32}\mu(\omega\phi)^2\left(\frac{\omega}{2\nu}\right)^{1/2} D^4$$

$$= \frac{\pi}{32}(1.0\,Pa\,s)(1E(3)\,s^{-1})^2(1E(-3))^2\left(\frac{(1E(3)\,s^{-1})(9E(2)\,kg/m^3)}{2 \times 1.0\,Pa\,a}\right)^{1/2}(0.5\,m)^4$$

$$= 4.116\,W$$

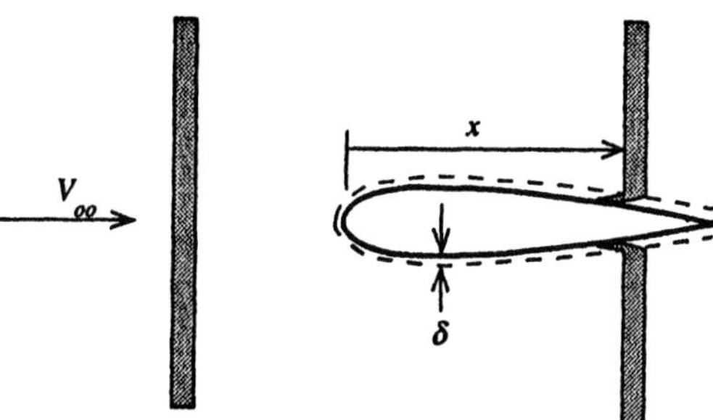

Figure 6.16 A thin slab of fluid flowing past an airfoil with speed V_∞ is retarded by viscous forces only within the thin boundary layer of thickness δ close to the surface of the airfoil.

6.6 Laminar Boundary Layers

We have been considering flows for which the viscous force is never small compared to the pressure or gravity force or to the fluid inertia. These have all been flows that are influenced by a solid surface causing the nearby fluid to move differently than more distant fluid. They are characterized by a shear stress in the fluid next to the solid surface. Such behavior will always be present when a real fluid moves with respect to a solid surface.

What is less clear from the cases that we have considered is that the region of the flow influenced by the fluid sticking to the solid surface, *i.e.*, the region of high shear stress and significant viscous force, is generally only a thin layer near the surface, called the *boundary layer*, at least under typical engineering flow conditions. To see why this is so, consider the steady flow of air past a stationary airfoil, as sketched in figure 6.16. A thin slab of upstream fluid will, as it flows downstream, suddenly find itself decelerated by contact with the surface of the airfoil, at least in a region close to the surface of thickness δ, which is related to the time of travel $t \simeq x/V_\infty$ from the front of the airfoil to a point x along the surface by:

$$\delta \simeq \sqrt{\nu t} \simeq \left(\frac{\nu x}{V_\infty}\right)^{1/2} \tag{6.79}$$

where we have used the relationship of equation 6.72 to estimate the thickness of the region that has been influenced by the mismatch of velocity between the fluid and the solid surface. Now let us compare the thickness δ of the boundary layer with the distance x from the leading edge of the airfoil:

$$\frac{\delta}{x} \simeq \left(\frac{\nu}{V_\infty x}\right)^{1/2} = \frac{1}{\sqrt{Re_x}} \tag{6.80}$$

where the length Reynolds number Re_x is based on the distance x and the free stream velocity V_∞:

$$Re_x \equiv \frac{V_\infty x}{\nu} \tag{6.81}$$

Thus the thickness δ is much smaller than the length x whenever $Re_x \gg 1$. The boundary layer thickness on most engineering structures—aircraft wings, ship's hulls, propellers, turbine blades, etc.—is small compared to the major dimension of the device because they ordinarily operate at large values of the Reynolds number. What is even more important, the viscous force in the flow field surrounding the device that lies *outside* the boundary layer is negligible, *i.e.* the external flow may be considered inviscid. For this reason, we can consider the flow field to be divided into two regions, the boundary near the solid surface that must be treated as a viscous flow and the remainder of the flow field that can be considered inviscid.[27]

6.6.1 The Boundary Layer Approximation

The existence of a thin boundary layer in a flow of high Reynolds number ensures a significant simplification of the Navier-Stokes equation as it applies to the flow within the boundary layer. To see how this comes about, consider the boundary layer flow along the surface of the airfoil shown in figure 6.16. Selecting axes x, y, with x measured in the flow direction along the surface and y normal to the surface, we may write the x-component of the Navier-Stokes equation 6.12 as:

$$u\frac{\partial u}{\partial x} + v\frac{\partial u}{\partial y} = -\frac{1}{\rho}\frac{\partial p^*}{\partial x} + \nu\left(\frac{\partial^2 u}{\partial x^2} + \frac{\partial^2 u}{\partial y^2}\right)$$

The magnitudes of the two second derivative terms on the right side of this equation are quite different from each other: the y derivative is much greater because the distance δ over which it changes is much less than the distance x from the leading edge:

$$\frac{\partial^2 u}{\partial y^2} \sim \frac{u}{\delta^2}; \qquad \frac{\partial^2 u}{\partial x^2} \sim \frac{u}{x^2} \simeq \frac{u}{\delta^2} \times \frac{1}{Re_x} \ll \frac{\partial^2 u}{\partial y^2}$$

where we have used 6.80 to replace x^2. It is therefore possible to neglect the term $\partial^2 u/\partial x^2$ compared with the other terms to arrive at the *boundary layer approximation*

[27] Ludwig Prandtl (1875–1953) developed these physical ideas in a remarkable but brief paper in 1905, greatly influencing the development of fluid mechanics in this century.

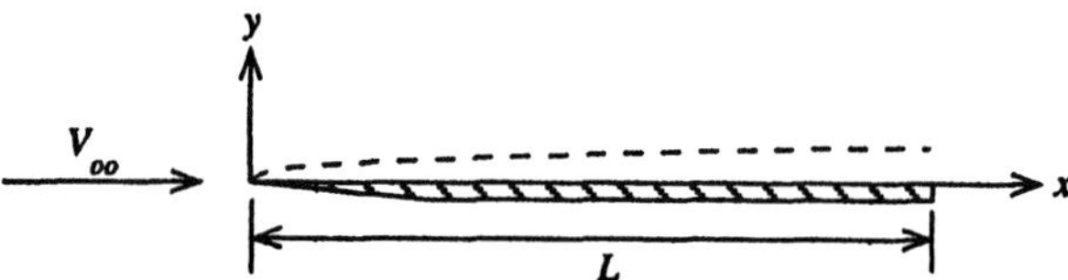

Figure 6.17 The boundary layer on the upper surface of a flat plate aligned with the flow in a wind tunnel grows in thickness with distance x from the leading edge.

to the Navier-Stokes equation:[28]

$$u\frac{\partial u}{\partial x}+v\frac{\partial u}{\partial y}=-\frac{1}{\rho}\frac{dp^*}{dx}+\nu\frac{\partial^2 u}{\partial y^2} \tag{6.82}$$

In addition, the boundary layer flow must satisfy the mass conservation equation for incompressible flow:

$$\frac{\partial u}{\partial x}+\frac{\partial v}{\partial y}=0 \tag{6.83}$$

These equations are the starting point for examining several kinds of boundary layer flows.

6.6.2 The Boundary Layer on a Flat Plate

The simplest kind of boundary layer is the one that develops on a flat plate aligned with the flow, as illustrated in figure 6.17. Because the upper surface is precisely aligned with the flow, there is no change of pressure in the x direction and $\partial p^*/\partial x = 0$. The boundary layer equation 6.82 is thereby simplified, and in combination with the incompressible mass conservation equation, the velocity components u and v must satisfy the momentum and mass conservation relations:

$$u\frac{\partial u}{\partial x}+v\frac{\partial u}{\partial y}=\nu\frac{\partial^2 u}{\partial y^2}$$

$$\frac{\partial u}{\partial x}+\frac{\partial v}{\partial y}=0 \tag{6.84}$$

This combination of one first order and one second order partial differential equations requires three boundary conditions: $u\{x, 0\} = 0$ and $v\{x, 0\} = 0$ to satisfy no slip at

[28] By considering mass conservation, it can be shown that the term $v(\partial u/\partial y)$ has the same magnitude as $u(\partial u/\partial x)$ and cannot be neglected in equation 6.82. Also, by considering the y-component of the Navier-Stokes equation, we find that $\partial p^*/\partial y = 0$ and thus $\partial p^*/\partial x$ is independent of y, which can then be written as the total derivative dp^*/dx.

the wall ($y = 0$) and $u\{x, \infty\} = V_\infty$ to ensure that the tangential flow speed reaches the free stream value far from the surface of the plate.

The boundary layer equations 6.84 are nonlinear, and there is no simple solution in terms of transcendental functions such as we found for the infinite plate, equation 6.71. Nevertheless, a simple numerical solution may be found if we assume that the form of the solution is:

$$u = V_\infty \left(\frac{df}{d\eta}\right)$$

$$v = \left(\frac{\nu V_\infty}{2x}\right)^{1/2} \left[\eta \frac{df}{d\eta} - f\right] \tag{6.85}$$

where the function f depends only upon the variable η, defined as:

$$\eta \equiv \left(\frac{V_\infty}{2\nu x}\right)^{1/2} y \tag{6.86}$$

We see that η is approximately the distance y from the wall divided by the boundary layer thickness δ estimated in equation 6.79. Assuming this form, equations 6.84 reduce to a third order total differential equation called the *Blasius equation*:

$$\frac{d^3 f}{d\eta^3} + f\left(\frac{d^2 f}{d\eta^2}\right) = 0 \tag{6.87}$$

where the boundary conditions on u and v become $f\{0\} = 0$, $(df/d\eta)_{\eta=0} = 0$, and $(df/d\eta)_{\eta=\infty} = 1$. The numerical solution to this equation for f and its derivatives may be substituted into equation 6.85 to find the velocity components u and v. These solutions are plotted in figure 6.18 as functions of the dimensionless distance η from the surface of the flat plate.[29]

Skin Friction and Drag

The u velocity profile in figure 6.18 has nearly the same shape as the velocity difference $U - u$ for the impulsively moved plane surface (see figure 6.14). From this velocity profile, we can calculate the value of the wall shear stress τ_w to be:

$$\tau_w = \mu\left(\frac{\partial u}{\partial y}\right)_{y=0} = \frac{1}{\sqrt{2}}\left(\frac{d^2 f}{d\eta^2}\right)_{\eta=0}\left(\frac{\rho\mu V_\infty^3}{x}\right)^{1/2} = 0.3321\left(\frac{\rho\mu V_\infty^3}{x}\right)^{1/2} \tag{6.88}$$

[29]Numerical values of f and its derivatives can be found in F. M. White, 1974, *Viscous Fluid Flow* (New York: McGraw-Hill Book Co.) and L. Rosenhead, ed., 1963, *Laminar Boundary Layers* (Oxford: Oxford University Press).

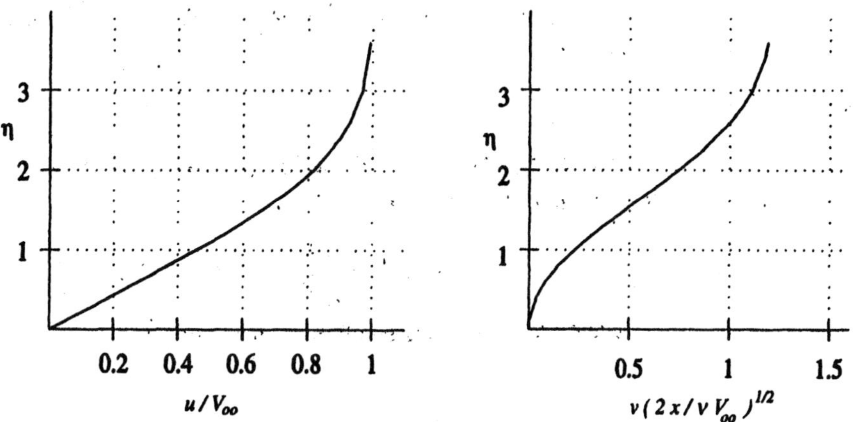

Figure 6.18 The velocity profiles (*left*) u/V_∞ and (*right*) $v\,(2x/\nu V_\infty)^{1/2}$ plotted as functions of $\eta \equiv y\,(V_\infty/2\nu x)^{1/2}$ for the boundary layer on a flat plate.

where the numerical coefficient is obtained from the numerical solution to the Blasius equation 6.87. It is customary to define a dimensionless wall shear stress, called the *skin friction coefficient* C_f, by dividing τ_w by the dynamic pressure $\rho V_\infty^2/2$ of the oncoming flow:

$$C_f \equiv \frac{2\tau_w}{\rho V_\infty^2} \tag{6.89}$$

The flat plate skin friction coefficient $(C_f)_{fp}$ can be found by substituting equation 6.88 into 6.89:

$$\left(C_f\right)_{fp} = 0.6641\left(\frac{\nu}{V_\infty x}\right)^{1/2} = \frac{0.6641}{\sqrt{Re_x}} \tag{6.90}$$

Engineers may be interested in the total drag force $\mathcal{D}$ on a flat plate. For a plate of length L in the streamwise direction and width W perpendicular to the flow, the drag $\mathcal{D}$ would be:

$$\mathcal{D} = 2\left(W\int_0^L \tau_w\,dx\right) = 0.6641(2WL)\left(\frac{\rho\mu V_\infty^3}{L}\right)^{1/2} \tag{6.91}$$

where the factor $2WL$ is the total plate area of the upper and lower surfaces exposed to the flow. The dimensionless form of the drag force, called the *drag coefficient* $C_{\mathcal{D}}$, is the ratio of the drag force $\mathcal{D}$ to the product of the exposed area A times the dynamic pressure $\rho V_\infty^2/2$:

$$C_\mathcal{D} \equiv \frac{2\mathcal{D}}{\rho V_\infty^2 A} \tag{6.92}$$

For the flat plate, the drag coefficient $(C_\mathcal{D})_{fp}$ is calculated by substituting equation 6.91 into 6.92 to obtain:

$$(C_\mathcal{D})_{fp} = \frac{2\mathcal{D}}{\rho V_\infty^2 (2WL)} = 1.3282 \left(\frac{\nu}{V_\infty L}\right)^{1/2} = \frac{1.3282}{\sqrt{Re_L}} \tag{6.93}$$

where Re_L is the plate length Reynolds number.

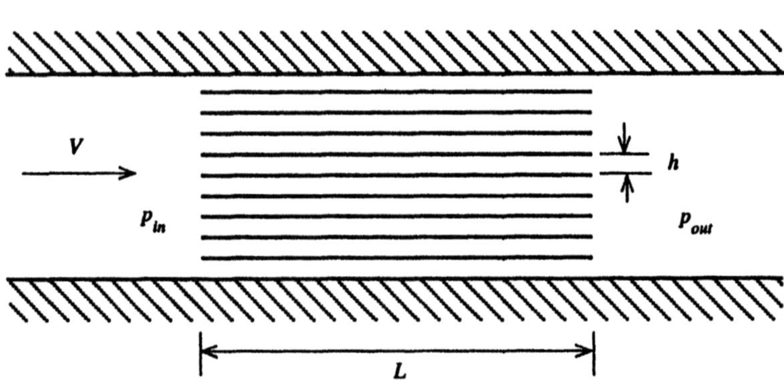

Figure E 6.14

Example 6.14

A flame arrestor in the intake duct of a gasoline engine, which prevents the propagation of a flame should there be fuel fumes in the intake air, consists of a series of thin parallel plates aligned with the intake flow, as depicted in figure E 6.14. The plate spacing is h, and the plate length is L.

Assuming the flow is incompressible, derive expressions for the pressure drop $p_{in} - p_{out}$ between the inflow and outflow streams for the limiting cases of (*a*) low velocity V, where the flow between each pair of plates is a plane Poiseuille flow, and (*b*) high velocity flow, where a boundary layer develops on each plate surface as though it were uninfluenced by the adjacent plates. (*c*) Calculate the Reynolds number Vh/ν at which the pressure drop in (*a*) and (*b*) are equal if $L = 10h$.

Solution

(a) For plane Poiseuille flow, the pressure drop may be found from equation 6.29:

$$V = \frac{h^2}{12\,\mu}\left(-\frac{\partial p^*}{\partial x}\right) = \frac{h^2}{12\,\mu}\frac{p_{in} - p_{out}}{L}$$

$$p_{in} - p_{out} = \frac{12\,\mu L V}{h^2}$$

(b) For boundary layer flow, the drag force $\mathcal{D}$ on a plate of width W is given by equation 6.91:

$$\mathcal{D} = 0.6641(2WL)\left(\frac{\rho\mu V^3}{L}\right)^{1/2}$$

If we now apply the momentum integral 5.11 to a control volume that encloses one channel, we find:

$$0 = (p_{in} - p_{out})(Wh) - \mathcal{D}$$

Combining these expressions and solving for the pressure drop:

$$p_{in} - p_{out} = 1.328\left(\frac{\rho\mu V^3 L}{h^2}\right)^{1/2}$$

(c) Setting the two expressions of (a) and (b) for the pressure drop equal to each other and solving for Vh/ν,

$$\frac{Vh}{\nu} = \left(\frac{L}{h}\right)\left(\frac{12}{1.328}\right)^2 = 10(8.165)^2 = 666.7$$

For flow velocities having $Vh/\nu \ll 666.7$, the pressure drop would be given by (a), while for $Vh/\nu \gg 666.7$ it would be given by (b).

Displacement Thickness

As can be seen on the right side of figure 6.18, the v component of velocity reaches a constant value $v\{x, \infty\}$ at the outside edge of the boundary layer. This value,

$$v\{x, \infty\} = 1.2168\left(\frac{\nu V_\infty}{2x}\right)^{1/2} \qquad (6.94)$$

is determined by the numerical solution of equation 6.87. This behavior means that the fluid at the outside edge of the boundary layer is moving outward at a small angle equal to $v\{x, \infty\}/V_\infty$ because the fluid within the boundary layer is moving

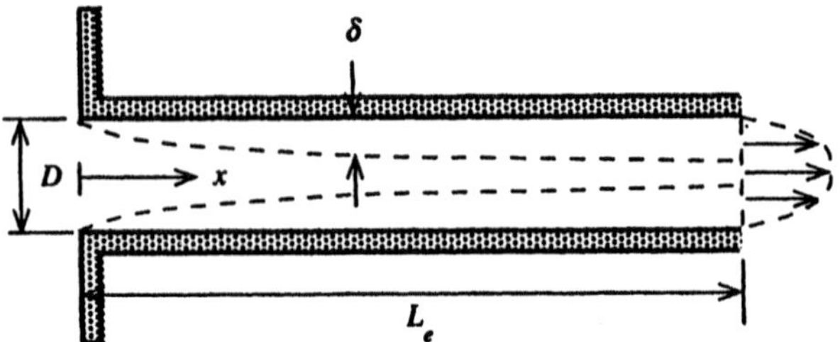

Figure 6.19 A viscous fluid flowing into the entrance of a long pipe creates boundary layers that grow toward the center, eventually leading to a Poiseuille flow at a distance L_e from the entrance.

more slowly than the free stream and cannot pass the same mass flow rate as the free stream. As a result, the external flow is displaced outward an amount δ^*, called the *displacement thickness*, by the slower moving fluid inside the boundary layer. The amount of this displacement is such that the volume flow rate between the wall and a distance y outside the boundary layer just equals the volume flow rate at the free stream velocity V_∞ between y and δ^*:

$$\int_0^y u\,dy \equiv V_\infty(y-\delta^*)$$

$$\delta^* \equiv y - \int_0^y \frac{u}{V_\infty}dy = \int_0^y \left(1-\frac{u}{V_\infty}\right)dy = \int_0^\infty \left(1-\frac{u}{V_\infty}\right)dy \quad (6.95)$$

For the flat plate boundary layer, the displacement thickness δ^*_{fp} can be evaluated by substitution of equation 6.85 into 6.95 and using equation 6.86:

$$\delta^*_{fp} = \left(\frac{2\nu x}{V_\infty}\right)^{1/2} \int_0^\infty \left(1-\frac{df}{d\eta}\right)d\eta = \left(\frac{2\nu x}{V_\infty}\right)^{1/2} |\eta - f|_{\eta=\infty}$$

$$\frac{\delta^*_{fp}}{x} = \frac{1.721}{\sqrt{Re_x}} \quad (6.96)$$

where the numerical value comes from the value of $\eta - f$ for large values of η in the numerical solution to the Blasius equation. Thus a thin flat plate appears to the surrounding inviscid flow as a plate of increasing thickness $2\delta^*_{fp}\{x\}$ along which the inviscid flow is slipping without friction. Of course, within the boundary layer, the fluid viscosity is causing a deceleration of more and more fluid as x increases.

Entrance Length in Pipe Flow

When fluid flows into the entrance of a long pipe or tube, it does not immediately achieve the parabolic velocity distribution of circular Poiseuille flow described by equation 6.40. Instead, a boundary layer begins to develop along the pipe walls, the slowed fluid growing in depth δ with increasing distance x along the pipe axis, as sketched in figure 6.19. When the boundary layer thickness grows to about half the pipe radius, the velocity profile in the pipe will have about reached its steady form for Poiseuille flow. We may estimate the distance L_e, called the *entrance length*, required for the velocity profile to reach this fully developed flow by setting the boundary layer thickness δ (equation 6.79) at $x = L_e$ equal to half the the pipe radius:

$$\delta \sim \left(\frac{\nu L_e}{V_\infty}\right)^{1/2} \sim \frac{D}{4}$$

$$\frac{L_e}{D} \sim \frac{V_\infty D}{16\nu} = \frac{1}{16} Re_D \tag{6.97}$$

In this entrance region, the growing boundary layer on the pipe wall displaces the fluid toward the pipe axis, narrowing the flow area near the entrance by an amount $\pi D \delta^*$ and causing a corresponding increase in the velocity u of the core flow, *i.e.*, the fluid near the axis but outside the boundary layer. The core flow is inviscid and experiences a pressure gradient $-dp/dx$ equal to $\rho u(du/dx)$ in the core flow. Thus there will be a pressure drop in the entrance region associated with the adjustment of the flow from a uniform velocity to that of circular Poiseuille flow which turns out to be greater than that of the Poiseuille flow along an equal length of pipe. This increment of pressure drop is called the *entrance loss* and will be discussed further in chapter 9.

6.6.3 Boundary Layer at a Stagnation Point

In what way would the boundary layer on a curved surface, where there would be a pressure gradient acting on the flow, differ from that on a flat plate where the pressure gradient is zero? When the pressure gradient is negative, *i.e.*, the pressure force is in the direction of motion (called a favorable pressure gradient), the boundary layer flow will tend to resemble a plane Poiseuille flow with a parabolic profile. However, when the pressure gradient is in the opposite direction, as we shall see below, the boundary layer cannot survive, and flow separation occurs. In this section, we consider an example of flow with a favorable pressure gradient: the flow at the stagnation point such as is found at the leading edge of an airfoil. The stagnation point marks the location where the oncoming flow divides to pass on either side of the airfoil.

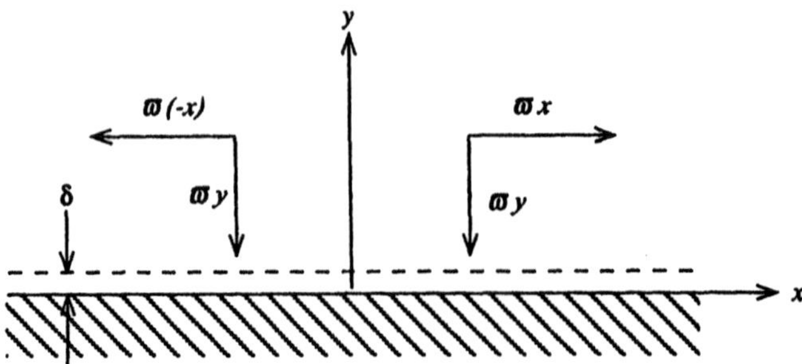

Figure 6.20 The boundary layer at a stagnation point on an airfoil has a constant thickness δ. The inviscid flow outside the boundary layer is symmetric about the plane $x = 0$.

In figure 6.20 we sketch the flow in the immediate vicinity of the stagnation point. The boundary layer thickness δ is so small as compared to the radius of curvature of the nose of the airfoil that the surface of the airfoil is locally plane. The axes x, y measure the distances along and normal to the airfoil surface, the plane $x = 0$ separating the fluid that passes to the right and to the left of the stagnation point (0, 0).[30] Outside of the boundary layer, the velocity field is inviscid and has the same form as that of example 4.2:

$$u = \varpi x; \qquad v = -\varpi y \tag{6.98}$$

where the parameter ϖ has the dimensions of s^{-1}.[31] As a fluid particle approaches the surface, it stretches sideways and slows down.

For the flow within the boundary layer, we choose the form:

$$u = \varpi x \left(\frac{df}{d\eta}\right); \qquad v = -\sqrt{\varpi \nu} f \tag{6.99}$$

where the function f depends only upon the dimensionless variable η, defined as:

$$\eta \equiv \left(\frac{\varpi}{\nu}\right)^{1/2} y \tag{6.100}$$

If we now substitute equations 6.99 and 6.100 into the boundary layer equations 6.82

---

[30]The stagnation point in figure 6.20 is a point along the *stagnation line* that is normal to the plane of the figure and runs spanwise along the leading edge of the wing.

[31]The term ϖ is approximately equal to the ratio of the flow velocity approaching the airfoil to the foil's radius of curvature.

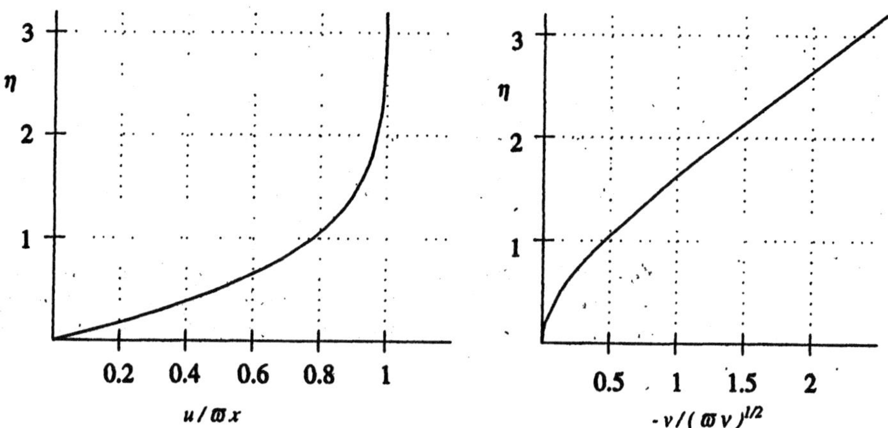

Figure 6.21 The velocity profiles (*left*) $u/\varpi x$ and (*right*) $-v/(\varpi\nu)^{1/2}$ plotted as functions of $\eta \equiv (\varpi/\nu)^{1/2}y$ for the stagnation point boundary layer.

and 6.83, we find that f is the solution to the total differential equation:[32]

$$\frac{d^3f}{d\eta^3} + f\left(\frac{df}{d\eta}\right) + 1 - \left(\frac{df}{d\eta}\right)^2 = 0 \tag{6.101}$$

with boundary conditions $f\{0\} = 0$, $(df/d\eta)_{\eta=0} = 0$ and $(df/d\eta)_{\eta=\infty} = 1$. The numerical solution to this equation[33] can be substituted into equation 6.99 to calculate the velocity components u and v. These are plotted in figure 6.21 as functions of the dimensionless distance from the airfoil surface, η.

Compared to the flat plate boundary layer, shown in figure 6.18, the stagnation point boundary layer shows a fuller, more curved profile for $u\{\eta\}$. The boundary layer displacement thickness δ^* is:

$$\delta^* = \int_0^\infty \left(1 - \frac{u}{\varpi x}\right) dy = \left(\frac{\nu}{\varpi}\right)^{1/2} \int_0^\infty \left(1 - \frac{df}{d\eta}\right) d\eta = \left(\frac{\nu}{\varpi}\right)^{1/2} (\eta - f)_{\eta=\infty}$$

$$= 0.6479\left(\frac{\nu}{\varpi}\right)^{1/2} \tag{6.102}$$

where the numerical coefficient comes from the numerical solution of equation 6.101.

---

[32]This solution is actually a solution to the Navier-Stokes equation because $\partial^2 u/\partial x^2 = \partial^2 v/\partial x^2 = 0$ identically in this case.

[33]See L. Rosenhead, ed., 1963, *Laminar Boundary Layers* (Oxford: Oxford University Press).

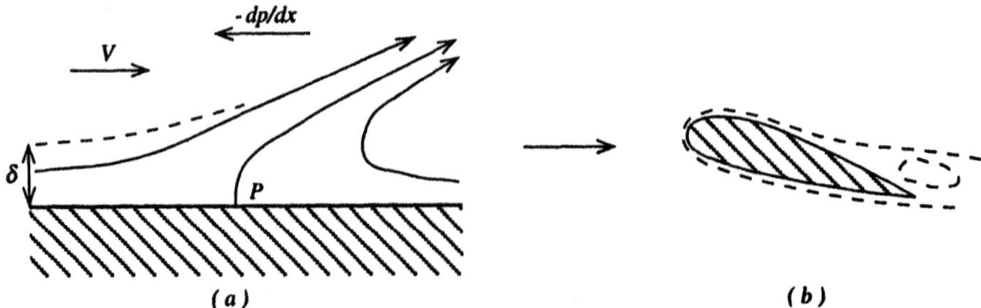

Figure 6.22 Above, (*a*) shows the flow near the separation point *P* of a boundary layer subject to an unfavorable pressure gradient, and in (*b*), boundary layer separation on the upper surface of a lifting airfoil produces a viscous wake region near the trailing edge that reduces lift and increases drag.

Far from the airfoil surface, outside the boundary layer, the velocity component v has the form:

$$v = -\sqrt{\varpi\nu}(\eta - 0.6479) = -\varpi(y - \delta^*) \tag{6.103}$$

To the external inviscid flow, the surface of the airfoil appears to be at $y = \delta^*$.

6.6.4 Boundary Layer Separation

In a flat plate boundary layer flow, the fluid within the boundary layer is decelerated by friction with the wall, *i.e.*, there is a balance between the inertia of the fluid, $\rho D\mathbf{V}/Dt$, and the viscous force, $\mu\partial^2 u/\partial y^2$. When a favorable (negative) pressure gradient is present, as in the case of the stagnation point boundary layer described above, the pressure force tends to accelerate the fluid within the boundary layer and must be balanced by an increase in the viscous force. However, if an unfavorable (positive) pressure gradient exists, the pressure force may be sufficient by itself to decelerate the fluid within the boundary layer, even causing the flow and the shear stress to reverse direction. If this happens, the boundary layer will separate from the solid wall because its momentum is too small to overcome the adverse pressure force acting upon it.

The flow in the neighborhood of a separation point is sketched in figure 6.22(*a*). Upstream of the separation point *P*, the boundary layer thickens rapidly as the separation point is approached, and the wall shear stress declines toward zero. At the separation point, the fluid closest to the wall moves away from it toward the free stream as it runs head on into upstream-moving fluid being pushed by the increasing pressure. The reverse flow downstream of the separation point forms a separated flow, *i.e.*, a region of slower fluid strongly influenced by pressure and viscous forces. The subsequent motion of the boundary layer, once it has left the solid surface, is determined by its interaction with the external flow, which itself is modified by the separated region. Needless to say, this complicated flow downstream of the separation point is very difficult to calculate.

Boundary layer separation may seriously degrade a flow in regions where the pressure rises too rapidly. For example, consider the flow over an airfoil placed at an angle to the oncoming flow so as to develop lift, as sketched in figure 6.22(*b*). Lift is created in part by a region of low pressure on the upper surface of the airfoil. The boundary layer on the upper surface first experiences a falling pressure (favorable pressure gradient) over the forward portion of the airfoil and then a rising pressure as it approaches the trailing edge. If this pressure rise is too abrupt, the boundary layer may separate from the upper surface as shown and create a large wake of recirculating flow behind the airfoil. The flow external to the boundary layer along the upper surface is substantially altered, as is the pressure distribution on the airfoil surface, reducing the lift on the airfoil and increasing its drag. For the efficient operation of aircraft wings, propellers and axial flow compressors, boundary layer separation must be avoided by careful design of the airfoil sections.

A boundary layer may avoid separation even in the presence of an adverse pressure gradient provided that the latter is small enough. This criterion may be quantified by requiring that the pressure force $-dp/dx$ (which equals the free stream inertia $\rho V dV/dx$) be much less than the viscous force, $\mu \partial^2 u/\partial y^2$:

$$-\frac{dp}{dx} = \rho V \frac{dV}{dx} \ll \mu \frac{\partial^2 u}{\partial y^2} \simeq \mu \frac{V}{\delta^2}$$

$$\frac{dV}{dx} \ll \frac{\nu}{\delta^2} \simeq \frac{V}{x} \tag{6.104}$$

where we have assumed that the thickness δ of the boundary layer at any point is related to the distance x from the leading edge by equation 6.79. Equation 6.104 may be interpreted to say that the length over which the pressure rises must be much larger than the length over which the external flow was accelerated or, alternatively, that the boundary layer growth during the pressure rise phase must be greater than that during the pressure fall phase. It is for this reason that streamline shapes, such as airfoils, have a rounded nose with a radius of curvature that is small compared to the length of the airfoil.

6.7 Problems

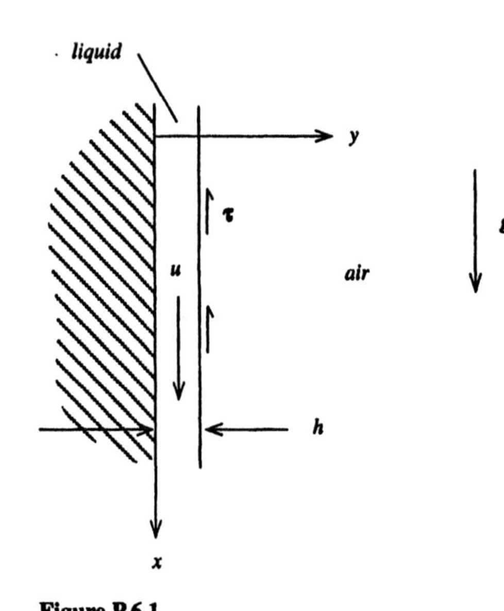

Figure P 6.1

Problem 6.1

Under the influence of gravity a viscous liquid (density ρ and viscosity μ) flows down a stationary vertical wall, forming a thin film of constant thickness h. An upflow of air next to the film exerts an upward constant shear stress τ on the surface of the liquid layer, as shown in figure P 6.1. The pressure in the flow is uniform.

Derive expressions for (a) the film velocity $u\{y\}$ as a function of y and the flow parameters ρ, μ, h and τ, and (b) the shear stress τ that would result in a zero net volume flow rate in the film.

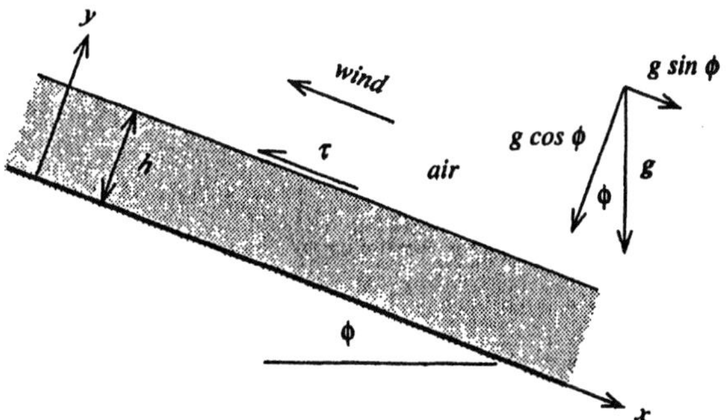

Figure P 6.2

Problem 6.2

When rain flows down a roof, it forms a layer whose velocity u and volume flow rate Q/W are given by equations 6.32 and 6.34, respectively. As illustrated in figure P 6.2, consider the case of an upward wind flow that exerts a shear stress τ on the upper surface ($y = h$) of the liquid layer. The wind flow tends to push the rain layer upward while gravity pulls the layer downward. Derive an expression for the thickness h of the liquid layer for which there is no net volume flow Q/W along the roof, expressing your answer in terms of the parameters τ, ρ, g and ϕ.

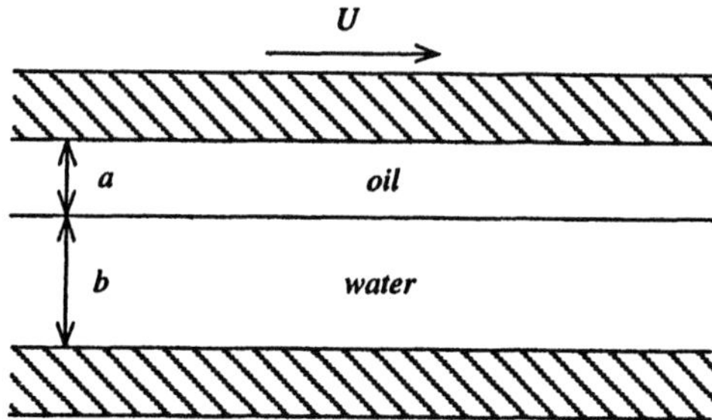

Figure P 6.3

Problem 6.3

A layer of oil of thickness a and viscosity μ_o floats on top of a layer of water of thickness b and viscosity μ_w. Both layers are contained between two large flat plates, the lower of which is stationary and the upper of which moves at a speed U in the x-direction (see figure P 6.3). Derive expressions for (a) the speed V_i of the water-oil interface and (b) the volumetric flow rates of oil and water, Q_0/W and Q_w/W, per unit distance normal to the direction of the flow. Express your answers in terms of the known parameters a, b, μ_w, μ_o and U.

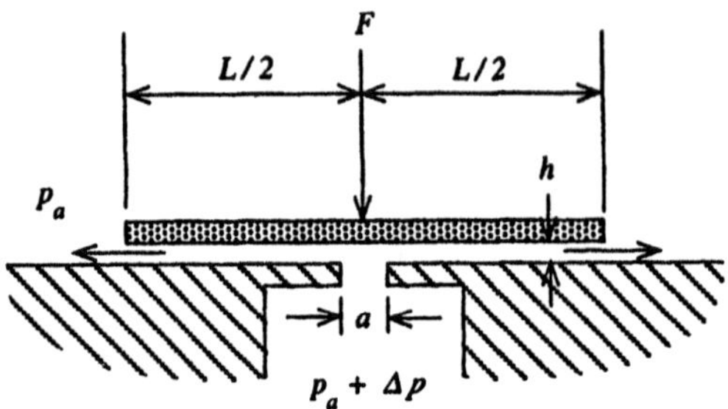

Figure P 6.4

Problem 6.4

In figure P6.4, a plate of width $L = 10\,cm$ and length $W = 1\,m$ supports a load $F = 1E(5)\,N$ by a flow of oil of viscosity $\mu = 1E(-2)\,Pa\,s$ between the plate and a support structure. The oil, supplied at a pressure $p_a + \Delta p$ and volume flow rate Q, is introduced at the center through a slot of width $a = 1\,cm$ and flows out horizontally to the atmosphere, whose pressure is p_a. Calculate the pressure Δp and flow rate Q required to maintain a clearance height $h = 0.1\,mm$ between the plate and the support structure.

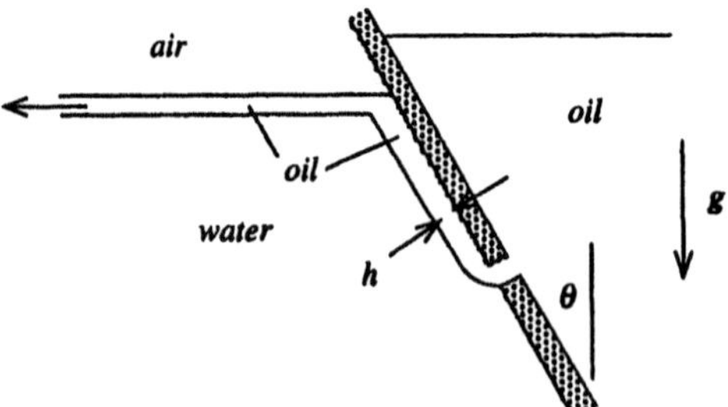

Figure P 6.5

Problem 6.5

An oil-filled barge has developed a narrow longitudinal crack in its side which extends a distance $W = 3\,m$ in a direction perpendicular to the sketch of figure P 6.5. Oil leaks out of the crack and, being less dense than water, runs up the side of the barge (inclined at an angle $\theta = 30°$ from the vertical) in a thin layer of constant thickness $h = 1\,cm$. Upon reaching the air-water interface, it flows laterally away from the barge. The oil density is $\rho_o = 0.9E(3)\,kg/m^3$, and the water density is $\rho_w = 1.03E(3) kg/m^3$. The oil viscosity, $\mu_o = 0.1\,Pa\,s$, is very much greater than that of the water. Calculate the value of the volumetric flow rate Q from the barge.

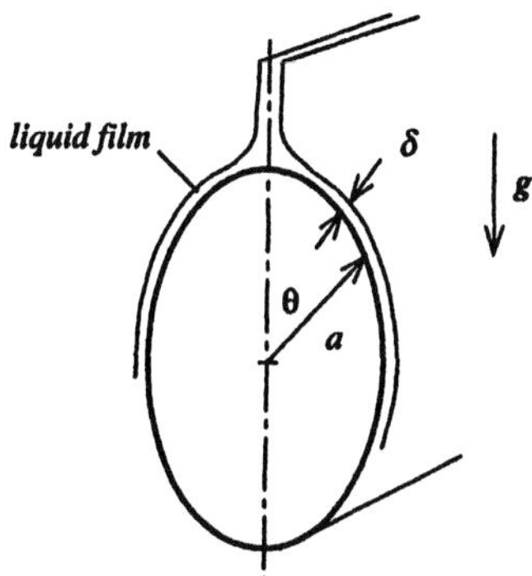

Figure P 6.6

Problem 6.6

A stream of viscous liquid, of density ρ and viscosity μ, is discharged from a slot onto the upper surface of a horizontal circular cylinder of radius a at a known volume flow rate Q/W per unit length of the cylinder. The fluid flows around the circumference of the cylinder under the action of gravity, forming a thin film as shown in figure P 6.6.

(a) Assuming that the inertia of the fluid can be neglected, derive an expression for the thickness δ of the layer as a function of the angle θ from the vertical and the parameters a, ρ, μ, g and Q/W. (b) Assuming the flow were inviscid, derive an expression for the thickness δ. (c) Derive an expression of inequality that ensures that the flow of (a) rather than (b) will be observed.

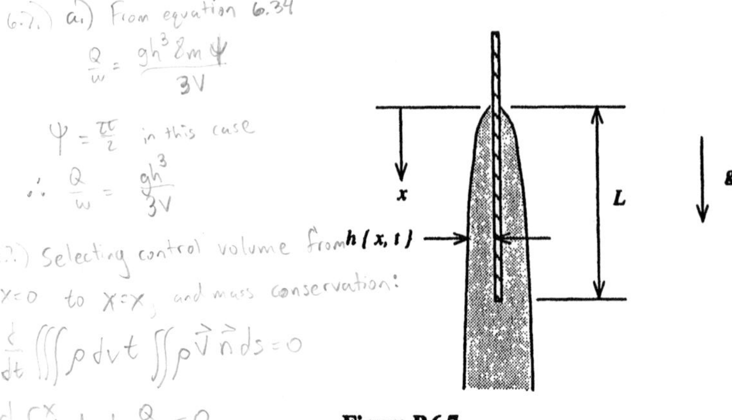

Figure P 6.7

• *Problem 6.7*

A thin flat plate of width W is dipped into a viscous fluid to a depth L and subsequently pulled out. It is then held stationary in a vertical position while the fluid drains from it, as shown in figure P 6.7. The thickness $h\{x, t\}$ of the fluid film increases from top ($x = 0$) to bottom ($x = L$) and decreases with time as the fluid drains away. Assuming that the fluid inertia is negligible and that at any point there is a balance of gravity and viscous forces in the fluid film on the plate, (a) derive an expression for the volume flow rate per unit width, Q/W, in the film on one side of the plate as a function of the film thickness $h\{x, t\}$ and the fluid properties. (b) Using the principle of mass conservation, show that there is a solution of the form:

$$h\{x, t\} = x^{1/2} f\{t\}$$

where $f\{t\}$ is a function of t only, and determine the value of the function $f\{t\}$ if $h_0 \equiv h\{L, 0\}$ is known.

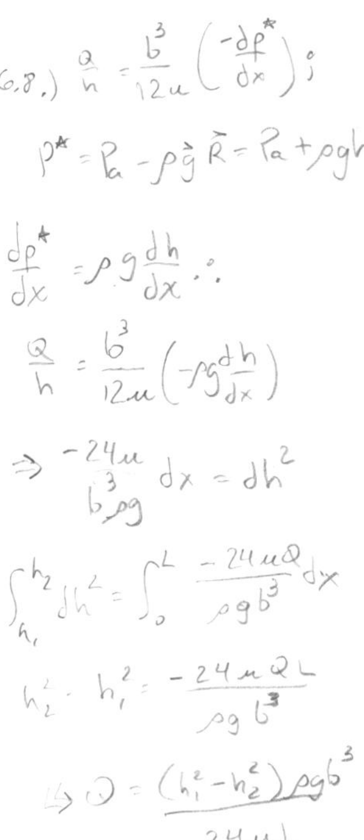

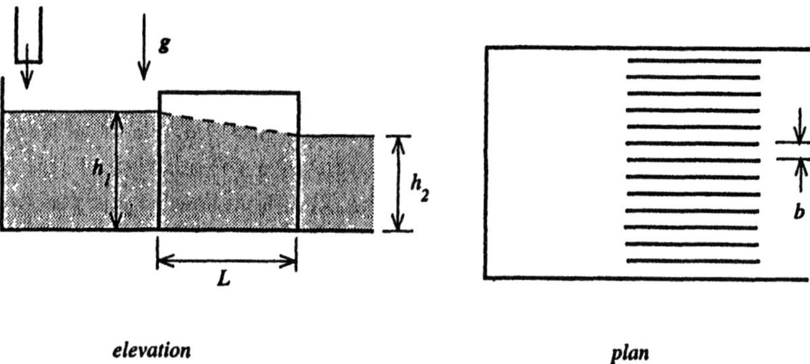

Figure P 6.8

Problem 6.8

Consider the problem of measuring the volume flow rate of a very viscous liquid flowing in an open channel. A novel method of calculating the flow rate involves measuring the height change in the fluid as it passes between a set of closely spaced vertical plates. In figure P 6.8, fluid enters the channel from above through a spigot at the left and flows steadily to the right with a uniform depth h_1. At some distance downstream, the fluid passes between a set of evenly spaced splitter plates oriented vertically in the channel. The spacing between the plates is b, and the length in the flow direction is L. When the flow emerges from between the plates, the fluid depth is h_2. The geometry is such that $b \ll L, h_1, h_2$, which allows one to neglect all end effects. Assuming the flow between the plates is locally a plane Poiseuille flow, derive an expression for the volume flow rate Q between two plates in terms of the fluid properties ρ and μ and the geometric parameters b, L, h_1 and h_2.

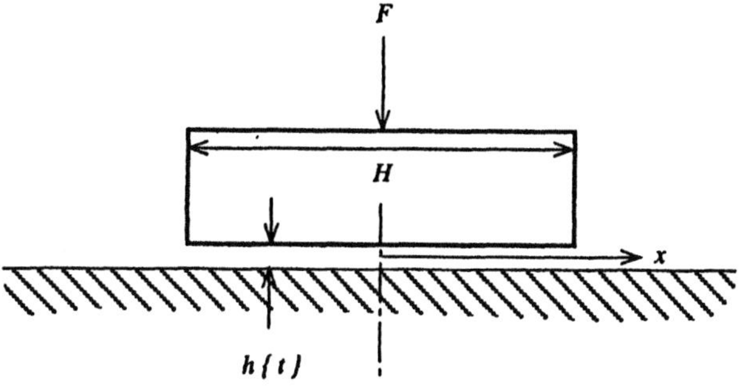

Figure P 6.9

Problem 6.9

Figure P 6.9 shows a rectangular bearing pad of width L and length $W \gg L$ which rests on a flat horizontal surface but which is separated from that surface by a thin film of oil of instantaneous thickness $h\{t\}$ that is much less than L. A constant force F causes the pad to sink downward at a variable speed $S\{t\}$. (a) Derive an expression for the volume flow rate $Q\{x, t\}$ of fluid flowing past a section x in terms of W, x and $h\{t\}$. Assuming that the inertia of the oil is negligible, derive expressions for (b) the pressure $p\{x, t\}$ at any position x and (c) the speed $S\{t\}$ as a function of F, L, W, h and the fluid viscosity μ. (d) Show how it will take an infinite time for the pad to come into contact with the underlying surface.

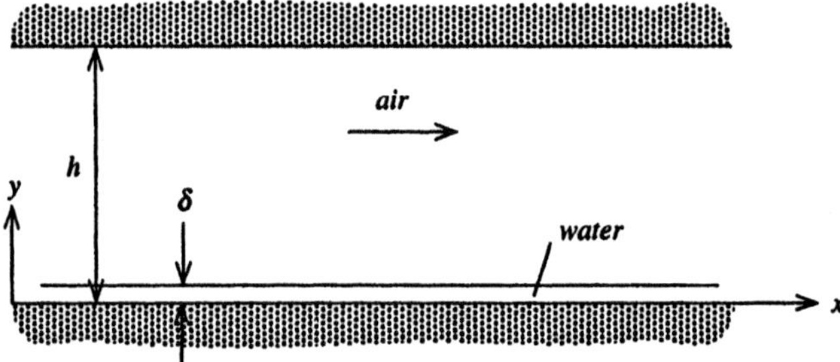

Figure P 6.10

Problem 6.10

A long horizontal two-dimensional channel of height h carries a steady fully-developed laminar flow of air in the x direction caused by a constant negative pressure gradient dp/dx. (There is no flow or change in the z direction.) A very thin layer of water of constant thickness δ flows along the bottom of the channel, as shown in figure P 6.10. The water layer thickness δ is much less than the channel height h, and the viscosity μ_w of water is much greater than that of air, μ_a. In terms of the flow parameters h, δ, μ_a, μ_w and dp/dx, derive expressions for (a) the shear stress on the air-water interface, assuming the water layer has no effect on the air flow, and (b) the velocity profile $u\{y\}$ in the water layer, assuming the shear stress on the water surface has the value of (a). (c) Show that the direct effect of the pressure gradient on $u\{y\}$ is small.

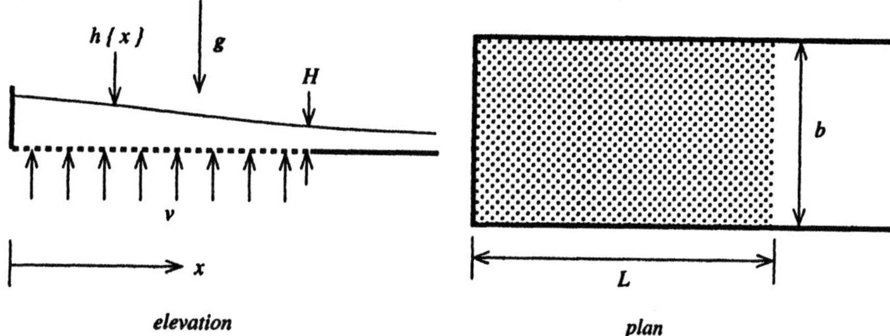

Figure P 6.11

Problem 6.11

A viscous liquid of density ρ and viscosity μ slowly filters up through the porous floor of a rectangular channel open to the atmosphere above, at a speed v that is uniform throughout the porous area of length L and width b, as shown in figure P 6.11. The channel is closed at the left end $x = 0$ and extends to the right beyond the end of the porous floor, $x = L$. The fluid thickness $h\{x\}$ decreases with increasing x, having the value H at $x = L$. In terms of the known parameters v, b, L, H, ρ and μ and the unknown layer thickness $h\{x\}$, derive expressions for (a) the volume flow rate $Q\{x\}$, (b) the pressure distribution $p\{x, y\}$ and (c) the horizontal velocity distribution $u\{x, y\}$, at any section a distance x from the left end. (d) From these expressions derive an expression for $h\{x\}$.

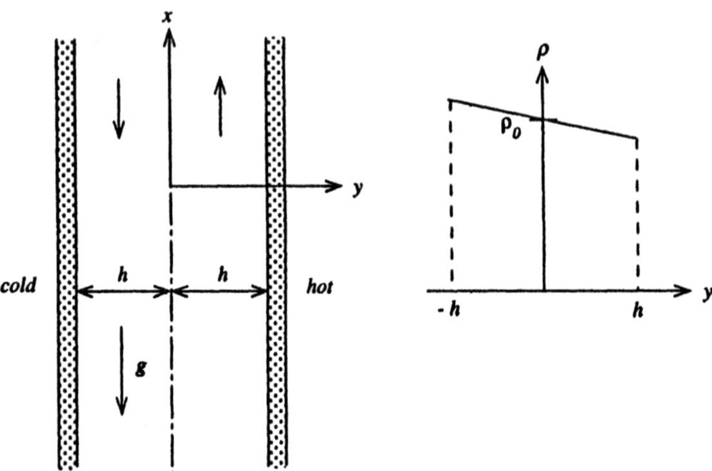

Figure P 6.12

Problem 6.12

A layer of air between two sheets of glass in a double pane insulating window has a density ρ varying linearly with distance y from the center of the air layer,

$$\rho = \rho_0(1 - \alpha y)$$

when heat is conducted from the hot inside to the cold outside, as shown in figure P 6.12. (However, the viscosity μ is uniform within the air layer.) As a consequence, air moves upward on the hot side ($0 \leq y \leq h$) and downward on the cold side ($-h \leq y \leq 0$), the streamlines being parallel to the glass surface. By symmetry, the upward volume flow on the hot side equals the downward volume flow on the cold side. (a) Sketch the vertical velocity profile $u\{y\}$. State the boundary conditons for u at $y = 0$ and $y = \pm h$. (b) Derive an expression for the pressure gradient $\partial p/\partial x$ along the center plane $y = 0$, noting that the pressure force and gravity force just balance there. (c) Derive an expression for the vertical velocity profile $u\{y\}$ in terms of the parameters ρ_0, α, h and μ.

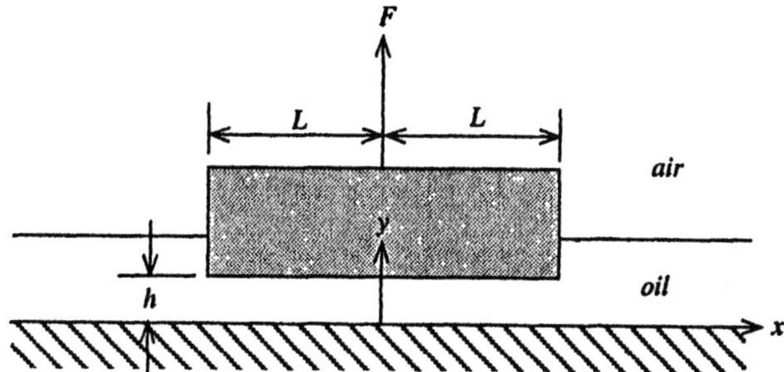

Figure P 6.13

Problem 6.13

A solid rectangular block of width $2L$ and large length $W \gg L$ is partially immersed in a layer of oil in a pan, separated from the bottom of the pan by a distance h that is much smaller than L (see figure P 6.13). It is pulled upward at a constant speed v by a force F. (a) Derive an expression for the volumetric flow rate $Q\{x\}$ at any section a distance x from the plane of symmetry as a function of x, v and W. (b) Neglecting gravity and assuming that the oil inertia can be disregarded, derive an expression for the force F as a function of the gap height h and the parameters μ, v and L.

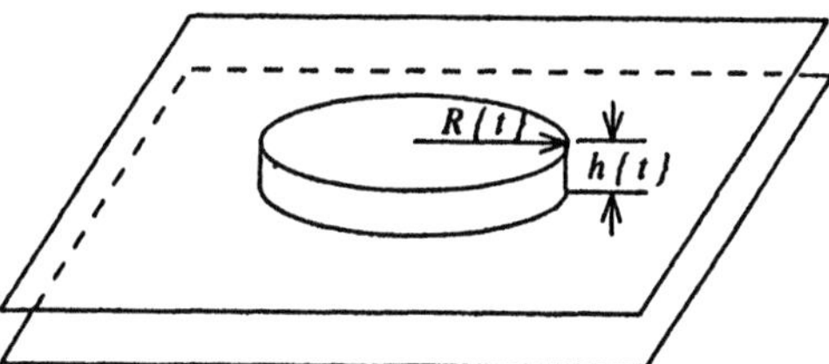

Figure P 6.14

Problem 6.14

A fixed volume $\mathcal{V}$ of viscous oil is trapped between two parallel plates. A constant force F is applied normal to the plates so that the oil forms a cylindrical volume of radius $R\{t\}$ and height $h\{t\}$ that vary with time as the oil is squeezed between the two plates (see figure P 6.14).

(a) By applying mass conservation to a cylindrical volume of oil of radius r, derive an expression for the average radial velocity $\bar{V}_r$ in terms of r, the plate separation h and its time derivative dh/dt. (b) Assuming that the radial flow of the oil between the plates is locally like a plane Poiseuille flow, derive an expression for the force F as a function of the volume $\mathcal{V}$, the fluid viscosity μ, h and dh/dt. (c) By integration of (b), derive an expression for $h\{t\}$.

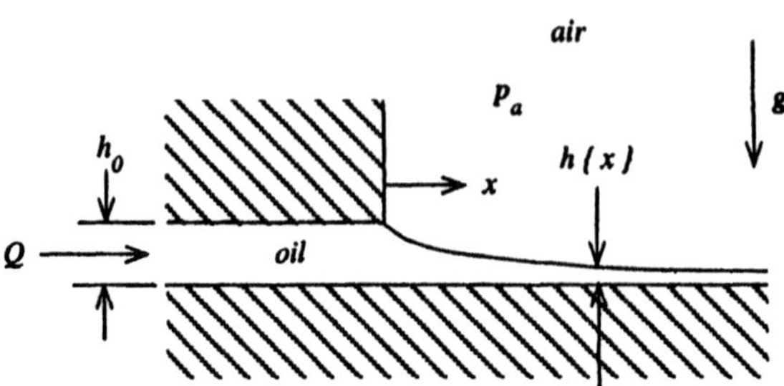

Figure P 6.15

Problem 6.15

In figure P6.15, a viscous oil leaks at a volumetric flow rate Q to the atmosphere through a crack of height h_0 onto a horizontal surface, where it continues to flow horizontally but with diminishing thickness $h\{x\}$. The crack width W in the direction normal to the plane of the flow is much greater than h_0. Assuming that the initial thickness of the layer at $x = 0$ is h_0, derive an expression for $h\{x\}$ as a function of the parameters h_0, Q, W and the oil properties ρ and μ.

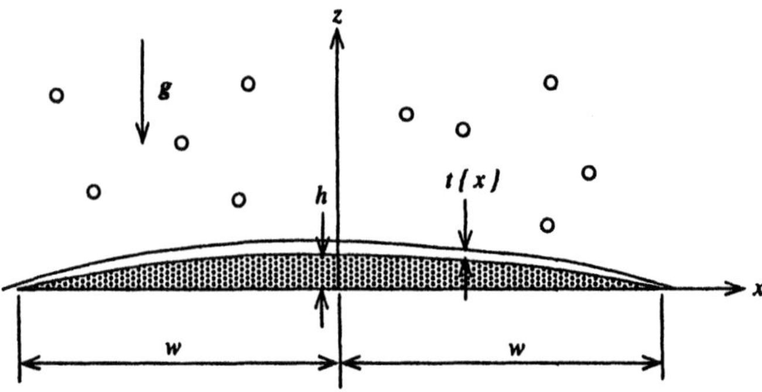

Figure P 6.16

Problem 6.16

A roadway of half-width w is crowned in the center so that rain will run off to either side. The surface elevation z (measured above the level of the roadside) is a parabola of height h ($\ll w$) at the road centerline:

$$z = h\left(1 - \frac{x^2}{w^2}\right)$$

as shown in figure P 6.16. During a rainstorm a water layer of thickness $t\{x\}$ forms on the roadway surface. The rain mass flow rate per unit of road area is measured to be m. In terms of the known parameters h, w, m and the rain density ρ and viscosity μ, derive expressions for (a) the volume flow rate $q\{x\}$ of water draining from the roadway per unit of roadway length normal to the plane of figure P 6.16, at any distance x from the centerline and (b) the thickness $t\{x\}$ of the layer of water that forms on the road surface as a function of the distance x from the roadway centerline. (c) Calculate t at $x = 0$ if $m = 1E(-2)\,kg/m^2s$, $h = 0.1\,m$, $w = 5\,m$, $\mu = 1.31E(-3)\,Pa\,s$ and $\rho = 1E(3)\,kg/m^3$.

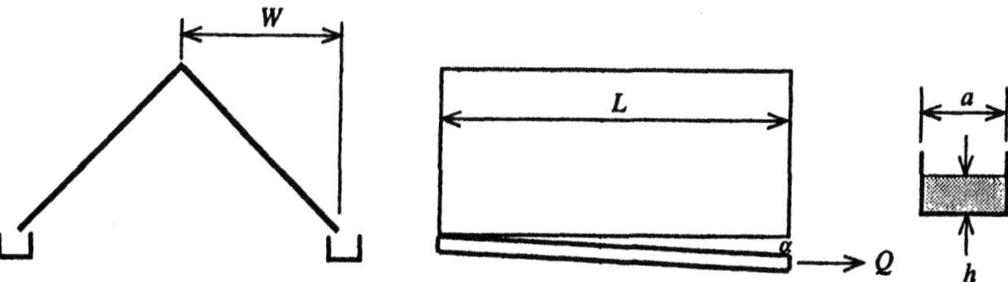

Figure P 6.17

Problem 6.17

An architect is designing rain gutters to carry away the rain that falls on a simple pitched roof, as shown in figure P 6.17. The roof has a length $L = 20\,m$ and a half width $W = 8\,m$. Each gutter will have a square cross-section of width $a = 0.1\,m$. They are to be installed at a slight angle $\alpha = 1E(-3)$ *radian* from the horizontal so that all the rain falling on the roof will drain from one end of each gutter. The architect wants to know whether the gutter will overflow under the highest expected rainfall rate of $p = 1$ *in/hr*. (This is the rate at which rainewater would accumulate on a flat, level, undrained surface.)

(a) Calculate the numerical value of the volumetric flow rate $Q\,[m^3/s]$ leaving the lower end of one gutter during the design rainstorm. (b) Assuming laminar viscous flow, derive an expression for the height h of the rainwater in the gutter at the lower end in terms of the volume flow rate Q, the width a, the water density ρ and the water viscosity μ. (c) Calculate the value of h using your answers to (a) and (b) with $\rho = 1E(3)\,kg/m^3$ and $\mu = 1E(-3)\,Pa\,s$. (d) What would be the height of the rainwater in the gutter at a point halfway along the length of the gutter?

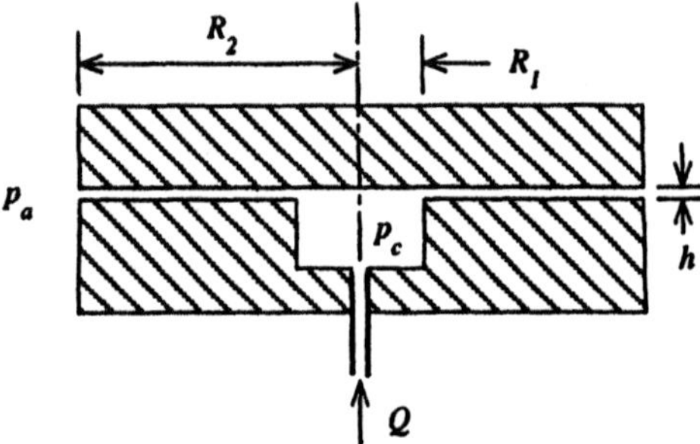

Figure P 6.18

Problem 6.18

A makeshift flowmeter consists of an upper circular plate of radius R_2 and a lower circular plate of the same outer radius but containing a central cavity of radius R_1, a cross-section of which is shown in figure P 6.18. The two plates are separated by a small gap of thickness h ($\ll R_1$). The cavity is supplied with a steady flow of gas at an unknown volumetric flow rate Q, and its pressure p_c is measured. The gas moves radially outward through the gap between the plates, exhausting to the atmosphere where the pressure is p_a. The gas flow may be regarded as incompressible.

(a) Assuming the flow were inviscid, derive an expression for the volumetric flow rate Q in terms of the known quantities p_c, p_a, ρ and the dimensions shown in figure P 6.18. (b) Assuming that the flow is very viscous and that at any radius r the velocity profile is that of a plane Poiseuille flow, derive an expression for Q in terms of p_c, p_a, ρ, μ and the dimensions shown in figure P 6.18. (c) Express the requirement that the flow would be governed by the conditions of (b) rather than (a) in terms of a dimensionless parameter.

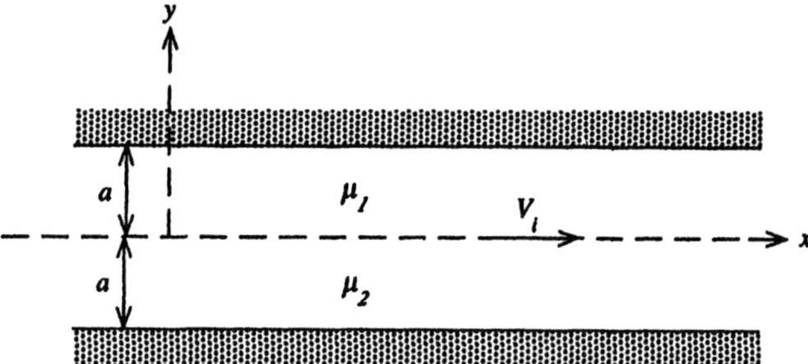

Figure P 6.19

Problem 6.19

Two immiscible liquids of viscosities μ_1 and μ_2 are flowing one above the other in a horizontal channel of fixed height $2a$ and of infinite extent in the flow direction under the influence of a constant negative pressure gradient dp^*/dx. (The channel dimension W in the direction normal to the flow plane is much larger than $2a$.) Each of the fluids occupies half of the channel, as shown in figure P 6.19. Considering the flow in both the upper and lower fluids to be a combined Poiseuille and Couette flow with the pressure gradient dp^*/dx and an interfacial velocity V_i, derive expressions for (a) V_i, (b) the volume flow rates Q_1/W and Q_2/W for the two fluids and (c) the velocity profile $u_1\{y\}$ in fluid 1 in terms of the parameters a, μ_1, μ_2 and dp^*/dx.

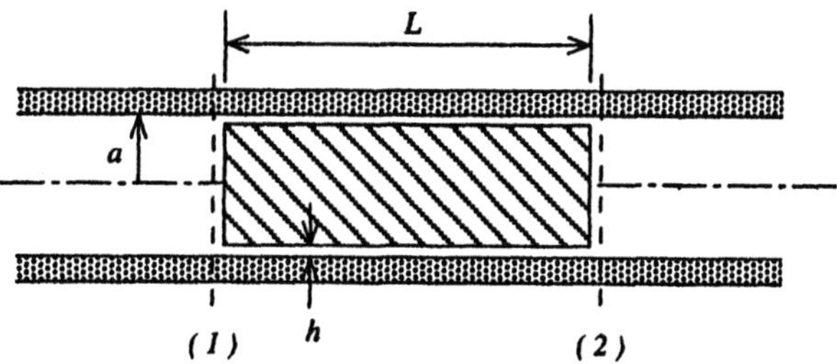

Figure P 6.20

Problem 6.20

A "pig" is a solid circular cylinder inserted in a pipe of slightly larger diameter to separate two different fluids being pumped in series through the pipe. Figure P 6.20 shows a pig of length L inside a pipe of radius a and having a radial clearance $h \ll a$ between its surface and the inner surface of the pipe. When the pressure p_1 at 1 exceeds the pressure p_2 at 2, the pig will move to the right at a constant velocity V.

Assuming that the flow between the pig and the pipe wall can be considered to be a steady plane Couette plus Poiseuille flow in a reference frame attached to the pig, (a) derive an expression for the pig velocity V in terms of the parameters p_1, p_2, L, h, a and the fluid viscosity μ. (b) If Q is the volume flow rate of fluid leaking through the clearance gap, relative to the pig, derive an expression for the ratio $Q/\pi a^2 V$, which is the ratio of leakage rate to the flow rate of fluid through the pipe.

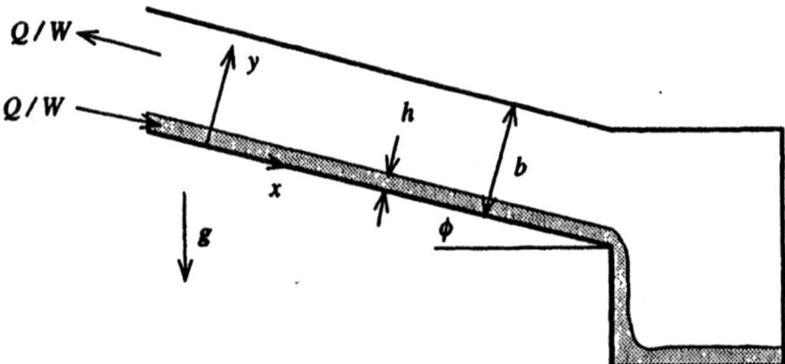

Figure P 6.21

Problem 6.21

As shown in figure P 6.21, a rectangular channel of height b and width W that is tilted above the horizontal by an angle ϕ, is supplied with a fluid of density ρ_1 at at volumetric flow rate Q. The channel is long and wide enough that a steady plane viscous flow ensues. The dense fluid forms a layer of constant thickness h as it flows down the channel and into a closed chamber at the lower end of the channel. In this chamber it displaces (without mixing) a lighter fluid, of density ρ_2, which then moves upward along the channel at the same volumetric flow rate Q. The lighter fluid has a viscosity μ_2 that is much less than that of the dense fluid, μ_1.

(a) Derive expressions for the thickness h of the layer of dense fluid and its velocity u_i at the interface between the two fluids, noting that it is much less viscous than the light fluid. (b) Derive an expression for the pressure gradient dp/dx along the channel.

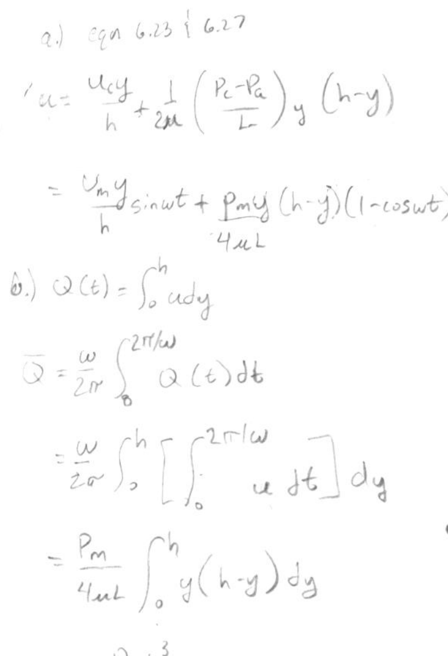

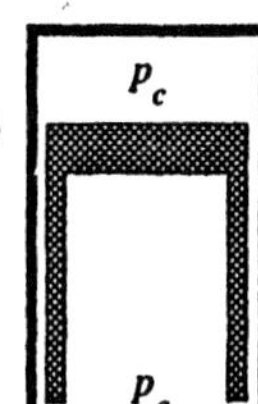

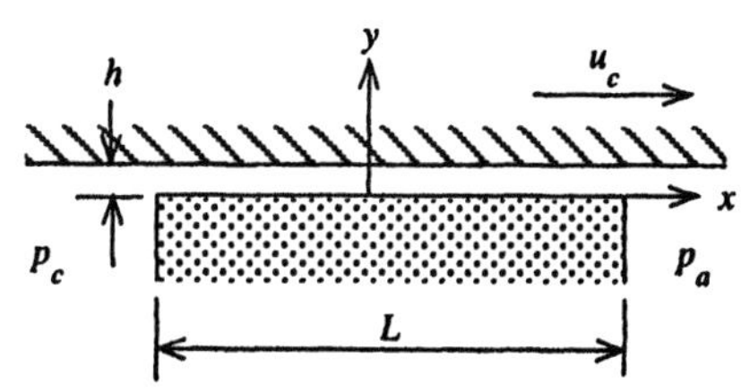

Figure P 6.22

Problem 6.22

In figure P 6.22, a piston in the cylinder of an internal combustion engine slides back and forth on a thin film of oil that fills the gap between the piston surface and the cylinder wall. The gap of thickness h is uniform along the length L and circumference of the piston and is much smaller than both of these dimensions.

We wish to determine the velocity of the oil in the gap. Assume that the inertia of the oil is negligible compared to the pressure and viscous forces, and choose a reference system fixed in the moving piston as shown in figure P 6.22. In this coordinate system, the cylinder wall moves with the speed u_c:

$$u_c = u_m \sin \omega t$$

where u_m is the maximum piston speed and where ω is the angular frequency of the crankshaft. The pressure below the piston is atmospheric pressure p_a while that above the piston in the combustion chamber, p_c, is:

$$p_c = p_a + \frac{p_m}{2}(1 - \cos \omega t)$$

where p_m is the maximum pressure increase in the cylinder, which occurs at $\omega t = \pi$.

Assuming that the oil flow in the gap is a combination of plane Couette and Poiseuille flows, (a) derive an expression for the velocity distribution $u\{y, t\}$ in terms of the parameters p_m, u_m, ω, h, L and the viscosity μ. (b) Derive an expression for the time-averaged volumetric flow rate $\overline{Q}$ of oil past the piston.

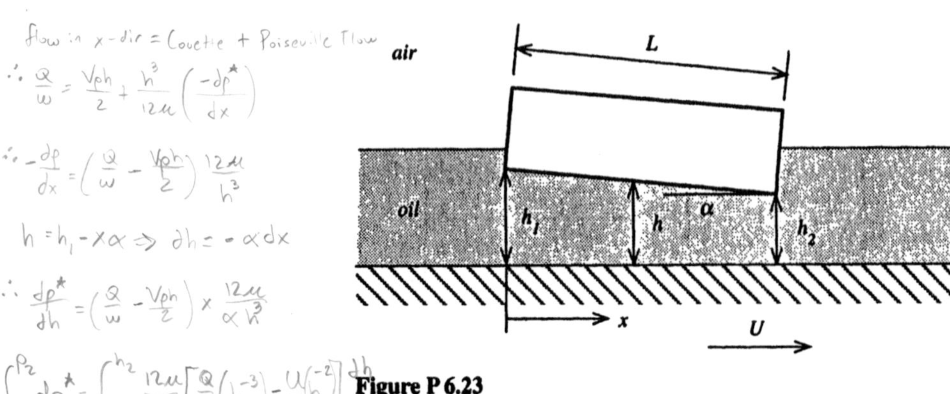

Figure P 6.23

Problem 6.23

It is everyday experience that, if you lubricate sliding solid surfaces with oil, the friction is greatly reduced. The situation is illustrated in figure P 6.23, in which a block of large length W (perpendicular to the plane of the figure) is sliding with relative velocity U on an oil film of thickness $h\{x\}$ that is much smaller than the block width $L\,(\ll W)$. Figure P 6.23 has been drawn in the reference frame of the block so that the bottom surface is moving with a velocity U while the block is stationary. The block is tilted with an inclination angle α to the horizontal that is small so that $\sin\alpha \simeq \alpha$ and $\cos\alpha \simeq 1$. Neglecting gravity so that $p\{0\} = p\{L\}$ and assuming the inertia of the fluid is negligible, derive an expression for the volumetric flow rate Q of oil through the gap between the block and the horizontal surface as measured in the reference frame of figure P 6.23. Express your answer as a function of U, W, h_1 and h_2.

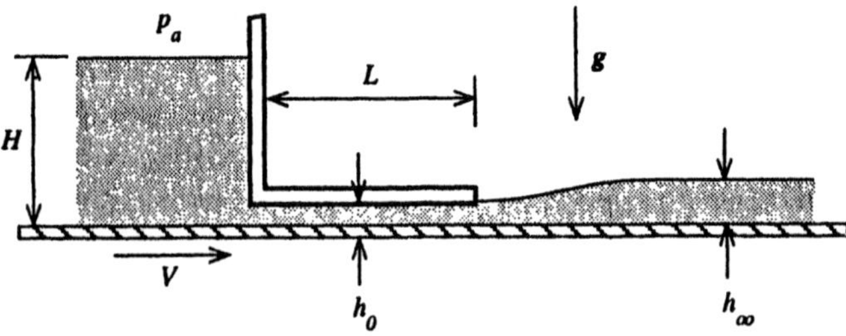

Figure P 6.24

Problem 6.24

In a manufacturing operation, a continuous polymer film is made by drawing a belt at fixed speed V under a reservoir which contains the liquid polymer (see figure P 6.24). The liquid is dragged onto the belt via a narrow slot of height h_0, and then, once the liquid has attained its final equilibrium thickness h_∞ on the belt, the polymer hardens and is eventually rolled up far to the right of figure P 6.24.

The polymer liquid has a viscosity μ and density ρ, and the slot height h_0 is much smaller than its length L. Depending upon the liquid elevation H in the reservoir, the final film thickness h_∞ can be either greater or less than the slot height h_0. We wish to determine the reservoir height H that will make $h_\infty = h_0$. Derive expressions for H that will ensure this equality when (a) the fluid is very viscous and (b) when the fluid may be considered inviscid. Express your answers in terms of the known parameters V, L, h_0, μ and ρ. (c) Derive a criteria that ensures that either (a) or (b) is valid.

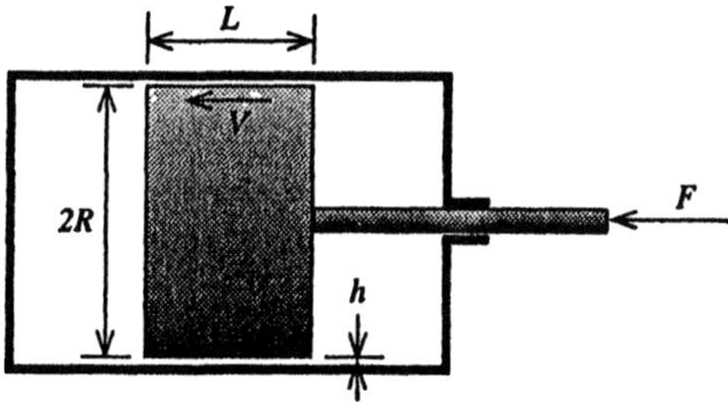

Figure P 6.25

Problem 6.25

A shock absorber is a device installed between the wheel axle and the chassis of a vehicle to dampen vibration. It consists of a piston of radius R inside a cylinder, with a clearance h between the piston and the cylinder wall that is much smaller than the length L of the piston, as shown in figure P 6.25. When the piston moves at a speed V, it displaces oil from one side to the other of the piston through the clearance passage.

Neglecting the inertia of the oil, derive expressions for (a) the volumetric flow rate Q past a section of the clearance space, (b) the pressure gradient dp/dz in that space, and (c) the force F applied to the piston in terms of the parameters V, h, L, R and the viscosity μ of the oil.

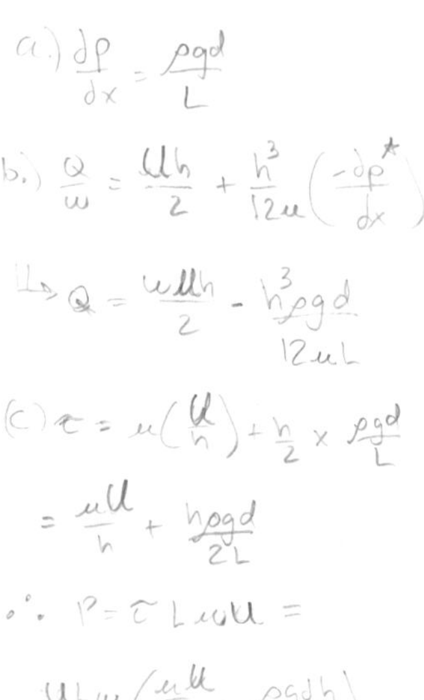

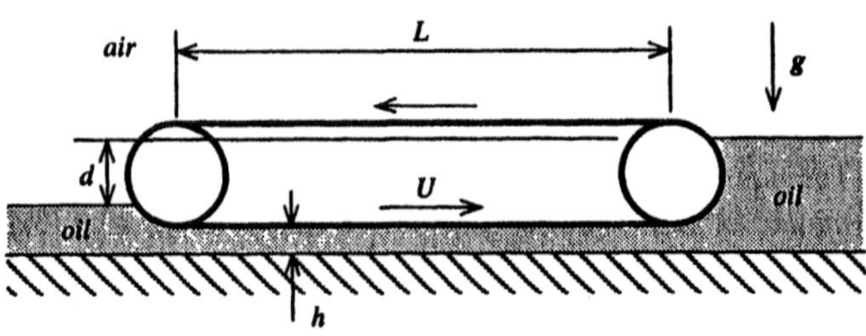

Figure P 6.26

Problem 6.26

Oil (viscosity μ and density ρ) is pumped from one reservoir to another, both of which are open to the atmosphere, by means of a moving belt pump, a cross-section of which is sketched in figure P 6.26. The fluid velocity components lie entirely in the plane of figure P 6.26 and do not depend upon the distance normal to this plane. The difference in the levels of the free surface in the two reservoirs is d. The belt moves at a velocity U toward the deeper reservoir, carrying along the oil in a thin layer of depth h. The length of this layer, L, is so great that a fully developed unidirectional laminar viscous flow exists within the layer, having a velocity profile independent of the horizontal distance along the length L. The width of the pump, normal to the plane of figure P 6.26, is W.

In terms of the known parameters D, L, h, W, U, μ and ρ, derive expressions for (a) the horizontal component of the pressure gradient $\partial p/\partial x$ in the layer, (b) the volume flow rate Q through the pump and (c) the power $\mathcal{P}$ required to run the pump.

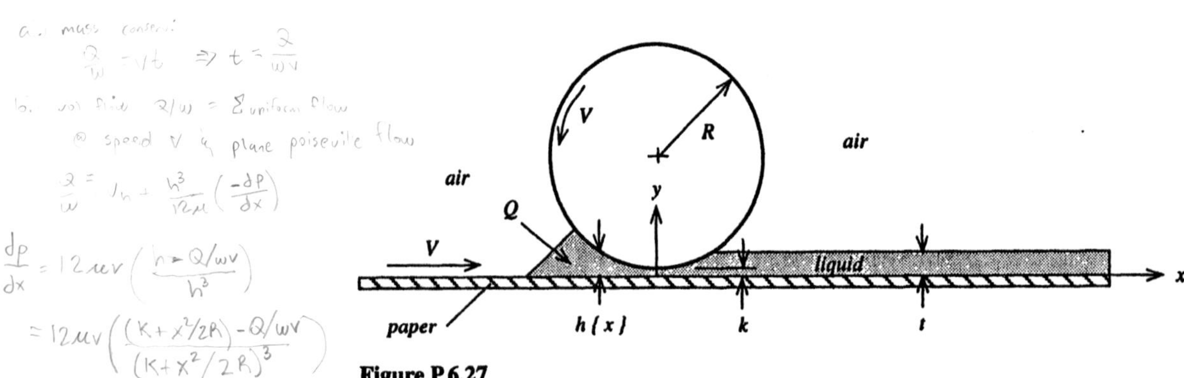

Figure P 6.27

Problem 6.27

A continuous sheet of paper of width W is coated with a layer of viscous liquid by the process illustrated in figure P 6.27. A stationary cylindrical roller of radius R rotates counterclockwise at a peripheral speed V that equals the speed V with which the paper moves past the roller from left to right. The very viscous fluid, supplied from the left of the roller at a volume flow rate Q, is squeezed into a thin film on the paper by the roller. Downstream of the roller, where the fluid thickness t is uniform, the fluid moves at the same speed V as does the paper. The thickness t is greater than the minimum clearance k between the roller and the paper. Both t and k are much less than R.

In the region of the flow where $x \ll R$, the thickness h of the viscous liquid film may be approximated as:

$$h\{x\} = k + \frac{x^2}{2R}$$

At any value of x in this region, you may assume that the fluid flow *relative to the paper sheet* is locally a plane Poiseuille flow.

In answering the following questions, express your answers only in terms of the known parameters V, Q, W, R, k, x and the fluid viscosity μ.

Utilizing the principle of mass conservation, derive expressions for (a) the fluid thickness t and (b) the pressure gradient dp/dx. Employing these results, derive expressions for (c) the value of x at which the pressure is a maximum, (d) the shear stress τ_w acting on the paper sheet and (e) the values of x at which τ_w is zero. (f) Show that the maximum (gage) pressure in the fluid is less than $24\mu V\sqrt{2R(t-k)}(t-k)/k^3$.

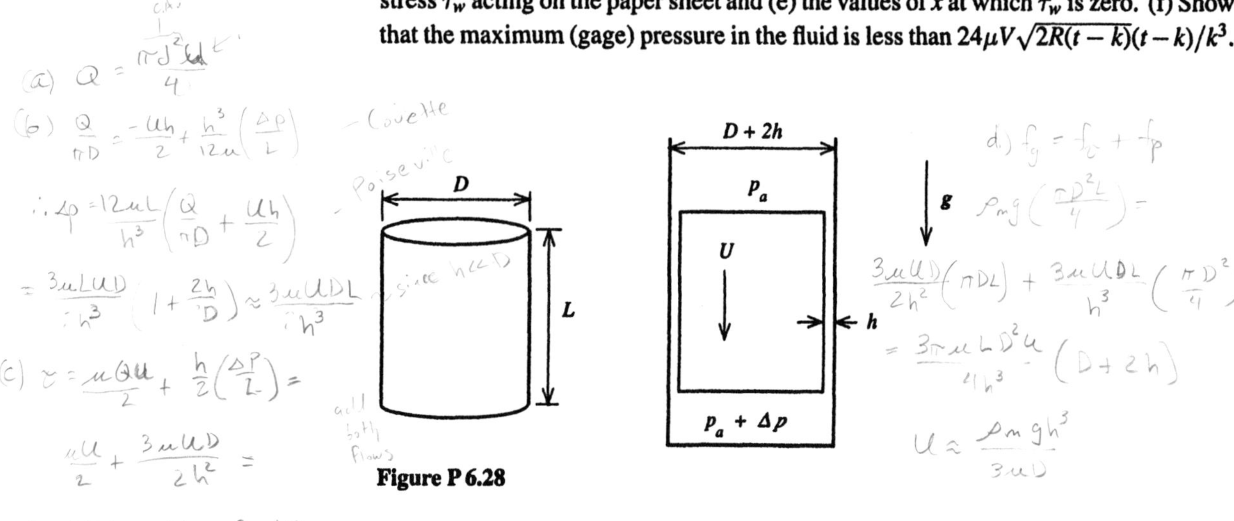

Figure P 6.28

Problem 6.28

A solid metal circular cylinder of diameter D, length L and density ρ_m is placed inside a hollow circular cylinder of slightly large inside diameter, $D+2h$. When released, the metal cylinder is observed to fall slowly at a constant speed U (see figure P 6.28). We want to calculate U, assuming that the air flow through the narrow gap of thickness h between the two cylinders is a steady laminar flow. Denoting the air density by ρ and viscosity by μ and considering the air to be incompressible, (a) derive an expression for the volumetric flow rate Q of air through the gap between the two cylinders. (b) If the pressure above the metal cylinder is the atmospheric pressure p_a and that below is $p_a + \Delta p$, derive an expression for Δp in terms of μ, L, D, U and h. (c) Derive an expression for the vertical shear stress τ on the outside lateral surface of the solid metal cylinder in terms of the quantities μ, L, D, U and h. (d) Using a force balance on the metal cylinder, derive an expression for the speed of fall U in terms of the quantities ρ_m, h, μ and D.

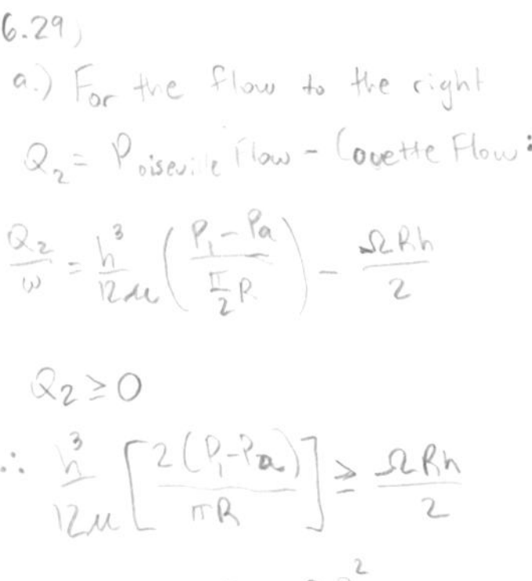

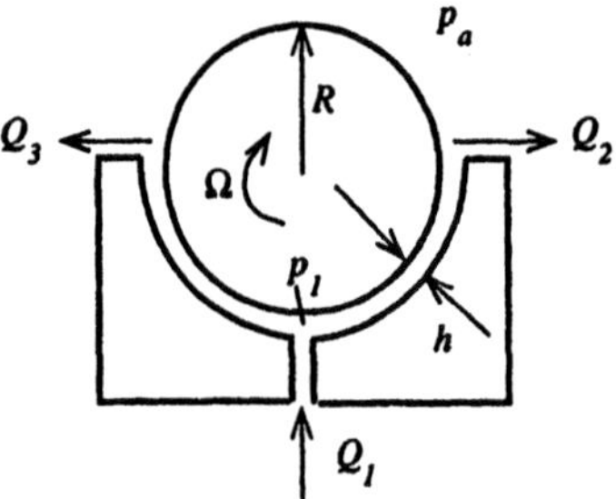

Figure P 6.29

Problem 6.29

A shaft of radius $R = 0.1\,m$, rotating in a clockwise direction at an angular speed $\Omega = 1.0E(2)\,rad/s$, is supported in a semicircular bearing having a constant clearance distance $h = 1E(-4)\,m$, as shown in figure P 6.29. To keep a film of oil in the bearing, a supply of oil of volume flow rate Q_1 is pumped into the bearing through a narrow slot at the bottom, part of this flow exiting the bearing at the right (Q_2) and the rest at the left (Q_3). The pressure of the oil at the inlet, p_1, is greater than the atmospheric pressure, p_a, at the outlets. The bearing length, $W = 0.5\,m$, is sufficiently long that the oil flow is two dimensional, with no flow velocity component normal to the plane of figure P 6.29. The oil viscosity is $\mu = 1.0E(-2)\,Pa\,s$.

The flow in the clearance space may be considered to be a linear combination of plane Poiseuille and plane Couette flow. (a) Calculate the minimum pressure difference, $p_1 - p_a$, that will just ensure a positive outflow of oil on the right ($Q_2 \geq 0$). For the condition of (a), calculate (b) the flow rate Q_1 and (c) the torque T required to rotate the shaft against friction in the bearing.

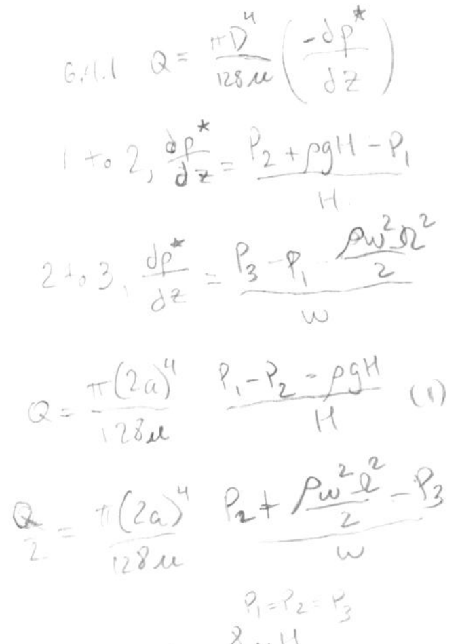

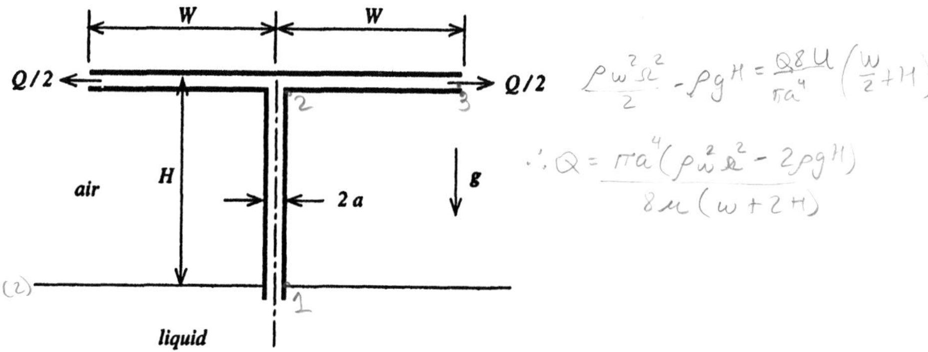

Figure P 6.30

Problem 6.30

In figure P 6.30, a T-shaped glass tube of inside radius a open at the ends is inserted just below the surface of a viscous liquid (density ρ and viscosity μ). The tube, initially filled with the same liquid, is spun about its vertical axis at a high angular speed Ω. The centrifugal force on the fluid in the tube arms pumps the viscous fluid slowly up the central column and out through both arms.

Selecting a steadily rotating reference frame fixed to the tube, using equation 4.27 and neglecting the Coriolis acceleration, the Navier-Stokes equation may be written as:

$$\frac{D\mathbf{V}}{Dt} = -\frac{p}{\rho} + \mathbf{g} + \Omega^2 r\mathbf{i}_r + \nabla^2\mathbf{V}$$

$$= -\frac{1}{\rho}\nabla\left(p + \rho g z - \frac{\rho\Omega^2 r^2}{2}\right) + \nabla^2\mathbf{V}$$

where z is the vertical axis and r is the radial distance from the axis in figure P 6.30. For Poiseuille flow in the tube sections, dp^*/dx may therefore be replaced by $d(p + \rho g z - \Omega^2 r^2/2)/ds$, where s is the distance along a streamline. Using this information, derive an expression for the volumetric flow rate Q through the tube in terms of the known parameters ρ, μ, Ω, H, W and a.

Problem 6.31

A very long porous pipe of diameter D is used to irrigate crops. It is supplied with water at one end at a gage pressure p_0. At each section along the pipe axis z, the water leaks to the atmosphere through the pipe wall with a radial flow speed $V_r = \alpha p\{z\}$, where $p\{z\}$ is the pipe internal gage pressure and α is a constant. Assuming circular Poiseuille flow at each section along the pipe axis, derive expressions for (a) the derivative dQ/dz

of the volumetric flow rate, (b) the pressure $p\{z\}$ and (c) the volumetric flow rate Q_0 at the pipe inlet. Express these answers in terms of the parameters D, α, μ, p_o and the axial distance z.

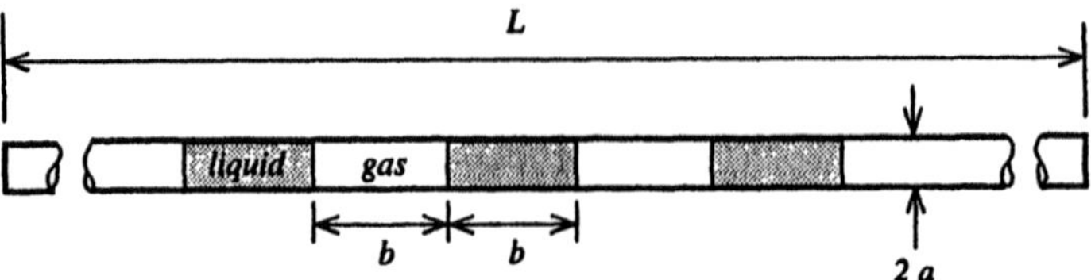

Figure P 6.32

Problem 6.32

A long horizontal circular tube of length L and internal radius a is subject to a pressure difference Δp between the entrance and exit. The fluid in the tube consists of alternate slugs of gas and liquid of length b that move at constant speed U along the tube, as sketched in figure P 6.32. The tube radius a is much smaller than the slug length b which in turn is much smaller than the tube length L (*i.e.*, $a \ll b \ll L$). The viscosity μ_l of the liquid is much less than that of the gas, μ_g.

Express your answers to the following questions in terms of the known quantities a, b, L, μ_l and μ_g. (a) Consider the flow in the liquid as observed by someone moving at the slug speed U. In such a moving reference frame r, z, where z is measured from the midlength of one of the liquid slugs, sketch the axial *relative* velocity profile $V_z\{0, r\}$ assuming that a steady laminar flow exists in the liquid slug. Indicate in your sketch the value of $V_z\{0, a\}$ at the tube wall. (b) Determine the value of the axial *relative* flow speed $V_z\{0, 0\}$ at the tube centerline, assuming a parabolic velocity distribution. (c) Derive an expression for the pressure difference, Δp.

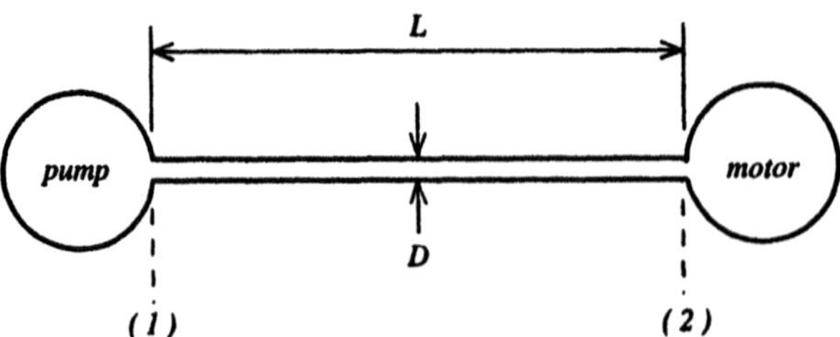

Figure P 6.33

Problem 6.33

An hydraulic system for transmitting power to a robot consists of a pump connected to an hydraulic motor by a very long tube of diameter D and length L, filled with an incompressible fluid of viscosity μ and density ρ. The volumetric flow rate $Q\{t\}$ of the pump is sinusoidal in time with an amplitude Q_0 and period T:

$$Q = Q_0 \sin\left(\frac{2\pi t}{T}\right)$$

(a) Assuming that the viscous force term in the equation of motion is negligible compared to the inertia term, derive an expression for the pressure difference $p_1\{t\} - p_2\{t\}$ between the pump and the motor, in terms of the parameters D, L, Q_0, T and ρ. (b) Now assume that the reverse is true, namely, that the viscous term greatly exceeds the inertia term so that the flow can be considered to be a circular Poiseuille flow at the instantaneous flow rate $Q\{t\}$. Derive an expression for $p_1\{t\} - p_2\{t\}$ in terms of the parameters D, L, Q_0, T and μ. (c) If the diameter D is small enough, the viscous term will be as important as the inertia term in determining the pressure difference. Derive an expression for the value of D for which these two terms are about equal. (d) The time-averaged power $\overline{\mathcal{P}}$ lost between the pump and the motor is:

$$\overline{\mathcal{P}} = \frac{1}{T}\int_0^T (p_1 - p_2)Q\,dt$$

For case (b) above, calculate the value of $\overline{\mathcal{P}}$ if $D = 2E(-2)\,m$, $L = 10\,m$, $Q_0 = 1E(-3)\,m^3/s$, $\mu = 1\,Pa\,s$ and $\rho = 1E(3)\,kg/m^3$.

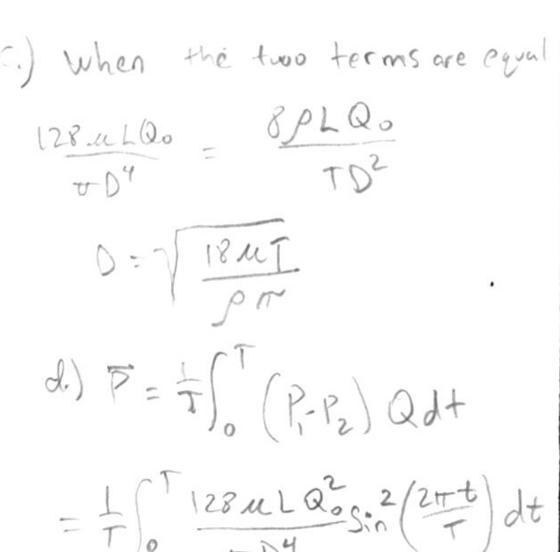

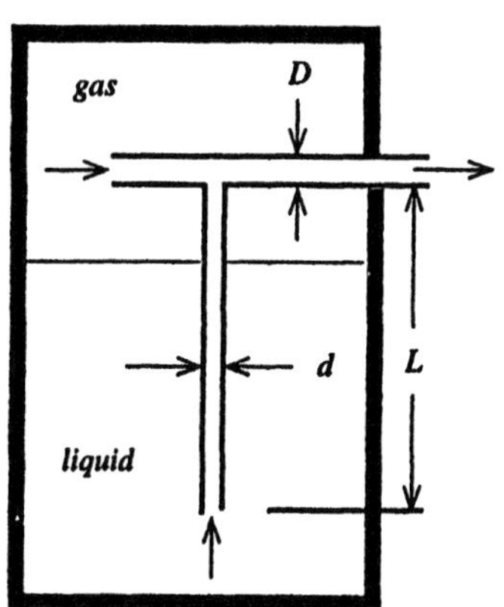

Figure P 6.34

Problem 6.34

An aerosol-generating device consists of a pressurized vessel containing a gas of density $\rho_g = 2\,kg/m^3$ and gage pressure $p_g = 5E(4) Pa$ and a liquid of density $\rho_l = 1.0E(3)\,kg/m^3$ and viscosity $\mu = 1.0E(-2)\,Pa\,s$. As shown in figure P 6.34, when the gas discharge tube of diameter $D = 1.0E(-3)\,m$ is opened to the atmosphere, the lower pressure in the tube induces both an outward flow of gas and an upward flow of liquid through the liquid tube of diameter $d = 1.0E(-4)\,m$ and length $L = 0.1\,m$.

In answering the following questions, neglect the effect of gravity on the flow. (a) Assuming that the gas flow is incompressible and inviscid and that the liquid flow is also inviscid, calculate the ratio Q_l/Q_g of the liquid volumetric flow rate Q_l to the gas volumetric flow rate Q_g. (b) Again assuming that the gas flow is incompressible and inviscid but that the liquid flow is a laminar viscous Poiseuille flow, calculate the ratio Q_l/Q_g. (c) Explain which alternative model of the flow, (a) or (b), is better and why it is so.

Problem 6.35

The designer of a machine has incorporated in his rough sketches a long, circular tube of diameter D_A and length L_A through which a viscous liquid will flow steadily at a volume flow rate Q_A, supplied by a positive displacement pump. The designer determines that the flow will be laminar and calculates the pressure drop is Δp_A, which is within acceptable limits. Subsequently, another designer finds a way to reduce the length of the tube by a factor of 2 so that $L_B = L_A/2$, but she must use a tube of rectangular cross-section of height b and width $10\,b$, yet having a flow area equal to that of the circular tube of diameter D_A. Because design B has the same flow area but only half the length, the second designer believes its pressure drop Δp_B will be less than Δp_A. Is the second designer correct?

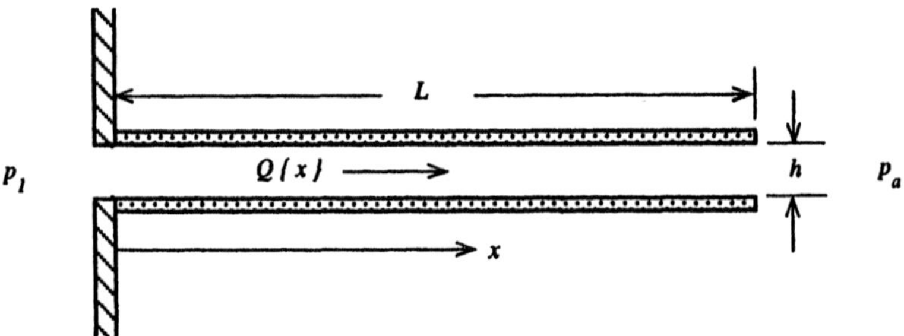

Figure P 6.36

Problem 6.36

An incompressible fluid of viscosity μ flows through a rectangular channel of length L, width W and height h, where $h \ll L \ll W$. The fluid enters the channel from a reservoir at pressure p_1 and emerges from the channel outlet to the atmosphere, whose pressure is p_a (see figure P 6.36). The upper and lower plates of the channel are slightly porous. The fluid leaks through the plates at a speed $w = (k/\mu)(p\{x\} - p_a)/t$ (Darcy's law, equation 6.65), where k is the permeability and t the thickness of the plates and where $p\{x\} - p_a$ is the pressure difference between the fluid in the channel and the atmosphere surrounding the plates.

(a) By using the principle of mass conservation, derive an expression for $dQ\{x\}/dx$, where $Q\{x\}$ is the volume flow rate at any station x along the channel, expressing your result in terms of the quantities k, μ, t, W and $p\{x\} - p_a$. (b) Assuming that the flow locally is a plane Poiseuille flow, derive a second order differential equation for $p\{x\} - p_a$ in terms of k, h and t. (c) Show that the solution to this equation is:

$$p - p_a = \frac{(p_1 - p_a)e^{\alpha x}}{e^{2\alpha L} - 1}\left(e^{2\alpha(L-x)} - 1\right)$$

where $\alpha \equiv 24k/h^3t$. (d) Using the solution of (c), derive an expression for the inlet volume flow rate $Q\{0\}$ in terms of W, k, h, μ, t, L and $p\{x\} - p_a$.

Problem 6.37

A square plate of side dimension $L = 1E(-2)\,m$, thickness $h = 1E(-3)\,m$ and specific gravity $SG = 3$ falls steadily through water of density $\rho_w = 1E(3)\,kg/m^3$ and viscosity $1E(-6)\,Pa\,s$, the plane of the plate being vertical. Assuming laminar flow in the plate boundary layer, calculate (a) the steady fall velocity V and (b) the plate Reynolds number VL/ν.

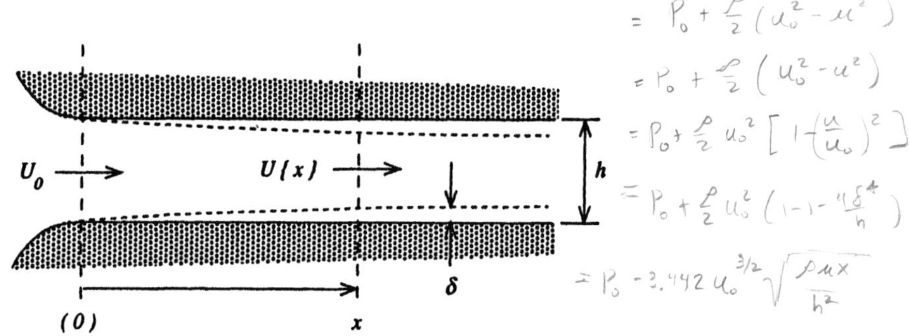

Figure P 6.38

Problem 6.38

Figure P 6.38 is a sketch of the test section of a wind tunnel having a rectangular cross-section of height h and width which is very much larger than h, so that the tunnel flow lies in the plane of figure P 6.38. The tunnel air speed U is low enough so that the air flow in the tunnel is incompressible. The flow into the tunnel entrance, section 0, is uniform in pressure (p_0) and velocity (U_0). A boundary layer develops on the tunnel wall, beginning at section (0), where $x = 0$. This growing boundary layer displaces the core flow, changing its velocity $U\{x\}$ and pressure $p\{x\}$ as the air moves down the tunnel. Assuming laminar flow in this boundary layer, derive expressions for $U\{x\}$ and $p\{x\}$ in terms of the parameters U_0, h and the air properties ρ and μ.

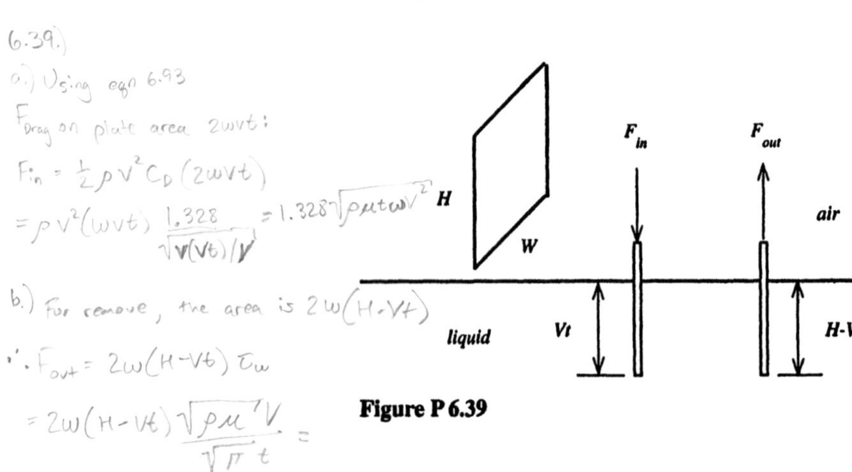

Figure P 6.39

Problem 6.39

In a manufacturing process, a thin rectangular plate of width W and vertical height H is to be dipped into a liquid of viscosity μ and density ρ at a constant speed V, as illustrated in figure P 6.39, until its upper edge is just submerged. Later on in the process, it is removed at the same constant speed. We need to estimate the viscous drag force F required to insert and remove the plate.

(a) In estimating the insertion force F_{in}, assume that F_{in} is equal to the drag force on a flat plate in steady flow moving at a speed V and having a length in the flow direction of Vt, where t is the time since the insertion began. Derive an expression for F_{in} as a function of time t and the parameters W, H, ρ, V and μ. (b) In estimating F_{out}, assume that the shear stress τ_w on the submerged portion of the plate is independent of position on the plate and has a value equal to that for an infinite plate suddenly set into motion at the speed V:

$$\tau_w = \mu \frac{V}{\sqrt{\pi \nu t}} = \frac{\sqrt{\rho\mu}\, V}{\sqrt{\pi t}}$$

where t is the time since the plate removal commenced. Derive an expression for F_{out} as a function of t, neglecting the effect of any liquid that might adhere to the plate above the liquid surface. (c) For the time $t = H/2V$, when the plate is half in the liquid, calculate the ratio F_{in}/F_{out}.

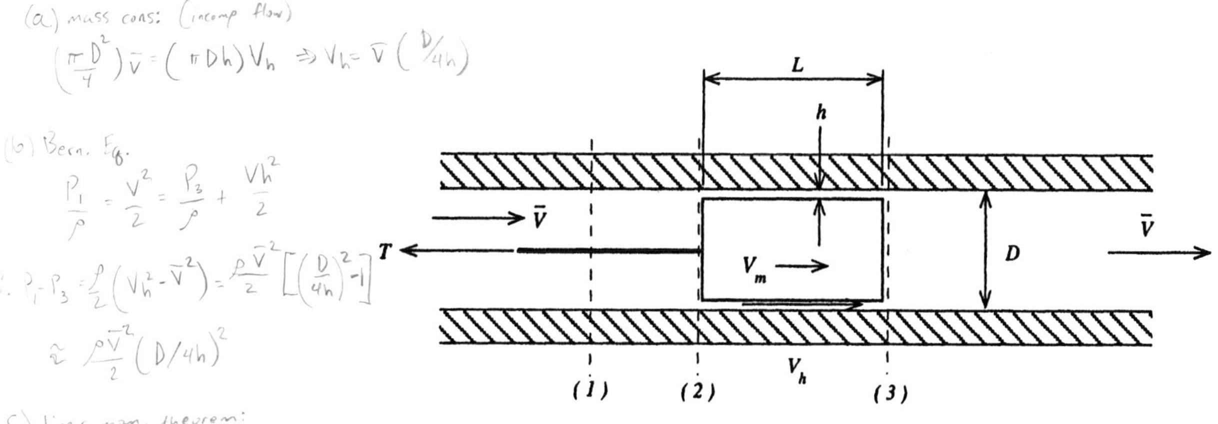

Figure P 6.40

Problem 6.40

A "mouse" is a device for pulling cable inside a pipe. As shown in the sketch, it consists of a solid cylinder of length L and having a radial clearance distance h between the cylinder and the inside of the pipe that is much smaller than the pipe's inside diameter D. When a fluid is pumped through the pipe at a mean speed $\bar{V}$, the mouse will move at a slower speed V_m while exerting a tension T on the cable.

Consider the case where the mouse is stationary ($V_m = 0$) and the flow may be considered incompressible and inviscid. Derive expressions for (a) the velocity V_h in the clearance annulus and (b) the pressure change $p_1 - p_3$ between the sections 1 and 3 shown in the sketch. (c) By utilizing the linear momentum theorem, derive an expression for the tension T in the cable. (These expressions should contain only the known parameters D, h, $\bar{V}$ and the fluid density ρ.) (d) Derive an expression for the tension T if the mouse moves at a steady speed V_m.

If the fluid is viscous, having a viscosity μ, there will be a viscous pressure drop from 2 to 3 in the clearance annulus. (e) Assuming that the flow in the annulus is a plane Poiseuille flow, derive an expression for the tension T, for the case of $V_m = 0$.

Bibliography

Abramowitz, Milton, and Irene A. Stegun. 1964. *Handbook of Mathematical Functions*. Washington, D.C: U.S. Department of Commerce.

Brodkey, Robert S. 1967. *The Phenomena of Fluid Motions*. Reading, Mass: Addison-Wesley Publishing Co.

Carslaw, H. S., and J.C. Jaeger. 1947. *Conduction of Heat in Solids*. Oxford: Oxford University Press.

Rosenhead, L., ed. 1963. *Laminar Boundary Layers*. Oxford: Oxford University Press.

Sherman, Frederick S. 1990. *Viscous Flow*. New York: McGraw-Hill Publishing Co.

White, Frank M. 1974. *Viscous Fluid Flow*. New York: McGraw-Hill Publishing Co.

7 Turbulent Viscous Flow

7.1 Introduction

In the previous chapter, we found that the effects of viscosity on a fluid flow can be taken into account by adding a viscous force term to the equation of motion, obtaining the Navier-Stokes equation. Several steady and unsteady solutions of this equation showed how the friction of the fluid against a solid surface slows the fluid down, preventing it from moving as fast as it otherwise would if the viscosity were zero. This type of flow was called laminar because the fluid particles moved along smooth paths in a predictable manner. However, there is another kind of viscous flow, obeying the same Navier-Stokes equation but having a distinctly different character—*turbulent flow*. In a viscous turbulent flow, the motion of an individual fluid particle is no longer predictable, except in a statistical sense, because the fluid motion is partially random. We must rely much more on experimental measurement than upon theory to predict the properties of turbulent flows.

This chapter describes the principal features of turbulent flows that determine how we utilize empirical information about them to solve engineering problems. Our interest here is more in the physics than the mathematics of turbulent flow because the former is a better guide to understanding how a turbulent flow is likely to behave in an engineering context. Even when armed with basic information about turbulent flows, we may still have to resort to experiment to find precise values for the engineering quantities of interest. In this sense, turbulent fluid flow is still principally an empirical science.

7.2 Characteristics of Turbulent Flow

Turbulent flow is one of the unsolved problems of classical physics. Despite more than a hundred years of research by the world's leading hydrodynamicists, we still lack a complete understanding of turbulent flow. Nevertheless, the principle physical features of turbulent flows, especially in their engineering applications, are by now well determined. In this section, we describe these aspects, providing a basis for a qualitative understanding of how turbulent flow behaves in common engineering circumstances.

7.2.1 The Onset of Turbulent Flow

When viscous fluids move slowly through small channels or around small bodies, the flow is laminar, even when unsteady, but when the flow speed of a fluid is increased enough, the flow will become randomly unsteady in part, usually where it comes into contact with solid surfaces. We can characterize any flow by its Reynolds number, $Re \equiv VL/\nu$, where V and L are a characteristic velocity and length scale of the flow field. What we observe empirically is that the flow becomes turbulent whenever the Reynolds number exceeds a certain value, $(Re)_{tr}$, called the *transition Reynolds number*. By speeding up the flow sufficiently, the flow Reynolds number can be increased to the point where it exceeds the value of the transition Reynolds number for that flow geometry, and the flow changes from laminar to turbulent.

A classical example of transition to turbulent flow occurs in a pipe flow when the volumetric flow rate is gradually increased from zero. For low values of the pipe Reynolds number $Re_D \equiv \bar{V}D/\nu$, the flow is steady and laminar, and the pressure drop is related to the volume flow rate by equation 6.41. However, as the flow approaches the transition Reynolds number, which has the value:

$$(Re_D)_{tr} \equiv \left(\frac{\bar{V}D}{\nu}\right)_{tr} = 2300 \qquad (7.1)$$

the flow begins to become unsteady, although it is still laminar. As the Reynolds number is further increased, bursts of turbulent flow are encountered until, at higher Reynolds numbers, the flow is entirely turbulent. Thus there is a range of Reynolds number, centered about the transition value, over which the flow changes from a completely steady laminar flow to a completely turbulent flow. Transition is not an instantaneous process.

A similar phenomena can be observed in the development of a boundary layer on a flat plate. Near the leading edge, where the Reynolds number $Re_x \equiv V_\infty x/\nu$ is small enough, a steady laminar boundary layer develops having a wall shear stress given by equation 6.88. Further downstream, where Re_x is greater, the laminar flow begins to show some unsteadiness and then becomes turbulent a little further along the plate. The value of the transition Reynolds number dividing these flow regimes is:

$$(Re_x)_{tr} \equiv \left(\frac{V_\infty x}{\nu}\right)_{tr} = 3E(5) \qquad (7.2)$$

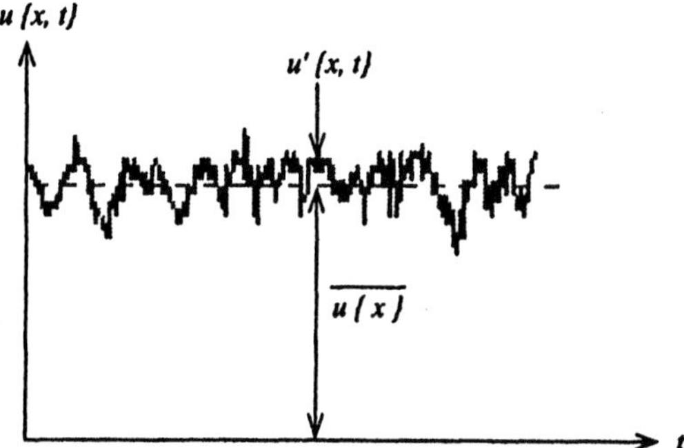

Figure 7.1 The turbulent velocity component $u\{x, t\}$ at position x as a function of time t, showing its time-averaged value $\overline{u\{x\}}$ and random increment $u'\{x, t\}$.

Again, the transition from steady laminar flow to a completely turbulent flow takes place over a finite length interval.

7.2.2 The Random Character of Turbulent Flow

Even though it may be generated by a globally steady process, such as a steady volume flow through a pipe, turbulent flow is never a locally steady flow. Nevertheless, the unsteadiness of the flow is not necessarily an overwhelming feature but more nearly a small disturbance of the average flow.

To illustrate this point, consider the time history of the velocity measured at a point in a turbulent pipe flow, as sketched in figure 7.1, where the x component of velocity $u\{x, t\}$ measured at a fixed location x is plotted as a function of time t. We can see that $u\{x, t\}$ can be considered to be the sum of a time-averaged value, $\overline{u\{x\}}$, and a time variable increment, $u'\{x, t\}$, that is usually quite a bit smaller than the time-averaged value:

$$u\{x, t\} \equiv \overline{u\{x\}} + u'\{x, t\} \tag{7.3}$$

Here the time-averaged value of $u\{x, t\}$, denoted by $\overline{u\{x\}}$, is defined as the integral on time over a time interval T, divided by the time interval:[1]

$$\overline{u\{x\}} \equiv \frac{1}{\mathrm{T}} \int_t^{t+\mathrm{T}} u\{x, t'\}\, dt' \tag{7.4}$$

[1]The time interval T need only be long enough to average out the fluctuations in $u\{x, t\}$.

Note that the time-averaged vaue of $u'\{x, t\}$ is automatically zero:

$$\overline{u'\{x, t\}} = \frac{1}{T}\int_t^{t+T} [u\{x, t'\} - \overline{u\{x\}}]\, dt' = \overline{u\{x\}} - \frac{\overline{u\{x\}}}{T}\int_t^{t+T} dt' = 0 \tag{7.5}$$

The random component of the velocity, $u'\{x, t\}$, has some of the characteristics of random noise signals, such as electrical noise in electronic circuits. If we decompose this velocity record for a pipe flow into its component frequencies, we find that the lowest frequency is approximately $\bar{V}/D$ while the highest is about $(Re_D)^{3/4}(\bar{V}/D)$, a frequency range that is usually more than a factor of a hundred. There is obviously quite a lot of small amplitude, high frequency, random motion involved in a turbulent flow, the details of which are very difficult to calculate or to predict.

The time average of the kinetic energy per unit mass of fluid at a point in the flow may be calculated in terms of the time-averaged velocity $\overline{\mathbf{V}}$ and the fluctuating velocity $\mathbf{v}'$ by expressing the velocity $\mathbf{V}$ as the sum $\overline{\mathbf{V}} + \mathbf{v}'$:

$$\mathbf{V} \cdot \mathbf{V} = (\overline{\mathbf{V}} + \mathbf{v}') \cdot (\overline{\mathbf{V}} + \mathbf{v}') = \overline{\mathbf{V}} \cdot \overline{\mathbf{V}} + 2\overline{\mathbf{V}} \cdot \mathbf{v}' + \mathbf{v}' \cdot \mathbf{v}'$$

and then performing a time averaging, noting that the time average of the term involving $\mathbf{v}'$ to the first power is zero:

$$\overline{\mathbf{V} \cdot \mathbf{V}} = \overline{\mathbf{V}} \cdot \overline{\mathbf{V}} + \overline{\mathbf{v}' \cdot \mathbf{v}'}$$

$$\frac{\overline{|\mathbf{V}|^2}}{2} = \frac{|\overline{\mathbf{V}}|^2}{2} + \frac{\overline{|\mathbf{v}'|^2}}{2} \tag{7.6}$$

The time-averaged kinetic energy is thus the sum of the kinetic energy of the average flow plus the kinetic energy associated with the turbulent fluctuations, called the *turbulent kinetic energy*. In cartesian coordinates, the turbulent kinetic energy can be expressed as:

$$\frac{\overline{|\mathbf{v}'|^2}}{2} = \frac{\overline{(u')^2} + \overline{(v')^2} + \overline{(w')^2}}{2} \tag{7.7}$$

The turbulent kinetic energy is a measure of how much kinetic energy has been invested in the random motion of the flow turbulence. In general it amounts to only a few percent of the kinetic energy of the time-averaged flow, $\overline{V}^2/2$.[2] Even so, the random velocity field produces shear stresses in the flow that are much larger than those that would exist if the flow were laminar.

[2] We denote the time average of the velocity by $\overline{V}$, whereas the spatial average of the velocity in a pipe flow is denoted by $\bar{V}$.

7.2.3 Eddy Description of Turbulent Flow

When a turbulent flow is visualized by injecting a stream of dye into the flow at a point, we can see that the fluid distorts in patterns of great complexity, containing both coarse and fine features. The flow is said to contain *eddies*, regions of swirling flow that, for a time, retain their identities as they drift with the flow but which ultimately break up into smaller eddies. The velocity field of a turbulent flow can be regarded as the superposition of a large number of eddies of various sizes, the largest being limited by the transverse dimension of the flow, such as the pipe diameter or boundary layer thickness, and the smallest being those that are rapidly damped out by viscous forces.

Where do the turbulent eddies come from? Mathematical analyses of steady laminar viscous flows show that infinitesimal disturbances to the flow can grow exponentially with time whenever the Reynolds number of the flow is sufficiently large. Such flows are unstable under these conditions and cannot remain steady under practical circumstances because there are always some disturbances present which may then grow spontaneously. The most rapidly growing disturbances are those whose size is comparable to the transverse dimension of the flow. These disturbances grow to form the largest eddies. Their velocity amplitude is generally about 10% of the average flow speed. These large eddies are themselves unstable, breaking down into smaller eddies and being replaced by new large eddies that are being generated continually.

The generation and breakup of eddies provides a mechanism for converting the energy of the mean flow into the random energy of molecules by viscous dissipation in the smallest eddies. Compared to a laminar flow of the same Reynolds number, a turbulent flow is like a short circuit in the flow field; it increases the rate at which energy is lost. As a consequence, a turbulent flow produces higher drag forces and pressure losses than would a laminar flow under the same flow conditions.

7.2.4 The Turbulent Energy Spectrum

The time-averaged turbulent energy per unit mass, $\overline{|\mathbf{v}'|^2}/2$, is the combination of the kinetic energies of many eddies of different sizes. The allocation of this kinetic energy to motions of different length scales is called the *turbulent energy spectrum.* A knowledge of this spectrum can tell us something about the details of the turbulent motion.

An energy spectrum with which we are undoubtedly familiar is that of sunlight. If sunlight is decomposed into its constituents of different wavelengths by reflection from a diffraction grating, the energy flux in each wavelength, or frequency band, can be measured. When this is done, it is found that the peak energy flux occurs at about

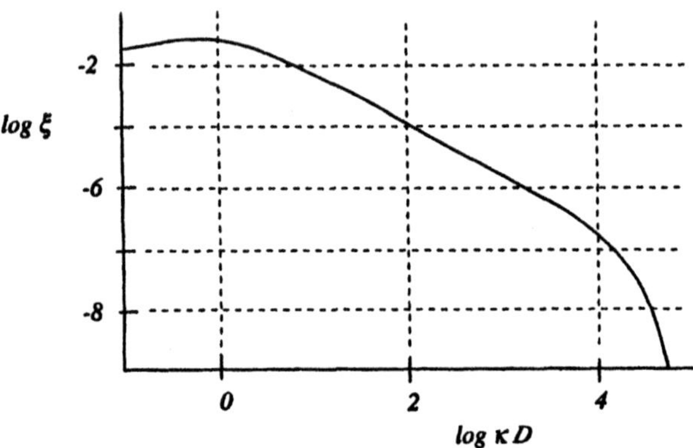

Figure 7.2 A log-log plot of the dimensionless turbulent energy spectrum $\xi \equiv E\{\kappa\}/\overline{|\mathbf{v}'|^2}D$ as a function of the dimensionless wave number κD for a pipe flow at a Reynolds number of $Re_D = 1E(6)$.

a wavelength of 0.5 μm, with much lower values encountered at longer (infrared) and shorter (ultraviolet) wavelengths. This distribution reflects the thermal equilibrium in the sun's photosphere, the origin of sunlight.

The distribution of turbulent kinetic energy among motions of different length scales is commonly given as a function of the wave number κ (which is inversely proportional to eddy size[3]) defined as:

$$\kappa \equiv \frac{\varpi}{\bar{V}} \tag{7.8}$$

where ϖ is the angular frequency of the components of the turbulent velocity history illustrated in figure 7.1. Denoting this turbulent energy per unit wave number by $E\{\kappa\}$, the total turbulent energy is:[4]

$$\overline{|\mathbf{v}'|^2} = \int_0^{\infty} E\{\kappa\}\, d\kappa \tag{7.9}$$

Figure 7.2 shows a sketch of the turbulent energy spectrum $E\{\kappa\}$ for a pipe flow at a Reynolds number of $Re_D = 1E(6)$. The peak value of $E\{\kappa\}$ occurs at $\kappa \sim 1/D$ at the low wave number end of the spectrum, where most of the turbulent energy is associated with the biggest eddies (lowest wave number). Over most of the spectral range $E\{\kappa\}$ varies as $\kappa^{-5/3}$, smaller eddies (greater wave number) contributing less to the total turbulent energy. At wave numbers greater than about $(Re_D)^{3/4}/D$, there is negligible

[3] For a sinusoidal disturbance, the wave number equals 2π divided by the wavelength.

[4] By tradition, $\overline{|\mathbf{v}'|^2}$ is called the turbulent energy although it is twice the turbulent kinetic energy.

energy because viscous dissipation causes the rapid decay of such small eddies.[5]

The amount of turbulent energy in a flow is limited by the loss of energy due to viscous dissipation. We can calculate the rate at which turbulent energy is lost from a knowledge of the velocity derivatives in the form of the dissipation function Φ_i for incompressible flow, equation 8.61. When averaged over time, the rate of loss of turbulent energy, denoted by ϵ, can be expressed in the form:

$$\epsilon = \frac{\overline{\Phi_i}}{\rho} = 2\nu \int_0^\infty \kappa^2 E\{\kappa\}\, d\kappa \tag{7.10}$$

Notice that the principal contribution to the integral on the right side of 7.10 comes at the high wave number end of the spectrum of figure 7.2 where $\kappa^2 E\{\kappa\}$ reaches its maximum. Thus the smallest eddies contribute the most to the energy dissipation while the largest eddies contribute most to the turbulent energy.

Kolmogoroff showed that the smallest eddy has a wave number κ_K that is related to ϵ and the flow Reynolds number by:[6]

$$\kappa_K = \left(\frac{\epsilon}{\nu^3}\right)^{1/4} \simeq \frac{Re_D^{3/4}}{D} \tag{7.11}$$

The smallest eddies are therefore several orders of magnitude smaller than the largest, energy containing eddies since the Reynolds number of a turbulent flow is necessarily large. This fundamental physical property of turbulent flow makes it extremely difficult to describe completely a turbulent flow, even using the largest and fastest computers available today. Instead, we must compromise by using much less information, such as $\overline{|\mathbf{v}'|^2}$ and ϵ, to characterize the effects of turbulence and be content to obtain approximate solutions to the mean flow field.

7.2.5 Turbulent Reynolds Stress

In a laminar pipe flow, the wall shear stress τ_w equals $8\mu\bar{V}/D$ (equation 6.43), but in a turbulent pipe flow, the wall shear stress is much greater than this value. Why is the shear stress much greater in a turbulent flow than in a laminar flow? The answer to this question, as we shall see below, is that the turbulent flow is unsteady and thereby capable of transferring momentum to the pipe wall much more rapidly than in a steady laminar flow.

---

[5]The $-5/3$ power law of $E\{\kappa\}$ and the scale length $D/Re_D^{3/4}$ were first derived by Kolmogoroff using dimensional arguments.

[6]The first of these relations implies that the Reynolds number of the Kolmogoroff eddy, $\sqrt{\kappa_K E\{\kappa_K\}}/\kappa_K \nu = 1$. If this is so, then $E\{\kappa_K\} = \kappa_K \nu^2$ and $\epsilon \simeq \nu \kappa_K^2 E\{\kappa_K\}\kappa_K = \nu^3 \kappa_K^4$.

In many engineering applications involving turbulent flows, we are satisfied if we can describe the time-averaged flow field, $\overline{\mathbf{V}}\{\mathbf{R}\}$, while ignoring the time-dependent component, $\mathbf{v}'\{\mathbf{R}, t\}$, whose time-average is zero. If we replace $\mathbf{V}\{\mathbf{R}, t\}$ in the incompressible Navier-Stokes equation 6.12 by the sum of its constituent components, $\overline{\mathbf{V}} + \mathbf{v}'$, we find the form:

$$\frac{\partial \mathbf{v}'}{\partial t} + [(\overline{\mathbf{V}} + \mathbf{v}') \cdot \nabla](\overline{\mathbf{V}} + \mathbf{v}') = -\frac{1}{\rho}\nabla(\overline{p^*} + p') + \nu\nabla^2(\overline{\mathbf{V}} + \mathbf{v}')$$

where we have replaced p^* by the sum of a time-averaged value, $\overline{p^*}$, and a fluctuating component, p'. If we now time average this equation, all the terms involving $\mathbf{v}'$ and p' to the first power average to zero, leaving:

$$(\overline{\mathbf{V}} \cdot \nabla)\overline{\mathbf{V}} = -\frac{1}{\rho}\nabla(\overline{p^*}) + \nu\nabla^2\overline{\mathbf{V}} - \overline{(\mathbf{v}' \cdot \nabla)\mathbf{v}'} \tag{7.12}$$

The time average of the incompressible mass conservation equation 3.17 gives:

$$\overline{\nabla \cdot (\overline{\mathbf{V}} + \mathbf{v}')} = 0$$

$$\nabla \cdot \overline{\mathbf{V}} = 0 \tag{7.13}$$

Next we multiply the incompressible mass conservation equation by $\mathbf{v}'$ and then time average, finding:

$$\overline{\mathbf{v}'(\nabla \cdot [\overline{\mathbf{V}} + \mathbf{v}'])} = \overline{\mathbf{v}'(\nabla \cdot \overline{\mathbf{V}})} + \overline{\mathbf{v}'(\nabla \cdot \mathbf{v}')} = \overline{\mathbf{v}'(\nabla \cdot \mathbf{v}')} = 0$$

Subtracting this from equation 7.12 and then multiplying by ρ, we have Reynolds' expression for the time-averaged equation of motion in a globally steady turbulent flow:

$$\rho(\overline{\mathbf{V}} \cdot \nabla)\overline{\mathbf{V}} = -\nabla(\overline{p^*}) + \mu\nabla^2\overline{\mathbf{V}} - \rho[\overline{(\mathbf{v}' \cdot \nabla)\mathbf{v}'} + \overline{\mathbf{v}'(\nabla \cdot \mathbf{v}')}] \tag{7.14}$$

The last term on the right side of equation 7.14 involves the derivatives of quantities that have the dimension of stress. For example, choosing the x-component of 7.14, we find, after rearranging terms:

$$\rho\left(\overline{u}\frac{\partial \overline{u}}{\partial x} + \overline{v}\frac{\partial \overline{u}}{\partial y} + \overline{w}\frac{\partial \overline{u}}{\partial z}\right) = -\frac{\partial \overline{p^*}}{\partial x} + \frac{\partial}{\partial x}\left(\mu\frac{\partial \overline{u}}{\partial x} - \rho\overline{u'u'}\right) + \frac{\partial}{\partial y}\left(\mu\frac{\partial \overline{u}}{\partial y} - \rho\overline{u'v'}\right)$$
$$+ \frac{\partial}{\partial z}\left(\mu\frac{\partial \overline{u}}{\partial z} - \rho\overline{u'w'}\right) \tag{7.15}$$

The terms in parentheses on the right side of equation 7.15 are the time averages of the shear stresses τ_{xx}, τ_{xy} and τ_{xz} acting at a point in the flow, as can be seen by comparing with equation 6.7. The first term in each parenthesis is the laminar shear stress due to the time-averaged flow $\overline{\mathbf{V}}$, while the second term, called the *Reynolds stress*, is the contribution from the unsteady fluctuating component $\mathbf{v}'$. In turbulent flows, the Reynolds stress greatly exceeds the laminar stress component, except very close to a solid surface. In a steady laminar flow, $\mathbf{v}' = 0$ and $\overline{\mathbf{V}} = \mathbf{V}$, so that equation 7.14 reduces to the familiar incompressible Navier-Stokes equation.

7.3 Turbulent Skin Friction and Drag

In a turbulent flow, the wall shear stress is much larger than it would be if the flow were laminar at the same Reynolds number. In addition, it is much less dependent upon the flow Reynolds number—in some circumstances even being independent of Reynolds number. For some simple flows, such as pipe flow and flat plate boundary layers, the relationship between the wall shear stress and the flow variables of a turbulent flow have been determined by extensive experiments whose results have been correlated by simple algebraic formulae. Unlike laminar flow, we have no analytic solutions to the flow field derivable from first principles that can universally and accurately match these measurements.

In this section, we consider two elementary turbulent flows that are applicable to many engineering systems and for which we have already determined the laminar flow solutions: flow in a circular tube and the boundary layer on a flat plate. While we are mostly interested in the accompanying pressure drop in pipe flow and the drag force on a flat plate, we will remark on the difference in the velocity profiles between turbulent and laminar flow for these two cases.

7.3.1 Turbulent Pipe Flow

The pressure drop in a turbulent pipe flow may be expressed in the dimensionless form of the Darcy friction factor f defined in equation 6.47. For turbulent flow in pipes with smooth walls, the measured friction factor is a function of the diameter Reynolds number Re_D that may be represented by the formula:[7]

$$\frac{1}{\sqrt{f}} = 2.0 \log \left(\frac{Re_D \sqrt{f}}{2.51} \right) \tag{7.16}$$

The friction factor f is only weakly dependent upon the Reynolds number, varying approximately as $[\log(Re_D)]^{-2}$. This dependence is illustrated by the lowest curve of

[7]This transcendental equation for $f\{Re_D\}$ may be solved readily by numerical iteration. Estimate a value for f, say f_0, insert in the right side of 7.16 and then solve 7.16 for the next approximation, f_1. Repeat the iterations, which converge rapidly to the correct value.

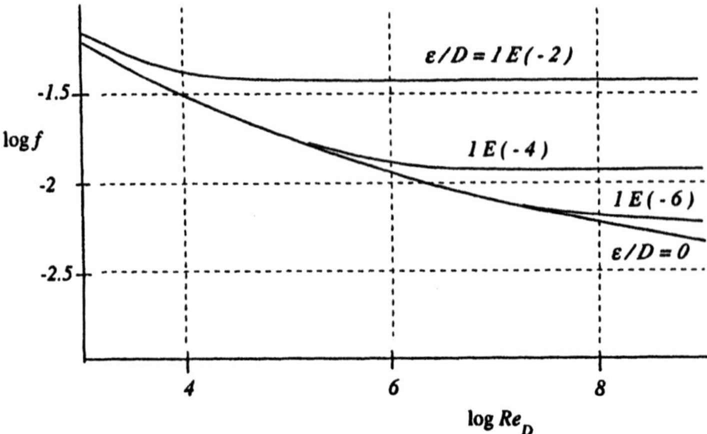

Figure 7.3 A log-log plot of the friction factor f for turbulent pipe flow versus the diameter Reynolds number Re_D. The lowest curve is for a smooth-wall pipe, while the progressively higher curves are for rough pipes of roughness ratios ε/D equal to $1E(-6)$, $1E(-4)$ and $1E(-2)$, respectively.

figure 7.3 for $Re_D > 2E(3)$, the minimum value for which the flow is turbulent. Note how the friction factor declines only by a factor of ten for a millionfold increase in the Reynolds number.

Because random motions close to the wall are strongly damped, turbulent pipe flow becomes laminar there, within what is called the *laminar sublayer*. This layer is responsible for the slight dependence of f on the fluid viscosity as embodied in the Reynolds number dependence of 7.16.

Few pipes have very smooth walls. Many have a surface roughness derived from their manufacture that distorts the laminar sublayer and generates additional turbulence in the flow near the wall that would not be present if the wall were completely smooth. Turbulent flow in a rough-wall pipe has a higher friction factor than flow in smooth-wall pipe at the same Reynolds number. In experiments on pipes lined with sandpaperlike roughness of physical height ε, the measured friction factor f was found by Colebrook to be related to the Reynolds number Re_D and the ratio ε/D by the formula:

$$\frac{1}{\sqrt{f}} = -2.0 \log\left(\frac{\varepsilon/D}{3.7} + \frac{2.51}{Re_D\sqrt{f}}\right) \tag{7.17}$$

Curves of f versus Re_D for several values of ε/D are shown in figure 7.3. Note that for a smooth wall $\varepsilon = 0$, and equation 7.17 reduces to 7.16.

For very rough pipes such that $\varepsilon/D \gg 1/(Re_D\sqrt{f})$, the friction factor depends only

Table 7.1 Wall roughness ε of pipe materials

Material	ε (m)
Concrete	$0.3E(-3)$–$3.0E(-3)$
Cast iron	$3E(-4)$
Galvanized iron	$1.5E(-4)$
Commercial steel	$5E(-5)$
Drawn tubing	$1.5E(-6)$
Glass, plastic	0

upon ε/D and not upon the Reynolds number. This occurs when the laminar sublayer is completely disrupted by the surface roughness, and the fluid viscosity plays no role in the pressure drop.

The equivalent wall roughness ε for commercial pipes has been determined from flow tests. Values are listed in table 7.1.

The velocity profile in a turbulent pipe flow is quite different from the parabolic profile of a laminar flow, equation 6.40. For laminar flow, the axial flow velocity $V_z\{r\}$ as compared to its mean value $\bar{V}$ may be found from equations 6.40 and 6.42 to be:

$$\frac{V_z}{\bar{V}} = 2\left[1 - \left(\frac{r}{a}\right)^2\right] \tag{7.18}$$

where a is the pipe radius. In contrast, the turbulent time-averaged velocity profile $\overline{V_z\{r\}}$ can be approximated by the expression:

$$\frac{\overline{V_z}}{\bar{V}} = \frac{60}{49}\left(1 - \frac{r}{a}\right)^{1/7} \tag{7.19}$$

As illustrated in figure 7.4, turbulent pipe flow has a much more uniform velocity profile than does the laminar flow, undergoing rapid decline only near the wall. This difference reflects the much higher shear stress in the turbulent flow, compared with laminar flow at the same flow conditions.

Example 7.1

A steel pipe of diameter $D = 6\,in$ carries water at a volume flow rate of 2,000 gallons per minute. (a) Is the flow turbulent? (b) Calculate the pressure drop Δp^* in a length $L = 1.0\,km$ and (c) the power $(\Delta p^*)Q$ required to maintain the flow.

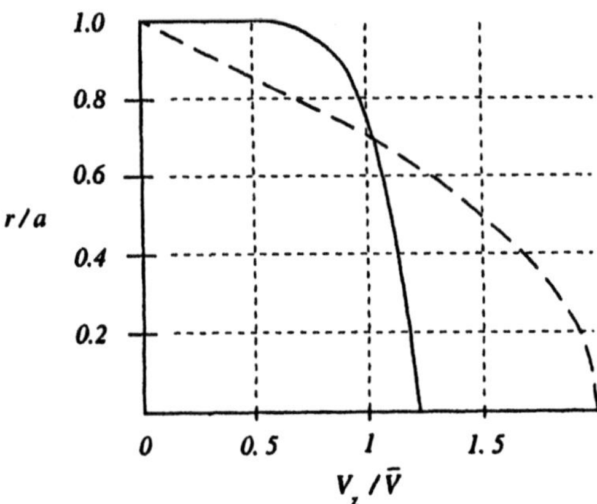

Figure 7.4 A comparison of the axial velocity $V_z/\bar{V}$ as a function of radial distance r/a for turbulent flow (solid line) and laminar flow (dashed line) in a circular pipe of radius a and mean velocity $\bar{V}$.

Solution

(a) To calculate the Reynolds number, find D, Q and $\bar{V}$ in SI units, using tables 1.1 and 1.6:

$$D = 6\,in \times \frac{2.54E(-2)\,m}{in} = 0.1524\,m$$

$$Q = \frac{2000\,gal}{min} \times \frac{min}{60\,s} \times \frac{3.782E(-3)\,m^3}{gal} = 1.262E(-1)\,m^3/s$$

$$\bar{V} = \frac{4Q}{\pi D^2} = \frac{4(1.262E(-1)\,m^3/s)}{\pi(0.1524\,m)^2} = 6.918\,m/s$$

$$Re_D = \frac{\bar{V}D}{\nu} = \frac{(6.918\,m/s)(0.1524\,m)}{1.0E(-6)\,m^2/s} = 1.054E(6)$$

The flow is turbulent since the Reynolds number exceeds the transition value of 2300.

(b) Selecting $\varepsilon = 5E(-5)\,m$ from table 7.1 and inserting this value in equation 7.17, the friction factor f is the solution to:

$$\frac{1}{\sqrt{f}} = -2.0\log\left(\frac{5E(-5)\,m}{3.7(0.1524\,m)} + \frac{2.51}{1.054E(6)\sqrt{f}}\right)$$

$$= -2.0\log\left(8.867E(-5) + \frac{2.381E(-6)}{\sqrt{f}}\right)$$

Guessing a value of $f = 1E(-2)$ and inserting this on the right side of the equation above, we solve to find $f = 1.603E(-2)$. Repeating this iteration twice more, we find $f = 1.588E(-2)$. To find Δp^*, substitute in equation 6.48:

$$\Delta p^* = f\left(\frac{1}{2}\rho\bar{V}^2\right)\frac{L}{D} = 1.588E(-2)\left(\frac{1E(3)\,kg/m^3(6.918\,m/s)^2}{2}\right)\frac{1E(3)\,m}{0.1524\,m}$$
$$= 2.493E(6)\,Pa$$

(c)

$$(\Delta p^*)Q = 2.493E(6)\,Pa \times 1.262E(-2)m^3/s = 3.146E(4)\,W$$

7.3.2 Turbulent Flat Plate Boundary Layer

At length Reynolds numbers $Re_L \equiv V_\infty L/\nu$ exceeding 3E(5), the boundary layer on a flat plate is turbulent, except for a small region near the leading edge. As in pipe flow, the turbulent wall shear stress τ_w is much greater than its laminar (Blasius) value given in equation 6.88. The empirical measurement of the flat plate drag force, which is the cumulative effect of the shear stress acting on the surface, is usually expressed in terms of the drag coefficient $C_\mathcal{D}$, as defined in equation 6.92 and based upon the plate area A exposed to the flow and the plate length L in the flow direction. For turbulent flow over a smooth flat plate, the drag coefficient $(C_\mathcal{D})_{fp}$ is related to the plate length Reynolds number $Re_L \equiv V_\infty L/\nu$ by:[8]

$$(C_\mathcal{D})_{fp} = \frac{0.455}{[\log(Re_L)]^{2.58}} \tag{7.20}$$

Like turbulent pipe flow, there is a weak dependence of drag coefficient on Reynolds number. Figure 7.5 shows $(C_\mathcal{D})_{fp}$ as a function of Re_L. Flat plate turbulent drag coefficients are of the order of $1E(-3)$ and are at least five times the Blasius value of equation 6.93 for Reynolds numbers exceeding the transition value of $3E(5)$.

The flat plate turbulent boundary layer is affected by wall roughness in the same manner as turbulent pipe flow, *i.e.*, the drag coefficient is increased above its smooth plate value, equation 7.20. When the wall is rough enough to disrupt the laminar sublayer, the drag coefficient becomes independent of the length Reynolds number and depends only upon the dimensionless ratio ε/L according to:

$$(C_\mathcal{D})_{fp} = \frac{0.30}{\left[\log\left(14.7\,L/\varepsilon\right)\right]^{2.5}} \tag{7.21}$$

---

[8]See H. Schlichting, 1979, *Boundary-Layer Theory*, 7th ed. (New York: McGraw-Hill Publishing Co.).

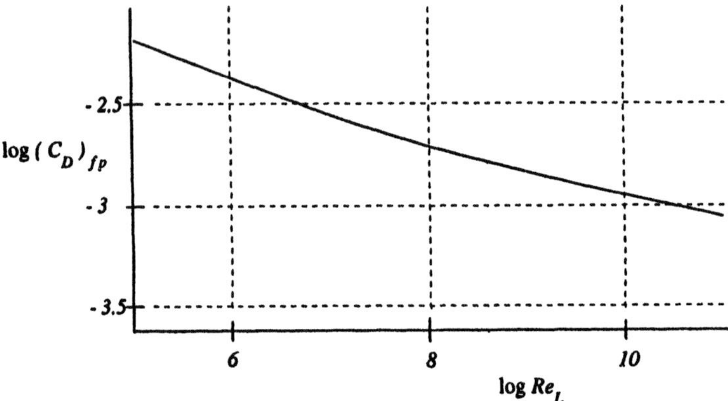

Figure 7.5 A log-log plot of the drag coefficient $(C_D)_{fp}$ versus length Reynolds number Re_L for turbulent flow on a smooth flat plate.

For a flat plate with a non-zero roughness height ε, the drag coefficient will be the larger of the values computed from equations 7.20 and 7.21.

The flat plate turbulent boundary layer is much thicker than its laminar counterpart. An approximate expression for the thickness $\delta_{fp}\{x\}$ to the outer edge of the turbulent boundary layer on a smooth flat plate at a distance x from the leading edge is:

$$\delta_{fp} \simeq 0.14 \left(\frac{\nu}{V_\infty}\right)^{1/7} x^{6/7} = 0.14 \left[\frac{x}{(Re_x)^{1/7}}\right] \tag{7.22}$$

Note that the turbulent flat plate boundary layer grows as $x^{6/7}$, which is much more rapid then the $x^{1/2}$ growth of a laminar boundary layer (see equation 6.79). This is a consequence of the much greater shear stress in the turbulent boundary layer.

Example 7.2

Derive an expression for the value of Re_L for which the drag coefficient of a smooth flat plate, equation 7.20, just equals the drag coefficient of a rough flat plate, equation 7.21.

Solution

Equating the two expressions for the drag coefficient and solving for $\log(Re_L)$,

$$\frac{0.455}{[\log(Re_L)]^{2.58}} = \frac{0.30}{[\log(14.7L/\varepsilon)]^{2.5}}$$

$$\log(Re_L) = 1.175 \left[\log\left(\frac{14.7L}{\varepsilon}\right)\right]^{1-0.03}$$

In the term on the right, we may consider the factor raised to the power −0.03 as unity and solve for Re_L:

$$Re_L = \left(\frac{14.7L}{\varepsilon}\right)^{1.175}$$

For Reynolds numbers less than this value, the plate may be considered smooth, and vice versa.

Example 7.3

A merchant ship of length $L = 100\,m$ and surface area $A = 3E(3)\,m^2$ exposed to sea water ($\rho = 1.03E(3)\,kg/m^3$) moves at a speed $V = 8\,m/s$. Calculate the ship's frictional drag force $\mathcal{D}$ and the power $\mathcal{D}V$ required to overcome this force, assuming $\varepsilon = 1E(-4)\,m$ and $\nu = 1E(-6)\,m^2/s$.

Solution

The length Reynolds number Re_L is:

$$Re_L = \frac{VL}{\nu} = \frac{(8\,m/s)(100\,m)}{1E(-6)\,m^2/s} = 8E(8)$$

which is well within the turbulent flow regime. If the ship surface were smooth, the drag coefficient from equation 7.20 would be:

$$(C_{\mathcal{D}})_{fp} = \frac{0.455}{(\log[8E(8)])^{2.58}} = 1.615E(-3)$$

For a rough surface, the drag coefficient of equation 7.21 is:

$$(C_{\mathcal{D}})_{fp} = \frac{0.30}{[\log(14.7[100\,m]/[1E(-4)\,m])]^{2.5}} = 2.181E(-3)$$

Since the latter is larger, the surface roughness determines the frictional drag. The drag $\mathcal{D}$ is found from equation 6.92:

$$\mathcal{D} = \left(\frac{1}{2}\rho V^2\right) AC_{\mathcal{D}} = 0.5(1.03E(3)\,kg/m^3)(8\,m/s)^2(3E(3)\,m^2)[2.181E(-3)]$$

$$= 2.094E(5)\,N$$

$$\mathcal{D}V = (2.094E(5)\,N)(8\,m/s) = 1.675\,MW$$

7.4 Simple Models of Turbulent Mean Flow

In this section, we give a few examples of how the mean flow $\overline{\mathbf{V}}$ can be calculated in simple turbulent flows if the Reynolds stress can be modeled. These examples emphasize the great difference between turbulent and laminar flows.

7.4.1 The Constant Stress Layer

Close to the wall of a pipe or the surface of a flat plate, the shear stress will be nearly the same as that at the wall, τ_w. In this region, called the constant stress layer, we can find the mean velocity profile $\bar{u}\{y\}$ by assuming a simple model of how the turbulent Reynolds stress $-\rho\overline{u'v'}$ depends upon the flow properties. In analogy with a laminar flow, we assume that the turbulent shear stress is proportional to the derivative of the mean velocity:[9]

$$\frac{\tau_w}{\rho} = \nu_T \frac{d\bar{u}}{dy} \tag{7.23}$$

where the kinematic viscosity coefficient ν_T, called the *eddy viscosity*, is not a constant property of the fluid but a variable property of the flow field that depends upon the properties of the eddies at each point in the flow. In a laminar flow of a gas, ν is approximately equal to the mean thermal speed of the molecules times the mean distance between molecular collisions. For ν_T we need a product of a local flow speed times a flow length. The characteristic flow speed we choose is called the *friction velocity* u_*, defined as:

$$u_* \equiv \sqrt{\frac{\tau_w}{\rho}} \tag{7.24}$$

and the length scale is the distance from the wall, y.[10] Thus we choose:

$$\nu_T = \mathrm{k} u_* y \tag{7.25}$$

Substituting equations 7.25 and 7.24 in 7.23 and solving for $d\bar{u}/dy$, we find:

$$\frac{d\bar{u}}{dy} = \frac{u_*}{\mathrm{k} y}$$

---

[9]This relationship was first suggested by T.V. Boussinesq in 1877.

[10]This form of ν_T was first proposed by L. Prandtl in 1933, and the length y is called the *mixing length*. The constant k is called the *von Karman constant*.

which may be integrated to give:

$$\overline{u} = \frac{u_*}{k} \ln\left(\frac{y}{y_0}\right) \tag{7.26}$$

where y_0, called the *roughness height*, identifies the height at which $\overline{u}$ extrapolates to zero. Contrasted with laminar (Poiseuille) flow, where u is proportional to y, turbulent flow shows a much weaker dependence of $\overline{u}$ on y.

The logarithmic velocity profile of equation 7.26 agrees quite well with measurements, the von Karman constant k having the empirical value of 0.4. It is suitable for flow over rough surfaces, where y_0 is related to the geometric surface roughness. For flow over smooth surfaces, equation 7.26 is only accurate if $yu_*/\nu > 30$; for this case the equivalent roughness height $y_0 = \nu/7.39\,u_*$:

$$\overline{u} = 2.5\,u_* \ln\left(\frac{7.39\,u_* y}{\nu}\right) = 5.75\,u_* \log\left(\frac{7.39\,u_* y}{\nu}\right); \quad \frac{yu_*}{\nu} \geq 30 \tag{7.27}$$

Example 7.4

Close to the earth's surface, the atmospheric boundary layer is a region of constant shear stress. A measurement of time-averaged horizontal wind speed $\overline{u}\{z\}$ at heights $z = 10\,m$ and $z = 1\,m$ yields values of $\overline{u}\{10\,m\} = 9\,m/s$ and $\overline{u}\{1\,m\} = 6\,m/s$, respectively. Calculate the numerical values of u_*, y_0 and the friction coefficient $C_f \equiv 2\tau_w/\rho(\overline{u}\{10\,m\})^2$.

Solution

We have two equations of the form 7.26 to be solved for u_* and y_0. Subtracting one from the other and setting k= 0.4,

$$\overline{u}\{10\,m\} - \overline{u}\{1\,m\} = 2.5\,u_* \left[\ln\left(\frac{10\,m}{y_0}\right) - \ln\left(\frac{1\,m}{y_0}\right)\right] = 2.5\,u_* \ln(10)$$

$$u_* = \frac{3\,m/s}{2.5\ln(10)} = 0.546\,m/s$$

Using this value of u_*, we solve for y_0:

$$\overline{u}\{10\,m\} = 2.5\,u_* \ln\left(\frac{10\,m}{y_0}\right)$$

$$\ln\left(\frac{10\,m}{y_0}\right) = \frac{9\,m/s}{2.5(0.546\,m/s)} = 6.593$$

$$y_0 = 1.37E(-2)\,m$$

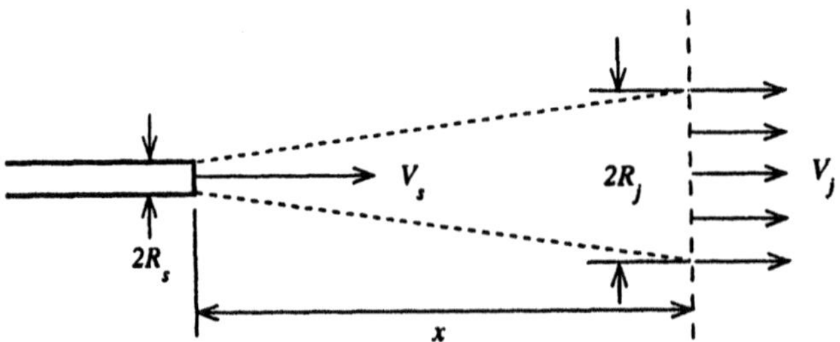

Figure 7.6 The turbulent flow of a jet into a stationary atmosphere entrains surrounding fluid.

The value of C_f is:

$$C_f = \frac{2u_*^2}{[\bar{u}\{10\,m\}]^2} = 2\left(\frac{0.546}{9}\right)^2 = 7.36E(-3)$$

7.4.2 Turbulent Jets and Wakes

In section 5.4.3, we applied the linear momentum theorem to flow in wakes and jets, finding how the flow speed varied with the cross-sectional area of the jet or wake (equations 5.22 and 5.23). In this section, we will show how to calculate the radius R of an axisymmetric jet or wake as a function of the distance from the source, assuming a simple law for the rate of entrainment of surrounding fluid into the jet or wake caused by the flow turbulence. These are both examples of turbulent mixing induced by a *free shear layer, i.e.*, flows in which parallel fluid streams, having different velocities, mix with each other in the absence of a solid wall. In such cases, we can expect that the time-averaged flow is independent of the fluid viscosity, provided the flow Reynolds number is sufficiently high to ensure turbulent flow.[11]

Turbulent Jet in a Stationary Fluid

Consider a turbulent jet of air issuing from a circular pipe and directed into a stationary atmosphere, as sketched in figure 7.6. As the fluid moves along the jet axis, the flow area of the jet, πR_j^2, increases as surrounding air is entrained into the jet, slowing it down. According to the analysis of section 5.4.3, the momentum flux in the jet is constant at all locations x downstream of the jet exit and equal to that of the source

---

[11] Typically, the Reynolds number based on the relative velocity and the jet or wake width must exceed about 100 for the flow to be turbulent.

flow:

$$\rho(\pi R_j^2)V_j^2 = \rho(\pi R_s^2)V_s^2$$

$$R_j V_j = R_s V_s \tag{7.28}$$

assuming the jet source and atmospheric density ρ to be the same.

The volume flow rate in the jet, $\pi R_j^2 V_j$, increases with x as the jet ingests surrounding air. Assume that the rate of increase of volume flow rate, $d(\pi R_j^2 V_j)/dx$, is proportional to the jet circumference $2\pi R_j$ times the jet velocity V_j:

$$\frac{d}{dx}(\pi R_j^2 V_j) = \alpha_j(2\pi R_j)V_j \tag{7.29}$$

where α_j is an empirical constant of order unity to be determined from experiments on jet flows. Combining 7.29 with 7.28 and integrating,

$$\frac{dR_j}{dx} = 2\alpha_j$$

$$R_j = R_s + 2\alpha_j x = R_s\left[1 + 2\alpha_j\left(\frac{x}{R_s}\right)\right] \tag{7.30}$$

where we use the boundary condition that $R_j = R_s$ at $x = 0$. Note that the jet radius increases linearly with x, as sketched in figure 7.6.

By combining equations 7.30 and 7.28, we find how V_j varies with x:

$$V_j = \frac{V_s}{1 + 2\alpha_j(x/R_s)} \tag{7.31}$$

At large distances ($x \gg R_s/2\alpha_j$), V_j varies as x^{-1}. If V_j is measured on the jet centerline, then $\alpha_j = 0.04$ best correlates the dependence of V_j on x/R_s.

Example 7.5

An air jet contains a small amount of toxic gas whose partial density (mass of toxic gas per unit volume of jet fluid) at the source, ρ_{ts}, is much less than the density ρ of the jet and the atmosphere. Derive an expression for the density ρ_{tj} of the toxic gas as a function of the distance x from the source.

Solution

The mass flow rate of the toxic gas at any axial position x must equal its rate at the source:

$$\rho_{ts}\pi R_s^2 V_s = \rho_{tj}\pi R_j^2 V_j$$

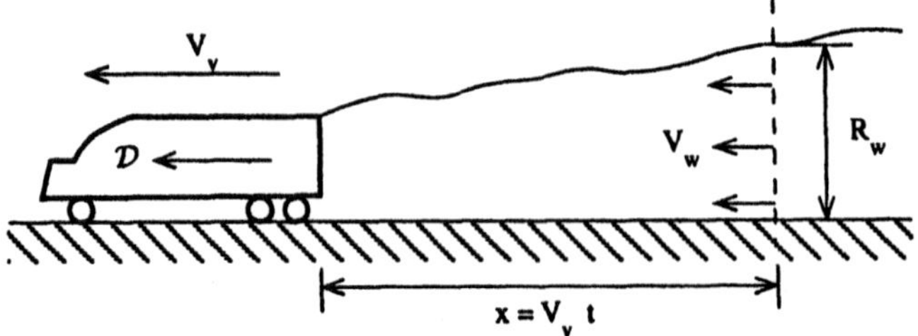

Figure 7.7 An instantaneous picture of the wake behind a truck moving along a highway shows the wake fluid moving in the direction of vehicle travel and the wake radius increasing with distance.

However, the jet radius and velocity are related by equation 7.28. Substituting 7.28 in the above and solving for ρ_{tj}/ρ_{ts},

$$\frac{\rho_{tj}}{\rho_{ts}} = \left(\frac{R_s}{R_j}\right)^2 \frac{V_s}{V_j} = \frac{V_j}{V_s} = \frac{1}{1 + 2\alpha_j(x/R_s)}$$

after using equation 7.31 for V_j.

Turbulent Wake of a Moving Vehicle

When a vehicle moves along a highway, it imparts momentum to the air passing by it, leaving a wake of fluid moving in the direction of travel of the vehicle. If we model this wake as having a semicircular cross-section of radius R_w and flow speed V_w relative to the ground, as shown in figure 7.7, we would observe that R_w increases and V_w decreases with time $t = x/V_v$ since the passage of the vehicle. Such effects are quite noticeable to an observer standing beside a highway, especially when a truck or bus passes.

We can relate R_w and V_w to the drag force $\mathcal{D}$ exerted on the air by the vehicle (see figure 7.7) by utilizing the solution in section 5.4.3 of the linear momentum integral given by equation 5.22. Noting the difference in reference frames between figures 5.4 and 7.7, the vehicle speed V_v and wake speed V_w of figure 7.7 are related by the momentum balance:[12]

$$(V_v - V_w)V_w = \frac{\mathcal{D}}{\rho(\pi R_w^2/2)}$$

---

[12] We are neglecting friction between the wake fluid and the road surface compared with the vehicle drag.

$$\left(\frac{\pi R_w^2}{2}\right) V_w = \frac{\mathcal{D}}{\rho V_v} \quad \text{if } V_w \ll V_v \tag{7.32}$$

If we make the same assumption about entrainment into the wake as for the jet, then:

$$\frac{d}{dx}\left(\frac{\pi R_w^2}{2}\right) V_v = \alpha_w(\pi R_w) V_w$$

$$\frac{dR_w}{dx} = 2\alpha_w \tag{7.33}$$

which has the same form as equation 7.30, but α_w is not necessarily the same as α_j. Integrating 7.33,

$$R_w = 2\alpha_w x \tag{7.34}$$

whenever x is large compared to the wake radius close behind the vehicle. It follows from 7.32 and 7.34 that V_w is:

$$V_w = \frac{2\mathcal{D}}{\pi \rho V_v (2\alpha_w x)^2} \tag{7.35}$$

so that V_w varies as x^{-2} far behind the vehicle.

If we define a vehicle drag coefficient $C_{\mathcal{D}}$ by:

$$C_{\mathcal{D}} \equiv \frac{2\mathcal{D}}{\rho V_v^2 A_v} \tag{7.36}$$

where A_v is the frontal area of the vehicle, then:

$$V_w = V_v \left[\frac{C_{\mathcal{D}} A_v}{\pi (2\alpha_w x)^2}\right] \tag{7.37}$$

The quantity $C_{\mathcal{D}} A_v$ is sometimes called the drag area of the vehicle. Since $C_{\mathcal{D}}$ depends mostly upon the shape and not the size of the vehicle, equation 7.37 shows that the wake speed is proportional to the vehicle speed V_v times the frontal area A_v. Thus, at the same distance x, buses and trucks have greater wake velocities than automobiles.

Example 7.6

An automobile travels at a fixed distance $x = 40\,m$ behind a truck, both at a vehicular speed of $V_v = 100\,km/h$. If the truck drag coefficient is $C_{\mathcal{D}} = 1.0$ and its frontal area is $A = 9\,m^2$, calculate the relative air speed $V_v - V_w$ experienced by the automobile, assuming $\alpha_w = 0.05$.

Solution

Inserting these values in equation 7.37 to calculate V_w:

$$V_w = (100\,km/h)\left(\frac{1.0(9\,m^2)}{\pi[2(0.05)40\,m]^2}\right) = 17.9\,km/h$$

$$V_v - V_w = 82.1\,km/h$$

The jet and wake growth laws, equations 7.33 and 7.30, which follow from the entrainment hypothesis, are manifestations of the concept of a turbulent eddy diffusivity ν_T. For laminar flow, the thickness δ of a shear layer, such as a boundary layer on a flat plate, is (equation 6.79):

$$\delta \simeq \sqrt{\nu \frac{x}{V}}$$

If we replace the laminar viscosity ν by $\nu_T = \alpha\delta V$, then:

$$\delta \simeq \sqrt{\alpha\delta V \frac{x}{V}}$$

$$\delta \simeq \alpha x \tag{7.38}$$

The linear growth of axisymmetric jet and wake widths with x are consistent with the existence of large turbulent shear stresses and eddy viscosities that are proportional to the local velocity and length scales.

7.5 Problems

Problem 7.1

An oil well consists of a pipe of diameter $D = 3.5\,in = 8.89\,cm$, vertical depth $L = 2{,}000\,m$ and roughness $\epsilon/D = 5E(-4)$. When not producing oil ($\bar{V} = 0$), the gage pressure p_{t0} at the top of the oil-filled pipe (called the *shut-in pressure*) is $4E(7)\,Pa = 5{,}800\,psig$. The oil has a viscosity $\mu = 0.1\,Pa\,s$ and a density $\rho = 950\,kg/m^3$.

(a) Calculate the gage pressure p_b in the oil reservoir at the level of the bottom of the well (called the *reservoir pressure*) when the well is not producing. (b) When the well is producing oil at a volumetric flow rate $Q = 20{,}000$ barrels per day ($= 3.768E(-2)\,m^3/s$), the gage pressure p_t at the top of the well is lower than the shut-in pressure while the pressure p_b in the reservoir at the level of the bottom of the well remains at the reservoir pressure. Calculate the value of p_t at this flow rate. (c)

Due to an accident, there is a blow out of the well, and the oil flows freely into the atmosphere, the well top gage pressure p_t being zero. Calculate the flow rate Q in barrels per day for this condition.

Problem 7.2

A light aircraft has a wing span $s = 12\,m$ and an average chord $c = 2\,m$. Assuming that the wing skin friction is the same as that on a flat plate of the same dimensions, calculate (a) the frictional drag $\mathcal{D}$ when the airplane flies at a speed $V = 200\,km/h$ in air having a density $\rho = 1.2\,kg/m^3$ and a kinematic viscosity $\nu = 1.5E(-5)\,m^2/s$ and (b) the power $\mathcal{P} = \mathcal{D}V$ required to overcome this frictional drag.

Problem 7.3

The wind speed at a height $y = 10\,m$ above the ground, upwind of a high-rise building, is measured to be $\bar{u}\{10\,m\} = 10\,m/s$. Assuming the roughness height of the surrounding urban terrain is $y_0 = 0.5\,m$, calculate the wind speed $\bar{u}\{100\,m\}$ at the top of the building, $y = 100\,m$.

Problem 7.4

An automobile having a drag area $C_{\mathcal{D}}A = 5\,m^2$ travels along a highway at a speed $V_v = 55\,mi/h$. It emits an air pollutant at a rate of $10\,g$ per mile traveled. Assuming $\alpha_w = 0.05$, calculate (a) the mass flow rate $\dot{m}_p$ of the pollutant emitted by the vehicle and (b) the mass density $\rho_p\,[\mu g/m^3]$ of the pollutant in the vehicle wake at a distance $x = 100\,m$ behind the automobile.

Problem 7.5

Air is discharged to the atmosphere from a building's ventilating system at a speed of $V_s = 3\,m/s$ from a circular opening of radius $R_s = 0.2\,m$. The discharge stream contains an air pollutant gas at a partial density of $(\rho_p)_s = 0.01\,kg/m^3$. Calculate the value of the pollutant density ρ_p at a distance $x = 30\,m$ from the point of discharge.

Bibliography

Brodkey, Robert S. 1967. *The Phenomena of Fluid Motions*. Reading, Mass.: Addison-Wesley Publishing Co.

Gerhart, Philip M., Richard J. Gross, and John I. Hochstein, 1992. *Fundamentals of Fluid Mechanics*. 2nd ed. Reading, Mass.: Addison-Wesley Publishing Co.

Monin, A. S., and A. M. Yaglom. 1975. *Statistical Fluid Mechanics*. Edited by John L. Lumley. Vol. 2. Cambridge, Mass.: MIT Press.

Schetz, Joseph A. 1980. *Injection and Mixing in Turbulent Flow*. New York: American Institute of Aeronautics and Astronautics.

Schlichting, H. 1968. *Boundary-Layer Theory*. 6th ed. New York: McGraw-Hill Book Co.

8 Conservation of Energy

8.1 Introduction

In treating fluid flows thus far, we have made use of two general laws, the conservation of mass and the conservation of momentum. Expressed in integral form, they apply to all fluid flows, but when expressed in differential form, the momentum conservation principle is restricted to inviscid flow (Euler's equation 4.5) or incompressible viscous flow (incompressible Navier-Stokes equation 6.11). Nevertheless, we are able to treat a range of important, practical flows within these restrictions.

There are other types of flows, also of great practical interest, which cannot be explained solely on the basis of the conservation of mass and momentum. Flow with heat transfer or chemical reaction (*e.g.*, flames), flow with two or more phases present and high speed flow (*i.e.*, flow with velocities comparable to the speed of sound) involve an interaction of the motion with the thermodynamic behavior of the fluid that needs to be taken into account. For these flows, we must invoke the principle of conservation of energy.

The engineering applications that require the use of energy conservation include, for example, steam and gas turbines, compressors, internal combustion engines, burners, heat transfer equipment, supersonic flow, acoustics, and many others. We cannot cover these interesting and important applications within a single chapter of this text. Nevertheless, because it is important to understand how these flows differ from what we have been considering thus far and to provide a link with other mechanical engineering subjects such as thermodynamics and heat transfer, which consider some of these applications in greater depth, we will introduce the principle of the conservation of energy and apply it, together with the conservation of mass and momentum, to a few simple fluid flows.

Our approach will be to derive an integral form of the energy conservation equation that is universally applicable and then to express it in differential form. This methodology will provide us with forms of the energy conservation principle that are compatible with the forms of the mass and momentum conservation expressions we have previously used.

However, before proceeding down this lengthy but necessary path, we will consider first a limited form of the energy conservation principal that is restricted to incompressible flows. While this version of energy conservation will not suffice for all examples of incompressible flow and certainly not for compressible flow, it is quite helpful in explaining some limitations on viscous incompressible flows.

8.2 Incompressible Viscous Flow

The principles that govern the motion of an incompressible viscous flow are embodied in the conservation of mass, equation 3.17:

$$\nabla \cdot \mathbf{V} = 0 \tag{8.1}$$

or the alternate form, equation 3.18:

$$\frac{\partial \rho}{\partial t} + (\mathbf{V} \cdot \nabla)\rho = 0 \tag{8.2}$$

and the Navier-Stokes equation 6.11, which we write here in the form:

$$\frac{\partial \mathbf{V}}{\partial t} + (\mathbf{V} \cdot \nabla)\mathbf{V} = -\frac{1}{\rho}\nabla p + \mathbf{g} + \frac{\mathbf{f}}{\rho} \tag{8.3}$$

where $\mathbf{f}$ is the viscous force per unit mass.[1] (If $\rho,\ \mu = const$, $\mathbf{f} = \mu\nabla^2\mathbf{V}$.)

In principle, we can find p and $\mathbf{V}$ from the solution of these equations, although in practice that proves very difficult in all but the simplest of cases. However, by expressing some of the information contained in these equations in an alternate form, we are able to see more clearly some of their consequences.[2] That is the objective of the next section.

8.2.1 The Conservation of Kinetic Energy

The kinetic energy per unit mass of fluid, $\mathbf{V} \cdot \mathbf{V}/2 = V^2/2$, is a property of each fluid particle in the flow field. Can we determine how rapidly this changes?

---

[1] As written here, this form is applicable to compressible flow as well.

[2] For example, in studying incompressible inviscid flow, we found that some of the information embodied in Euler's equation could be cast in the form of Bernoulli's integral.

To do so, multiply equation 8.3 by $\mathbf{V}\cdot$:

$$\mathbf{V}\cdot\left(\frac{\partial \mathbf{V}}{\partial t}\right)+\mathbf{V}\cdot[(\mathbf{V}\cdot\nabla)\mathbf{V}] = -\frac{1}{\rho}\mathbf{V}\cdot\nabla p+\mathbf{V}\cdot\mathbf{g}+\frac{\mathbf{V}\cdot\mathbf{f}}{\rho}$$

$$\frac{\partial}{\partial t}\left(\frac{V^2}{2}\right)+(\mathbf{V}\cdot\nabla)\left(\frac{V^2}{2}\right) = -\frac{1}{\rho}\mathbf{V}\cdot\nabla p+\mathbf{V}\cdot\mathbf{g}+\frac{\mathbf{V}\cdot\mathbf{f}}{\rho} \tag{8.4}$$

Equation 8.4 is the differential form of the conservation of kinetic energy in a viscous flow. The left side of 8.4 is the rate of increase of kinetic energy per unit mass, $V^2/2$, of a fluid particle as it moves along in a fluid flow. The right side of 8.4 is the rate at which the pressure, gravity and viscous forces are increasing this kinetic energy per unit mass.[3]

As it stands, equation 8.4 is not all that convenient for practical use. Instead, its integral form is much more useful. To derive it, we add equation 8.1 multiplied by $(V^2/2 - \mathbf{g}\cdot\mathbf{R})$ to 8.4, obtaining (after using equations 1.38 and 2.7):

$$\frac{\partial}{\partial t}\left(\frac{V^2}{2}\right)+\nabla\cdot\left[\left(\frac{V^2}{2}\right)\mathbf{V}\right] = -\frac{1}{\rho}\mathbf{V}\cdot\nabla p+\nabla\cdot[(\mathbf{g}\cdot\mathbf{R})\,\mathbf{V}]+\frac{\mathbf{V}\cdot\mathbf{f}}{\rho}$$

Next multiply by ρ, and then add equation 8.1 multiplied by p plus equation 8.2 multiplied by $(V^2/2 - \mathbf{g}\cdot\mathbf{R})$ to obtain (after using 1.38 again):

$$\frac{\partial}{\partial t}\left[\rho\left(\frac{V^2}{2}-\mathbf{g}\cdot\mathbf{R}\right)\right]+\nabla\cdot\left[\left(\frac{V^2}{2}\right)\rho\mathbf{V}\right] = -\nabla\cdot\left[\left(\frac{p}{\rho}-\mathbf{g}\cdot\mathbf{R}\right)\rho\mathbf{V}\right]+\mathbf{V}\cdot\mathbf{f} \tag{8.5}$$

Now integrate equation 8.5 over a fixed control volume, converting the divergence terms to surface integrals using the divergence theorem, equation 1.44, to obtain:

$$\frac{d}{dt}\iiint_{\mathcal{V}}\left[\rho\left(\frac{V^2}{2}-\mathbf{g}\cdot\mathbf{R}\right)\right]d\mathcal{V}+\iint_{S}\left(\frac{p}{\rho}+\frac{V^2}{2}-\mathbf{g}\cdot\mathbf{R}\right)\rho\mathbf{V}\cdot\mathbf{n}\,dS$$
$$= \iiint_{\mathcal{V}}(\mathbf{V}\cdot\mathbf{f})\,d\mathcal{V} \tag{8.6}$$

In this form, the energy conservation principle states that the rate of accumulation of kinetic plus potential energy within the fixed control volume plus the net outflow rate of the quantity $(p/\rho+V^2/2-\mathbf{g}\cdot\mathbf{R})$—a quantity that is conserved along a streamline in steady inviscid flow and has the same dimensions as kinetic energy—is equal to the rate at which viscous forces are doing work on the fluid inside the control volume.

In a subsequent section (8.5), we show that the viscous work term of equation 8.6

[3] Although we tend to think that viscous forces will reduce the kinetic energy of the flow, this is not true in general.

may be rewritten as:

$$\iiint_{\mathcal{V}} (\mathbf{V} \cdot \mathbf{f})\, d\mathcal{V} = \iint_{S} \boldsymbol{\tau} \cdot \mathbf{V}\, dS - \iiint_{\mathcal{V}} \Phi_i\, d\mathcal{V}$$

where $\boldsymbol{\tau}$ is the viscous stress. The function Φ_i, called the *incompressible dissipation function* and defined in equation 8.61, is proportional to the viscosity μ and is always a non-negative quantity. Combining this with equation 8.6, we have the integral form of the incompressible energy equation:

$$\frac{d}{dt} \iiint_{\mathcal{V}} \rho \left(\frac{V^2}{2} - \mathbf{g} \cdot \mathbf{R} \right) d\mathcal{V} + \iint_{S} \left(\frac{p}{\rho} + \frac{V^2}{2} - \mathbf{g} \cdot \mathbf{R} \right) \rho \mathbf{V} \cdot \mathbf{n}\, dS$$

$$= \iint_{S} \boldsymbol{\tau} \cdot \mathbf{V}\, dS - \iiint_{\mathcal{V}} \Phi_i\, d\mathcal{V} \tag{8.7}$$

In this form, the energy conservation principle balances the accumulation and flow of energy on the left side with the sum of two viscous terms on the right: the rate of work done by the viscous stress, $\boldsymbol{\tau} \cdot \mathbf{V}$, acting on the control surface and a *negative* term, minus the rate of dissipation of energy Φ_i integrated throughout the control volume.

This equation is most useful for steady flows in pipes, ducts and channels. For steady flow, it may be written as:

$$\iint_{S} \left[\left(\frac{p}{\rho} + \frac{V^2}{2} - \mathbf{g} \cdot \mathbf{R} \right) \rho \mathbf{V} \cdot \mathbf{n} - \boldsymbol{\tau} \cdot \mathbf{V} \right] dS = - \iiint_{\mathcal{V}} \Phi_i\, d\mathcal{V} \qquad \textit{if } \frac{\partial}{\partial t} = 0 \tag{8.8}$$

Example 8.1

For the steady plane Couette flow illustrated in figure 6.4, derive an expression for the wall shear stress τ_w by solving equation 8.8.

Solution

For a control volume consisting of a base of unit area and height equal to the channel height h, the only contribution to the left side of 8.8 comes from the term $\boldsymbol{\tau} \cdot \mathbf{V}$ on the upper surface of unit area since there is no change of any of the variables in the flow direction. Consequently 8.8 becomes:

$$-\tau_w V = - \int_0^h \Phi_i dy$$

Of the components of Φ_i in equation 8.61, only $\partial u/\partial y = V/h$ is non-zero. Hence:

$$\tau_w V = \int_0^h \mu\left(\frac{V}{h}\right)^2 dy = \mu\left(\frac{V}{h}\right)^2 \int_0^h dy = \mu\frac{V^2}{h}$$

$$\tau_w = \mu\frac{V}{h}$$

8.2.2 Steady Flow through a Pipe or Chamber

Now consider a flow through a pipe or heat transfer apparatus. On the walls of the pipe or container, $\mathbf{V} = 0$, and thus the walls make no contribution to the surface integral of 8.8. At the inflow and outflow surfaces of the control volume, the shear stress $\boldsymbol{\tau}$ is much smaller than p so that $\boldsymbol{\tau} \cdot \mathbf{V}$ can be neglected compared to $p\mathbf{V} \cdot \mathbf{n}$. Denoting the inflow and outflow integrals separately, and noting that $\mathbf{V} \cdot \mathbf{n}$ is positive for outflow and negative for inflow, we may write 8.8 as:

$$\iint_{in}\left(\frac{p}{\rho} + \frac{V^2}{2} - \mathbf{g}\cdot\mathbf{R}\right)(-\rho\mathbf{V}\cdot\mathbf{n})\,dS - \iint_{out}\left(\frac{p}{\rho} + \frac{V^2}{2} - \mathbf{g}\cdot\mathbf{R}\right)(\rho\mathbf{V}\cdot\mathbf{n})\,dS$$

$$= \iiint_{\mathcal{V}} \Phi_i \, d\mathcal{V} \geq 0 \tag{8.9}$$

where the inequality follows from the fact that Φ_i is always non-negative. Equation 8.9 states that the magnitude of the quantity $(p/\rho + V^2/2 - \mathbf{g}\cdot\mathbf{R})\rho\mathbf{V}\cdot\mathbf{n}$, integrated across the cross-section of the flowing fluid, declines in the downstream direction because of the effects of viscosity, *i.e.*, the upstream value exceeds the downstream value by an amount that is proportional to the fluid viscosity.

In chapter 3, we found that the quantity, $p/\rho + V^2/2 - \mathbf{g}\cdot\mathbf{R}$, which has the dimensions of energy per unit mass, is conserved along a streamline in steady, incompressible inviscid flow. Some texts refer to this quantity as the flow energy since it is conserved in inviscid flow, but we can see from 8.9 that the conservation of energy requires that this quantity will change in a viscous flow. Along an individual streamline, it can either increase or decrease in a steady viscous flow, although the summation of the changes from inlet to outlet of a flow-containing structure with non-moving walls will always experience a decrease in the flow direction.

The quantity $p+\rho(V^2/2-\mathbf{g}\cdot\mathbf{R})$ may be called the *total pressure* of the incompressible flow. It is the pressure that the fluid would have if it were brought to rest in equilibrium with stationary fluid at $\mathbf{g}\cdot\mathbf{R} = 0$. Equation 8.9 implies that the average total pressure of the flow through a pipe or chamber declines in the downstream direction.

Head

In treating incompressible flows, it is customary to define a flow variable called the *total head*, signified by the symbol h. It has the dimensions of length and is defined as:

$$\mathsf{h} \equiv \frac{p}{\rho g} + \frac{V^2}{2g} - \frac{\mathbf{g} \cdot \mathbf{R}}{g} \tag{8.10}$$

The terms on the right side of equation 8.10 are called the *static pressure head*, the *velocity head* and the *elevation head*, respectively. This usage originated in applications to flows of water in open channels and through hydroturbines and sluice gates.

In terms of the total head h, equation 8.9 may be written as:

$$\iint_{in} g\mathsf{h}(-\rho \mathbf{V} \cdot \mathbf{n})\, dS - \iint_{out} g\mathsf{h}(\rho \mathbf{V} \cdot \mathbf{n})\, dS = \iiint_{\mathcal{V}} \Phi_i \, d\mathcal{V} \geq 0 \tag{8.11}$$

Consequently, the value of $g\mathsf{h}(\rho\mathbf{V}\cdot\mathbf{n})$ integrated across the flow cross-section decreases in the flow direction.

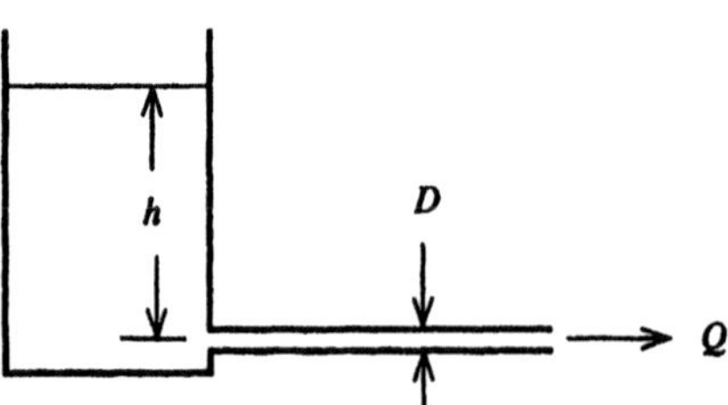

Figure E 8.2

Example 8.2

A tank of water open to the atmosphere at its top is being drained through a pipe of diameter $D = 2\, in$, as sketched in figure E 8.2. The pipe outlet, also open to the atmosphere, is at an elevation $h = 10\, m$ below that of the water level in the tank at the moment when the volumetric flow rate Q is measured to be $425\, gal/min$. What is the reduction in head $\mathsf{h}_{in} - \mathsf{h}_{out}$ between the static fluid in the tank and the fluid at the pipe outlet?

Solution

From the definition of h, equation 8.10, the reduction $\mathsf{h}_{in} - \mathsf{h}_{out}$ is:

$$\mathsf{h}_{in} - \mathsf{h}_{out} = \left(\frac{p}{\rho g} + \frac{V^2}{2g} + z\right)_{in} - \left(\frac{p}{\rho g} + \frac{V^2}{2g} + z\right)_{out}$$

$$= \frac{p_{in} - p_{out}}{\rho g} + \frac{V_{in}^2 - V_{out}^2}{2g} + z_{in} - z_{out}$$

$$= 0 - \frac{V_{out}^2}{2g} + h$$

where we take the inflow point at the air/water interface in the tank. Evaluating V_{out}:

$$Q = \frac{425\, gal}{min} \times \frac{min}{60\, s} \times \frac{3.785E(-3)\, m^3}{gal} = 2.681E(-2)\, m^3/s$$

$$V = \frac{4Q}{\pi D^2} = \frac{4(2.681E(-2)\, m^3/s)}{\pi[2 \times 2.54E(-2)]^2\, m^2} = 13.23\, m/s$$

$$\mathsf{h}_{in} - \mathsf{h}_{out} = 10\, m - \frac{(13.23\, m)^2}{2(9.807\, m/s^2)} = 1.076\, m$$

8.2.3 Energy Conservation in Steadily Rotating Machines

Rotating machines, such as pumps, turbines, compressors, fans, blowers, propellers, etc., are important devices that produce or absorb mechanical work while interacting with a flowing fluid. They power transportation vehicles and produce nearly all electrical power. In this section, we consider the special form of the principle of conservation of energy that applies to this class of devices whenever the flow can be considered to be incompressible.[4]

While the flow into or out of these machines may be a steady flow, the internal flow is far from steady because of the motions induced by the moving blades of the rotor.[5] In chapter 4, section 4.6.3, we noted that the flow in the noninertial reference frame of the rotor was steady and therefore it would be simpler to describe the flow relative to the rotor. In the noninertial reference frame of the rotor, we found that the Bernoulli integral along a streamline of the rotor flow had the same form as that for

---

[4]The general condition for incompressible flow is that the flow speed be much less than the speed of sound. We can consider air flow through fans and propellers as incompressible flow.

[5]The rotating element is called a *rotor* while the stationery structure is called a *stator*. Propellers and wind turbines have no stator.

an inertial reference frame except that the term $\mathbf{g}\cdot\mathbf{R}$ is replaced by $\mathbf{g}\cdot\tilde{\mathbf{R}}+(\Omega\tilde{r})^2/2$. In the following sections, we explore the changes in the expression for the conservation of energy in incompressible flow that result from this extra term.

For steady incompressible flow relative to the rotor, the conservation of energy equation 8.9 becomes:[6]

$$\iint_{out}\left(\frac{p}{\rho}+\frac{\tilde{V}^2}{2}-\mathbf{g}\cdot\tilde{\mathbf{R}}-\frac{(\Omega\tilde{r})^2}{2}\right)(\rho\tilde{\mathbf{V}}\cdot\mathbf{n})\,dS$$
$$-\iint_{in}\left(\frac{p}{\rho}+\frac{\tilde{V}^2}{2}-\mathbf{g}\cdot\tilde{\mathbf{R}}-\frac{(\Omega\tilde{r})^2}{2}\right)(-\rho\tilde{\mathbf{V}}\cdot\mathbf{n})\,dS=-\iiint_{\mathcal{V}}\tilde{\Phi}_i\,d\mathcal{V}\leq 0 \tag{8.12}$$

We may express the integrands of the surface integrals in terms of the velocity and coordinates of the stationary inertial reference frame by using equation 4.25, to find:[7]

$$\iint_{out}\left(\frac{p}{\rho}+\frac{V^2}{2}-\mathbf{g}\cdot\mathbf{R}\right)(\rho\mathbf{V}\cdot\mathbf{n})\,dS-\iint_{in}\left(\frac{p}{\rho}+\frac{V^2}{2}-\mathbf{g}\cdot\mathbf{R}\right)(-\rho\mathbf{V}\cdot\mathbf{n})\,dS$$
$$=\iint_{out}\Omega rV_\theta(\rho\mathbf{V}\cdot\mathbf{n})\,dS-\iint_{in}\Omega rV_\theta(-\rho\mathbf{V}\cdot\mathbf{n})\,dS-\iiint_{\mathcal{V}}\tilde{\Phi}_i\,d\mathcal{V} \tag{8.13}$$

If we now apply the conservation of angular momentum integral, equation 5.46, to the axially symmetric control volume, neglecting any viscous stress on the inlet and outlet flows, we find the torque $\mathbf{T}$ *applied to the rotor*:

$$\mathbf{T}=\mathbf{i}_z\iint_{out}rV_\theta(\rho\mathbf{V}\cdot\mathbf{n})\,dS-\mathbf{i}_z\iint_{in}\Omega rV_\theta(-\rho\mathbf{V}\cdot\mathbf{n})\,dS \tag{8.14}$$

Substituting in equation 8.13, we find the energy conservation equation applied to a rotating machine:

$$\iint_{out}\left(\frac{p}{\rho}+\frac{V^2}{2}-\mathbf{g}\cdot\mathbf{R}\right)(\rho\mathbf{V}\cdot\mathbf{n})\,dS-\iint_{in}\left(\frac{p}{\rho}+\frac{V^2}{2}-\mathbf{g}\cdot\mathbf{R}\right)(-\rho\mathbf{V}\cdot\mathbf{n})\,dS$$
$$=\mathbf{\Omega}\cdot\mathbf{T}-\iiint_{\mathcal{V}}\tilde{\Phi}_i\,d\mathcal{V} \tag{8.15}$$

An alternate form, assuming that the flow variables across the inlet and outlet flow streams are constant, is:

[6]See equation 4.28. Velocities and coordinates measured in the rotating reference frame are denoted by a tilde over the quantity (*e.g.*, $\tilde{V}$). The relations between velocities and coordinates in the inertial and noninertial reference frames are given by equation 4.25.

[7]We have chosen an axially-symmetric control volume so that $\tilde{\mathbf{V}}\cdot\mathbf{n}=\mathbf{V}\cdot\mathbf{n}$.

$$\dot{m}\left(\frac{p}{\rho}+\frac{V^2}{2}-\mathbf{g}\cdot\mathbf{R}\right)_{out}-\dot{m}\left(\frac{p}{\rho}+\frac{V^2}{2}-\mathbf{g}\cdot\mathbf{R}\right)_{in}=\boldsymbol{\Omega}\cdot\mathbf{T}-\iiint_{\mathcal{V}}\bar{\Phi}_i\,d\mathcal{V} \quad (8.16)$$

where $\dot{m}$ is the mass flow rate through the device.

In terms of the total head h defined by equation 8.10, equation 8.16 assumes the form:

$$\dot{m}g(\mathsf{h}_{out}-\mathsf{h}_{in})=\boldsymbol{\Omega}\cdot\mathbf{T}-\iiint_{\mathcal{V}}\bar{\Phi}_i\,d\mathcal{V} \quad (8.17)$$

Expressed in this form, the conservation of energy states that the increase in head through the device, $\mathsf{h}_{out}-\mathsf{h}_{in}$, multiplied by the mass flow rate $\dot{m}$ times g equals the power $\boldsymbol{\Omega}\cdot\mathbf{T}$ required to rotate the shaft minus the viscous power dissipated inside the machine, $\iiint_{\mathcal{V}}\bar{\Phi}_i\,d\mathcal{V}$.

For a pump, the power $\boldsymbol{\Omega}\cdot\mathbf{T}$ is positive since $\boldsymbol{\Omega}$ and $\mathbf{T}$ have the same direction. The magnitude of the pump power $(\Omega T)_p$ is always greater than the increase in the flux of head. The ratio of these two terms is defined as the pump efficiency η_p:

$$\eta_p\equiv\frac{\dot{m}g(\mathsf{h}_{out}-\mathsf{h}_{in})}{(\Omega T)_p}=1-\frac{1}{(\Omega T)_p}\iiint_{\mathcal{V}}\bar{\Phi}_i\,d\mathcal{V}\leq 1 \quad (8.18)$$

If the efficiency of a pump is known, then the power required to pump fluid at a mass flow rate $\dot{m}$ with an increase in head $\mathsf{h}_{out}-\mathsf{h}_{in}$ is:

$$(\Omega T)_p=\frac{\dot{m}g(\mathsf{h}_{out}-\mathsf{h}_{in})}{\eta_p} \quad (8.19)$$

For a turbine, on the other hand, $\boldsymbol{\Omega}\cdot\mathbf{T}$ is negative, and the turbine delivers power. For this case, the energy equation 8.17 may be written in the form:

$$-\boldsymbol{\Omega}\cdot\mathbf{T}=\dot{m}g(\mathsf{h}_{in}-\mathsf{h}_{out})-\iiint_{\mathcal{V}}\bar{\Phi}_i\,d\mathcal{V}$$

showing that the turbine power is always less than the reduction in the flux of head because of the viscous losses in the turbine. It is customary to define the turbine efficiency as:

$$\eta_t\equiv\frac{(\Omega T)_t}{\dot{m}g(\mathsf{h}_{in}-\mathsf{h}_{out})}=1-\frac{1}{\dot{m}g(\mathsf{h}_{in}-\mathsf{h}_{out})}\iiint_{\mathcal{V}}\bar{\Phi}_i\,d\mathcal{V}\leq 1 \quad (8.20)$$

Thus the power $(\Omega T)_t$ available from a turbine of known efficiency η_t that experiences a reduction of head of $\mathsf{h}_{in}-\mathsf{h}_{out}$ is:

$$(\Omega T)_t=\eta_t\dot{m}g(\mathsf{h}_{in}-\mathsf{h}_{out}) \quad (8.21)$$

Example 8.3

A hydroturbine delivers 8 MW of mechanical power to an electric generator. The change in head $\mathsf{h}_{in} - \mathsf{h}_{out}$ between the turbine inlet and outlet is 10 m, and the volumetric flow rate $Q = 100\, m^3/s$. Calculate the turbine efficiency η_t.

Solution

Using equation 8.20,

$$\eta_t = \frac{(\Omega T)_t}{\rho Q g(\mathsf{h}_{in} - \mathsf{h}_{out})} = \frac{8E(6)\, W}{(1E(3)\, kg/m^3)(100\, m^3/s)(9.807\, m/s^2)(10\, m)} = 81.57\,\%$$

8.3 The First Law of Thermodynamics

The principle of the conservation of energy in a fluid flow is an application of the first law of thermodynamics to a flowing fluid. This law defines the relationship between the internal energy (U) of a thermodynamic substance and the transfer of heat (Q) and work (W) to or from the substance during a process of change. When applied to a moving fluid element, the first law states that the element's total energy (E), which is the sum of its internal energy, kinetic energy and potential energy in the earth's gravitational field, is *increased by the net amount of heat added to the element and decreased by the amount of work done by the element on its environment.*[8] (The thermodynamic convention is that heat added to a fluid element or work done by it on its surroundings is a positive quantity.) Because the first law applies to a fixed amount of matter, it is expressed in Lagrangian form, *i.e.*, it describes the changes to an identified fluid element as it moves through the flow field. Using our previous notation for the material time derivative, we can express the first law in the form: the rate of increase of total energy equals the rate of heat addition minus the rate at which work is done by the fluid element, or

$$\frac{DE}{Dt} = \frac{DQ}{Dt} - \frac{DW}{Dt} \tag{8.22}$$

In this form, the first law is not readily applicable to a fluid flow problem. We need to express it in an Eulerian form that is compatible with our mass and momentum conservation equations.

[8]The quantities involved in the expression of energy conservation are those measured in an inertial reference frame.

Consider a control volume fixed in an inertial reference frame (*e.g.*, a laboratory reference frame). The energy E of the fluid inside the control volume is simply the integral of the product $\rho(\hat{u} + V^2/2 - \mathbf{g} \cdot \mathbf{R})$ over the control volume $\mathcal{V}$, where $\hat{u}$ is the internal energy per unit mass of fluid. Making use of the Reynolds transport theorem (equation 5.5) for the property, total energy per unit mass $\hat{u} + V^2/2 - \mathbf{g} \cdot \mathbf{R}$, the material derivative of the total energy E becomes:

$$\frac{DE}{Dt} = \frac{d}{dt} \iiint_{\mathcal{V}} \rho \left(\hat{u} + \frac{V^2}{2} - \mathbf{g} \cdot \mathbf{R} \right) d\mathcal{V} + \iint_{\mathcal{S}} \rho \left(\hat{u} + \frac{V^2}{2} - \mathbf{g} \cdot \mathbf{R} \right) \mathbf{V} \cdot \mathbf{n} \, d\mathcal{S} \quad (8.23)$$

The first term on the right side of equation 8.23 is the rate of accumulation of total energy E within the control volume, and the second is the net rate of transport of E out of the control volume by the fluid flowing across its boundaries.

We next turn to the determination of the heat transfer to the fluid in the control volume. Heat is transferred within a fluid because of temperature differences between adjacent locations in the fluid. For most fluids, the rate of heat flow per unit area across a surface in the fluid, $\mathbf{q}$, is proportional to the temperature gradient in accordance with *Fourier's Law*:[9]

$$\mathbf{q} = -\lambda \nabla T \quad (8.24)$$

where λ is the thermal conductivity of the fluid, a transport property like viscosity. The heat flows in the direction of decreasing temperature. The net rate of heat addition to the fluid inside the control volume from its surroundings is thus the integral of the heat flux $\mathbf{q}$ over the control surface $\mathcal{S}$:

$$\frac{DQ}{Dt} = \iint_{\mathcal{S}} \mathbf{q} \cdot (-\mathbf{n}) d\mathcal{S} \quad (8.25)$$

(The minus sign results from the fact that, when the heat flows into the control volume, $\mathbf{q}$ has the opposite direction from the outward normal $\mathbf{n}$ and $\mathbf{q} \cdot (-\mathbf{n})$ is positive, as required by the thermodynamic convention for Q.)

Now consider the rate at which work is done on the environment by the fluid in the control volume. The differential force exerted by the fluid on its surroundings is the stress acting on the surroundings times the surface area $d\mathcal{S}$. We can divide the stress into two components, the pressure $p\mathbf{n}$ which acts in the direction of the normal $\mathbf{n}$ pointing outward from the fluid towards the surroundings and the viscous stress $-\boldsymbol{\tau}$, which is opposite to the stress $\boldsymbol{\tau}$ acting on the fluid. (The viscous stress $\boldsymbol{\tau}$ was determined in section 6.2.) The rate at which the fluid does work is the scalar product

[9] In SI units, $\mathbf{q}$ has the dimension of W/m^2, and λ, the dimension W/mK.

of the force times the velocity of the fluid at the control surface:

$$\frac{D\mathcal{W}}{Dt} = \iint_S (p\mathbf{n} - \boldsymbol{\tau}) \cdot \mathbf{V}\, dS = \iint_S (p\mathbf{V} \cdot \mathbf{n} - \boldsymbol{\tau} \cdot \mathbf{V})\, dS \tag{8.26}$$

Notice that the work term contains two parts: that due to the pressure and that due to the viscous stress. In an inviscid flow, the latter term is zero, and only the pressure term contributes to the work done by the fluid. Also, if a portion of the control surface is a stationary wall, then its contribution to the work term is zero, both for inviscid flow ($\mathbf{V} \cdot \mathbf{n} = 0,\ \boldsymbol{\tau} = 0$) and for viscous flow ($\mathbf{V} = 0$).

The integral form of the conservation of energy in a flowing fluid can now be expressed by substituting equations 8.23, 8.25 and 8.26 in 8.22:

$$\frac{d}{dt}\iiint_{\mathcal{V}} \rho\left(\hat{u} + \frac{V^2}{2} - \mathbf{g}\cdot\mathbf{R}\right) d\mathcal{V} + \iint_S \rho\left(\hat{u} + \frac{V^2}{2} - \mathbf{g}\cdot\mathbf{R}\right)\mathbf{V}\cdot\mathbf{n}\, dS$$
$$= \iint_S \mathbf{q}\cdot(-\mathbf{n})dS - \iint_S (p\mathbf{V}\cdot\mathbf{n} - \boldsymbol{\tau}\cdot\mathbf{V})\, dS \tag{8.27}$$

In words, this conservation equation requires that the rate of increase of total energy of the fluid inside the control volume plus the outward rate of flow of total energy across the control surface are equal to the sum of the inward rate of heat flow across the control surface minus the rate at which the fluid is doing work on the environment as it moves across the control surface.

In applying the first law of thermodynamics to various flow processes, it is customary to introduce the property enthalpy $\hat{h}$, which is the sum of the internal energy $\hat{u}$ and the ratio p/ρ:

$$\hat{h} \equiv \hat{u} + \frac{p}{\rho} \tag{8.28}$$

Using equation 8.28 to replace $\rho\hat{u}$ by $\rho\hat{h} - p$ in the second term of equation 8.27, we find that the surface integral on the right involving $p\mathbf{V}\cdot\mathbf{n}$ cancels and that we have:

$$\frac{d}{dt}\iiint_{\mathcal{V}} \rho\left(\hat{u} + \frac{V^2}{2} - \mathbf{g}\cdot\mathbf{R}\right) d\mathcal{V} + \iint_S \rho\left(\hat{h} + \frac{V^2}{2} - \mathbf{g}\cdot\mathbf{R}\right)\mathbf{V}\cdot\mathbf{n}\, dS$$
$$= \iint_S \mathbf{q}\cdot(-\mathbf{n})dS + \iint_S \boldsymbol{\tau}\cdot\mathbf{V}\, dS \tag{8.29}$$

If we call the sum $(\hat{h} + V^2/2 - \mathbf{g}\cdot\mathbf{R})$ the *total enthalpy*, then equation 8.29 expresses the conservation of energy in terms of the total energy and the total enthalpy: the rate of accumulation of total energy inside the control volume plus the outflow of total enthalpy is equal to the rate of heat addition to the fluid in the control volume minus the rate at which viscous forces on the control surface do work on the environment.

Note that, if the flow were *adiabatic*, *i.e.* the heat flux $\mathbf{q} = 0$, then the first term on the right of equation 8.29 would be zero. Also, if the flow were inviscid ($\tau = 0$), then the second term would be zero. In the case of inviscid, adiabatic flow, the energy equation 8.29 has zero on the right side, a simpler form. If the flow is steady, then the first term on the left side is zero.

We have previously derived differential forms of mass conservation and momentum conservation equations, the latter for the special cases of inviscid flow (Euler equation 4.5) or incompressible viscous flow (incompressible Navier-Stokes equation 6.11). What is the corresponding differential form of the equation for conservation of energy? It turns out there is no simple path for deriving this equation because the viscous work term of equation 8.29 is not readily convertible to a volume integral. Instead, by applying equation 8.27 to an element of fluid volume and utilizing the Navier-Stokes equation, we arrive at the following differential form of the energy equation:[10]

$$\rho\frac{D\hat{u}}{Dt} - \frac{p}{\rho}\frac{D\rho}{Dt} = -\nabla\cdot\mathbf{q} + \Phi \tag{8.30}$$

where the last term of this equation, Φ, called the *dissipation function*, is proportional to the viscosity and is always positive or zero ($\Phi \geq 0$) because it contains the squares of sums of terms involving the spatial derivatives of $\mathbf{V}$. Note that the energy equation, like the mass conservation equation, is a scalar equation.

An alternate form of equation 8.30 may be obtained by replacing $\hat{u}$ by $\hat{h} - p/\rho$, as given in equation 8.28:

$$\rho\frac{D\hat{h}}{Dt} - \frac{Dp}{Dt} = -\nabla\cdot\mathbf{q} + \Phi \tag{8.31}$$

Example 8.4

In a journal bearing, such as that in example 6.4, the lubricating oil is locally a couette flow, as illustrated in figure 6.4. If the power dissipated in the bearing is removed by cooling the journal bearing material ($y \leq 0$), (a) calculate the heat flux q to the wall ($y = 0$) for the bearing described in example 6.4. (b) Calculate the temperature difference between the bearing surfaces if $\lambda = 4\ W/mK$.

Solution

(a) Referring to equation 8.31, and noting that there are no changes in the direction of flow, x, but only in the normal direction, y, of figure 6.4, the energy conservation

---

[10]See section 8.5 of this chapter for this derivation.

has the form:

$$0 = -\nabla \cdot \mathbf{q} + \Phi_i = -\frac{dq}{dy} + \Phi_i$$

which may be integrated, noting that $q = 0$ at the shaft surface $y = h$, to give:

$$\int_{q_w}^{0} dq = \int_0^h \Phi_i \, dy$$

$$-q_w = \int_0^h \mu \left(\frac{V}{h}\right)^2 dy$$

$$q_w = -\mu \frac{V^2}{h}$$

where we have used the Couette value of $\mu(V/h)^2$ for Φ_i. The negative value of q_w corresponds to heat being removed from the oil. Using the numerical values in example 6.4 of $\mu = 6.7E(-5)\,Pa\,s$, $V = \Omega D/2 = 18.85\,m/s$ and $h = 1E(-4)\,m$, q_w is:

$$q_w = -(6.7E(-5)\,Pa\,s)\frac{(18.85\,m/s)^2}{1E(-4)\,m} = -2.381E(2)\,W/m^2$$

(b) Noting that $q = -\lambda dT/dy$, energy conservation requires:

$$0 = \frac{d^2T}{dy^2} + \Phi_i$$

$$\frac{d^2T}{dy^2} = -\frac{\mu}{\lambda}\left(\frac{V}{h}\right)^2$$

Integrating twice on y with the boundary conditions that $q = \lambda dT/dy = 0$ at $y = h$ and $T = T_0$ at $y = 0$ gives:

$$T = -\frac{\mu}{2\lambda}\left(\frac{V}{h}\right)^2 y^2 + c_1 y + c_2 = T_0 + \frac{\mu}{\lambda}\left(\frac{V}{h}\right)^2 \left(y - \frac{y^2}{2h}\right)$$

The temperature difference $T_h - T_0$ across the oil gap is:

$$T_h - T_0 = \frac{\mu}{\lambda}\frac{V^2}{2h} = \frac{6.7E(-5)\,Pa\,s}{4\,W/mK} \times \frac{(18.85\,m/s)^2}{2[1E(-4)\,m]} = 29.98\,K$$

Thus far we have derived four alternative forms for expressing the conservation of energy—equations 8.27, 8.29, 8.30 and 8.31. Which form is the best to use? The answer to that question depends upon the kind of flow we are analyzing. In what follows, we consider several common kinds of flows and the the corresponding energy conservation expression that is easiest to use.

8.3.1 Steady Compressible Flow through a Duct or Chamber

In compressible flow, there can be a close coupling of the transport of heat with the change in density so that there is a linkage between the conservation of mass, momentum and energy. Consequently, we use equation 8.29 for the most general case of compressible flow. For steady flow, it has the form:

$$\iint_S \rho\left(\hat{h} + \frac{V^2}{2} - \mathbf{g}\cdot\mathbf{R}\right)\mathbf{V}\cdot\mathbf{n}\,dS = \iint_S \mathbf{q}(-\mathbf{n}\,dS + \iint_S \boldsymbol{\tau}\cdot\mathbf{V}\,dS \qquad (8.32)$$

If we apply this to the case of the flow of a compressible fluid in a duct or heat exchanger, neglecting the term $\boldsymbol{\tau}\cdot\mathbf{V}$ as compared to $\rho\hat{h}\mathbf{V}\cdot\mathbf{n}$ at the inlet and outlet surfaces of the control volume, then we find the compressible version of equation 8.9:

$$\iint_{out} \rho\left(\hat{h} + \frac{V^2}{2} - \mathbf{g}\cdot\mathbf{R}\right)\mathbf{V}\cdot\mathbf{n}\,dS - \iint_{in} \rho\left(\hat{h} + \frac{V^2}{2} - \mathbf{g}\cdot\mathbf{R}\right)(-\mathbf{V}\cdot\mathbf{n})\,dS$$
$$= \iint_S \mathbf{q}\cdot(-\mathbf{n})\,dS \qquad (8.33)$$

The quantity $\hat{h}+V^2/2-\mathbf{g}\cdot\mathbf{R}$ is called the *total enthalpy*.[11] It is the value of the enthalpy that the fluid would have if it were brought to rest in equilibrium with stationery fluid at $\mathbf{g}\cdot\mathbf{R} = 0$. Equation 8.33 reveals that the increase in the integrated flux of total enthalpy of the flow passing through the control volume is equal to the rate of heat addition through the control surface. In contrast to the incompressible case (equation 8.9), the change in total enthalpy flux is affected only by heat transfer, not viscosity.

We can find a differential form of the energy conservation equation 8.32 for the case of steady, adiabatic (λ, $\mathbf{q} = 0$), inviscid (μ, $\boldsymbol{\tau} = 0$) flow by applying the divergence theorem and steady mass conservation. Equation 8.32, with these restrictions, has the form:

$$\iint_S \rho\left(\hat{h} + \frac{V^2}{2} - \mathbf{g}\cdot\mathbf{R}\right)\mathbf{V}\cdot\mathbf{n}\,dS = 0$$

Applying the divergence theorem (equation 1.44), equation 1.38 and the conservation

[11]In most compressible flows, the change in $\mathbf{g}\cdot\mathbf{R}$ between the inlet and outlet stream is much smaller than the change in $\hat{h} + V^2/2$, and the term $\mathbf{g}\cdot\mathbf{R}$ may be dropped from the energy conservation equation as insignificant. The sum $\hat{h} + V^2/2$ is called the *stagnation enthalpy*.

of mass for a steady flow,

$$\iiint_{\mathcal{V}} \nabla \cdot \left[\rho\left(\hat{h} + \frac{V^2}{2} - \mathbf{g}\cdot\mathbf{R}\right)\mathbf{V}\right] d\mathcal{V} = 0$$

$$\iiint_{\mathcal{V}} \left[\left(\hat{h} + \frac{V^2}{2} - \mathbf{g}\cdot\mathbf{R}\right)\nabla\cdot(\rho\mathbf{V}) + \rho\mathbf{V}\cdot\nabla\left(\hat{h} + \frac{V^2}{2} - \mathbf{g}\cdot\mathbf{R}\right)\right] d\mathcal{V} = 0$$

$$\mathbf{V}\cdot\nabla\left(\hat{h} + \frac{V^2}{2} - \mathbf{g}\cdot\mathbf{R}\right) = 0 \qquad (\textit{if } \mu, \lambda, \frac{\partial}{\partial t} = 0) \tag{8.34}$$

The total enthalpy is thus conserved along a streamline in a steady, adiabatic, inviscid flow. This is the compressible flow equivalent to Bernoulli's equation.

In unsteady flows encountered in compressors and turbines, the flow relative to the rotor can be considered steady, as described below in section 8.4.1. In reciprocating compressors and engines, the time dependence of all terms of equation 8.29 can be substantial.

8.3.2 Flows with Heat Transfer

There are many practical flows in which a fluid is heated or cooled significantly. Indeed, it may be the purpose of the flow to exchange heat, as in the radiator of an automobile or the cooling coil of a refrigerator. In such flows, the expression of the conservation of energy may be simplified because some terms can be disregarded.

We begin by determining the conditions for which the dissipation term Φ in equation 8.30 or 8.31 is small compared to the divergence of the heat flux, $\nabla\cdot\mathbf{q}$. In order of magnitude, $\Phi \simeq \mu(V/L)^2$, where V is the characteristic velocity of the flow and L is its characteristic length scale. If ΔT is the characteristic temperature change in the flow, then by Fourier's Law (8.24), $q \simeq \lambda\Delta T/L$ and $\nabla\cdot\mathbf{q} \simeq \lambda\Delta T/L^2$. For the heat conduction term to dominate,

$$\lambda\Delta T/L^2 \gg \mu(V/L)^2$$

$$\frac{c_p\Delta T}{V^2} \gg \frac{c_p\mu}{\lambda} \tag{8.35}$$

where we have multiplied by the constant-pressure specific heat c_p to obtain the dimensionless form in the second line.

The term on the right side of equation 8.35, $c_p\mu/\lambda$, is called the *Prandtl number* and is a dimensionless ratio of fluid properties. For gases and many liquids, the Prandtl number is of order unity, but it can be quite small for liquid metals. In any case, it is sufficient for the ratio $c_p\Delta T/V^2$ to be very large compared to unity to justify the neglect of Φ in equation 8.30 or 8.31. For example, a typical flow velocity in a heat exchanger might be $1\,m/s$, and a typical specific heat is $1E(4)\,J/kg\,K$ so that equation 8.35 would be satisfied if $\Delta T \gg 1E(-4)\,K$. However, this condition cannot be satisfied for compressible flows when V^2 is the same order as $c_p\Delta T$.

A similar argument can be used to show that $\rho D\hat{h}/Dt \gg Dp/Dt$ in a heat transfer flow. In order of magnitude, $d\hat{h} \simeq d(c_p\Delta T)$ and $dp \simeq d(\rho V^2)$ so that:

$$\frac{\rho D\hat{h}/Dt}{Dp/Dt} \simeq \frac{c_p\Delta T}{V^2}.$$

Thus both Φ and Dp/DT may be neglected in equation 8.31, and the conservation of energy assumes the form:

$$\rho\frac{D\hat{h}}{Dt} = \nabla\cdot(\lambda\nabla T) \qquad (if\ c_p\Delta T \gg V^2) \tag{8.36}$$

Whenever the temperature change is small enough that λ and c_p do not vary much at all over the temperature range ΔT, then equation 8.36 may be simplified to:

$$\frac{DT}{Dt} = \frac{\lambda}{\rho c_p}\nabla^2 T \qquad (if\ \lambda,\ c_p = const.,\ \ c_p\Delta T \gg V^2) \tag{8.37}$$

The ratio $\lambda/\rho c_p$ is called the *thermal diffusivity* and is usually denoted by the symbol κ. It has the same dimensions as the viscous diffusivity ν and, for gases, about the same magnitude.[12]

In most heat transfer applications, the velocity field is first determined from the solution to the Navier-Stokes equation, after which the conservation of energy equation 8.37 can be solved for the temperature field. However, in natural convection problems, where the flow is affected by the buoyant force on less dense (heated) portions of the fluid, the momentum and energy conservation equations are closely coupled and cannot be treated separately. Such flows are more difficult to treat than forced convection flows where gravity forces are small compared with other forces or inertia.

Finally, we may obtain an integral form of the conservation of energy for steady forced convection flows with heat transfer by using equation 8.33, noting that the change in $\mathbf{g}\cdot\mathbf{R} \simeq gL$ is usually small compared to $c_p\Delta T$:

$$\int_{out} \rho\hat{h}\mathbf{V}\cdot\mathbf{n}\,dS - \int_{in} \rho\hat{h}(-\mathbf{V}\cdot\mathbf{n})\,dS = \iint_S \mathbf{q}\cdot(-\mathbf{n})\,dS$$

$$(if\ c_p\Delta T \gg V^2,\ gL;\ \frac{\partial}{\partial t} = 0) \tag{8.38}$$

[12]Note that ν/κ is equal to the Prandtl number.

Example 8.5

Water flowing through a pipe at a rate of $Q = 5\,gal/min$ is heated by an electric resistance heater at a rate of $10\,kW$. Calculate the temperature rise in the water, whose specific heat is $c_p = 4.18\,J/kg\,K$.

Solution

Applying equation 8.38 to the flow,

$$\rho Q c_p (T_{out} - T_{in}) = \iint_S \mathbf{q} \cdot (-\mathbf{n})\, dS$$

$$[1E(3)\,kg/m^3]\left(\frac{5\,gal}{min} \times \frac{min}{60\,s} \times \frac{3.785E(-3)\,m^3}{gal}\right)[4.18E(3)\,J/kg\,K](T_{out} - T_{in})$$

$$= 1E(4)\,W$$

$$(T_{out} - T_{in}) = 7.585\,K$$

8.4 The Second Law of Thermodynamics

The second law of thermodynamics defines two thermodynamic properties, absolute temperature T and entropy S. These are related to the heat Q by the inequality of Clausius,

$$\frac{DS}{Dt} \geq \frac{1}{T}\frac{DQ}{Dt} \tag{8.39}$$

which states that the entropy of a fluid element can only be decreased by cooling but can increase even in the absence of heat exchange. The limiting case of the equality of equation 8.39 is called a thermodynamically reversible flow. It is an idealization of a real flow, comparable to incompressible or inviscid flow, which is never exactly realized but may be closely approximated in many practical cases. For this reason, we want to express the second law in the more usual Eulerian form.

The integral form of equation 8.39 is found by applying Reynolds' transport theorem, equation 5.5, to a fixed control volume, obtaining:

$$\frac{d}{dt}\iiint_{\mathcal{V}} \rho \hat{s}\, d\mathcal{V} + \iint_S \rho \hat{s} \mathbf{V} \cdot \mathbf{n}\, dS \geq \iint_S \left(\frac{\mathbf{q} \cdot (-\mathbf{n})}{T}\right) dS \tag{8.40}$$

where $\hat{s}$ is the entropy per unit mass of fluid. In a steady adiabatic flow, where $\mathbf{q} \cdot \mathbf{n} = 0$, the second law requires that the entropy flux out of the control volume must equal, or exceed, that into the control volume; the entropy production within the control volume is never negative for this case.

By applying the divergence theorem and equation 5.4 to equation 8.40, we find the differential form of the second law:

$$\rho \frac{D\hat{s}}{Dt} \geq -\nabla \cdot \left(\frac{\mathbf{q}}{T}\right) \tag{8.41}$$

For an adiabatic flow ($\mathbf{q} = 0$), the entropy $\hat{s}$ of a fluid particle will either increase or remain the same, but it will never decrease.

The entropy change of a particle is related to the changes in internal energy and density by equation 1.9:

$$T\frac{D\hat{s}}{Dt} = \frac{D\hat{u}}{Dt} - \frac{p}{\rho^2}\frac{D\rho}{Dt} \tag{8.42}$$

However, the right side of equation 8.42 is related to $\mathbf{q}$ and Φ by the first law equation 8.30 so that equation 8.42 becomes:

$$\begin{aligned} \rho \frac{D\hat{s}}{Dt} &= \frac{\Phi}{T} - \frac{1}{T}\nabla \cdot \mathbf{q} \\ &= \frac{\Phi_T}{T} - \nabla \cdot \left(\frac{\mathbf{q}}{T}\right) \end{aligned} \tag{8.43}$$

where the function Φ_T is defined as:

$$\Phi_T \equiv \Phi - \frac{\mathbf{q} \cdot \nabla T}{T} \geq 0 \tag{8.44}$$

and is never negative because Φ and $-\mathbf{q} \cdot \nabla T$ are each never negative.[13]

If we compare equations 8.43 and 8.41, we can see that the source of the inequality in 8.41 and the irreversibility in the flow is the term Φ_T/T, which has contributions from both viscous dissipation (Φ) and heat conduction ($-\mathbf{q} \cdot \nabla T/T$). If the velocity and temperature gradients are small enough, Φ_T/T may be negligible compared to the other terms in the equation and the flow may be considered to be thermodynamically reversible.

By integrating equations 8.43 over the control volume, we find its integral form:

$$\frac{d}{dt}\iiint_{\mathcal{V}} \rho\hat{s}\, d\mathcal{V} + \iint_S \rho\hat{s}\mathbf{V}\cdot\mathbf{n}\, dS = \iint_S \left(\frac{\mathbf{q}\cdot(-\mathbf{n})}{T}\right) dS + \iiint_{\mathcal{V}} \left(\frac{\Phi_T}{T}\right) d\mathcal{V} \tag{8.45}$$

The first term on the right of equation 8.45 is the contribution of heat flow across the control surface to the entropy creation within the control volume, and the second term is the irreversible entropy production within the control volume.

[13]By Fourier's Law, equation 8.24, $\mathbf{q}$ has the opposite direction of ∇T so that $-\mathbf{q} \cdot \nabla T$ is positive.

8.4.1 Compressible Flow in Steadily Rotating Machines

In section 8.2.3, we considered incompressible flow in a steadily rotating machine such as a pump or propeller. For compresible flow in similar circumstances, which would be encountered in gas turbines, for example, we again consider the steady flow relative to the rotor. In this noninertial reference frame, the conservation of energy equation 8.33 has the form:

$$\iint_{out}\left(\hat{h}+\frac{\tilde{V}^2}{2}-\mathbf{g}\cdot\tilde{\mathbf{R}}-\frac{(\Omega\tilde{r})^2}{2}\right)(\rho\tilde{\mathbf{V}}\cdot\mathbf{n})\,dS$$
$$-\iint_{in}\left(\hat{h}+\frac{\tilde{V}^2}{2}-\mathbf{g}\cdot\tilde{\mathbf{R}}-\frac{(\Omega\tilde{r})^2}{2}\right)(-\rho\tilde{\mathbf{V}}\cdot\mathbf{n})\,dS=\iint_S\mathbf{q}\cdot(-\mathbf{n})\,dS \tag{8.46}$$

which is the compressible equivalent of equation 8.12. Following the same steps as for the previous section on incompressible flow, we arrive at the equivalent of equation 8.16:[14]

$$\dot{m}\left(\hat{h}+\frac{\tilde{V}^2}{2}\right)_{out}-\dot{m}\left(\hat{h}+\frac{\tilde{V}^2}{2}\right)_{in}=\mathbf{\Omega}\cdot\mathbf{T}+\iint_S\mathbf{q}\cdot(-\mathbf{n})\,dS \tag{8.47}$$

Note that this is not simply the incompressible energy equation 8.16 with p/ρ replaced by $\hat{h}$. There is no viscous dissipation term indicating a flow irreversibility through the machine nor any obvious way to define an efficiency which takes into account the viscous effects.

We must resort to the second law of thermodynamics to rewrite equation 8.47 in a form that explicitly acknowledges the effects of viscous dissipation. For a globally adiabatic flow ($\mathbf{q} = 0$ on the control surface), which is generally the case for compressors and turbines, the steady flow version of the second law equation 8.45 is:

$$\dot{m}\hat{s}_{out}-\dot{m}\hat{s}_{in}=\iiint_{\mathcal{V}}\left(\frac{\Phi_T}{T}\right)d\mathcal{V} \qquad \textit{if}\ \mathbf{q}\cdot(-\mathbf{n})\,dS=0 \tag{8.48}$$

We can conclude that the outflow entropy $\hat{s}_{out}$ is never less than the inflow entropy $\hat{s}_{in}$ because $\Phi_T \geq 0$. The entropy increase is the fluid property that measures the debilitating effects of viscous dissipation and internal heat conduction in a globally adiabatic compressible flow.

It is possible to introduce the entropy change implicitly into the compressible energy equation 8.47 by defining an isentropic outflow enthalpy $\hat{h}_s$ as the enthalpy of

---

[14] In equation 8.47, we have dropped the gravity term $\mathbf{g}\cdot\mathbf{R}$ since it is invariably small (compared to other terms) in compressible flows through turbomachines.

the fluid at the *outlet* pressure and the *inlet* entropy:

$$\hat{h}_s \equiv \hat{h}\{p_{out}, \hat{s}_{in}\} \le \hat{h}_{out} \tag{8.49}$$

where the inequality follows directly from the thermodynamic relation that $(\partial\hat{h}/\partial\hat{s})_p = T$ is always positive. Substituting in equation 8.47, we arrive at the adiabatic compressible flow equivalent of equation 8.17:

$$\dot{m}\left(\hat{h}_s + \frac{\tilde{V}^2}{2}\right)_{out} - \dot{m}\left(\hat{h} + \frac{\tilde{V}^2}{2}\right)_{in} = \mathbf{\Omega}\cdot\mathbf{T} - \dot{m}(\hat{h}_{out} - \hat{h}_s) \tag{8.50}$$

where the last term, which is always positive, is the analog of the dissipation integral of equation 8.17.

Continuing with the analogy to the incompressible flow, we define *adiabatic* compressor and turbine efficiencies by:

$$\eta_c \equiv \frac{1}{(\Omega T)_c}\left[\dot{m}\left(\hat{h}_s + \frac{\tilde{V}^2}{2}\right)_{out} - \dot{m}\left(\hat{h} + \frac{\tilde{V}^2}{2}\right)_{in}\right] \le 1$$

$$\eta_t \equiv (\Omega T)_t \left[\dot{m}\left(\hat{h} + \frac{\tilde{V}^2}{2}\right)_{in} - \dot{m}\left(\hat{h}_s + \frac{\tilde{V}^2}{2}\right)_{out}\right]^{-1} \le 1 \tag{8.51}$$

and the compressor and turbine powers can then be given in terms of the component efficiencies:

$$(\Omega T)_c = \frac{1}{\eta_c}\left[\dot{m}\left(\hat{h}_s + \frac{\tilde{V}^2}{2}\right)_{out} - \dot{m}\left(\hat{h} + \frac{\tilde{V}^2}{2}\right)_{in}\right]$$

$$(\Omega T)_t = \eta_t\left[\dot{m}\left(\hat{h} + \frac{\tilde{V}^2}{2}\right)_{in} - \dot{m}\left(\hat{h}_s + \frac{\tilde{V}^2}{2}\right)_{out}\right] \tag{8.52}$$

8.5 Derivation of the Differential Form of the First Law

In this section, we derive the differential form of the first law, equation 8.30, for any flow, whether compressible or incompressible. To do so, we will have to utilize the relationships between the viscous stress $\boldsymbol{\tau}$ and the velocity derivatives previously used in deriving the Navier-Stokes equation.

If we apply equation 5.4 to the left side of the first law equation 8.27 and also apply the divergence theorem to the heat flux and pressure terms on the right, we obtain:

$$\iiint_{\mathcal{V}} \rho\frac{D}{Dt}\left(\hat{u} + \frac{V^2}{2} - \mathbf{g}\cdot\mathbf{R}\right) d\mathcal{V} = -\iiint_{\mathcal{V}} [\nabla\cdot\mathbf{q} + \nabla\cdot(p\mathbf{V})]\, d\mathcal{V} + \iint_{S} \boldsymbol{\tau}\cdot\mathbf{V}\, dS$$

If we now select as our control volume a unit volume of fluid, these integrals may be evaluated to obtain:

$$\rho \frac{D}{Dt}\left(\hat{u} + \frac{V^2}{2} - \mathbf{g}\cdot\mathbf{R}\right) = -\nabla\cdot\mathbf{q} - p\nabla\cdot\mathbf{V} - \mathbf{V}\cdot\nabla p + \psi \tag{8.53}$$

where the viscous work term ψ, expressed in cartesian coordinates, is:

$$\begin{aligned}\psi \equiv\; & \frac{\partial}{\partial x}(u\tau_{xx}) + \frac{\partial}{\partial y}(u\tau_{xy}) + \frac{\partial}{\partial z}(u\tau_{xz}) \\ & + \frac{\partial}{\partial x}(v\tau_{xy}) + \frac{\partial}{\partial y}(v\tau_{yy}) + \frac{\partial}{\partial z}(v\tau_{yz}) \\ & + \frac{\partial}{\partial x}(w\tau_{xz}) + \frac{\partial}{\partial y}(w\tau_{yz}) + \frac{\partial}{\partial z}(w\tau_{zz})\end{aligned} \tag{8.54}$$

The negative of the term ψ is the rate at which the viscous stress $\boldsymbol{\tau}$ acting on a unit volume of fluid does work on the surrounding environment.

Next we consider the equation of conservation of kinetic energy, equation 8.4, in the form:

$$\rho \frac{D}{Dt}\left(\frac{V^2}{2}\right) = -\mathbf{V}\cdot\nabla p + \rho\mathbf{V}\cdot\mathbf{g} + \mathbf{V}\cdot\mathbf{f} \tag{8.55}$$

The components of $\mathbf{f}$ are given in equation 6.7, and the term $\mathbf{V}\cdot\mathbf{f}$ may be expressed in cartesian coordinates as:

$$\begin{aligned}\mathbf{V}\cdot\mathbf{f} \equiv\; & u\left(\frac{\partial\tau_{xx}}{\partial x} + \frac{\partial\tau_{xy}}{\partial y} + \frac{\partial\tau_{xz}}{\partial z}\right) \\ & + v\left(\frac{\partial\tau_{xy}}{\partial x} + \frac{\partial\tau_{yy}}{\partial y} + \frac{\partial\tau_{yz}}{\partial z}\right) \\ & + w\left(\frac{\partial\tau_{xz}}{\partial x} + \frac{\partial\tau_{yz}}{\partial y} + \frac{\partial\tau_{zz}}{\partial z}\right)\end{aligned} \tag{8.56}$$

The term $\mathbf{V}\cdot\mathbf{f}$ is the rate at which the viscous force per unit volume is doing work on moving the fluid element through space.

Finally, we note that the material derivative of the gravity term $\mathbf{g}\cdot\mathbf{R}$ may be written as:

$$\begin{aligned}\rho \frac{D}{Dt}(-\mathbf{g}\cdot\mathbf{R}) &= \rho\mathbf{V}\cdot\nabla(-\mathbf{g}\cdot\mathbf{R}) \\ &= -\rho\mathbf{V}\cdot\mathbf{g}\end{aligned} \tag{8.57}$$

where we have made use of the time independence of $\mathbf{g}$ and equation 2.7.

By subtracting equations 6.5 and 8.57 from the first law equation 8.53 and noting that $p\nabla\cdot\mathbf{V} = -(p/\rho)D\rho/Dt$ by mass conservation, we find the following form of the first law:

$$\rho\frac{D\hat{u}}{Dt} - \frac{p}{\rho}\frac{D\rho}{Dt} = -\nabla \cdot \mathbf{q} + \Phi \tag{8.58}$$

where the dissipation function, Φ, is related to ψ and $\mathbf{V} \cdot \mathbf{f}$ by:

$$\Phi \equiv \psi - \mathbf{V} \cdot \mathbf{f} = \tau_{xx}\frac{\partial u}{\partial x} + \tau_{yy}\frac{\partial v}{\partial y} + \tau_{zz}\frac{\partial w}{\partial z} + \tau_{xy}\left(\frac{\partial v}{\partial x} + \frac{\partial u}{\partial y}\right) + \tau_{yz}\left(\frac{\partial w}{\partial y} + \frac{\partial v}{\partial z}\right) + \tau_{xz}\left(\frac{\partial w}{\partial x} + \frac{\partial u}{\partial z}\right) \tag{8.59}$$

If we now replace the viscous stress components by their appropriate values for a Newtonian fluid, equations 6.4 and 6.5, we find the following form for Φ, expressed in cartesian coordinates:

$$\Phi = \frac{2}{3}\mu\left[\left(\frac{\partial u}{\partial x} - \frac{\partial v}{\partial y}\right)^2 + \left(\frac{\partial v}{\partial y} - \frac{\partial w}{\partial z}\right)^2 + \left(\frac{\partial u}{\partial x} - \frac{\partial w}{\partial z}\right)^2\right] + \mu\left[\left(\frac{\partial v}{\partial x} + \frac{\partial u}{\partial y}\right)^2 + \left(\frac{\partial w}{\partial y} + \frac{\partial v}{\partial z}\right)^2 + \left(\frac{\partial w}{\partial x} + \frac{\partial u}{\partial z}\right)^2\right] + \mu_B(\nabla \cdot \mathbf{V})^2 \tag{8.60}$$

where μ_B is the bulk viscosity.

Note that Φ is always positive or zero since it is the sum of squared terms. If U is the magnitude of the velocity of a flow and L is the magnitude of the length over which the velocity changes, then the magnitude of Φ is $\mu(U/L)^2$.

For an incompressible fluid, $\nabla \cdot \mathbf{V} = 0$ and Φ assumes a somewhat simpler form, denoted by Φ_i:

$$\Phi_i = 2\mu\left[\left(\frac{\partial u}{\partial x}\right)^2 + \left(\frac{\partial v}{\partial y}\right)^2 + \left(\frac{\partial w}{\partial z}\right)^2\right] + \mu\left[\left(\frac{\partial v}{\partial x} + \frac{\partial u}{\partial y}\right)^2 + \left(\frac{\partial w}{\partial y} + \frac{\partial v}{\partial z}\right)^2 + \left(\frac{\partial w}{\partial x} + \frac{\partial u}{\partial z}\right)^2\right] \quad \text{if} \quad \nabla \cdot \mathbf{V} = 0 \tag{8.61}$$

8.6 Problems

Problem 8.1

For the flow through an hydraulic jump, sketched in figure 5.10, derive an expression for the loss of head, $h_{in} - h_{out}$, along the surface streamline in terms of h_{in} and h_{out}. Explain why $h_{out} \geq h_{in}$ in the hydraulic jump.

Problem 8.2

Air enters a centrifugal compressor at a speed $V_{in} = 100\,m/s$ through an inlet of diameter $D_{in} = 15\,cm$. The air is drawn from the ambient atmosphere at a density $\rho_a = 1.204 kg/m^3$ and temperature $20°\,C$. The air leaves the compressor at a temperature of $300°\,C$ and a speed $V_{out} = 80\,m/s$. In this temperature range, the enthalpy $\hat{h} = \hat{c}_p T$, where the specific heat $\hat{c}_p = 1.005E(3)\,J/kg\,K$.

(a) Assuming inviscid flow from the atmosphere to the compressor inlet, calculate the value of $(\hat{h} + V^2/2g)_{out} - (\hat{h} + V^2/2g)_{in}$. (b) Assuming that the inflow density is the same as that of the atmosphere, calculate the mass flow rate $\dot{m}$ through the compressor. (c) Calculate the power $\mathcal{P}$ required to run the compressor.

Problem 8.3

For the turbulent jet of figure 7.6, derive an expression for the kinetic energy flux $\iint (V^2/2)(\rho\mathbf{V} \cdot \mathbf{n})\,dS$ at any station x in terms of the parameters x, R_s, V_s and α_j. Does this expression agree with the inequality of equation 8.11 for the decline of the head flux in a viscous flow?

Problem 8.4

A water pump with equal inflow and outflow areas increases the water pressure from inlet to outlet by $100\,psi$ when the flow rate is $50\,gal/min$. If the pump efficiency $\eta_p = 78\%$, calculate the power $\mathcal{P}$ required to operate the pump under these flow conditions.

Problem 8.5

A perfect gas ($\hat{c}_p = 1E(3)\,J/kg\,K$, $\hat{h} = \hat{c}_p T$) flows adiabatically ($\mathbf{q} = 0$) through a long pipe of constant diameter. The upstream temperature $T_1 = 300\,K$, pressure $p_1 = 1E(6)\,Pa$ and velocity $V_1 = 10\,m/s$, and the downstream pressure $p_2 = 1E(5)\,Pa$. Calculate the downstream temperature T_2 and velocity V_2.

Bibliography

Sabersky, Rolf H., Allan J. Acosta, and Edward G. Hauptman. 1989. *Fluid Flow, A First Course in Fluid Mechanics.* 3rd ed. New York: Macmillan Publishing Co.

Batchelor, G. K. 1967. *An Introduction to Fluid Mechanics.* Cambridge: Cambridge University Press.

Howarth, L., ed. 1953. *Modern Developments in Fluid Dynamics. Vol. 1, High Speed Flows.* Oxford: Oxford University Press.

9 Flow in Fluid Systems

9.1 Introduction

We are surrounded by examples of systems that convey fluids inside pipes and ducts for various uses. A home has a domestic water system that supplies water for drinking, cooking, washing, and operating toilets. It also has a house drain system for collecting waste water and sending it to a treatment system, either an individual septic system or a community sewage treatment plant. Cities and most towns have potable water supply systems that pump water from wells or reservoirs and distribute the water to residences, businesses, etc. These systems must also provide emergency water supplies for fighting fires. Natural gas fuel is supplied to homes, factories and power plants by a national network connecting thousands of wells to the consumers of gaseous fuel. Fuel oil is pumped from wells to storage tanks and then pipelined to a refinery for processing. Nearly every building has a heating system (and many have a cooling system as well) that moves heat from a central furnace to peripheral areas by means of a stream of hot air, water or steam. Similar systems carry heat or cold in transportation vehicles. In steam power plants, steam and water flow through a cycle that converts some of the heat from the fuel to mechanical work. As you read this paragraph, your heart is pumping blood throughout a very complex system that supplies the cells in your body with oxygen and nutrients while simultaneously removing waste products. Clearly, both we and our society depend upon the successful functioning of many fluid systems for our physical support.

This chapter is concerned with the problem of making fluids move through systems in a desired way. We know from our previous study that energy must be expended to make fluids move through pipes and ducts, so we also need to determine how much power must be supplied to make our fluid system function the way we want it to. In recent years, the efficiency of energy utilization has become of increasing importance, affecting the design and operation of many engineering systems. We will study in this chapter the methods for analyzing the efficient flow of fluids in systems.

9.2 Head Loss in Pipes and Ducts

In chapter 8, equation 8.10, we defined a quantity called *head*, used in incompressible flow and denoted by the symbol h:

$$\mathsf{h} \equiv \frac{p}{\rho g} + \frac{V^2}{2g} - \frac{\mathbf{g} \cdot \mathbf{R}}{g} = \frac{p^*}{\rho g} + \frac{V^2}{2g} \tag{9.1}$$

We found that the head, when averaged across the cross-section of a duct in a steady flow, decreases in the direction of flow because of viscous dissipation (equation 8.11).[1] Between the inlet and outlet from a pipe or duct, there will be a reduction in head, $\Delta \mathsf{h}_l$, which we call the *head loss*. In a pipe, the amount of this head loss depends upon the pipe size and length and the volume flow rate of fluid through the pipe. There will be additional head losses if the pipe is not straight but contains sharp bends or if the pipe contains valves or other restrictions. In this section, we will be concerned with the quantitative evaluation of these head losses.

9.2.1 Head Loss in Straight Pipes

In chapters 6 and 7, we learned how viscous shear on a pipe wall causes a pressure drop in the flow. For a straight section of pipe of length L and diameter D, the quantitative relationship between the pressure drop[2] from inlet to outlet, $p^*_{in} - p^*_{out}$, and the flow rate was expressed in equation 6.48 in terms of the Darcy friction factor f (defined in equation 6.47) and the average flow velocity $\bar{V}$ as:

$$p^*_{in} - p^*_{out} = f \left(\frac{1}{2} \rho \bar{V}^2 \right) \frac{L}{D} \tag{9.2}$$

[1]But, in pumps and turbines, the head can be increased or decreased, respectively, because the flow is unsteady within these devices and power is added or extracted (see equation 8.17).

[2]In this and following sections, we will call the sum, $p^* \equiv p - \rho \mathbf{g} \cdot \mathbf{R} = p + \rho g z$, the "pressure," although it is the sum of the static pressure p and what is sometimes called the hydrostatic pressure, $-\rho \mathbf{g} \cdot \mathbf{R} = \rho g z$. This simplifies the analysis of piping systems for which there is a change of elevation between inlet and outlet because only the change in p^*, not p, determines the flow rate.

Since there is no change in $\bar{V}$ between inlet and outlet, the loss in head, $\mathsf{h}_{in} - \mathsf{h}_{out}$, is simply the pressure difference $p^*_{in} - p^*_{out}$ divided by ρg:

$$\Delta \mathsf{h}_f \equiv \mathsf{h}_{in} - \mathsf{h}_{out} = \frac{p^*_{in} - p^*_{out}}{\rho g} = f\left(\frac{\bar{V}^2}{2g}\right)\frac{L}{D} \tag{9.3}$$

where we have used 9.2 to eliminate the pressure drop.

It is customary to express the head loss between two points in a flow in terms of a dimensionless *head loss coefficient* K:

$$K \equiv \frac{\Delta \mathsf{h}}{\bar{V}^2/2g} = \frac{2g(\Delta \mathsf{h})}{\bar{V}^2} \tag{9.4}$$

so that the head loss becomes:

$$\Delta \mathsf{h} = K\left(\frac{\bar{V}^2}{2g}\right) \tag{9.5}$$

By comparing equation 9.4 with 9.3, we can see that the head loss coefficient K_f in a pipe flow is simply:

$$K_f = f\left(\frac{L}{D}\right) \tag{9.6}$$

Thus, in pipe flow, the head loss coefficient is proportional to the friction factor and the pipe length-to-diameter ratio.

The pipe friction factor f is a function of the diameter Reynolds number $Re_D \equiv \bar{V}D/\nu$ and, if the flow is turbulent, the wall roughness height ratio, ε/D. For laminar and turbulent flow, these relationships are, respectively, given in equations 6.49 and 7.17:

$$f = \frac{64}{Re_D} \qquad \text{if } Re_D < 2300$$

$$\frac{1}{\sqrt{f}} = -2.0\log\left(\frac{\varepsilon/D}{3.7} + \frac{2.51}{Re_D\sqrt{f}}\right) \qquad \text{if } Re_D > 2300 \tag{9.7}$$

In the scheme of equation 9.7, the Darcy friction factor f increases discontinuously at $Re_D = 2300$ from $2.783E(-2)$ for laminar flow to $4.728E(-2)$ for turbulent flow in a smooth pipe. Actually, the increase in friction factor is spread over a range in Reynolds number near 2300. This numerical discontinuity seldom causes trouble in pipe flow calculations.

The friction factor f as a function of the diameter Reynolds number $\bar{V}D/\nu$ and the wall roughness ratio ϵ/D is plotted from equations 9.7 in figure 9.1, using a logarithmic scale for both axes. This plot is called a *Moody diagram* and was commonly used for graphic solutions of 9.7 before the advent of digital computers. It is convenient for estimating approximate values of the relevant variables.

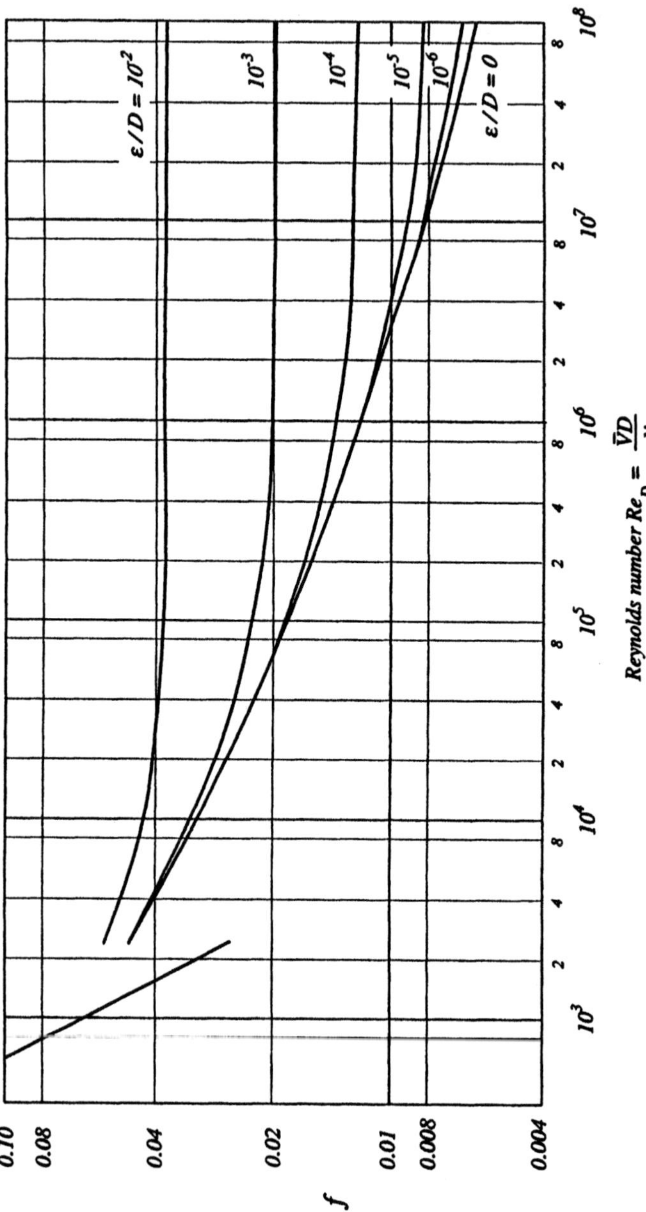

Figure 9.1 The Darcy friction factor f as a function of diameter Reynolds number Re_D and pipe wall roughness ratio ϵ/D for laminar ($Re_D < 2300$) and turbulent ($Re_D > 2300$) flow in circular pipes.

Head Loss for a Given Flow Rate

If the volumetric flow rate Q through a given pipe is known, it is a straightforward task to calculate the head loss and pressure drop by following these steps:

1. Calculate the Reynolds number $Re_D = \bar{V}D/\nu = 4Q/\pi D\nu$.
2. Depending upon whether the Reynolds number is greater or less than 2300, calculate f from the turbulent or laminar form of equation 9.7.
3. Using this value of f, calculate the loss coefficient, head loss and pressure drop from equations 9.6, 9.5 and 9.2.

Example 9.1

An 8 *in* diameter commercial steel pipe carries water from a storage tank outlet (elevation 100 *m*) to a town water distribution system (elevation 22 *m*) at a distance $L = 2.2\,km$ from the storage tank. If the flow rate is 1,000 gallons per minute, calculate (a) the head loss coefficient, (b) the head loss and (c) the *static* pressure change $p_{in} - p_{out}$ between the pipe inlet and outlet.

Solution

(a) First calculate D, $\bar{V}$ and Re_D using data from tables 1.1 and 1.6:

$$D = 8\,in \times \frac{2.54E(-2)\,m}{in} = 0.2032\,m$$

$$Q = \frac{1{,}000\,gal}{min} \times \frac{min}{60\,s} \times \frac{3.782E(-3)\,m^3}{gal} = 6.303E(-2)\,m^3/s$$

$$\bar{V} = \frac{4Q}{\pi D^2} = \frac{4(6.303E(-2)\,m^3/s)}{\pi(0.2032\,m)^2} = 1.944\,m/s$$

$$Re_D = \frac{\bar{V}D}{\nu} = \frac{(1.944\,m/s)(0.2032\,m)}{1.0E(-6)\,m^2/s} = 3.950E(5)$$

The flow is turbulent since Re_D exceeds 2300. Selecting $\varepsilon = 5E(-5)\,m$ from table 7.1, f must be found from the solution to equation 9.7 for turbulent flow:

$$\frac{1}{\sqrt{f}} = -2.0\log\left(\frac{5E(-5)\,m}{3.7(0.2032\,m)} + \frac{2.51}{3.950E(5)\sqrt{f}}\right)$$

$$= -2.0\log\left(6.650E(-5) + \frac{6.354E(-6)}{\sqrt{f}}\right)$$

Begin by assuming $f = 1E(-2)$ on the right side of this equation, and then calculate $f = 1.656E(-2)$ by evaluating the right side. After three more iterations, we find $f = 1.616E(-2)$. The head loss coefficient from equation 9.6 is:

$$K_f = f\left(\frac{L}{D}\right) = 1.616E(-2)\left(\frac{2,000\,m}{0.2032\,m}\right) = 1.590E(2)$$

(b) The head loss is calculated from equation 9.5:

$$\Delta h_f = K_f\left(\frac{\bar{V}^2}{2g}\right) = 1.590E(2)\left(\frac{(1.944\,m)^2}{2(9.807\,m/s^2)}\right) = 30.64\,m$$

(c) Noting that $p^*_{in} - p^*_{out} = p_{in} - p_{out} + \rho g(z_{in} - z_{out})$,

$$p_{in} - p_{out} = p^*_{in} - p^*_{out} - \rho g(z_{in} - z_{out}) = [(h_{in} - h_{out}) - (z_{in} - z_{out})]\rho g$$

$$= [30.64\,m - (100 - 22)\,m](1E(3)\,kg/m^3)(9.807\,m/s^2) = -4.645E(5)\,Pa$$

Note that there is a static pressure increase despite the head loss because the exit is at a sufficiently lower elevation that $z_{in} - z_{out}$ exceeds $h_{in} - h_{out}$.

Flow Rate for a Given Head Loss

In many cases of pipe flow, the flow rate through the pipe is not controlled but grows to a level that matches the pressure drop available. For example, when a sink water tap is opened wide, the flow rate increases until the pressure drop in the pipe equals the pressure difference between the water supply and atmospheric pressure.

The procedure for calculating the flow rate Q given the head loss $\Delta h_f = h_{in} - h_{out}$ available to overcome the wall friction depends upon whether the flow is laminar or turbulent. If the flow is laminar, an expression for $\bar{V}$ in terms of known quantities may be derived by eliminating f between equations 9.7 and 9.3:

$$\bar{V} = \frac{g(\Delta h_f)D^2}{32\,\nu L} \tag{9.8}$$

On the other hand, if the flow is turbulent, we first determine the value of $\sqrt{f}Re_D$ using equation 9.3 and $Re_D \equiv \bar{V}D/\nu$ to find:

$$\sqrt{f}Re_D = \left(\frac{2g(\Delta h_f)D^3}{\nu^2 L}\right)^{1/2} \tag{9.9}$$

and, using this value, solve for Re_D from equation 9.7 in the form:

$$Re_D = -2.0(\sqrt{f}Re_D)\log\left(\frac{\varepsilon/D}{3.7} + \frac{2.51}{\sqrt{f}Re_D}\right) \tag{9.10}$$

from which $\bar{V}$ may be determined.

The procedure for calculating the flow rate Q follows these steps:

1. Assuming laminar flow, calculate $\bar{V}$ from equation 9.8 and $Re_D = \bar{V}D/\nu$.
2. If $Re_D < 2300$, the flow is laminar, and $Q = \pi D^2 \bar{V}/4$.
3. If the flow is not laminar, calculate $\sqrt{f}Re_D$ from equation 9.9 and Re_D from 9.10.
4. Calculate $Q = \pi D^2 \bar{V}/4 = \pi D \nu Re_D/4$.

Example 9.2

For the pipe flow system of example 9.1, calculate the volume flow rate Q when the static pressure drop $p_{in} - p_{out}$ is zero.

Solution

If $p_{in} - p_{out} = 0$, then $p^*_{in} - p^*_{out} = p_{in} - p_{out} + \rho g(z_{in} - z_{out}) = \rho g(z_{in} - z_{out})$, and the head loss $\Delta \mathsf{h}_f = \mathsf{h}_{in} - \mathsf{h}_{out} = (p^*_{in} - p^*_{out})/\rho g = z_{in} - z_{out} = 100\,m - 22\,m = 78\,m$. It is very likely that the flow is turbulent, so we calculate $\sqrt{f}Re_D$ from 9.9:

$$\sqrt{f}Re_D = \left(\frac{2(9.807\,m/s^2)(78\,m)(0.2032\,m)^3}{[1E(-6)\,m^2/s]^2(2.2E(3)\,m)}\right)^{1/2} = 7.638E(4)$$

and then solve for Re_d from 9.10:

$$Re_D = -2[7.638E(4)]\log\left(6.65E(-5) + \frac{2.51}{7.638E(4)}\right) = 6.115E(5)$$

confirming that the flow is turbulent. Now calculate Q:

$$Q = \frac{\pi D \nu Re_D}{4} = \frac{\pi(0.2032\,m)(1E(-6)\,m^2/s)[6.115E(5)]}{4} = 9.759E(-2)\,m^3/s$$

$$= 9.759E(-2)\,m^3/s \times \frac{gal}{3.782E(-3)\,m^3} \times \frac{60\,s}{min} = 1,548\,gal/min$$

Selecting a Pipe Diameter

Sometimes a design engineer must select the size of pipe needed to provide a given level of service. For example, the maximum flow rate Q may be specified, together with the maximum allowable pressure drop or head loss at this flow rate. The engineer must select the smallest pipe diameter that will allow this flow without exceeding the limiting head loss.

Pipes are manufactured in standard thicknesses and diameters. For a given diameter, a thicker pipe will withstand a higher pressure. The cost per unit length of pipe will increase with the amount of material, which is proportional to the diameter times the thickness. The designer will choose the smallest diameter that will meet the flow requirements and the lowest wall thickness that will withstand the highest pressure that might be encountered in the system, so as to minimize the cost of the piping.

Sometimes the pipe selection is governed by more complicated economic considerations. While the cost of a pipe is certainly an important factor, the cost of energy to pump the required flow through the pipe is also an item of expense. By selecting a larger pipe, a lower pressure drop will be experienced, and less power will be required. The savings in reduced energy expense may more than offset the increased cost of the larger diameter pipe. *Life cycle costing* is a method of economic/engineering analysis that takes into account the cost of constructing and operating an engineered system and selecting the components to accomplish the technical objectives while minimizing the total costs.

Example 9.3

A factory is to be supplied with city water from a large main whose gage pressure is 100 *psi*. A horizontal pipe line 100 *m* long, connecting the city main with the factory, must supply a maximum of 2,000 *gal/min* of water for fire fighting purposes, with the gage pressure at the end of the pipe not less than zero. Commercial steel pipe of even diameters (in inches) is available. What is the minimum diameter pipe that will meet these requirements?

Solution

The available pressure drop $p_{in} - p_{out}$ of 100 *psi* equals a head loss $\mathsf{h}_{in} - \mathsf{h}_{out}$ of:

$$\mathsf{h}_{in} - \mathsf{h}_{out} = \frac{p_{in} - p_{out}}{\rho g} = \frac{100\,psi}{(1E(3)\,kg/m^3)(9.807\,m/s^2)} \times \frac{6.895E(3)\,Pa}{psi} = 70.31\,m$$

Now begin by assuming that a 4 *in* pipe will do the job. Follow the steps of example 9.1 to find the pressure drop:

$$D = 4\,in \times \frac{2.54E(-2)\,m}{in} = 0.1016\,m$$

$$Q = \frac{2000\,gal}{min} \times \frac{min}{60\,s} \times 3.782E(-3)\,m^3 gal = 1.261E(-1)\,m^3/s$$

$$\bar{V} = \frac{4Q}{\pi D^2} = \frac{4(1.261E(-1)\,m^3/s)}{\pi(0.1016\,m)^2} = 15.55\,m/s$$

$$Re_D = \frac{\bar{V}D}{\nu} = \frac{(15.55\,m/s)(0.1016\,m)}{1E(-6)\,m^2/s} = 1.580E(6)$$

$$\frac{1}{\sqrt{f}} = -2.0\log\left(\frac{5E(-5)\,m}{3.7(0.1016\,m)} + \frac{2.51}{1.580E(6)\sqrt{f}}\right)$$

$$f = 1.697E(-2)$$

$$K_f = f\left(\frac{L}{D}\right) = 1.697E(-2)\left(\frac{100\,m}{0.1016\,m}\right) = 16.70$$

$$\mathsf{h}_{in} - \mathsf{h}_{out} = K_f\left(\frac{\bar{V}^2}{2g}\right) = 16.70\left(\frac{(15.55\,m/s)^2}{2(9.807\,m/s^2)}\right) = 205.9\,m$$

The head loss of 205.9 m is greater than the allowable loss of 70.31 m. If we try the next size pipe, $D = 6\,in$, and repeat the process we will find that $\bar{V} = 6.913\,m/s$, $Re_D = 1.053E(6)$, $f = 1.530E(-2)$, $K_f = 10.04$ and $\mathsf{h}_{in} - \mathsf{h}_{out} = 24.46$. This is smaller than the allowable head loss so that a 6 in diameter pipe is acceptable.

9.2.2 Head Loss in Straight Noncircular Ducts

Noncircular ducts are commonly used in heating, ventilating and air conditioning systems because of ease of manufacture, assembly and fitting to construction space in walls and ceilings. There is no problem in withstanding the internal pressure of these systems, which is only slightly above atmospheric pressure. However, to select properly the fans needed to deliver air to all parts of the system we first need to know how to determine the pressure drop in noncircular ducts.

In treating laminar viscous flow in noncircular tubes in section 6.5.5, we introduced in equation 6.46 the definition of the *hydraulic diameter* $\mathcal{D}$ of a noncircular duct, which depended upon its cross-sectional area and perimeter:

$$\mathcal{D} \equiv \frac{4 \times Area}{Perimeter} \tag{9.11}$$

For a circular duct of diameter D, $\mathcal{D} = D$, but for other shapes, $\mathcal{D}/b$ has values given in table 6.1, where b is the shorter dimension of the duct cross-section (see figure 6.9).

If the Reynolds number based upon the hydraulic diameter, $Re_{\mathcal{D}} \equiv \bar{V}\mathcal{D}/\nu$, is less than about 2300, the flow will be laminar, and the Darcy friction factor f is related to the Reynolds number by:

$$f = \frac{const.}{Re_{\mathcal{D}}} \quad (laminar) \tag{9.12}$$

where the constant in the numerator of the right side of equation 9.12 depends upon the duct shape, as listed in table 6.1. But, for turbulent flow, the Darcy friction factor is nearly independent of duct shape and and can be expressed in a form similar to that of equation 9.7:

$$\frac{1}{\sqrt{f}} = -2.0 \log\left(\frac{\varepsilon/\mathcal{D}}{3.7} + \frac{2.51}{Re_{\mathcal{D}}\sqrt{f}}\right) \qquad \textit{(turbulent)} \qquad (9.13)$$

Equations 9.11–9.13 comprise a set that permits the determination of head loss in noncircular ducts.

Example 9.4

A vertical ventilation duct of rectangular cross-section 6 *in* × 12 *in*, having a length $L = 20\,ft$, will carry air at a flow rate of 1000 *cubic feet per minute* (*cfm*) from the basement to the upper floor of a house. Calculate the head loss and pressure drop $p^*_{in} - p^*_{out}$ for this flow if the duct roughness $\varepsilon = 1E(-5)\,m$.

Solution
First express the flow quantities in SI units, using equation 9.11 and table 1.6:

$$\mathcal{D} = \frac{4(6 \times 12)\,in^2}{2(6+12)\,in} = 8\,in \times \frac{2.54E(-2)\,m}{in} = 0.2032\,m$$

$$Q = \frac{1000\,ft^3}{min} \times \frac{min}{60\,s} \times \frac{2.832E(-2)\,m^3}{ft^3} = 0.472\,m^3/s$$

$$A = 72\,in^2 \times \left(\frac{2.54E(-2)\,m}{in}\right)^2 = 4.645E(-2)\,m^2$$

$$\bar{V} = \frac{Q}{A} = \frac{0.472\,m^3/s}{4.645E(-2)\,m^2} = 10.16\,m/s$$

$$L = 20\,ft \times \frac{0.3048\,m}{ft} = 6.096\,m$$

Next calculate the Reynolds number and the head loss, selecting $\nu = 1.51E(-5)\,m^2/s$ from table 1.1 and using equation 9.13:

$$Re_{\mathcal{D}} = \frac{\bar{V}\mathcal{D}}{\nu} = \frac{(10.16\,m/s)(0.2032\,m)}{1.51E(-5)\,m^2/s} = 1.367E(5)$$

$$\frac{1}{\sqrt{f}} = -2\log\left(\frac{1E(-5)\,m}{3.7(0.2032\,m)} + \frac{2.51}{1.367E(5)\sqrt{f}}\right)$$

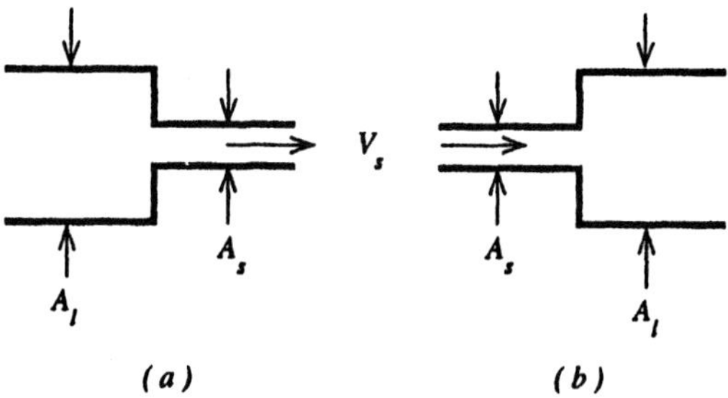

Figure 9.2 In (*a*) a sudden contraction or (*b*) an expansion of flow area in a pipe flow, a loss of head results.

$$= -2\log\left(1.33E(-5) + \frac{1.836E(-5)}{\sqrt{f}}\right)$$

$$f = 1.718E(-2)$$

$$\mathrm{h}_{in} - \mathrm{h}_{out} = f\left(\frac{L}{D}\right)\left(\frac{\bar{V}^2}{2g}\right) = 1.718E(-2)\left(\frac{6.096\,m}{0,2032\,m}\right) \times \frac{(10.16\,m)^2}{2(9.807\,m/s^2)} = 2.713\,m$$

Assuming that the air density is $1.204\,kg/m^3$ (see table 1.1), the pressure drop is:

$$p^*_{in} - p^*_{out} = \rho g(\mathrm{h}_{in} - \mathrm{h}_{out}) = (1.204\,kg/m^3)(9.807\,m/s^2)(2.713\,m) = 32.03\,Pa$$

The difference $p^*_{in} - p^*_{out}$ will equal the difference in gage pressure between inlet and outlet if the air density surrounding the duct is the same as that inside the duct.

9.2.3 Other Losses

Piping systems include more than straight sections of pipe. Pipes must bend around obstructions, they are equipped with valves and other fittings that cause local flow deviations, and those of different diameters are joined together. Whenever the parallel flow in a straight pipe is disturbed, there will be a head loss caused by this localized change that adds to the head loss in the straight pipe section. In this section, we show how to estimate these additional losses.

Loss from an Area Change

Sometimes coaxial pipes of different diameters are joined together as shown in figure 9.2. If the area of the larger pipe is denoted by A_l and that of the smaller pipe by A_s, then the head loss Δh caused by a change in area may be defined in terms of a loss coefficient K:

$$\Delta h \equiv K\left(\frac{\bar{V}_s^2}{2g}\right) \tag{9.14}$$

where the velocity $\bar{V}_s$ is that in the *smaller* pipe and where the loss coefficient K depends upon the ratio of the two areas. For a contraction, the loss coefficient K_c is:

$$K_c = 0.4\left(1 - \frac{A_s}{A_l}\right) \tag{9.15}$$

while that for a sudden expansion, K_e, is:

$$K_e = \left(1 - \frac{A_s}{A_l}\right)^2 \tag{9.16}$$

These values may be reduced by making the transition in area between the two pipes more gradual, but such improvements are generally not justified economically.

When a pipe is supplied from, or discharges to, a large tank or reservoir, the inflow or outflow corresponds to a contraction or expansion from, or to, a pipe of infinite area so that $A_s/A_l = 0$ and the loss coefficients for inflow and outflow become:

$$K_{in} = 0.4; \qquad K_{out} = 1.0 \tag{9.17}$$

by virtue of equations 9.15 and 9.16.

Loss at a Bend

When a pipe bends through a right angle, the higher speed flow at the centerline moves to the outside of the bend while the lower speed flow near the wall flows toward the inside of the bend due to the imbalance in the centrifugal acceleration of the higher and lower speed fluids. The bend thus induces flow in the plane of the pipe or duct cross-section. This flow extracts energy from the axial flow, causing a head loss. Such a flow is called a *secondary flow*, to distinguish it from the primary axial flow.

The loss coefficient for a typical right angle bend, such as a pipe elbow, is approximately 1.0. If the bend is made more gradual by increasing the radius of curvature compared with the pipe diameter, the loss coefficient may be reduced, approximately in inverse proportion to the radius of curvature.

Loss at Fittings

Piping systems often contain valves for regulating or stopping the flow, as well as T intersections with branch pipes. Any fitting inserted into a straight section of piping will produce an additional flow loss, depending upon how intense is its disturbance to the axial flow in the pipe. Tabulations of loss coefficients for pipe fittings are available[3] and typically fall within the range of $0.1 - 1.0$.

9.2.4 Total Head Loss

For flow through a pipe or duct, the total head loss Δh_l will be the sum of the loss Δh_f due to wall friction plus the entrance and exit losses and losses due to valves and fittings. For any pipe or duct section of length L_i, hydraulic diameter $\mathcal{D}_i$ and mean velocity $\bar{V}_i$, the total head loss $(\Delta h_l)_i$ would then be:

$$(\Delta h_l)_i = \left[f_i \left(\frac{L_i}{\mathcal{D}_i} \right) + \sum_j (K_j)_i \right] \left(\frac{\bar{V}_i^2}{2g} \right) \tag{9.18}$$

where $(K_j)_i$ is the loss coefficient for each loss element j. Equation 9.18 may be used to determine the pressure drop $(p^*_{in} - p^*_{out})_i$ in the section i:

$$(p^*_{in} - p^*_{out})_i = \rho g (\Delta h_l)_i \tag{9.19}$$

Figure E 9.5

Example 9.5

A heat collecting circuit for a solar panel consists of a 1 cm diameter drawn copper tube of length $L = 22\,m$ arranged in the pattern shown in figure E 9.5. The water flow

---

[3]See, for example, V. L., Streeter, ed., 1961, *Handbook of Fluid Dynamics* (New York: McGraw-Hill Book Co.)

in the circuit, $Q = 4\,l/min$, is supplied from a header of much larger diameter than the tube and collected by another header of equal size. Assuming that the loss coefficient of each of the 180° turns is 1.0, calculate the head loss Δh_l between the supply and collector headers.

Solution

First calculate the friction loss in a straight pipe of length $L = 22\,m$:

$$Q = \frac{4\,l}{min} \times \frac{min}{60\,s} \times \frac{1E(-3)\,m^3}{l} = 6.667E(-5)\,m^3/s$$

$$A = \frac{\pi D^2}{4} = \frac{\pi(1E(-2)\,m)^2}{4} = 7.854E(-5)\,m^2$$

$$\bar{V} = \frac{Q}{A} = \frac{6.667E(-5)\,m^3/s}{7.854E(-5)\,m^2} = 0.8489\,m/s$$

$$Re_D = \frac{\bar{V}D}{\nu} = \frac{(0.8489\,m/s)(1E(-2)\,m)}{1E(-6)\,m^2/s} = 8.489E(3)$$

The flow is turbulent. Choosing $\varepsilon = 1E(-6)\,m$ from table 7.1, we calculate f from equation 9.13:

$$\frac{1}{\sqrt{f}} = -2.0\log\left(\frac{1E(-6)\,m}{3.7[1E(-2)\,m]} + \frac{2.51}{8.489E(3)\sqrt{f}}\right)$$

$$= -2.0\log\left(2.703E-5) + \frac{2.837E(-4)}{\sqrt{f}}\right)$$

$$f = 3.205E(-2)$$

$$f\left(\frac{L}{D}\right) = 3.205E(-2)\left(\frac{22\,m}{1E(-2)\,m}\right) = 70.53$$

The loss coefficients for the ten turns sum to 10, while the inlet and exit loss coefficients add an additional 1.4 (equation 9.17) for a total loss of $\Sigma K = 11.4$. The total head loss Δh_l becomes:

$$\Delta h_l = \left(f\left(\frac{L}{D}\right) + \Sigma K\right)\left(\frac{\bar{V}^2}{2g}\right)$$

$$= (70.53 + 11.4)\frac{(0.8489\,m)^2}{2(9.807\,m/s^2)} = 3.01\,m$$

9.3 Head Changes in Systems with Pumps and Turbines

Viscous flow through pipes and ducts results in a decrease in head h by the amount Δh_l given in equation 9.18. As explained in section 8.2.3, other head changes occur in pumps and turbines: a head increase $(\Delta h)_p$ in pumps (see equation 8.19) and a head decrease of amount $(\Delta h)_t$ in turbines (see equation 8.21). The total change in head for the system, $h_{out} - h_{in}$, from inflow to outflow, must be the algebraic sum of these changes:

$$h_{out} - h_{in} = -\Delta h_l + (\Delta h)_p - (\Delta h)_t \tag{9.20}$$

9.3.1 Matching Pumps and Hydroturbines to Flow Systems

For fluids to flow through pipes or ducts, power must be supplied to overcome the flow losses. A pump, fan, compressor, or gravity source must provide an increase in head to overcome the head loss caused by viscosity. In designing such fluid flow systems, the engineer must choose a source having the proper characteristics to provide the necessary flow through the system. This matching of the flow and head change requirements holds equally well for hydroturbines or hydraulic motors that extract power from a fluid flow. In this section, we investigate the nature of the problem of matching power machinery to the fluid system to which they are connected.

Pumping Fluid

Liquid centrifugal pumps and air ventilating fans ordinarily operate at a fixed speed, providing an increase in head Δh that varies with the flow rate Q, in the manner sketched in figure 9.3. At zero flow, such as would be experienced if the discharge line were closed off, the maximum head gain Δh_m is experienced. As more fluid is allowed to flow through the pump, the head increase diminishes until it reaches zero at the maximum flow rate Q_m. The shape of the curve of Δh versus Q_m depends upon the design of the pump or fan but, for centrifugal devices, is approximately parabolic, as shown in figure 9.3.

The ideal power delivered to the fluid stream passing through a pump or fan is the product $\dot{m}g\Delta h = \rho Q\Delta h$, while the power required to turn the pump is larger than this by the factor $1/\eta_p$, where η_p is the pump efficiency (see equation 8.19). The product $Q\Delta h$ is plotted in figure 9.3, showing how the maximum power is produced when the head increase Δh and flow rate Q are less than their maximum values, Δh_m and Q_m. The design of pumps is such that the maximum efficiency is reached near the flow condition of maximum power. If possible, a pump should be operated at maximum efficiency.

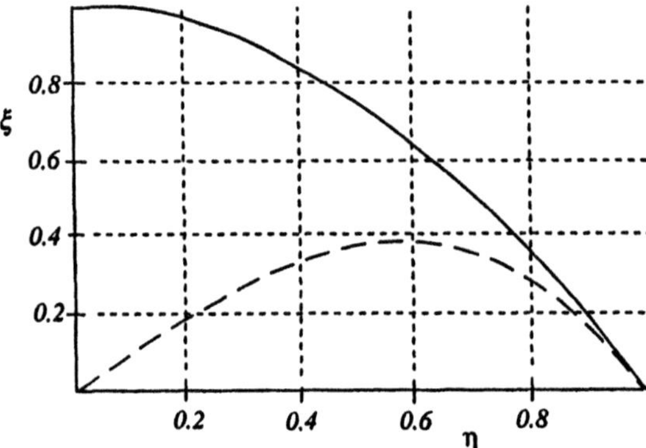

Figure 9.3 A plot of (*solid line*) the dimensionless pump head $\xi \equiv \Delta h/\Delta h_m$ versus dimensionless flow rate $\eta \equiv Q/Q_m$ and (*dashed line*) the dimensionless product $\xi \equiv Q\Delta h/Q_m\Delta h_m$ versus dimensionless flow rate $\eta \equiv Q/Q_m$, where Δh_m and Q_m are the maximum values of the head and flow rate, shows typical centrifugal pump characteristics.

The characteristics of a pump, for which the head increase Δh declines as the flow rate Q increases, is the inverse of the characteristic of a piping system, for which the head loss Δh_l rises with increasing flow rate. When supplying fluid to a piping system, a given pump will only supply that flow for which the pump head increase Δh equals the head loss Δh_l of the piping system. To illustrate this principle, figure 9.4 shows a pump Δh versus Q curve (solid line) and a corresponding Δh versus Q curve (dashed line) for the head loss in a piping system supplied by the pump. The flow rate through the system will be that at which these two curves intersect.

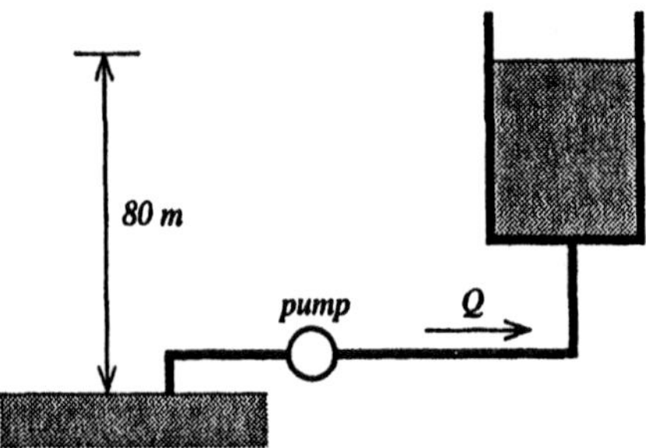

Figure E 9.6

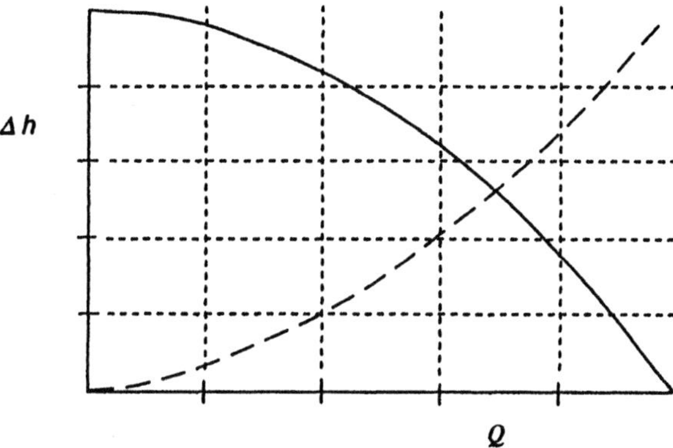

Figure 9.4 The head versus flow rate curves for (*solid line*) a pump and (*dashed line*) a piping system supplied by the pump intersect at the point where the head change and flow rates match.

Example 9.6

A pump extracts water from a reservoir and delivers it to an elevated holding tank whose surface lies 80 m above the surface of the reservoir. The commercial steel supply pipe, 1.5 km in length, is 6 in in diameter. The pump head rise Δh is related to the flow rate $Q\,(GPM)$ by:

$$\Delta h = (150\,m)\left[1 - \left(\frac{Q\,(GPM)}{1000\,(GPM)}\right)^2\right]$$

Calculate the flow rate $Q\,(GPM)$ through the system.

Solution

The fluid delivered from the reservoir to the tank experiences an increase of head equal to 80 m. The pump head rise Δh must supply this increase plus the viscous head loss Δh_l:

$$\Delta h = 80\,m + \Delta h_l$$

Begin by assuming a head loss $\Delta h_l = 20\,m$ and calculate the flow rate Q_{20} that would give rise to this loss. Following the procedure of example 9.2,

$$D = 6\,in \times \frac{2.54E(-2)\,m}{in} = 0.1524\,m$$

$$\sqrt{f}Re_D = \left(\frac{2(9.807\,m/s^2)(20\,m)(0.1524\,m)^3}{[1E(-6)\,m^2/s]^2(1500\,m)}\right)^{1/2} = 3.043E(4)$$

$$Re_D = -2[3.043E(4)]\log\left(\frac{5E(-5)\,m}{3.7(0.1524\,m)} + \frac{2.51}{3.043E(4)}\right) = 2.292E(5)$$

$$Q = \frac{\pi(0.1524\,m)(1E(-6)\,m^2/s)[2.292E(5)]}{4} = 2.744E(-2)\,m^3/s$$

$$Q_{20} = \frac{2.744E(-2)\,m^3}{s} \times \frac{60\,s}{min} \times \frac{gal}{3.785E(-3)\,m^3} = 435\,GPM$$

At this flow rate, the pump head Δh would be:

$$\Delta h = (150\,m)\left[1 - \left(\frac{435}{1000}\right)^2\right] = 121.6\,m$$

which is $21.6\,m$ higher than the total of $80\,m + 20\,m = 100\,m$ needed for the flow rate of 435 (*GPM*). Thus the flow rate will be greater than 435 (*GPM*). As a second guess, assume that $\Delta h = 40\,m$ and repeat the calculation above to find $Q_{40} = 626\,(GPM)$ and $\Delta h = 91.22$, which is $28.8\,m$ below the value of $120\,m$ needed to provide the flow of 626 (*GPM*). Thus the flow rate will be less than 626 (*GPM*). We can estimate the flow rate by linear interpolation between these values so as to reduce the discrepancy to zero:

$$Q = \left(\frac{21.6}{21.6 - (-28.8)} \times (626 - 435) + 435\right)(GPM) = 517\,(GPM)$$

Using this flow rate, we calculate $\Delta h = 109.9\,m$ and $\Delta h_l = 29.91\,m$. Continuing this process to improve our estimate of Q and Δh we finally arrive at:

$$Q = 527.7(GPM); \qquad \Delta h = 108.3\,m$$

Hydroturbines

Hydropower is produced by damming a river to provide a reservoir of water at a level that is higher than that downstream. Water is piped from the reservoir to a hydroturbine that discharges into the downstream flow. Since river flow may exceed the hydroturbine capacity in some seasons of the year, a spillway is provided to bypass the hydroturbines whenever necessary.

Hydroturbines ordinarily run at a fixed speed because they generate electricity of required frequency. The available head is nearly constant, although the reservoir level may change with the river flow. The power generated by the hydroturbine can be regulated by varying the flow rate through the turbine which is accomplished by adjusting the inflow passage area. The mechanical power of the hydroturbine must match the electrical power of the alternator it drives, which in turn is tied into an electric power distribution network. Hydroturbines should be protected from overspeeding if the electrical load is suddenly removed.

The head losses in the pipes that supply water from the reservoir to the turbine and carry the discharge from the turbine to the river downstream will reduce the head available to power the turbine and therefore are energy losses to the power producing system. Designers make these pipes large enough so that the cost of reducing these power losses further is no more than the cost of increasing the power from the turbine.

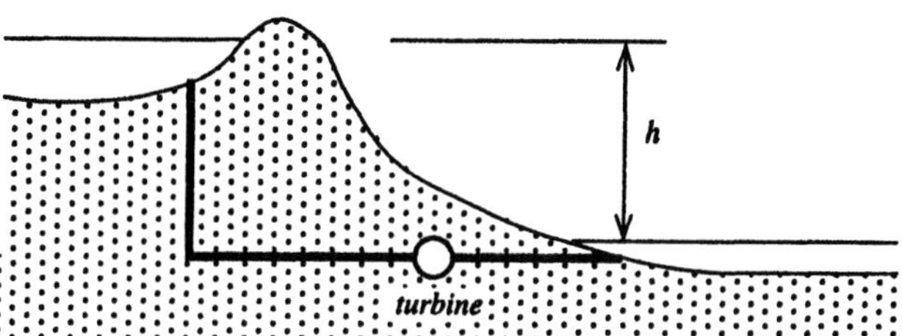

Figure E 9.7

Example 9.7

A hydroturbine draws water from a lake whose surface is a distance $h = 100\,m$ above that of a river to which the turbine discharges, as sketched in figure E 9.7. The turbine flow rate Q is $10\,m^3/s$ and its efficiency η_t is 85%. It is supplied by inflow and outflow pipes of diameter $D = 1.5\,m$, total length $L = 300\,m$ and surface roughness $\varepsilon = 1E(-4)\,m$. Calculate the head loss in the piping and the power produced by the turbine.

Solution

We first calculate the friction loss coefficient in the piping using the method of example 9.1:

$$\bar{V} = \frac{4Q}{\pi D^2} = \frac{4(10\,m^3/s)}{\pi(1.5\,m)^2} = 5.659\,m/s$$

$$Re_D = \frac{\bar{V}D}{\nu} = \frac{(5.659\,m/s)(1.5\,m)}{1E(-6)\,m^2/s} = 8.488E(6)$$

$$\frac{1}{\sqrt{f}} = -2.0\log\left(\frac{1E(-4)\,m}{3.7(1.5\,m)} + \frac{2.51}{8.488E(6)\sqrt{f}}\right)$$

$$= 2.0 \log \left(1.802E(-5) + \frac{2.957E(-7)}{\sqrt{f}} \right)$$

$$f = 1.140E(-2)$$

$$K_f = f\left(\frac{L}{D}\right) = 1.140E(-2)\left(\frac{300\,m}{1.5\,m}\right) = 2.281$$

To this loss coefficient must be added inlet and outlet loss coefficients of 0.4 and 1.0 (equation 9.17) for a total of 3.681:

$$\Delta h_l = \Sigma K \left(\frac{\bar{V}^2}{2g}\right) = 3.681 \left(\frac{(5.659\,m/s)^2}{2(9.807\,m/s^2)}\right) = 6.010\,m$$

The head change across the turbine, Δh_t, is the difference between the head supplied by the reservoir, $h = 100\,m$, and the head loss in the piping, $\Delta h_l = 6.010\,m$, or $\Delta h_t = 93.99\,m$. Using equation 8.21, the turbine power becomes:

$$(\Omega T)_t = \eta_t(\rho Q)g(\Delta h_t) = 0.85(1E(3)\,kg/m^3)(10\,m^3/s)(9.807\,m/s^2)(6.010\,m)$$

$$= 5.010E(5)\,W = 501\,kW$$

9.4 Complex Networks

As illustrations of piping systems, we have thus far considered pipes or ducts without branches or divisions of the flow, but many piping systems, such as those that deliver potable water to homes and offices in cities and towns, contain pipes of different sizes and lengths connected to each other in a system somewhat resembling a tree. How do we analyze the flow in such systems? In this section, we will show how to apply the principles previously used for a single pipe to such complex systems.

9.4.1 Series and Parallel Components

As a starting point, consider two pipes, A and B, arranged in series as in figure 9.5(a). The volume flow rate Q is the same in each pipe, *i.e.*, $Q_A = Q_B = Q$. If Q is known, then the head losses Δh_A and Δh_B may be separately calculated using the method of example 9.1 and added together to obtain the overall head loss. On the other hand, if the overall head loss Δh_l is known and if the volume flow rate Q is desired, we must resort to an iterative solution. The easiest method is to guess a head loss Δh_A in pipe A and its corresponding head loss $\Delta h_l - \Delta h_A$ in pipe B and to calculate the flow rates Q_A and Q_B. If the guess is correct, then $Q_A = Q_B = Q$. Invariably the first

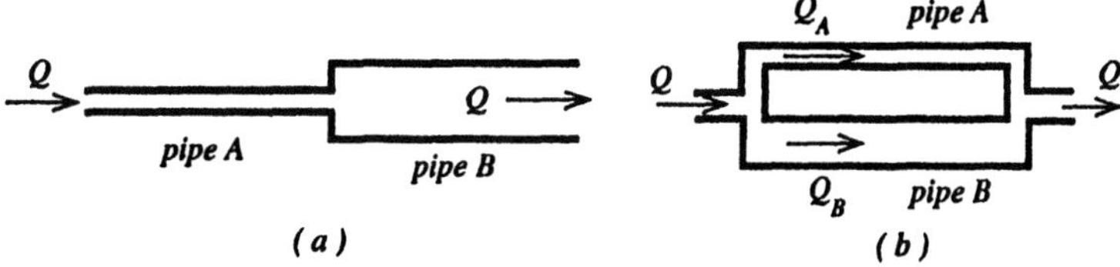

Figure 9.5 Two different pipes arranged in (*a*) series and (*b*) parallel experience different head loss and volume flow rate, respectively.

guess is wrong, and $Q_A - Q_B \neq 0$. We need to find a value for Δh_A that will make $Q_A - Q_B = 0$. By successive guesses and linear interpolation, we can home in on the correct answer.

Example 9.8

Two $50 ft$ lengths of garden hose, one $3/4\, in$ (A) and the other $1/2\, in$ (B) in diameter, are attached in series to a water tank pressurized to $40\, psig$. Calculate the flow rate through the hoses when discharging to the atmosphere at the same level as the tank. (Assume $\varepsilon = 0$.)

Solution

In SI units, the diameters and length are:

$$D_A = 0.75\, in \times \frac{2.54E(-2)\, m}{in} = 1.905E(-2)\, m;$$

$$D_B = 0.5\, in \times \frac{2.54E(-2)\, m}{in} = 1.270E(-2)\, m$$

$$L = 50 ft \times \frac{0.3048\, m}{ft} = 15.24\, m$$

while the available head loss Δh_l is:

$$\Delta h_l = \frac{40\, psi}{\rho g} = \frac{40\, psi}{(1E(3)\, kg/m^3)(9.807\, m/s^2)} \times \frac{6.895E(3)\, Pa}{psi} = 28.12\, m$$

Initially, neglect the inlet and outlet losses and the contraction loss at the connection of hoses A and B. Since hose A has twice the area of hose B, assume about one-third of the head loss will occur in A, or $\Delta h_A = 10\, m$ and $\Delta h_B = 18.12\, m$. Using the method of example 9.1, calculate Q_A and Q_B:

$$(\sqrt{f}Re_D)_A = \left(\frac{2g(\Delta h)D^3}{\nu^2 L}\right)_A^{1/2} = \left(\frac{2(9.807\,m/s^2)(10\,m)(1.905E(-2)\,m)^3}{(1E(-6)\,m^2/s)^2(15.24\,m)}\right)^{1/2}$$

$$= 9433$$

$$(Re_D)_A = -2.0(\sqrt{f}Re_D)\log\left(\frac{2.51}{\sqrt{f}Re_D}\right) = -2(9433)\log\left(\frac{2.51}{9433}\right) = 6.745E(4)$$

$$Q_A = \left(\frac{\pi D\nu Re_D}{4}\right)_A = \frac{\pi(1.905E(-2)\,m)(1E(-6)\,m^2/s)(6.745E(4))}{4}$$

$$= 1.009E(-3)\,m^3/s$$

$$(\sqrt{f}Re_D)_B = \left(\frac{2(9.807\,m/s^2)(18.12\,m)(1.27E(-2)\,m)^3}{(1E(-6)\,m^2/s)^2(15.24\,m)}\right)^{1/2} = 5135$$

$$(Re_D)_B = -2(5135)\log\left(\frac{2.51}{5135}\right) = 3.400E(4)$$

$$Q_B = \frac{\pi(1.27E(-2)\,m)(1E(-6)\,m^2/s)(3.4E(4))}{4} = 3.392E(-4)\,m^3/s$$

Now estimate the inlet, outlet and contraction losses from equations 9.15–9.17:

$$\bar{V}_A = \frac{4Q_A}{\pi D_A^2} = \frac{4(1.009E(-3)\,m^3/s)}{\pi(1.905E(-2)\,m)^2} = 3.540\,m/s$$

$$\bar{V}_B = \frac{4(3.392E(-3)\,m^3/s)}{\pi(1.27E(-2)\,m)^2} = 2.678\,m/s$$

$$\Sigma K\left(\frac{\bar{V}^2}{2g}\right) = \frac{0.4\bar{V}_A^2 + (0.2+1)\bar{V}_B^2}{2g} = \frac{0.4(3.540\,m/s)^2 + 1.2(2.678\,m/s)^2}{2(9.807\,m/s^2)} = 0.6943\,m$$

For the next round, we estimate the available friction head loss to be $\Delta h_f = 28.12\,m - 0.694\,m = 27.43\,m$, and we decide to allocate this total to $\Delta h_A = 2\,m$ and $\Delta h_B = 25.43$. Repeating the steps above, we arrive at:

$$Q_A = 4.072E(-4)\,m^3/s; \qquad Q_B = 4.108E(-4)\,m^3/s; \qquad \Sigma K\left(\frac{\bar{V}^2}{2g}\right) = 0.6851\,m$$

This second guess is quite close, the volume flow rates matching to within 0.8% and the head loss to within 0.03%. Should we need greater accuracy, we would interpolate linearly on the values of Δh_A *vs.* $Q_A - Q_B$ to find the value of Δh_A that would make $Q_A - Q_B$ equal zero. Indicating the first and second guesses by ' and ", we find a

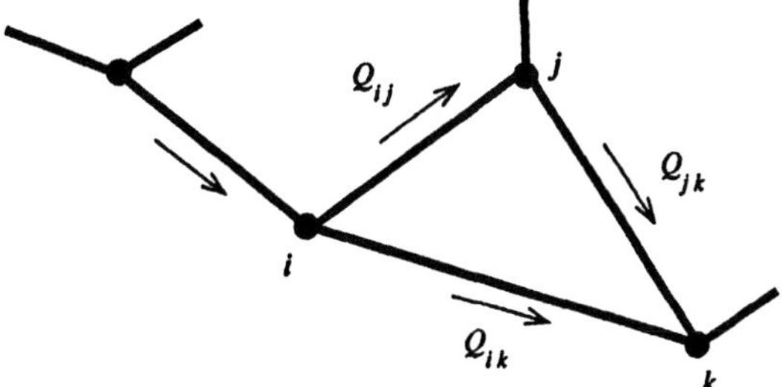

Figure 9.6 A redundant network identifying the nodal points i, j, k and the flow rates Q_{ij} and Q_{ik} between them.

third guess to be:

$$\Delta h = \Delta h''_A - (Q''_A - Q''_B)\left(\frac{\Delta h'_A - \Delta h''_A}{(Q'_A - Q'_B) - (Q''_A - Q''_B)}\right)$$
$$= 2.043\,m$$

For parallel pipes, as sketched in figure 9.5(b), if the head loss is known then Q_A and Q_B can be calculated by the method of example 9.2 and added to give Q. If Q is known, then the head loss Δh_l may be guessed, and Q_A and Q_B calculated, improving the guess until $Q-(Q_A+Q_B)$ becomes zero. Linear interpolation between the guessed values may be used to predict an improved value of Δh_l for use in the next iteration.

9.4.2 Redundant Networks

A redundant fluid network is one for which there are multiple paths for the fluid to flow between two points in the system. The simplest example is that of figure 9.5(b), in which the fluid may flow through both pipes A and B. A common example is the distribution system for water to urban dwellings. Water mains follow the street system and are interconnected in the same pattern as are the streets. Withdrawing water at one location may induce flows in many nearby pipes in the network. Another example is the human circulatory system, which is very redundant with respect to the small blood vessels but not the large ones.

Figure 9.6 shows several nodes (points where pipes connect to each other) of a complex network. A node i is connected to a few nearby nodes $j, k, \ldots$ by pipes $ij, ik, \ldots$. If the values of the head h_i, h_j, h_k, ... at $i, j, k, \ldots$ are known, then the flow rates $Q_{ij}, Q_{ik}, \ldots$ may be computed using the method of Example 9.2.[4] If the node i is a point where fluid is not withdrawn, then:

$$\sum_j Q_{ji} = 0 \tag{9.21}$$

On the other hand, if fluid is withdrawn from the system at i at a known rate Q_{iw}, then:

$$\sum_j Q_{ji} = Q_{iw} \tag{9.22}$$

The analysis of a network flow requires the determination of the set of nodal head values $h_i, h_j, \ldots$ that will give rise to flows that satisfy equation 9.21 or 9.22 at all nodal points. Such a solution can always be found when either h_i or Q_i is known at a node where fluid leaves or enters the system or when a functional relation exists between them, as for a pump or turbine flow to or from the system at that node.

The simplest way to solve for the flow through a redundant system is to guess a set of values for $h_i, h_j, \ldots$, calculate the flow rates Q_{ij} for all combinations of i and j connected by pipes and determine the amounts ΔQ_i by which $\sum Q_i$ fail to satisfy the conditions 9.21 or 9.22:

$$\Delta Q_i \equiv \sum_j Q_{ji} - Q_{iw} \tag{9.23}$$

Subsequent guesses are made by changing h_i so as to reduce ΔQ_i at each node. By linear interpolation of h_i *vs.* ΔQ_i, we can select a trial value of h_i to make $\Delta Q_i = 0$. By continuing this process, a precise solution will be reached eventually.

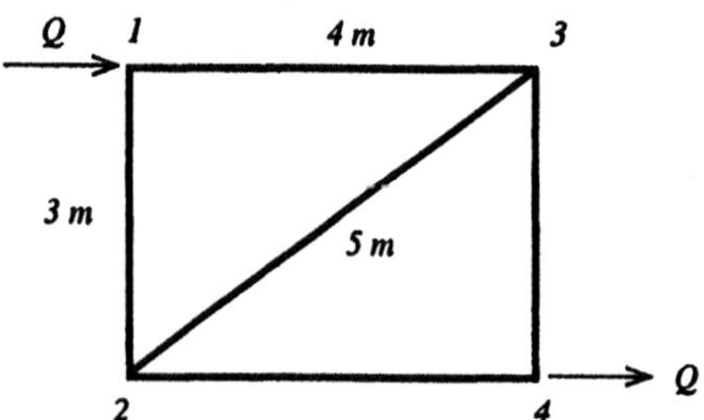

Figure E 9.9

[4] The flow rate Q_{ij} is the rate of flow from i to j, and is positive if the flow is in that direction. Thus $Q_{ji} = -Q_{ij}$.

Example 9.9

The pipe network shown in figure E 9.9 receives water at node 1 and discharges it at node 4. The head at 1 is $h_1 = 10\,m$ and that at 4 is $h_4 = 0\,m$. The segment lengths are shown as $3\,m$, $4\,m$, and $5\,m$, and the tube dimensions are $D = 1\,cm$, $\varepsilon = 0$ for all segments. Calculate the flow rate Q.

Solution

Because of the symmetry of the network, $h_1 - h_2 = h_3 - h_4$, $Q_{12} = Q_{34}$ and $Q_{13} = Q_{24}$. Thus we have only to calculate Q_{12}, Q_{23} and Q_{13} to solve for the network flow. To start the calculation, assume $h_2 = h_3 = 5\,m$ so that $Q_{23} = 0$. Then calculate Q_{12} and Q_{13} using the method of example 9.2, noting that $\Delta h_{12} = \Delta h_{13} = 5\,m$:

$$(\sqrt{f}Re_D)_{12} = \left(\frac{2g(\Delta h)D^3}{\nu^2 L}\right)^{1/2}_{12} = \left(\frac{2(9.807\,m/s^2)(5\,m)(1E(-2)\,m)^3}{(1E(-6)\,m^2/s)^2(3\,m)}\right)^{1/2} = 5718$$

$$(Re_D)_{12} = -2(\sqrt{f}Re_D)\log\left(\frac{2.51}{(\sqrt{f}Re_D)}\right) = -2(5718)\log\left(\frac{2.51}{5718}\right) = 3.840E(4)$$

$$Q_{12} = \frac{\pi D\nu Re_D}{4} = \frac{\pi(1E(-2)\,m)(1E(-6)\,m^2/s)(3.84E(4))}{4} = 3.016E(-4)\,m^3/s$$

$$(\sqrt{f}Re_D)_{13} = \left(\frac{2(9.807\,m/s^2)(5\,m)(1E(-2)\,m)^3}{(1E(-6)\,m^2/s)^2(4\,m)}\right)^{1/2} = 4952$$

$$(Re_D)_{13} = -2(4952)\log\left(\frac{2.51}{4952}\right) = 3.263E(4)$$

$$Q_{13} = \frac{\pi(1E(-2)\,m)(1E(-6)\,m^2/s)(3.263E(4))}{4} = 2.563E(-4)\,m^3/s$$

Now calculate ΔQ_2:

$$\Delta Q_2 = Q_{12} - Q_{23} - Q_{24} = (3.016 - 0 - 2.563)E(-4)\,m^3/s = 0.453E(-4)\,m^3/s$$

This corresponds to an outflow at 2 and an equal inflow at 3. If we increase h_2 and decrease h_3 by an equal amount, ΔQ_2 will decrease. Trying $h_2 = 6\,m$ and $h_3 = 4\,m$, we follow the steps above to calculate:

$$Q_{12} = 2.896E(-4)\,m^3/s; \qquad Q_{13} = 2.842E(-4)\,m^3/s$$

$$Q_{23} = 1.341E(-4)\,m^3/s; \qquad \Delta Q_2 = -1.287E(-4)\,m^3/s$$

We have obviously increased h_2 too much because ΔQ_2 has become negative. Thus h_2 must lie between $5\,m$ and $6\,m$. The two assumed values of head, $h_2 = 5\,m$, $6\,m$, have resulted in two values of the flow imbalance, $\Delta Q_2 = 0.453E(-4)\,m^3/s$, $-1.287E(-4)$

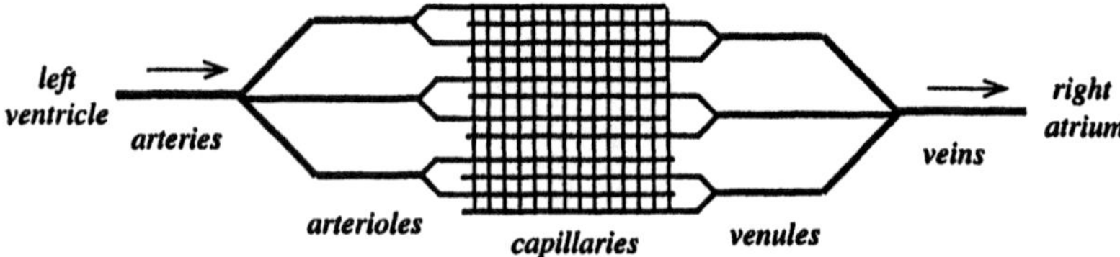

Figure 9.7 A diagram of the human systemic circulatory system, showing the components of the arterial supply and venous collection systems and the capillary bed.

m^3/s. Using linear interpolation,[5] the next guess for h_2 is:

$$h_2 = 6m - \frac{(6-5)m}{(-1.287 - 0.453)E(-4)\,m^3/s}(-1.287E(-4)\,m^3/s) = 5.26m$$

Proceeding in this manner for two more iterations, we converge to the solution:

$$h_2 = 5.11\,m; \qquad h_3 = 4.89\,m$$

$$Q_{12} = 2.978E(-4)\,m^3/s; \qquad Q_{13} = 2.595E(-4)\,m^3/s; \qquad Q_{23} = 0.375E(-4)\,m^3/s$$

$$Q = Q_{13} + Q_{12} = 5.573E(-4)\,m^3/s$$

9.4.3 The Human Blood Circulatory System

The heart circulates blood throughout the body by means of two loops in series. The first loop, called the *systemic circulatory system*, reaches all organs of the body except the lungs, which are supplied by the second loop, or *pulmonary circulatory system*. Although the volume flow rate of blood is the same for both systems since they are connected in series, the resistance to flow is greater in the systemic circulation because it serves a greater body volume at a larger distance. Consequently, the left ventricle, which supplies the systemic circulatory system, must produce more power than the right ventricle. The asymmetric heart structure reflects this difference in function.

Both circulatory systems are very complex, involving redundant paths within the smallest vessels that provide some protection against blockage or damage. A diagram of the systemic circulation is shown in figure 9.7. Blood leaves the left ventricle, passing in turn through the large arteries, small arteries, arterioles, capillaries, venules, small veins, and large veins and then returns to the right atrium of the heart. On the arterial side, there is increasing branching of the system into parallel paths until the

---

[5]The two calculations are fit to the straight line $h_2 = m(\Delta Q_2) + b$ to solve for the slope m and the intercept b, the latter being the value of h_2 when $\Delta Q_2 = 0$.

Table 9.1 Flow properties of the systemic circulation

	$\mathcal{V}$ (%)	A (cm^2)	$\bar{V}$ (cm/s)	$\bar{p}$ ($mm\,Hg$)	L (cm)
Heart	7				
Large arteries	8	4	20	100 ± 20	100
Small arteries	5	20	4	90 ± 20	13
Arterioles	2	40	2	60 ± 5	3
Capillaries	5	2500	0.03	20	0.1
Venules	10	250	0.3	10	0.4
Small veins	15	80	1	5	10
Large veins	39	20	4	1	100
Pulmonary vessels	9				

flow reaches the capillary bed, which is highly branched and interconnected so that blood is brought close to all the body cells. On the venous side, the structure is similar but reversed in function, serving to collect the flow from the capillary bed. Most of the surface area of the blood vessels, across which nutrients and waste products must diffuse to and from the body cells, lies within the capillary bed.

The arterial and venous systems, while providing distribution and collection of the blood flow to and from the capillary bed, are not symmetrical in form, the venous system having a greater volume and flow area than the arterial system. Table 9.1 lists the attributes of the components of the systemic circulatory system that affect the flow of blood through it. The first column lists the volume $\mathcal{V}$ of blood in each component, expressed as a percentage of the normal blood volume, 5 l. Note that the venous side (venules and small and large veins) comprise nearly two-thirds of the volume of the entire circulatory system, providing an elastic storage volume. The second column lists the total flow area A of the parallel channels in each component, with the capillary bed exhibiting the greatest flow area by far. Since a capillary diameter is about 10 μm, there are about a billion capillaries in parallel. The average velocity $\bar{V} = Q/A$ is listed in the next column. Only in the largest artery is the flow Reynolds number high enough to sustain turbulent flow, the other components experiencing laminar flow. The fourth column shows how the volume-averaged pressure $\bar{p}$ declines between the inflow and the outflow, the largest pressure drop occurring in the arterioles. The fluctuating component of the pressure, caused by the pulsatile operation of the heart as a pump, declines as the flow area and distance from the heart increase. Even though the inflow to the right atrium is also pulsatile, there is no noticeable fluctuating pressure because of the large venous volume and the great elasticity of the venous system. Finally, the last column lists the average vessel length L, which equals the volume V divided by the flow area A.

The human blood circulatory system is obviously very complex and difficult to model as a fluid mechanical system—especially those billion capillaries! It has an equally complex control system. The heart rate responds to several stimuli but principally to the demand for more oxygen. The distribution of blood flow within the system is also controlled. For example, blood flow is increased locally to damaged areas and is increased or decreased to the skin when heat evolution needs to be augmented or suppressed. Nevertheless, the principles discussed in this chapter apply to its components and can be used in the diagnosis and treatment of circulatory problems.

9.4.4 Computational Aids

The examples of this chapter show that the calculation of the flow behavior in fluid systems can become tedious when carried out on a hand calculator, even a programmable one, because often an iterative solution is required. Software for personal computers, workstations and minicomputers is available that treats piping systems of the kind commonly encountered in engineering design, greatly reducing the time and effort required to solve a problem. In special cases, it may be necessary to develop programs for particular applications that are not included in commercial software programs. The principles described above serve as a starting point for assembling such special purpose programs.

When faced with an occasional problem that is not too complex, the engineer may find it less time consuming to solve it by using the methods of this chapter rather than learning or relearning the use of the applicable software.

9.5 Problems

Problem 9.1

A town engineer has been asked to determine the size of a water main needed to supply a new residential development. The main will be supplied from a tank whose minimum water level is $200\,ft$ above the grade level in the development. The length L of the main will be 3 miles. To provide fire protection, a flow of $10,000$ gallons per minute at no less than atmospheric pressure in the main must be available at the development site. Available pipe sizes are $12\,in$, $18\,in$, $24\,in$, $36\,in$ and $48\,in$ inside diameter D, with a roughness height $\epsilon = 0.01\,in$.

(a) If a $12\,in$ diameter pipe were used, calculate the loss in head Δh_f between the supply tank and the development site. (b) What is the minimum size of pipe that should be used? (c) What will be the main pressure at the development site for this pipe size when carrying the required flow?

Problem 9.2

A fire hose of length $L = 300\,ft$ and inner diameter $D = 3\,in$ feeds a nozzle of exit diameter $D_e = 1\,in$. It is supplied by a hydrant having a gage pressure $p_h - p_a$ corresponding to a head of $100\,ft$. The effects of bends are negligible and the internal roughness ϵ is the same as that of cast iron pipe. Calculate the volume flow rate Q through the hose in gallons per minute (GPM).

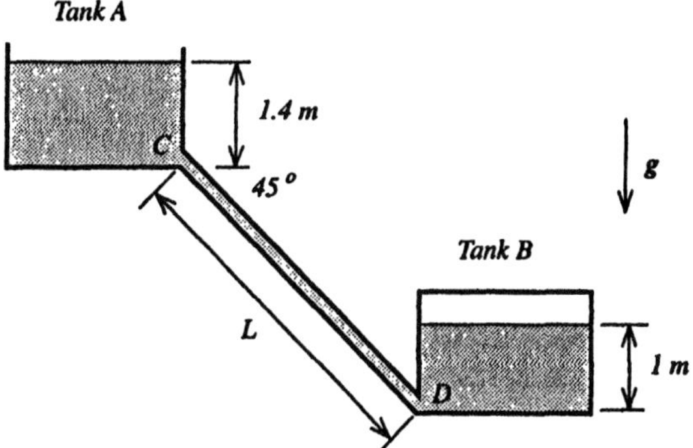

Figure P 9.3

Problem 9.3

A pressurized tank B is connected to an open tank A through a straight pipe of diameter $D = 6\,mm$ and length $L = 4.3\,m$ that forms an angle with the horizontal of 45°. The two tanks contain water of density $\rho = 1E(3)\,kg/m^3$ and viscosity $\mu = 1E(-3)\,Pa\,s$. The gage air pressure above the water surface in tank B is $p_B = 3.45E(4)\,Pa$. The water surfaces in the two tanks are higher than the pipe ends, C and D, by $1.4\,m$ and $1\,m$, respectively, as shown in figure P 9.3.

(a) Prove that the water flow is from tank B to tank A. (b) Assuming that the flow is laminar, calculate the mean flow speed $\bar{V}$ if entrance and exit losses are neglected. (c) Prove that the assumption of laminar flow is valid.

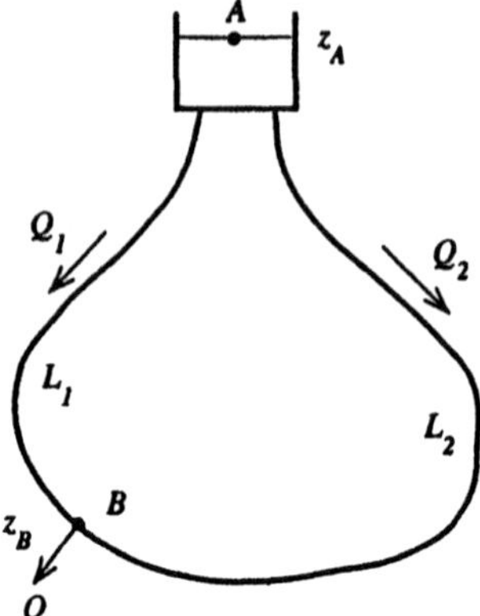

Figure P 9.4

Problem 9.4

A water supply for a small town consists of a storage tank and a 10 *km* long pipe of inner diameter $D = 0.2\,m$ ($\epsilon/D = 2E(-4)$) that forms a complete loop through the town and back to the storage tank, as illustrated in figure P 9.4.

At a point *B* along the water main, located a distance $L_1 = 3\,km$ from the tank, a fire engine is pumping water from the main to a fire at the volume flow rate $Q = 0.1\,m^3/s$. At *B*, the elevation z_B is 100 *m* lower than that of the free surface of the water in the storage tank, z_a. Part of the water, Q_1, flows along the shorter leg ($L_1 = 3\,km$) and part, Q_2, along the longer leg ($L_2 = 7\,km$), the total, $Q_1 + Q_2$, being equal to the pump flow *Q*.

In answering the following questions, use $\rho = 1E(3)\,kg/m^3$ and $\nu = 1E(-6)\,m^2/s$. (a) Assuming that the head loss in either leg is due entirely to frictional loss and that the friction factors f_1 and f_2 for each leg are equal, calculate the flow rates Q_1 and Q_2. (b) Under the assumptions of (a), calculate the pressure p_B at the hydrant from which the fire engine is drawing its water. (c) Justify the assumption that the inlet loss is negligible compared to the friction loss.

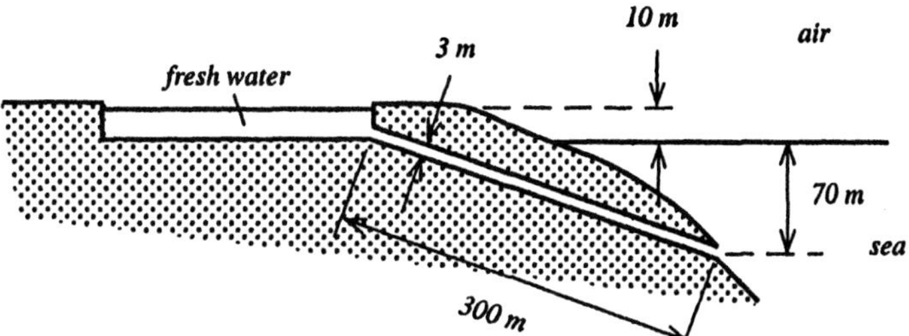

Figure P 9.5

Problem 9.5

The outfall from the settling lagoon of a sewage treatment plant discharges fresh water ($\rho_f = 1E(3)\, kg/m^3$; $\nu_f = 1E(-6)\, m^2/s$) to the sea ($\rho_s = 1.03E(3)\, kg/m^3$) at a depth $70\, m$ below sea level. The surface of the pond is $10\, m$ above sea level. The outfall pipe is $70\, m$ long and $3\, m$ in diameter and has a surface roughness of $3\, mm$.

(a) If the water flow through the pipe were inviscid, calculate what would be the velocity V in the pipe. (b) Assuming viscous flow, calculate $\bar{V}$ for the pipe flow.

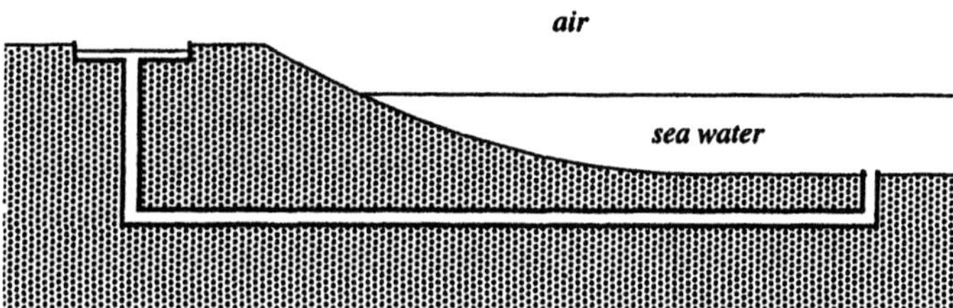

Figure P 9.6

Problem 9.6

The outfall from the sewage treatment plant in Boston harbor consists of a tunnel about 24 feet in diameter ($D = 7.3\, m$, $\varepsilon = 2E(-3)\, m$) and total length of about 9.5 miles ($L = 1.53E(4)\, m$) that conducts treated effluent from a settling tank to an outlet at the bottom of Massachusetts Bay, as sketched in figure P 9.6. The water surface of the settling tank is elevated above sea level by a distance $\Delta h = 20\, m$. Assuming that the treated sewage and sea water have the same density and kinematic viscosity $\nu = 1E(-6)\, m^2/s$, and further assuming that the loss of head in the tunnel flow is entirely due to friction, (a) calculate the volumetric flow rate Q in the tunnel. (b)

From your answer to (a), calculate the entrance head loss and a revised value of Q that takes this loss into account.

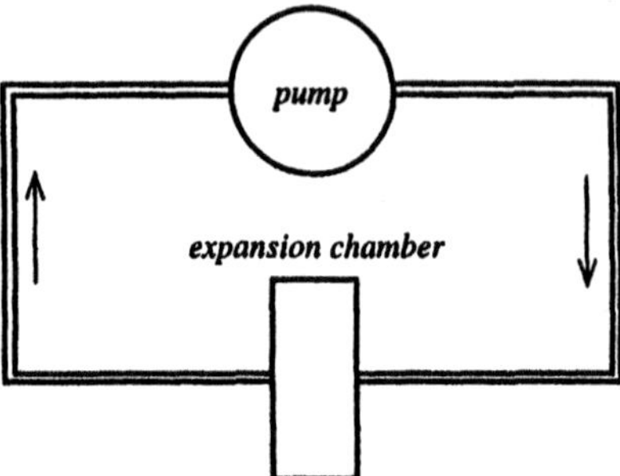

Figure P 9.7

Problem 9.7

In figure P 9.7, a cooling system consists of a pump driving liquid coolant through a pipe of length $L = 10\,m$, diameter $D = 5\,cm$ and roughness $\epsilon = 0$. The pipe has four right angle bends, for each of which the loss coefficient $K_e = 1.0$. The pump is provided with shaft power $\mathcal{P} = 1E(4)\,W$ and has an efficiency $\eta_p = 0.6$. The coolant density $\rho = 1E(3)\,kg/m^3$ and its viscosity $\mu = 2E(-3)\,Pa\,s$. Mounted in the line is an expansion chamber of area much larger than that of the pipe. The loss coefficients for flow into and out of the pipe connections to the expansion chamber are $K_{in} = 0.4$ and $K_{out} = 1.0$, respectively. Calculate the volumetric flow rate Q through the pipe.

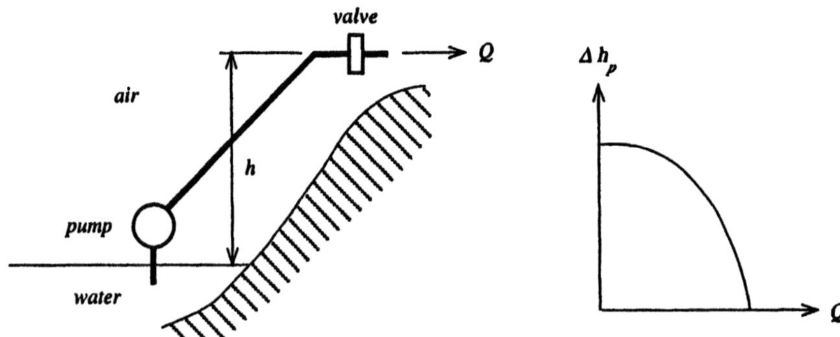

Figure P 9.8

Problem 9.8

A pump delivers water at a volume flow rate Q to the atmosphere through a pipe of diameter D that is equipped with a partially closed valve near its exit. The exit is located at a height h above the surface of the reservoir from which the water is drawn (see figure P 9.8). The pump head gain Δh_p is related to the volume flow rate Q by:

$$\Delta h_p = A - BQ^2$$

where A and B are dimensional constants. The pump efficiency η_p is independent of Q. The head loss in the piping system is predominantly due to the loss in the valve, whose loss coefficient is K_v, so that the frictional loss in the pipe may be neglected. (a) Derive an expression for the volumetric flow rate Q in terms of the parameters ρ, A, B, h, D and K_v. (b) Derive an expression for the pumping power $\mathcal{P}_p$ required in terms of the given parameters.

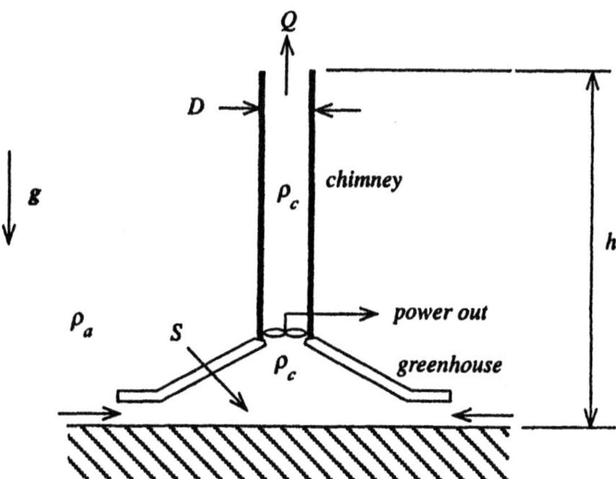

Figure P 9.9

Problem 9.9

Figure P 9.9 shows a device that produces mechanical power at a rate $\mathcal{P}$ from solar radiation. A circular greenhouse, open to the atmosphere at its periphery, transmits solar radiative energy at a rate S which is absorbed on a black floor. This energy heats the air in the greenhouse, the resulting density ρ_c being less than that of the outside atmosphere ρ_a. The warmed air passes through a turbine of efficiency η_t as it rises in a long chimney, inside of which the air retains its lower density ρ_c. The height h of the chimney is much greater than that of the greenhouse, and its diameter D is much smaller than that of the greenhouse. The chimney's inside walls have a roughness

ϵ, and the entrance to the chimney is characterized by a loss coefficient (based on the chimney gas velocity) of K_c. The air density decrement $\rho_a - \rho_c$ is related to the incident solar power S and the air volume flow rate Q by:

$$\rho_a - \rho_c = \alpha \frac{S}{Q}$$

where α is a dimensional constant.

(a) Derive an expression for the power $\mathcal{P}$ that might be extracted from this device as a function of the known parameters ρ, h, D, ϵ, K_c, η_t, S, α and the variable flow rate Q. (b) What is the maximum power that can be obtained from this device?

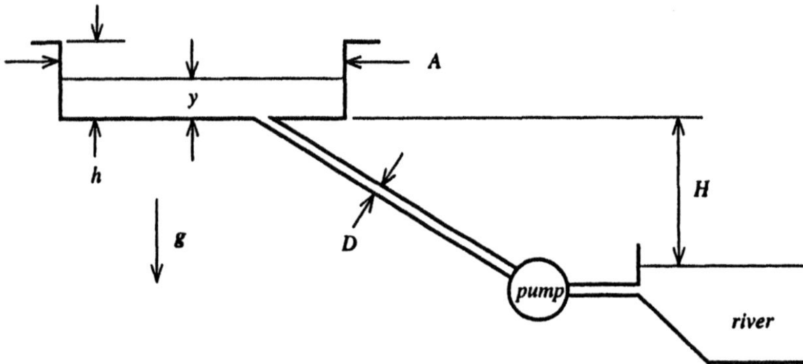

Figure P 9.10

Problem 9.10

A pumped storage system consists of a reservoir elevated above a river surface level by a height $H = 150\,m$. The reservoir can be filled to an additional height $h = 10\,m$ by a pump of flow rate $Q = 3\,m^3/s$ via a pipe of diameter $D = 1\,m$, roughness height $\epsilon = 1\,cm$ and total length $L = 1\,km$. The reservoir surface area $A = 1E(4)\,m^2$, and the reservoir is rectangular in volume, as shown in figure P 9.10.

(The pump is powered by an electric motor which extracts power from an electric utility grid. After filling the reservoir, the pump/motor is run at a later time in reverse as a turbine/generator and supplies electric power to the grid.)

In answering the following questions, express your answer in numerical form. (a) Starting with an empty reservoir, how long (*seconds*) will it take to fill the reservoir? (b) If you neglect all frictional and other head losses, what would be the ideal work (*Joules*) required to fill the reservoir? (c) Now consider the real flow of viscous water through the pipe into the reservoir. At the time when the reservoir is half full ($y = 5\,m$), what is the pressure difference (*Pascals*) between the exit and entrance to the pump?

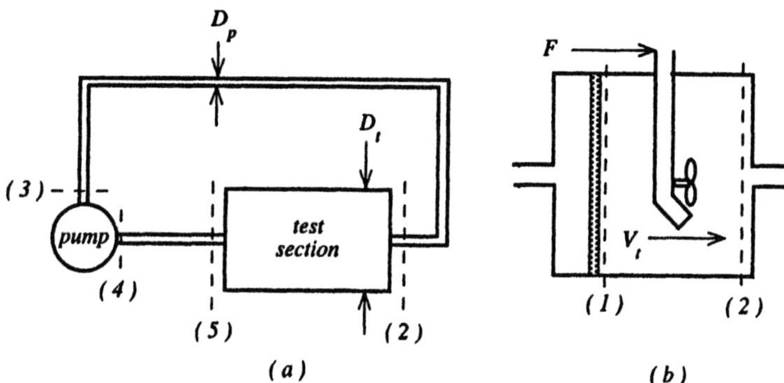

Figure P 9.11

Problem 9.11

An experimental facility for testing marine outboard engines consists of a pump, a closed loop of piping and a test section, as shown in figure P 9.11(a). The piping has a total length of $L = 10\,m$, a diameter $D_p = 0.15\,m$ and a roughness height $\epsilon = 0.01\,mm$. The test section, shown in figure P 9.11(b), has a diameter $D_t = 0.45\,m$. At the upstream end of the test section is a screen which ensures a uniform test section velocity V_t and pressure p_1 of the fluid at section 1. At section 2, the pressure p_2 is also uniform across the test section. The facility is filled with water of density $\rho = 1E(3)\,kg/m^3$ and viscosity $\mu = 1E(-3)\,Pa\,s$.

(a) A test is run in which the fluid test speed $V_t = 1\,m/s$. The restraining force F exerted on the outboard engine in the direction of flow is measured to be $1E(3)\,N$. Neglecting any frictional force on the walls of the test section, calculate the pressure difference $p_1 - p_2$. (b) In another test at the same test speed $V_t = 1\,m/s$, the pressure difference $p_5 - p_2$ is measured to be $1E(4)\,Pa$. Calculate the pressure rise $p_4 - p_3$ across the pump for this test. (Assume the loss coefficient for each elbow is $K_e = 0.25$.)

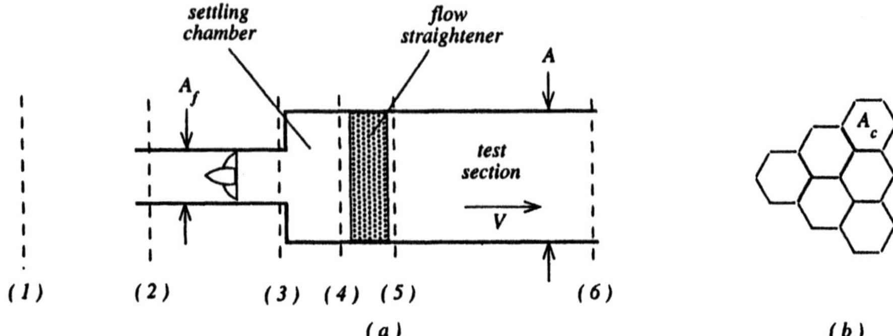

Figure P 9.12

Problem 9.12

A small low-speed wind tunnel is to be constructed from available materials. As shown in figure P 9.12(a), the tunnel consists of an axial flow fan of entrance and exit area $A_f = 0.5\,m^2$ and a settling chamber, flow straightener and test section, all of flow area $A = 1\,m^2$. The wind tunnel is to be designed to operate at a steady flow speed $V = 1\,m/s$. The axial flow fan will provide a pressure rise $p_3 - p_2 = 10\,Pa$ when delivering the volumetric flow rate $Q = 1\,m^3/s$ needed to operate the tunnel.

The flow straightener consists of a honeycomb structure, as shown in figure P 9.12(b), in which the individual channels have a flow area $A_c = 1E(-5)\,m^2$ and length L (to be determined by you). The viscous flow through the straightener will damp out all fluctuations, producing a smooth, uniform flow of velocity V in the test section. You may consider the flow in each of the flow straightener channels to be a steady viscous flow in a circular tube of area A_c, and the entire flow as incompressible with density $\rho = 1.204\,kg/m^3$ and viscosity $\mu = 1.82E(-5)\,Pa\,s$.

(a) Assuming an entrance loss coefficient of $K_c = 0.5$ for the flow into the inlet fan, calculate the pressure difference $p_3 - p_a$ when operating at the design conditions. (b) Assuming the appropriate loss coefficient K_e for a sudden enlargement, calculate the value of $p_4 - p_3$ for the same operating conditions. (c) Assuming negligible wall friction loss in the test section, calculate the value of the pressure drop $p_4 - p_5$ across the flow straightener. (d) Calculate the Reynolds number of the flow in the channels of the flow straightener. (e) Calculate the length L of the flow straightener that will provide the pressure drop of part (c).

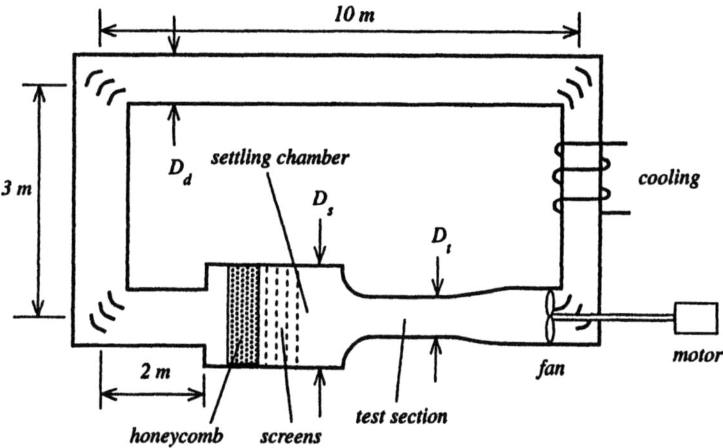

Figure P 9.13

Problem 9.13

Figure P 9.13 shows the preliminary design of a pressurized, closed-circuit wind tunnel which, for the sake of Reynolds number range, is to operate at a pressure level of 8 atmospheres absolute. The circuit lies in the horizontal plane.

The wind tunnel test section is fed by a smooth, well-shaped contraction upstream of which is a settling chamber. The latter contains a honeycomb section to remove swirl and large eddies and a series of four screens which reduce the length scale of turbulent motions so that the turbulence is dissipated by viscosity before the contraction. Downstream of the test section is a diffuser to decelerate the air stream to recover static pressure. At the corners of the duct circuit are turning vanes to minimize flow separation, non-uniform flow and head losses. Flow is maintained by a fan. To prevent the air temperature from rising to a high level, the power input is removed as heat in a cooling section.

The test section flow speed is to be $V_t = 30\,m/s$. The test section diameter is $D_t = 0.3\,m$, the duct diameter is $D_d = 0.6\,m$ and the settling chamber diameter is $D_s = 0.9\,m$, all sections having smooth walls ($\epsilon = 0$). The tunnel air density (which can be considered constant) is $\rho = 9.60\,kg/m^3$ and viscosity is $\mu = 1.77E(-5)\,Pa\,s$.

The head losses in the honeycomb, the contraction section and the test section may all be neglected because they are small compared to the other losses in the circuit. The loss coefficient for each vane section is $K_v = 0.3$ (based on the duct speed V_d), for each screen is $K_s = 2.0$ (based on the settling chamber speed V_s) and for the diffuser is $K_d = 0.2$ (based on the test section speed V_t).

(a) Calculate what would be the reading, in meters of water on a manometer, of a pitot tube pointing upstream in the center of the test section. (b) Calculate the friction head loss in the pipe portion of the circuit ducting (not counting the settling chamber, contraction, diffuser and turning corners), expressed in meters of air at the tunnel density. (c) Calculate the pressure rise in the fan. (d) Assuming the fan has an efficiency $\eta_f = 70\%$, calculate the power required to drive it.

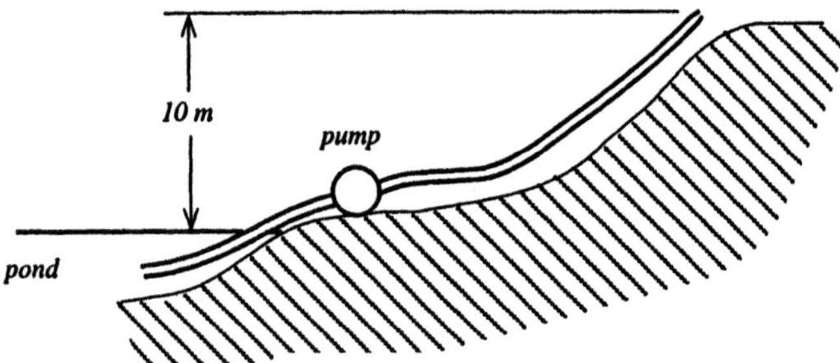

Figure P 9.14

Problem 9.14

A fire engine pumps water from a pond at a volume flow rate of $Q = 0.1\, m^3/s$ through a fire hose of total length $L = 100\, m$, diameter $D = 0.1\, m$ and roughness height $\epsilon = 1E(-3)\, m$. The elevation of the fire hose nozzle is $10\, m$ above that of the pond surface. The end of the intake hose is inserted below the water level of the pond, as shown in figure P 9.14. The nozzle exit diameter is $0.05\, m$.

Calculate (a) the head rise $(\Delta h)_p$ across the pump and (b) the pump power required if the pump efficiency η_p is 80%. You may neglect flow losses in the nozzle.

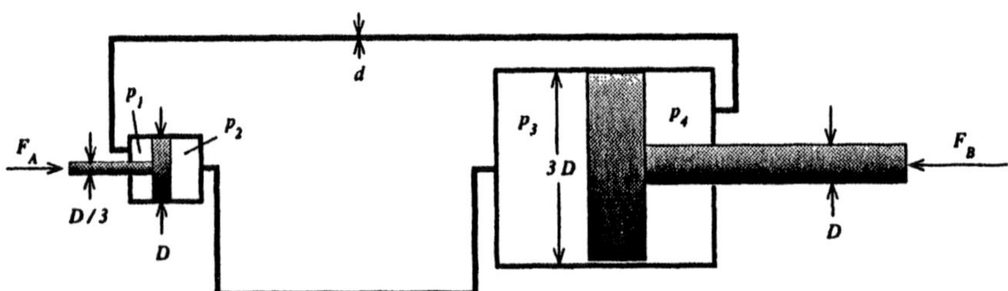

Figure P 9.15

Problem 9.15

An hydraulic press used in a manufacturing operation consists of an actuating piston and cylinder of diameter $D = 3\,cm$ connected to a press piston and cylinder of diameter $3D$ by two hydraulic lines, each of length $L = 10\,m$ and diameter $d = 5\,mm$, as shown in the sketch. The hydraulic fluid has a viscosity $\mu = 2E(-2)\,Pa\,s$.

(a) If a force $F_A = 100\,N$ is applied to the actuator piston, and no motion ensues, calculate the gage pressure $p_2 = p_3$ and the force F_B that restrains the press piston, assuming $p_1 = p_4$ equals atmospheric pressure. When the resistive force reaches $F_B = 1000\,N$, the press piston moves to the right at a speed $V_B = 1\,mm/s$. Calculate (b) the speed V_A of the actuator piston, (c) the pressure drop $\Delta p \equiv p_2 - p_3 \simeq p_4 - p_1$ assuming laminar viscous flow in the hydraulic lines, (d) the force F_A needed to cause this motion and (e) the power efficiency $F_B V_B / F_A V_A$ of the system at this operating condition.

Bibliography

Gerhart, Philip M., Richard J. Gross, and John I. Hochstein. 1992. *Fundamentals of Fluid Mechanics*. 2nd ed. Reading, Pa.: Addison-Wesley Publishing Company.

Guyton, Arthur C. 1976. *Textbook of Medical Physiology*. 5th ed. Philadelphia: W. B. Saunders Company.

Streeter, V. L., ed. 1961. *Handbook of Fluid Mechanics*. New York: McGraw-Hill Book Co.

White, Frank M. 1986. *Fluid Mechanics*. 2nd ed. New York: McGraw-Hill Book Co.

10 Dimensional Analysis and Modeling

10.1 Introduction

Dimensional analysis is an analytical procedure that helps us to organize empirical information about a fluid flow. Relying on the physical principle of *dimensional homogeneity*, it enables us to simplify somewhat the relationship between the things we want to know about a flow and the constraints we impose upon that flow. It cannot produce new information, but it shows us how to reorder what information is at hand to provide a clearer picture of phenomenological relationships.

In earlier chapters, we studied some simple flows that could be described by algebraic expressions, such as Bernoulli's equation or closed-form solutions to the Navier-Stokes equations for incompressible flow. Many engineering flows of great practical importance are more complex than these examples and have no such elegant mathematical solutions describing their behavior. Very often it is necessary to resort to an experiment to establish even the principal physical features of the flow. In designing such experiments and interpreting their results, dimensional analysis can be very helpful.

For over a hundred years, engineers have used small-scale models of engineering structures to help provide information that will make their design more effective. Airplanes, motor vehicles, ships and dam spillways are some examples of engineering structures that are tested in laboratories to determine the details of the fluid flow associated with their use. Physical modeling of these flows in a laboratory setting is often a necessary step in designing the full-scale device. Dimensional analysis provides the engineer with the tools to design, conduct and analyze the results of model testing and to predict the important flow properties that will be encountered by the full-scale structure.

In this chapter, we describe first how dimensional analysis can be applied to any fluid flow problem and used to simplify the expression of the dependence of important flow properties on the flow variables. Then we show how to derive useful information from model experiments by applying the principle of dimensional analysis. These both are essential tools of the engineer in designing complex systems.

We finish this chapter by including a discussion of the drag and lift forces acting on bodies immersed in a flow, an example of how to tie dimensional analysis to an understanding of important physical features of some interesting and practical flows.

10.2 Dimensional Analysis

While dimensional analysis may be applied to any physical experiment, in this section, we are concerned with its implications for fluid flows. Although there are several interesting ways to view the significance of dimensional analysis, we adopt the straightforward approach of showing how it simplifies the organizing of information about fluid flow experiments, thus reducing the amount of experimentation needed to produce a complete description of flow phenomena. Dimensional analysis becomes, not a magical sleight of hand but an exercise in rational analysis of a complex flow situation that can lead to useful physical and mathematical modeling techniques for the design of engineering systems.

10.2.1 Dimensions and Units of Measurement

From the discussion in section 1.3, recall that an observed physical quantity has a *dimension* and a magnitude that is measured in the *units* of that dimension. The fundamental dimensions (mass, length, time, temperature and electric current) are those that cannot be measured in terms of each other. All other physical quantities have dimensions that are products of these fundamental dimensions, *e.g.*, density has the dimensions of mass divided by (length)^3. To express the dimension of a physical quantity, we shall use a shorthand. The fundamental dimensions we use in fluid mechanics, mass, length, time, and temperature, will be designated by $\mathcal{M}$, $\mathcal{L}$, $\mathcal{T}$ and Θ, respectively. The dimension of a quantity, such as density ρ for example, will be designated by $[\rho]$ and may be expressed in terms of the fundamental dimensions as:

$$[\rho] = \frac{\mathcal{M}}{\mathcal{L}^3} = \mathcal{M}\mathcal{L}^{-3} \tag{10.1}$$

The dimensions of all the quantities we use in fluid mechanics can be expressed as multiples of the fundamental dimensions $\mathcal{M}$, $\mathcal{L}$, $\mathcal{T}$ and Θ. Since we will use such quantities extensively in dimensional analysis, we list many of them in table 10.1. For convenience, they are categorized as geometric quantities, fluid properties and flow quantities.

Table 10.1 Dimensions of fluid mechanical quantities

Quantity	Symbol	Dimension
Geometric quantities		
Length		$\mathcal{L}$
Area	$\mathcal{S}$	$\mathcal{L}^2$
Volume	$\mathcal{V}$	$\mathcal{L}^3$
Angle	θ, ϕ	1
Time	t	$\mathcal{T}$
Gradient, divergence, curl	$\nabla, \nabla\cdot, \nabla\times$	$1/\mathcal{L} = \mathcal{L}^{-1}$
Laplacian	∇^2	$1/\mathcal{L}^2 = \mathcal{L}^{-2}$
Time derivative	$\partial/\partial t, D/Dt$	$1/\mathcal{T} = \mathcal{T}^{-1}$
Fluid properties		
Density	ρ	$\mathcal{M}/\mathcal{L}^3 = \mathcal{M}\mathcal{L}^{-3}$
Viscosity	μ	$\mathcal{M}/\mathcal{L}\mathcal{T} = \mathcal{M}\mathcal{L}^{-1}\mathcal{T}^{-1}$
Kinematic viscosity	ν	$\mathcal{L}^2/\mathcal{T} = \mathcal{L}^2\mathcal{T}^{-1}$
Surface tension	Υ	$\mathcal{M}/\mathcal{T}^2 = \mathcal{M}\mathcal{T}^{-2}$
Thermal conductivity	λ	$\mathcal{M}\mathcal{L}/\mathcal{T}^3\Theta = \mathcal{M}\mathcal{L}\mathcal{T}^{-3}\Theta^{-1}$
Kinetic, potential energy per unit mass	$V^2/2, gz$	$\mathcal{L}^2/\mathcal{T}^2 = \mathcal{L}^2\mathcal{T}^{-2}$
Internal energy, enthalpy	$\hat{e}, \hat{h}$	$\mathcal{L}^2/\mathcal{T}^2 = \mathcal{L}^2\mathcal{T}^{-2}$
Specific heat, entropy, gas constant	$\hat{c}_p, \hat{s}, R$	$\mathcal{L}^2/\mathcal{T}^2\Theta = \mathcal{L}^2\mathcal{T}^{-2}\Theta^{-1}$
Flow quantities		
Velocity	$\mathbf{V}$	$\mathcal{L}/\mathcal{T} = \mathcal{L}\mathcal{T}^{-1}$
Acceleration	$D\mathbf{V}/Dt, \mathbf{g}$	$\mathcal{L}/\mathcal{T}^2 = \mathcal{L}\mathcal{T}^{-2}$
Angular velocity, frequency	Ω	$1/\mathcal{T} = \mathcal{T}^{-1}$
Vorticity	ω	$1/\mathcal{T} = \mathcal{T}^{-1}$
Pressure, stress	p, τ	$\mathcal{M}/\mathcal{L}\mathcal{T}^2 = \mathcal{M}\mathcal{L}^{-1}\mathcal{T}^{-2}$
Pressure gradient	∇p	$\mathcal{M}/\mathcal{L}^2\mathcal{T}^2 = \mathcal{M}\mathcal{L}^{-2}\mathcal{T}^{-2}$
Volumetric flow rate	$\mathcal{Q}$	$\mathcal{L}^3/\mathcal{T} = \mathcal{L}^3\mathcal{T}^{-1}$
Mass flow rate	$\dot{m}$	$\mathcal{M}/\mathcal{T} = \mathcal{M}\mathcal{T}^{-1}$
Force	$\mathbf{F}$	$\mathcal{M}\mathcal{L}/\mathcal{T}^2 = \mathcal{M}\mathcal{L}\mathcal{T}^{-2}$
Moment of force, torque	$\mathbf{R}\times\mathbf{F}, \mathbf{T}$	$\mathcal{M}\mathcal{L}^2/\mathcal{T}^2 = \mathcal{M}\mathcal{L}^2\mathcal{T}^{-2}$
Linear momentum	$\mathbf{M}$	$\mathcal{M}\mathcal{L}/\mathcal{T} = \mathcal{M}\mathcal{L}\mathcal{T}^{-1}$
Moment of momentum	$\mathbf{H}$	$\mathcal{M}\mathcal{L}^2/\mathcal{T} = \mathcal{M}\mathcal{L}^2\mathcal{T}^{-1}$
Momentum flux	$\dot{m}\mathbf{V}$	$\mathcal{M}\mathcal{L}/\mathcal{T}^2 = \mathcal{M}\mathcal{L}\mathcal{T}^{-2}$
Temperature	T	Θ
Head	h	$\mathcal{L}$
Work, heat	$\mathcal{W}, \mathcal{Q}$	$\mathcal{M}\mathcal{L}^2/\mathcal{T}^2 = \mathcal{M}\mathcal{L}^2\mathcal{T}^{-2}$
Power	$\mathbf{F}\cdot\mathbf{V}, \mathbf{\Omega}\cdot\mathbf{T}$	$\mathcal{M}\mathcal{L}^2/\mathcal{T}^3 = \mathcal{M}\mathcal{L}^2\mathcal{T}^{-3}$
Heat flux	$\mathbf{q}$	$\mathcal{M}/\mathcal{T}^3 = \mathcal{M}\mathcal{T}^{-3}$
Dissipation function	Φ	$\mathcal{M}/\mathcal{L}\mathcal{T}^3 = \mathcal{M}\mathcal{L}^{-1}\mathcal{T}^{-3}$

Example 10.1

Show that each term in the incompressible Navier-Stokes equation 6.11:

$$\frac{\partial \mathbf{V}}{\partial t} + (\mathbf{V} \cdot \nabla)\mathbf{V} = -\frac{1}{\rho}\nabla p + \mathbf{g} + \nu \nabla^2 \mathbf{V}$$

has the same dimension.

Solution

Using table 10.1, determine the dimension of each term:

$$\left[\frac{\partial \mathbf{V}}{\partial t}\right] = \frac{[V]}{[t]} = \frac{\mathcal{L}\mathcal{T}^{-1}}{\mathcal{T}} = \mathcal{L}\mathcal{T}^{-2}$$

$$[(\mathbf{V} \cdot \nabla)\mathbf{V}] = [V]^2[\nabla] = (\mathcal{L}\mathcal{T}^{-1})^2\mathcal{L}^{-1} = \mathcal{L}\mathcal{T}^{-2}$$

$$\left[\frac{1}{\rho}\nabla p\right] = \frac{[\nabla][p]}{[\rho]} = \frac{(\mathcal{L}^{-1})(\mathcal{M}\mathcal{L}^{-1}\mathcal{T}^{-2})}{\mathcal{M}\mathcal{L}^{-3}} = \mathcal{L}\mathcal{T}^{-2}$$

$$[\mathbf{g}] = \mathcal{L}\mathcal{T}^{-2}$$

$$[\nu \nabla^2 \mathbf{V}] = [\nu][\nabla^2][V] = (\mathcal{L}^2\mathcal{T}^{-1})(\mathcal{L}^{-2})(\mathcal{L}\mathcal{T}^{-1}) = \mathcal{L}\mathcal{T}^{-2}$$

10.2.2 The Principle of Dimensional Homogeneity

When we use algebra and calculus, we ordinarily deal with pure numbers. An equation such as:

$$y = x^2 \tag{10.2}$$

describes a parabola passing through the origin of (x, y)-space in this case a two-dimensional space, or plane. We could equally well write equation 10.2 as:

$$f\{x, y\} \equiv y - x^2 = 0 \tag{10.3}$$

where $f\{x, y\}$ denotes a function of the variables x and y. Geometrically speaking, we can say that $f\{x, y\} = 0$ is the equation of a line in the (x, y)-plane, *i.e.*, a "surface" in the two-dimensional space, (x, y). Since the variables x and y are pure numbers, there is an unlimited number of functions $f\{x, y\}$ that will describe such "surfaces."

Algebraic expressions that describe the relationships among physical variables in an experiment are more restricted than those of mathematical variables in an algebraic equation. For example, suppose we have a collection of square figures of different size cut from paper, and for each one, we measure its area S and the length s of a side. We would find that these two variables are related by:

$$\mathcal{S} = s^2 \tag{10.4}$$

While equation 10.4 looks identical to 10.2, these equations are really quite different. Unlike 10.2, the variables $\mathcal{S}$ and s of 10.4 represent physical quantities and have dimensions:

$$[\mathcal{S}] = \mathcal{L}^2; \qquad [s] = \mathcal{L} \tag{10.5}$$

More important, the dimensions of the two sides of equation 10.5 are the same, since $[s^2] = [s]^2 = \mathcal{L}^2 = [\mathcal{S}]$. This is a universal requirement of equations that represent physical reality. It is called the *principle of dimensional homogeneity*.

We might also express the results of our measurements in the form:

$$f\{\mathcal{S}, s\} \equiv \mathcal{S} - s^2 = 0 \tag{10.6}$$

and be tempted to describe this form as a surface in $(\mathcal{S}, s)$-space. But this doesn't make any physical sense because the dimensions of this space, $[\mathcal{S}] = \mathcal{L}^2$ and $[s] = \mathcal{L}$, respectively, are incommensurate.[1] To think of equation 10.6 as a surface in a space, we must put it in the form:

$$\mathcal{F}\left\{\frac{\mathcal{S}}{s^2}\right\} \equiv \left(\frac{\mathcal{S}}{s^2}\right) - 1 = 0 \tag{10.7}$$

where $\mathcal{F}$ is a dimensionless function. However, notice that there is only one variable in equation 10.7, $\mathcal{S}/s^2$, which is dimensionless. We have reduced the number of variables from two dimensional variables, $(\mathcal{S}, s)$, in equation 10.6 to one dimensionless variable, $\mathcal{S}/s^2$, in equation 10.7 by utilizing the principle of dimensional homogeneity. Furthermore, equation 10.7 has now the form of a surface in a space: the "surface" is the point (1) in the "space" $(\mathcal{S}/s^2)$, a line.

It might seem that there is no difference between equations 10.6 (or 10.4) and 10.7, and that other than its more pleasing geometrical interpretation, the latter is not preferable. But the power of expressing the results of our experiment in dimensionless form has been obscured because we started with the correlation of all our test results, equation 10.4. If we had begun only with the observation that the area $\mathcal{S}$ depended upon the side dimension s in some way not yet understood, then we would have said that:

$$\mathcal{S} = f\{s\} \tag{10.8}$$

[1]There is no way to calculate the distance between two points in such a space or to measure volumes and areas in it.

The dimensionless form of this equation, corresponding to equation 10.7 would be:

$$\left(\frac{S}{s^2}\right) - \frac{f\{s\}}{s^2} = 0 \tag{10.9}$$

Since this is a dimensionless equation, the second term on the left side must also be dimensionless, but the only way that $f\{s\}/s^2$ can be dimensionless is if f =(*constant*)s^2. Thus we can conclude from equation 10.9 that S/s^2 is a constant, whose value can be determined from a measurement of S and s for one square only (where it would be found to be 1). Knowing this, we can save much experimental time by making only one, rather than a large number, of measurements to arrive at the conclusion of equation 10.4.

Let us see how this approach might work in a fluid flow experiment. Suppose that we have a pitot tube installed in a flow channel as shown in figure 4.6 and that we think that the flow velocity V_1 that we want to measure depends upon the manometer column height h, the fluid density ρ, the manometer fluid density ρ_m and the value of g:

$$V_1 = f\{h, \rho, \rho_m, g\} \tag{10.10}$$

Can we deduce the form of f? Let us try to write equation 10.10 in a dimensionless form by dividing it by some combination of h, ρ, ρ_m and g. Now we know from Bernoulli's equation that gz and $V^2/2$ have the same dimension, so we will divide equation 10.10 by $\sqrt{gh}$ to find:

$$\frac{V_1}{\sqrt{gh}} - \frac{f\{h, \rho, \rho_m, g\}}{\sqrt{gh}} = 0 \tag{10.11}$$

Now consider the second term of equation 10.11. It is not only dimensionless, but it must be a function of dimensionless variables that are combinations of h, ρ, ρ_m and g. The only such function would be a dimensionless function of ρ/ρ_m, say $\mathcal{F}$:

$$\frac{V_1}{\sqrt{gh}} = \mathcal{F}\left\{\frac{\rho}{\rho_m}\right\} \tag{10.12}$$

This is a considerable advance over equation 10.10. We need only run a set of experiments in which the density ratio ρ/ρ_m is varied while V_1 is held fixed at a known value. If this were done, we would find that $\mathcal{F}\{\rho/\rho_m\} = \sqrt{2(\rho_m - \rho)/\rho}$.

Although we have considered only two particular examples, additional similar analyses would lead us to conclude that *the relationship among physical variables in a fluid flow may be reduced to a dimensionless relationship among fewer dimensionless variables that are formed by combinations of the physical variables.* The number of reduced variables can be determined from the Π-theorem, to be discussed next.

10.2.3 The Π-Theorem

We usually describe the results of a flow experiment as the determination of how a *dependent* variable, say v_1, depends upon the *independent* variables $v_2, v_3, \ldots, v_n$. The latter are those physical quantities that are under the control of the experimenter and can be varied one at a time to see what effect they will have on the dependent variable.[2] In a wind tunnel experiment, for instance, the dependent variable might be the drag force on an airplane model while the independent variables might be the tunnel flow speed, air density and air temperature. The object of the experiment is to find the relation:

$$v_1 = f\{v_2, v_3, \ldots, v_n\} \tag{10.13}$$

The principle of dimensional homogeneity assures us that this relation can be expressed in dimensionless form but with fewer (dimensionless) independent variables. But how many fewer dimensionless variables will there be? The Π-theorem determines that number:

> A relationship among n physical variables (such as equation 10.13) may be reduced to a dimensionless relationship among $n - r$ dimensionless variables, denoted by $\Pi_1, \Pi_2, \ldots, \Pi_{n-r}$, *e.g.*,
>
> $$\Pi_1 = \mathcal{F}\{\Pi_2, \Pi_3, \ldots, \Pi_{n-r}\} \tag{10.14}$$
>
> The amount r of the reduction in number of the variables is usually equal to the number d of fundamental dimensions required for the dimensions of the n physical variables.[3] The $n - r$ Π's are formed from products of the v_i's but are not uniquely determined. However, they are independent.[4]

The procedure for determining a set of Π's for a particular fluid flow for which there are n dimensional variables $v_1, v_2, \ldots, v_n$ is not set by the Π-theorem. However, the following steps are the easiest way to arrive at the desired result:

1. Prepare a list of the dimensions of the n physical variables, using table 10.1. Determine the number d of fundamental dimensions ($\mathcal{M}$, $\mathcal{L}$, $\mathcal{T}$, Θ) that appear in this list.

---

[2]Independent variables are those which cannot be expressed in terms of each other. For example, only two of the three variables ρ, μ and ν are independent since $\nu = \mu/\rho$.

[3]In rare cases, r may be less than d. In these instances, r is the maximum number of dimensional variables that will not form a dimensionless group.

[4]This means that no Π can be expressed in terms of the others.

2. Assuming the reduction r equals d, select from among the independent variables a number r that contain all of the d fundamental dimensions but which by themselves cannot form a dimensionless Π. Call these the selected variables $v_s, \ldots, v_{s+r-1}$.
3. For each of the remaining $n-r$ unselected variables, form a Π by finding a product of that variable and the selected variables that is dimensionless. To do this select the form:

$$\Pi_i = v_i(v_s)^{a_s} \cdots (v_{s+r-1})^{a_{s+r-1}} \tag{10.15}$$

where the exponents $a_s, \ldots, a_{s+r-1}$ are chosen so as to make the product dimensionless:

$$[v_i(v_s)^{a_s} \cdots (v_{s+r-1})^{a_{s+r-1}}] = 1 \tag{10.16}$$

Sometimes the exponents may be determined easily by inspection[5]. Otherwise the exponents are determined from a set of r algebraic equations.

4. Since the Π's are not unique, any product of the Π's is also a Π. It may be convenient to express equation 10.14 in an alternate form, such as:

$$\Pi_1 = \mathcal{F}\{\Pi_2(\Pi_3)^2, \sqrt{\Pi_3}, \ldots, \Pi_{n-r}\}$$

This does not change the number of Π's, and they are still independent of each other.

Applying the Π-Theorem

To see how this procedure leads to a dimensionless formulation of the physical variables, let us consider the experimental determination of the pressure drop in a steady flow through a straight horizontal pipe, which was treated in chapters 6 and 7. If we were starting with some limited knowledge of the experiments, we would find that the pressure gradient ∇p^* would depend upon the pipe diameter D, the pipe roughness height ϵ, the mean flow velocity $\bar{V}$, the fluid viscosity μ and density ρ, or:

$$\nabla p^* = f\{D, \epsilon, \bar{V}, \mu, \rho\}$$

Now proceed to step 1 and determine the dimensions of these variables using table 10.1:

$$[\nabla p^*] = \mathcal{M}\mathcal{L}^{-2}\mathcal{T}^{-2}; \qquad [D], [\epsilon] = \mathcal{L}; \qquad [\bar{V}] = \mathcal{L}\mathcal{T}^{-1};$$
$$[\mu] = \mathcal{M}\mathcal{L}^{-1}\mathcal{T}^{-1}; \qquad [\rho] = \mathcal{M}\mathcal{L}^{-3}$$

Since only $\mathcal{M}$, $\mathcal{L}$ and $\mathcal{T}$ appear in these dimensions, $d = 3 = r$.

[5] For example, if $[v_i] = \mathcal{L}$ and $[v_s] = \mathcal{L}$, then $\Pi_i = v_i/v_s$.

Proceeding to step 2, we must choose three of the independent variables D, ε, $\bar{V}$, μ, ρ as selected variables. It is usually a good idea to pick each variable from a different category of table 10.1, so we choose D, $\bar{V}$ and μ as the selected variables. Among them they include the three fundamental dimensions, $\mathcal{M}, \mathcal{L}$ and $\mathcal{T}$, and they cannot be combined to give a dimensionless variable.[6]

Following step 4, we next calculate the dimensionless Π's. Start by writing:

$$\Pi_1 = \nabla p^* D^a \mu^b \bar{V}^c$$

and then determine the dimensions of the product:

$$[\nabla p^* D^a \mu^b \bar{V}^c] = (\mathcal{M}\mathcal{L}^{-2}\mathcal{T}^{-2})(\mathcal{L})^a(\mathcal{M}\mathcal{L}^{-1}\mathcal{T}^{-1})^b(\mathcal{L}\mathcal{T}^{-1})^c$$
$$= (\mathcal{M}\mathcal{L}^{-2}\mathcal{T}^{-2})(\mathcal{M}^b\mathcal{L}^{a-b+c}\mathcal{T}^{-b-c}) = \mathcal{M}^{1+b}\mathcal{L}^{-2+a-b+c}\mathcal{T}^{-2-b-c}$$

Now we must choose a, b and c so that the exponents of $\mathcal{M}$, $\mathcal{L}$ and $\mathcal{T}$ are zero:

$$1 + b = 0; \qquad b = -1$$
$$-2 - b - c = 0; \qquad c = -(2 + b) = -(2 - 1) = -1$$
$$-2 + a - b + c = 0; \qquad a = 2 + b - c = 2 - 1 + 1 = 2$$

Thus we find the first Π is:

$$\Pi_1 = \frac{\nabla p^* D^2}{\mu \bar{V}}$$

There are two remaining unselected independent variables, ε and ρ. The first of these easily forms a Π with D:

$$\Pi_2 = \frac{\varepsilon}{D}$$

For the remaining variable ρ, we follow the steps above that we used for ∇p:

$$\Pi_3 = \rho D^a \mu^b \bar{V}^c$$
$$[\rho D^a \mu^b \bar{V}^c] = (\mathcal{M}\mathcal{L}^{-3})(\mathcal{L})^a(\mathcal{M}\mathcal{L}^{-1}\mathcal{T}^{-1})^b(\mathcal{L}\mathcal{T}^{-1})^c = \mathcal{M}^{1+b}\mathcal{L}^{-3+a-b+c}\mathcal{T}^{-b-c}$$

$$1 + b = 0; \qquad b = -1$$
$$-b - c = 0; \qquad c = -b = 1$$
$$-3 + a - b + c = 0; \qquad a = 3 + b - c = 1$$

[6]If we had chosen D, ε and μ, we would fail the last test because D/ε is dimensionless.

so that Π_3 becomes:

$$\Pi_3 = \frac{\rho\bar{V}D}{\mu} = Re_D$$

Π_3 is the familiar Reynolds number based upon pipe diameter. Equation 10.14 thus assumes the form:

$$\frac{\nabla p^* D^2}{\mu\bar{V}} = \mathcal{F}\left\{\frac{\rho\bar{V}D}{\mu}, \frac{\varepsilon}{D}\right\}$$

The dependent variable Π_1 is not a familiar dimensionless variable. If we divide it by Π_2, we have:

$$\frac{\Pi_1}{\Pi_2} = \frac{\nabla p^* D}{\rho\bar{V}^2} = \frac{f}{2}$$

where f is the Darcy friction factor defined in equation 6.47. Following step 4, we may thus substitute f for Π_1 to obtain:

$$f = \mathcal{F}\left\{\frac{\rho\bar{V}D}{\mu}, \frac{\varepsilon}{D}\right\} \tag{10.17}$$

This is precisely the form of equation 9.7 for the friction factor in laminar and turbulent pipe flow. Dimensional analysis has enabled us to specify the number and kind of the dimensionless variables but does not tell us the form of the function $\mathcal{F}$ that relates them. The latter must come from experiment, but as we have seen, the number and kind of experiments that are needed to find $\mathcal{F}$ are greatly reduced by having accomplished a dimensional analysis.

The choice of the selected variables (in this case, D, $\bar{V}$, μ) is not crucial to the outcome of the dimensional analysis. A different choice would lead to a different set of Π's that would be recognizable as products of those in equation 10.17 and could be rearranged to the same form as 10.17 if desired.[7] The results of a dimensional analysis is usually expressed in terms of conventional dimensionless variables, such as Reynolds number, drag coefficient, friction factor, etc.

Example 10.2

A solid sphere of radius a and density ρ_s is dropped into a container of liquid whose density and viscosity are ρ_l and μ_l. A short time after entering the liquid, the sphere

[7]There are seven possible choices of the selected variables for this example that satisfy the criteria of item 2 above.

is observed to descend into the liquid at a constant speed V_f. Derive a dimensionless expression for the dependence of V_f on the experimental variables a, ρ_s, ρ_l and μ_l.

Solution

A sphere falls through the liquid because the gravity force acting on the sphere exceeds the buoyant force. Even though we do not vary the value of g in this experiment, we must consider it to be an independent variable in principle because it characterizes an important force affecting the flow in this case. We therefore need to perform a dimensional analysis for the relation:

$$V_f = f\{\rho_s, \rho_l, \mu_l, a, g\}$$

among the $n = 6$ dimensional variables. The dimensions of these variables are:

$$[V_f] = \mathcal{L}\mathcal{T}^{-1}; \qquad [\rho_s, \rho_l] = \mathcal{M}\mathcal{L}^{-3}; \qquad [\mu] = \mathcal{M}\mathcal{L}^{-1}\mathcal{T}^{-1};$$

$$[a] = \mathcal{L}; \qquad [g] = \mathcal{L}\mathcal{T}^{-2}$$

Since $\mathcal{M}$, $\mathcal{L}$ and $\mathcal{T}$ appear in these dimensions, $r = d = 3$, and there are $n - r = 3$ Π's to be determined, one dependent (Π_1) and two independent (Π_2, Π_3). Choose as selected variables ρ_l, a and g, leaving ρ_s and μ_l as the remaining independent variables. Noting that $\sqrt{ga}$ has the dimensions of velocity, form the first Π from V_f by inspection:

$$\Pi_1 = \frac{V_f}{\sqrt{ga}}$$

The Π formed from ρ_s is clearly:

$$\Pi_2 = \frac{\rho_s}{\rho_l}$$

To form a Π from μ_l, divide it by a density ρ_l times a velocity $\sqrt{ga}$ times a length a, as in an inverse Reynolds number:

$$\Pi_3 = \frac{\mu_l}{\rho_l\sqrt{ga}(a)} = \frac{\mu_l}{\rho_l g^{1/2} a^{3/2}}$$

The general dimensionless relation becomes:

$$\frac{V_f}{\sqrt{ga}} = \mathcal{F}\left\{\frac{\rho_s}{\rho_l}, \frac{\rho_l g^{1/2} a^{3/2}}{\mu_l}\right\}$$

where we have gratuitously inverted Π_3 so as to put it in the form of a Reynolds number.

It is possible to determine $\mathcal{F}$ for a limiting case. For small diameter spheres falling through a very viscous liquid such that the Reynolds number $\rho_l V_f a/\mu_l$ is less than one, the velocity V_f is that for Stokes flow given in equation 6.63, which may be expressed in the dimensionless form:

$$\frac{V_f}{\sqrt{ga}} = \frac{2}{9}\left(\frac{\rho_s - \rho_l}{\rho_l}\right)\left(\frac{\rho_l g^{1/2} a^{3/2}}{\mu_l}\right)$$

This certainly has the general form of the dimensional analysis given above.

If we had choosen ρ_l, a and μ_l as the selected variables, the form of the dimensional analysis would have been:

$$\frac{\rho_l V_f a}{\mu_l} = \mathcal{F}'\left\{\frac{\rho_s}{\rho_l}, \frac{\rho_l g^{1/2} a^{3/2}}{\mu_l}\right\}$$

where the left side is simply the ratio Π_1/Π_3 of the previous dimensional analysis and where $\mathcal{F}' = \mathcal{F}/\Pi_3$. Other choices of the selected variables would lead to different, but equally valid, expressions of the dimensional analysis.

As example 10.2 shows, it is not always self-evident what are the true independent variables in a physical flow. Preliminary experiments may be used to determine whether changing the obvious variables will have some effect on the outcome. Some variables, like the magnitude of the gravitational acceleration g, may be impossible to alter while others, like the density of a liquid, may not readily be changed very much. Where empirical evidence is lacking or is incomplete, we must depend upon previous experience and understanding of the important physical features of the flow to guide us in selecting the significant variables of a given flow.

10.2.4 Dimensionless Form of Conservation Laws

The conservation of mass, momentum or energy in a fluid flow may be expressed in differential or integral form. In either form, the laws have dimensional terms, but by the principle of dimensional homogeneity, each additive term must have the same dimension. Since these expressions are equations with known terms, we do not apply the Π-theorem to them but use the principle of dimensional homogeneity to reduce them to dimensionless form.

Incompressible Flow

As an example, consider an unsteady, viscous incompressible flow with density ρ and viscosity ν constant throughout the flow field. The differential forms of the conservation of mass and momentum, equations 3.17 and 6.12, are:

$$\nabla \cdot \mathbf{V} = 0 \tag{10.18}$$

$$\frac{\partial \mathbf{V}}{\partial t} + (\mathbf{V} \cdot \nabla)\mathbf{V} = -\frac{1}{\rho}\nabla p^* + \nu \nabla^2 \mathbf{V} \tag{10.19}$$

Suppose the flow is characterized by a length scale L, a velocity scale V and a frequency Ω. (For example, the flow might be that through a propeller of diameter L and angular speed Ω advancing into a fluid with a speed V.) We can use the dimensional parameters L, V, Ω and ρ to nondimensionalize the variables of equations 10.18 and 10.19, the dimensionless form being identified by ($'$):

$$t' \equiv \Omega t; \quad \mathbf{R}' \equiv \frac{\mathbf{R}}{L}; \qquad \nabla' \equiv L\nabla; \qquad \mathbf{V}' \equiv \frac{\mathbf{V}}{V}; \qquad (p^*)' \equiv \frac{p^*}{\rho V^2} \tag{10.20}$$

If we substitute these expressions in equations 10.18 and 10.19, they assume the dimensionless form:

$$\begin{aligned} &\nabla' \cdot \mathbf{V}' = 0 \\ &(S)\frac{\partial \mathbf{V}'}{\partial t'} + (\mathbf{V}' \cdot \nabla')\mathbf{V}' = -\nabla'(p^*)' + \left(\frac{1}{Re}\right)(\nabla')^2\mathbf{V}' \end{aligned} \tag{10.21}$$

where the flow Reynolds number Re and *Strouhal number* S are:

$$Re \equiv \frac{VL}{\nu}; \qquad S \equiv \frac{\Omega L}{V} \tag{10.22}$$

The Reynolds number Re may be interpreted as the magnitude of the ratio of the acceleration of a fluid particle in steady flow, $(\mathbf{V} \cdot \nabla)\mathbf{V}$, to the viscous force per unit mass, $\nu\nabla^2\mathbf{V}$, since it is the ratio of the corresponding terms in 10.21. The Strouhal number S is the ratio of the time L/V for a fluid particle to flow a distance L at the speed V to the cycle time $1/\Omega$.

The principal advantage of the dimensionless conservation equations is that the differential terms are of order unity, which is very convenient for numerical computation. If we were to solve these equations, they would have the form:

$$\begin{aligned} \mathbf{V}' &= \mathcal{F}_V\{\mathbf{R}', t'; Re, S\} \\ p' &= \mathcal{F}_p\{\mathbf{R}', t'; Re, S\} \end{aligned} \tag{10.23}$$

where the independent Π's include both the parameters Re and S, as well as the independent variables $\mathbf{R}'$ and t'. From equations 10.23, we could also calculate the nondimensional thrust and torque of the propeller as functions of Re and S.

In some flows, the boundary conditions on equations 10.21 define additional dimensionless parameters that do not appear explicitly in the these equations. If the incompressible fluid has a free surface, such as water with air above it, the boundary condition on the pressure p at a point on the free surface ($z = z_B$) is that it should exceed the constant air pressure p_B by an amount proportional to the surface tension Υ:

$$p_B^* = (p + \rho g z)_B = p_B + \frac{\Upsilon}{r_1 + r_2} + \rho g z_B \tag{10.24}$$

where r_1 and r_2 are the radii of curvature of the surface at B. If we make equation 10.24 dimensionless by dividing it by ρV^2, we find:

$$(p_B^*)' = \frac{p_B}{\rho V^2} + \left(\frac{\Upsilon}{\rho V^2 L}\right) \frac{1}{r_1' + r_2'} + \left(\frac{gL}{V^2}\right) z_B' \tag{10.25}$$

The coefficients in parentheses of the last two terms of equation 10.25 are new dimensionless parameters. Their usual forms are called the *Weber number* and the *Froude number*:[8]

$$W \equiv \frac{\rho V^2 L}{\Upsilon}; \qquad F_r \equiv \frac{V}{\sqrt{gL}} \tag{10.26}$$

The Weber number W may be interpreted as the ratio of the surface energy of a fluid sample of dimension L to its kinetic energy, $\Upsilon L^2/\rho V^2 L^3 = W$. The square of the Froude number F_r is the ratio of the acceleration of a fluid particle in steady flow, $\mathbf{V} \cdot \nabla \mathbf{V} = V^2/L$, to the acceleration of gravity, g.

In some liquid flows, such as the flow around a ship's propeller or in a pump, the fluid may vaporize, forming little bubbles because the pressure p falls below the vapor pressure p_v of the liquid. This phenomenon is called *cavitation.* Over time, cavitation can cause erosion of a propeller or impeller as the vapor bubbles collapse against the sureface when the pressure increases above the vapor pressure. A dimensionless number, the *cavitation number Ca*, is used to characterize the appearance of vapor bubbles in the flow, *i.e.*:

$$Ca \equiv \frac{p_\infty - p_v}{\rho V^2} \tag{10.27}$$

where p_∞ is a reference pressure of the flow. The cavitation number is clearly related to the first term on the right side of equation 10.25.

In summary, we can now see that the solution of the dimensionless conservation equations 10.21 could involve all of these dimensionless parameters, depending upon the boundary conditions:

$$\mathbf{V}' = \mathcal{F}_V\{\mathbf{R}', t'; Re, S, W, Fr, Ca\}$$

$$p' = \mathcal{F}_p\{\mathbf{R}', t'; Re, S, W, Fr, Ca\} \tag{10.28}$$

[8]William Froude (1810–1879) showed how to use model experiments to predict the resistance of ships.

Only in unusual cases would all of these effects to be important simultaneously such as for example, the flow around a high speed ship's propeller close to the ocean surface.

Compressible Flow

If a flow is compressible, the mass and momentum conservation equations 10.18 and 10.19 do not completely describe the flow. Instead, the compressible flow form of the Navier-Stokes equation[9] and the mass conservation equation 3.13 must be used. However, even this is not enough because ρ is now a dependent variable, rather than a constant of the flow. To the conservation of mass and momentum, we must add also the conservation of energy, equation 8.31, together with a thermodynamic relation between the variables p, ρ and $\hat{h}$ and Fourier's law, equation 8.24. These compressible flow equations introduce additional dimensionless parameters, the *Mach number* M[10] and the *Prandtl number* Pr, defined by:

$$M \equiv \frac{V}{a}; \qquad Pr \equiv \frac{\hat{c}_p \mu}{\lambda} \tag{10.29}$$

where $a \equiv [(\partial p/\partial \rho)_s]^{1/2}$ is the speed of sound and λ is the thermal conductivity of the fluid. In addition to these two dimensionless parameters, Re and S also enter explicitly into the dimensionless conservation laws.

Integral Forms

We often use the integral form of a conservation law, such as mass conservation equation 3.9, linear momentum conservation equation 5.11, angular momentum conservation equation 5.46 or energy conservation equation 8.29. They are useful in enabling us to calculate the external force $\mathbf{F}_{ex}$ or torque $\mathbf{T}_{ex}$ applied to the fluid in a control volume or the rate of heat transfer from it. These integral expressions may be rendered nondimensional by the same procedures as were used to derive equations 10.21 above.

The conventional method of expressing the dimensionless external force is the *force coefficient* $C_{\mathcal{F}}$:

$$C_{\mathcal{F}} \equiv \frac{\mathcal{F}}{(1/2)\rho V^2 A_{\mathcal{F}}} \tag{10.30}$$

---

[9]See L. Howarth, ed., 1953, *Modern Developments in Fluid Dynamics*, Vol. I, *High Speed Flows* (Oxford: Oxford University Press) for the expression of the compressible Navier-Stokes equation.

[10]Ernst Mach (1838–1916) demonstrated the existence of supersonic flow and shock waves.

where the force $\mathcal{F}$ is usually the drag or lift force on a body immersed in a moving fluid. The area $\mathcal{A}_{\mathcal{F}}$ is determined by the circumstances of the flow, usually being the planform area of a wing for the lift force and the projected area of a blunt body for the drag force. For an external torque or moment of a force, the *torque coefficient* $C_{\mathcal{T}}$ is usually defined as:

$$C_{\mathcal{T}} \equiv \frac{\mathcal{T}}{\rho V^2 \mathcal{R}_{\mathcal{T}}^3} \tag{10.31}$$

where $\mathcal{R}_{\mathcal{T}}$ is a suitable length, usually the diameter of an impeller or propeller.

The wall shear stress τ_w and heat transfer rate q_w appear in the momentum and energy equations. They are made dimensionless in different ways. For the wall shear stress we define the *friction coefficient* C_f as:

$$C_f \equiv \frac{\tau_w}{(1/2)\rho V^2} \tag{10.32}$$

The dimensionless wall heat transfer rate is the *Nusselt number, Nu*:

$$Nu \equiv \frac{q_w L}{\lambda(\Delta T)} \tag{10.33}$$

where ΔT is the temperature difference between the fluid and the wall and where L is the characteristic dimension of the flow.

Table 10.2 lists the dimensionless parameters defined in this section.

10.3 Modeling

An engineering *model* is a physical or mathematical representation of a *prototype* engineering system design that is utilized to predict some aspects of the prototype behavior. Physical models of smaller size may be used in wind tunnels to obtain information on the prototype flow around airplanes, automobiles and skyscrapers. With the advent of supercomputers, flow over aircraft under conditions that cannot be simulated in a wind tunnel can now be calculated using mathematical models of such flow. Short range forecasts of weather conditions are aided by the exercising of mathematical models of the atmospheric circulation system. In this section, we will concentrate on the physical modeling of engineering systems as carried out in laboratory experiments.

The use of engineering models goes back at least to 1872 when William Froude built the first towing tank to test the resistance of ocean vessels. A small wind tunnel was utilized by the Wright brothers to develop their first aircraft. Nowadays wind tunnels, towing tanks and hydraulic models of dams, rivers and harbors are commonplace aids to the design of engineering systems. The usefulness of such models relies

Table 10.2 Dimensionless parameters of fluid flow

Parameter	Symbol	Value	Equation
Reynolds number	Re	$\frac{VL}{\nu}$	10.22
Strouhal number	S	$\frac{\Omega L}{V}$	10.22
Weber number	W	$\frac{\rho V^2 L}{\Upsilon}$	10.26
Froude number	Fr	$\frac{V}{\sqrt{gL}}$	10.26
Cavitation number	Ca	$\frac{p_\infty - p_v}{\rho V^2}$	10.27
Mach number	M	$\frac{V}{a}$	10.29
Prandtl number	Pr	$\frac{\hat{c}_p \mu}{\lambda}$	10.29
Force coefficient	$C_{\mathcal{F}}$	$\frac{\mathcal{F}}{(1/2)\rho V^2 \mathcal{A}_{\mathcal{F}}}$	10.30
Torque coefficient	$C_{\mathcal{T}}$	$\frac{\mathcal{T}}{\rho V^2 \mathcal{R}_{\mathcal{T}}^3}$	10.31
Friction coefficient	C_f	$\frac{\tau_w}{(1/2)\rho V^2}$	10.32
Nusselt number	Nu	$\frac{q_w L}{\lambda(\Delta T)}$	10.33

upon the principles that underlie the relation, called *similitude*, between the model characteristics and those of the prototype.

10.3.1 Similitude

In geometry we learned how to prove the similarity of plane figures, such as triangles and rectangles. Similar figures had the same shape but different sizes. Corresponding angles must be equal and the ratio of the lengths of corresponding sides also must be equal. This illustrates the principle of *geometric similitude*. As applied to modeling, it requires that the shapes of the model and the prototype be identical, but the size may be different by a geometric *scale ratio SR* that equals the ratio of their characteristic lengths:

$$SR \equiv \frac{L_m}{L_p} \tag{10.34}$$

where L_m and L_p are the same characteristic length (such as ship length or aircraft wing span) of the model and prototype, respectively.[11] If we were to take photographs

[11] A scale ratio of 0.1 is often expressed as $1/10$ or $1:10$.

of a model and its prototype, the photographs would be indistinguishable if taken at distances from the object that are in the same proportion to L_m and L_p.

If we are modeling an engineering structure that moves through a fluid, we must ensure that the model moves with time in a manner that is similar to that of the prototype. This will be achieved if the ratio of the acceleration of the structure to the acceleration of a fluid particle is the same for both model and prototype. For example, if we want to model a wind turbine of radius L rotating at an angular speed Ω in a wind of speed V_w, we must require that the ratio of the centrifugal acceleration of the turbine, $\Omega^2 L$, to a typical acceleration of a fluid element, $\mathbf{V} \cdot \nabla \mathbf{V} \simeq V_w^2/L$, should be the same for both model and prototype:

$$\left(\frac{\Omega^2 L}{V_w^2/L}\right)_m = \left(\frac{\Omega^2 L}{V_w^2/L}\right)_p$$

$$\left(\frac{\Omega L}{V_w}\right)_m = \left(\frac{\Omega L}{V_w}\right)_p \tag{10.35}$$

which is equivalent to the equality of the Strouhal number $\Omega L/V$ for both model and prototype. Note that, in combination with equation 10.34, this implies a constraint on the value of the flow length V/Ω:

$$\frac{(V_w/\Omega)_m}{(V_w/\Omega)_p} = SR \tag{10.36}$$

This illustrates the principle of *kinematic similitude*, where the acceleration of the fluid and its boundaries are maintained in the same ratio for both model and prototype. Using the photographic analogy, kinematic similarity between model and prototype implies that moving pictures of the model and prototype flows would appear identical if the framing rates are proportional to Ω_m and Ω_p and the flow speeds satisfy the relation of equation 10.36.

Although geometric and kinematic similitude are necessary conditions for a model experiment to portray a prototype flow, they are not sufficient. We need to ensure that the entire flow fields of the model and prototype flows are similar. One way to assure ourselves that this will be so is to require that the solutions to the dimensionless conservation laws, as described in section 10.2.4, are identical for both model and prototype flows. For this to be true, the dimensionless variables such as Re, S, F_r, W, *etc.*, must be the same for both model and prototype since they appear as parameters in the dimensionless expression of the conservation laws. However, these dimensionless variables are the same as those Π's that we derived in undertaking a dimensional analysis of a flow problem. We can then ensure a complete *dynamical similitude* between model and prototype flow by requiring that the model and prototype Π's be equal:[12]

$$(\Pi_1)_m = (\Pi_1)_p$$
$$(\Pi_2)_m = (\Pi_2)_p$$
$$\dots$$
$$(\Pi_{n-r})_m = (\Pi_{n-r})_p \tag{10.37}$$

The similarity of model and prototype flow fields is thus ensured if the length scales of all dimensions conform to geometric similarity (equation 10.34) and if the independent dimensionless Π's are the same for both model and prototype (equation 10.37). These conditions enable us to relate the dimensional model and prototype variables to each other, thereby providing a prediction of the prototype variables from a measurement of the model values. This is the objective of the model experiments.

As an example of modeling, suppose that we construct a laboratory scale model of a complex piping system in order to investigate the pressure drops and flow rates in the system. To ensure similitude, we need a geometrically similar model of scale ratio *SR* (all pipe diameters, lengths and roughness heights are in proportion to *SR*) and all independent Π's must be equal for both model and prototype. Using the dimensional analysis of equation 10.17 for pipe flow, the independent Π's may be chosen to be the Reynolds number $\rho\bar{V}D/\mu$ and the roughness ratio ε/D. Because we have a scale model, the roughness ratio equality is ensured:

$$\left(\frac{\varepsilon}{D}\right)_m = \left(\frac{(SR)\varepsilon}{(SR)D}\right)_p = \left(\frac{\varepsilon}{D}\right)_p$$

by virtue of equation 10.34. The equality of the Reynolds numbers requires:

$$\left(\frac{\rho\bar{V}D}{\mu}\right)_m = \left(\frac{\rho\bar{V}D}{\mu}\right)_p$$

$$\frac{\bar{V}_m}{\bar{V}_p} = \frac{D_p\nu_m}{D_m\nu_p} = \frac{1}{SR}\frac{\nu_m}{\nu_p} \tag{10.38}$$

If we decide to use the same fluid in the model as is used in the prototype flow (although this is not necessary[13]), the model flow speed must be higher than the prototype speed by the factor $(SR)^{-1}$.

---

[12]If the independent Π's are equal, then the dependent Π's will be equal also.

[13]It would be safer to use water in the laboratory experiment in place of toxic or flammable liquids in the prototype flow.

Changes in pressure p^* between any two points in the system will be proportional to the pressure gradient ∇p^* times the diameter D, but because the roughness ratio and Reynolds number of the model and prototype are each equal, so must be the friction factors (see equation 10.17):

$$\left(\frac{2\nabla p^* D}{\rho \bar{V}^2}\right)_p = \left(\frac{2\nabla p^* D}{\rho \bar{V}^2}\right)_m$$

$$\frac{(\nabla p^* D)_p}{(\nabla p^* D)_m} = \frac{\rho_p}{\rho_m}\frac{\bar{V}_p^2}{\bar{V}_m^2} = (SR)^2 \frac{(\mu\nu)_p}{(\mu\nu)_m} \tag{10.39}$$

where we have used equation 10.38 to eliminate the velocities in 10.39. If the fluids are the same for model and prototype, then the prototype differences in p^* are smaller than the model values by the factor $(SR)^2$, for corresponding points in the piping system.

10.3.2 Applications of Modeling

Modeling of engineering fluid flows is commonly used in the design of aircraft, ships, road vehicles, and civil works, but there are many special cases of modeling that are useful in research and development, such as blood flow in elastic vessels or the flow of a buoyant plume from a smoke stack. In this section, we consider several examples of modeling flows in engineering systems.

Aerodynamic Drag of Road Vehicles

In recent years, auto and truck manufacturers have sought improved energy efficiency for their vehicles. One component of the resistance to forward motion of such vehicles is the aerodynamic drag force exerted by the air motion relative to the vehicle. This source of resistance may be the major component of fuel consumption in steady horizontal travel at usual highway speeds. Vehicular shape, size and speed determine the amount of the drag force. Testing models in wind tunnels provides information on the effects of design variables on the aerodynamic energy consumption of road vehicles.[14]

To determine the conditions for testing a model of an automobile, we begin with a dimensional analysis of the relation between the aerodynamic drag force $\mathcal{D}$ and the

---

[14]In addition to overcoming vehicle drag, the vehicle engine must provide power for vehicle acceleration, friction losses in the drive train and wheels, engine cooling and vehicle ventilation. Stop-and-go driving is less energy efficient than steady driving because of energy lost to braking and because of the energy consumed while waiting at stoplights.

independent variables, vehicle speed V, vehicle size L, and the air properties ρ and μ:

$$\mathcal{D} = f\{V, L, \rho, \mu\} \tag{10.40}$$

Using table 10.1, we find the dimensions of the variables in equation 10.40:

$$[\mathcal{D}] = \mathcal{MLT}^{-2}; \qquad [V] = \mathcal{LT}^{-1}; \qquad [L] = \mathcal{L};$$
$$[\rho] = \mathcal{ML}^{-3}; \qquad [\mu] = \mathcal{ML}^{-1}\mathcal{T}^{-1} \tag{10.41}$$

There are $n = 5$ physical variables involving $r = 3$ dimensions; hence there are $n - r = 2$ dimensionless Π's. Choosing V, ρ and L as the selected variables (their dimensions involve $\mathcal{M}$, $\mathcal{L}$ and $\mathcal{T}$), the dimensionless independent variable formed from the remaining independent variable μ is clearly the Reynolds number:

$$\Pi_2 = \frac{\rho VL}{\mu} \tag{10.42}$$

Referring to table 10.2, a suitable form for the dimensionless drag force is:

$$\Pi_1 = \frac{\mathcal{D}}{(1/2)\rho V^2 L^2} \tag{10.43}$$

It is customary to replace L^2 in equation 10.43 by the frontal area $\mathcal{A}_f$ of the vehicle, *i.e.*, the vehicle area seen by an upstream observer. When we do so, Π_1 becomes the drag coefficient $C_\mathcal{D}$:

$$C_\mathcal{D} \equiv \frac{\mathcal{D}}{(1/2)\rho V^2 \mathcal{A}_f} \tag{10.44}$$

We can then express equation 10.40 in the dimensionless form:

$$C_\mathcal{D} = \mathcal{F}\{Re\}$$
$$\frac{\mathcal{D}}{(1/2)\rho V^2 \mathcal{A}_f} = \mathcal{F}\left\{\frac{\rho VL}{\mu}\right\} \tag{10.45}$$

The model and prototype Reynolds numbers must be equal. If atmospheric air is used in the wind tunnel tests so that $\rho_m = \rho_p$ and $\mu_m = \mu_p$, then:

$$\left\{\frac{\rho VL}{\mu}\right\}_m = \left\{\frac{\rho VL}{\mu}\right\}_p$$
$$V_m = V_p\left(\frac{L_p}{L_m}\right) = \frac{V_p}{SR} \tag{10.46}$$

The model wind speed must be larger than the prototype vehicle speed by the factor $(SR)^{-1}$. However, to avoid compressibility effects, V_m must be less than sound speed.

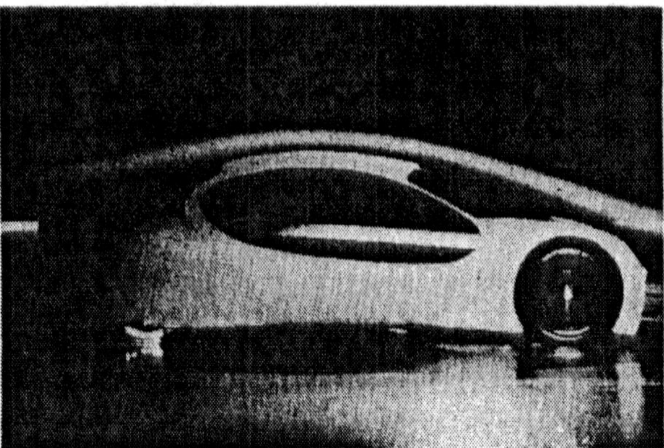

Figure 10.1 A model of an experimental fuel-efficient automobile in a wind tunnel flow, showing the streamline over the top of the model. The reported drag coefficient is $C_{\mathcal{D}} = 0.192$. (Photograph courtesy of the General Motors Technical Center)

This restricts the scale ratio SR to values not less than about 0.3 for modeling at usual highway speeds.

The relationship between the model and prototype drag forces when the model test is conducted at the speed given by equation 10.46 is found by equating the model and prototype drag coefficients:

$$\left(\frac{\mathcal{D}}{(1/2)\rho V^2 \mathcal{A}_f}\right)_p = \left(\frac{\mathcal{D}}{(1/2)\rho V^2 \mathcal{A}_f}\right)_m$$

$$\mathcal{D}_p = \mathcal{D}_m \left(\frac{V_p}{V_m}\right)^2 \frac{(\mathcal{A}_f)_p}{(\mathcal{A}_f)_m} = \mathcal{D}_m (SR)^2 \frac{1}{(SR)^2} = \mathcal{D}_m \qquad (10.47)$$

where we have used the geometric relation that the model area $\mathcal{A}_f$ is $(SR)^2$ times the prototype area. This surprising result, that the prototype drag is the same as the model drag, is due to two counterbalancing effects: the higher flow speed but smaller frontal area of the model as compared with the prototype vehicle.

Figure 10.1 shows a model of an automobile being tested in a wind tunnel. The flow over the top of the automobile is made visible by injecting smoke upstream of the model. Such tests do not completely simulate the air flow relative to the vehicle because the floor of the wind tunnel does not move at the same speed as the air in the wind tunnel and because the wheels do not rotate. By rotating the model about a vertical axis to a different position, the effects of a crosswind on the vehicle forces may be measured.

The Resistance of Ships

When ocean-going sailing vessels were first replaced by steam-powered ships in the latter half of the nineteenth century, little was understood about the amount of power needed to drive these steam vessels. However, by 1872, William Froude had developed the method of using models to predict the power requirements of ocean vessels that is still in use today.

A ship moving over the surface of the ocean creates surface gravity waves which carry energy and momentum away from the ship. The wave system alters the pressure distribution on the ship's hull, producing a horizontal force $\mathcal{D}_w$, called the wave resistance. In addition, viscous shear stresses on the surface of the hull add to give an additional horizontal force $\mathcal{D}_f$, called the frictional resistance. The thrust produced by the ship's propeller must counterbalance the total resistance, $\mathcal{D} \equiv \mathcal{D}_w + \mathcal{D}_f$.

Because the flow of water about the ship has a free surface, the acceleration of gravity g is an independent variable affecting the resistance $\mathcal{D}$, as well as the ship speed V, length L, and water properties ρ and μ:

$$\mathcal{D} = f\{V, L, \rho, \mu, g\} \tag{10.48}$$

If we choose V, L and ρ as the selected variables for a dimensional analysis, the two independent dimensionless variables formed from μ and g become the Reynolds number and the Froude number:

$$\Pi_2 = \frac{\rho VL}{\mu} \equiv Re_L; \quad \Pi_3 = \frac{V}{\sqrt{gL}} \equiv Fr \tag{10.49}$$

while the dimensionless resistance becomes a resistance coefficient:

$$\Pi_1 = \frac{\mathcal{D}}{(1/2)\rho VL^2} \equiv C_{\mathcal{D}} \tag{10.50}$$

The dimensionless form of equation 10.48 is thus:

$$C_{\mathcal{D}} = \mathcal{F}\{Re_L, Fr\}$$

$$\frac{\mathcal{D}}{(1/2)\rho VL^2} = \mathcal{F}\left\{\frac{\rho VL}{\mu}, \frac{V}{\sqrt{gL}}\right\} \tag{10.51}$$

Froude made a brilliant, practical suggestion. He hypothesized that the frictional component of the resistance, $\mathcal{D}_f$, depends only upon the Reynolds number and is equal to the drag force on a flat plate of length L and area $\mathcal{A}_w$ exposed to the fluid, where $\mathcal{A}_w$ equals the area of the ship's underwater surface, while the wave making component of the resistance, $\mathcal{D}_w$, depends only upon the Froude number. In essence, he assumed that there is negligible coupling between the viscous boundary layer on the ship's surface and the wave-generating flow produced by the ship's motion through the water. Expressed in the form of resistance coefficients, Froude's hypothesis is:

$$\frac{\mathcal{D}_f}{(1/2)\rho V^2 \mathcal{A}_w} = (C_{\mathcal{D}})_{fp}\left\{Re_L, \frac{\varepsilon}{L}\right\} \tag{10.52}$$

$$\frac{\mathcal{D}_w}{(1/2)\rho V^2 \mathcal{A}_w} = \mathcal{F}_w\left\{\frac{V}{\sqrt{gL}}\right\} \tag{10.53}$$

where $(C_{\mathcal{D}})_{fp}$ is the flat plate drag coefficient of equations 7.20–7.21 for turbulent flow.[15] By conducting experiments with flat plates and ship forms of various sizes, Froude confirmed the validity of his hypothesis, equations 10.52–10.53.

It is not possible to obtain complete similitude between ship model and prototype because Reynolds number and Froude number cannot both be made equal between the model and prototype. For equal Froude numbers,

$$\left(\frac{V}{\sqrt{gL}}\right)_m = \left(\frac{V}{\sqrt{gL}}\right)_p$$

$$V_m = V_p\sqrt{\frac{L_m}{L_p}} = (\sqrt{SR})\, V_p \tag{10.54}$$

while for equal Reynolds numbers,

$$\left(\frac{\rho VL}{\mu}\right)_m = \left(\frac{\rho VL}{\mu}\right)_p$$

$$V_m = V_p\sqrt{\frac{L_p}{L_m}} = \frac{V_p}{\sqrt{SR}} \tag{10.55}$$

Since the requirements of equations 10.54 and 10.55 are incompatible, only the Froude number or the Reynolds number can be matched. However, by Froude's hypothesis, equation 10.52, we can calculate the frictional resistance $\mathcal{D}_f$ and thereby need only require that the model and prototype Froude numbers be equal (equation 10.53) in order to determine the wave resistance (and total resistance) of the ship from a model experiment.

Figure 10.2 shows a ship model test in a towing tank. In such tests, the model drag force is measured while the model is moving down the tank at a steady speed. This measurement is used in the calculation of the prototype vessel resistance when moving at the same Froude number as the model.

Naval architects sometimes express the wave drag in terms of a resistance coefficient

---

[15]The flow around ships is invariably turbulent because of their size and operating speed. Model experiments are usually designed to insure turbulent flow because laminar flow tends to produce a dissimilar wave pattern emanating from the stern of the ship model.

Figure 10.2 A ship model being tested in the MIT towing tank, modeling the movement of a ship through ocean waves. The model is tested at the same Froude number as the prototype but at a smaller Reynolds number. (Photograph courtesy of Professor Michael S. Triantafyllou, MIT)

$\mathcal{R}_w$ defined as the ratio of the wave drag $\mathcal{D}_w$ to the buoyant force $\rho g \mathcal{V}$:

$$\mathcal{R}_w \equiv \frac{\mathcal{D}_w}{\rho g \mathcal{V}} = \mathcal{F}'_w \left\{ \frac{V}{\sqrt{gL}} \right\} \tag{10.56}$$

where $\mathcal{V}$ is the volume of water displaced by the vessel. The function $\mathcal{F}'$ of equation 10.56 is proportional to V^2/gL times $\mathcal{F}_w$ in 10.53. For a typical Froude number of 0.25, the resistance coefficient of ocean vessels is about $1E(-3)$ and varies approximately as Fr^4.

Example 10.3

A model of an ocean vessel of length $L_p = 100\,m$ is built to a scale of 1 : 50. The model length is $L_m = 2\,m$, displaced volume is $\mathcal{V}_m = 5E(-2)\,m^3$ and wetted area is $(\mathcal{A}_w)_m = 0.9\,m^2$. When tested in a towing tank ($\rho = 1E(3)\,kg/m^3$) at a speed $V_m = 1.1\,m/s$, the total drag force $\mathcal{D}$ is measured to be $2.66\,N$. Calculate (a) the corresponding speed V_p of the prototype ocean vessel, (b) its drag force $\mathcal{D}_p$ in ocean water ($\rho = 1.03E(3)\,kg/m^3$) and (c) its propulsive power $\mathcal{P}_p = \mathcal{D}_p V_p$.

Solution

(a) For equal Froude numbers, equation 10.55 requires:

$$V_p = \frac{V_m}{\sqrt{SR}} = \frac{1.1\,m/s}{\sqrt{0.02}} = 7.778\,m/s = 7.778\,m/s\left(\frac{3600\,s/h}{1.852E(3)\,m/naut.mi.}\right)$$
$$= 15.12\,naut.mi./h$$

A speed of one nautical mile per hour (*naut.mi./h*) is called a *knot*.

(b) To find the wave drag of the model, we first calculate the frictional drag on a flat plate of length $L_m = 2\,m$ and area $(\mathcal{A}_w)_m = 0.9\,m^2$ at a speed $V_m = 1.1\,m/s$, using equation 7.20 for the drag coefficient $C_\mathcal{D}$:

$$(Re_L)_m = \frac{V_m L_m}{\nu} = \frac{(1.1\,m/s)(2\,m)}{1E(-6)\,m^2/s} = 2.2E(6)$$

$$(C_\mathcal{D})_m = \frac{0.455}{[\log(Re_L)_m]^{2.58}} = \frac{0.455}{[\log(2.2E(6))]^{2.58}} = 3.874E(-3)$$

$$(\mathcal{D}_f)_m = \left(\frac{1}{2}\rho V^2 \mathcal{A}_w\right)_m (C_\mathcal{D})_m = [0.5(1E(3)\,kg/m^3)(1.1\,m/s)^2(0.9\,m^2)][3.874E(-3)]$$
$$= 2.11\,N$$

$$(\mathcal{D}_w)_m = \mathcal{D}_m - (\mathcal{D}_f)_m = 2.66\,N - 2.11\,N = 0.55\,N$$

Because the Froude numbers are the same for both model and prototype, their resistance coefficients will also be equal:

$$\left(\frac{\mathcal{D}_w}{\rho g \mathcal{V}}\right)_m = \left(\frac{\mathcal{D}_w}{\rho g \mathcal{V}}\right)_p$$

$$(\mathcal{D}_w)_p = (\mathcal{D}_w)_m \left(\frac{\rho_p}{\rho_m}\right)\left(\frac{\mathcal{V}_p}{\mathcal{V}_m}\right) = (0.55\,N)(1.03)(50)^3 = 7.081E(4)\,N$$

since $\mathcal{V}_m = (SR)^3 \mathcal{V}_p$. To this wave resistance must be added the frictional drag on a flat plate of length $L_p = 100\,m$ and area $(\mathcal{A}_w)_p = (0.9\,m^2)(50)^2 = 2250\,m^2$ at a speed $V_p = 7.778\,m/s$:

$$(Re_L)_p = \frac{(7.778\,m/s)(100\,m)}{1E(-6)\,m^2/s} = 7.778E(8)$$

$$(C_\mathcal{D})_p = \frac{0.455}{[\log(Re_L)_p]^{2.58}} = \frac{0.455}{[\log(7.788E(8))]^{2.58}} = 1.621E(-3)$$

$$(\mathcal{D}_f)_p = \left(\frac{1}{2}\rho V^2 \mathcal{A}_w\right)_p (C_\mathcal{D})_p$$

$$= [0.5(1.03E(3)\,kg/m^3)(7.778\,m/s)^2(2250\,m^2)][1.621E(-3)]$$

$$= 1.136E(5)\,N$$

$$\mathcal{D}_p = (\mathcal{D}_w)_p + (\mathcal{D}_f)_p = 7.081E(4)\,N + 1.136E(5)\,N = 1.844E(5)\,N$$

(c) The power $\mathcal{P}_P$ is:

$$\mathcal{P}_p = \mathcal{D}_p V_p = 1.844E(5)\,N \times 7.778\,m/s = 1.434\,MW$$

Propellers and Wind Turbines

Ship and aircraft propellers may be tested in water or wind tunnels on a smaller geometric scale than the prototypes. While a complete dynamic similitude is possible, usually model tests are conducted at a lower Reynolds number than the prototype to minimize tunnel power requirements.

For propellers, the dependent variables of interest are the thrust F_p and the power input P_p. These are functions of the independent variables, flight speed V_f, air density ρ, viscosity μ, propeller diameter D and angular velocity Ω:

$$F_p = f_F\{V_f,\ \rho,\ \mu,\ D,\ \Omega\}$$

$$P_p = f_P\{V_f,\ \rho,\ \mu,\ D,\ \Omega\} \tag{10.57}$$

Since the dimensions of the variables include the three dimensions $\mathcal{M}$, $\mathcal{L}$ and $\mathcal{T}$, we can reduce the number of independent variables in a dimensional analysis from five to two by choosing three of them as selected variables. If we choose ρ, Ω and D for this purpose, we find the following dimensionless forms of 10.57:

$$C_T \equiv \frac{F_p}{\rho\Omega^2 D^4} = \mathcal{F}_F\left\{\frac{\rho\Omega D^2}{\mu},\ \frac{V_f}{\Omega D}\right\}$$

$$C_P \equiv \frac{P_p}{\rho\Omega^3 D^5} = \mathcal{F}_P\left\{\frac{\rho\Omega D^2}{\mu},\ \frac{V_f}{\Omega D}\right\} \tag{10.58}$$

which define the *thrust coefficient* C_T and *power coefficient* C_P and in which the independent variables are a Reynolds number $\rho(\Omega D)D/\mu$ and an inverse Strouhal number $V_f/\Omega D$.

Aircraft propellers produce maximum thrust and power at take-off (small $V_f/\Omega D$), but less thrust and power are required under level flight at altitude; the thrust and power coefficients are generally decreasing functions of $V_f/\Omega D$. When operating near design conditions, these coefficients are only weakly dependent upon Reynolds number.

The propulsive efficiency η_{prop} is the ratio of the power delivered to the vehicle,

$F_p V_f$, to the propeller shaft power, P_p, and may be related to the thrust and power coefficients through equations 10.58:

$$\eta_{prop} \equiv \frac{F_p V_f}{P_p} = \left(\frac{V_f}{\Omega D}\right)\frac{C_T}{C_P} \tag{10.59}$$

Since η_{prop} cannot exceed unity, $C_P \geq (V_f/\Omega D)C_T$.

For complete dynamic similitude, the Reynolds and Strouhal numbers of model and prototype should be equal. Assuming equal ρ/μ, this requires that:

$$(\Omega D^2)_m = (\Omega D^2)_p$$
$$\left(\frac{V_f}{\Omega D}\right)_m = \left(\frac{V_f}{\Omega D}\right)_p \tag{10.60}$$

which may be satisfied by choosing Ω_m and $(V_f)_m$ so that:

$$\Omega_m = \Omega_p \left(\frac{D_p}{D_m}\right)^2$$
$$(V_f)_m = (V_f)_p \left(\frac{D_p}{D_m}\right) \tag{10.61}$$

The conditions 10.61 require that the model speed V_f be larger than the prototype speed by a factor SR^{-1}, which might introduce compressibility effects in the model experiment, and that the model power exceed the prototype power by the same factor, an impractical requirement. Usually the model test is conducted at the highest practical Reynolds number available in the tunnel, and only the prototype inverse Strouhal number $V_f/\Omega D$ is replicated in the model test.

The thrust and power coefficients and efficiency of a typical ship propeller are shown in figure 10.3 as a function of the dimensionless speed of advance. Both the thrust and power decline with increasing speed, but the efficiency increases to a maximum just short of the speed that results in zero thrust.

For ship propellers, cavitation may be important so the cavitation number $Ca \equiv (p_\infty - p_v)/\rho\Omega^2 D^2$ should be made equal for model and prototype. This is accomplished by adjusting the pressure level p_∞ in the test section of the propeller tunnel. Figure 10.4 is a snapshot of a cavitating flow around a model ship propeller.

For wind turbines, the axial thrust is of little interest. However, the turbine power P_{wt} is made dimensionless by dividing it by the kinetic energy flux through the propeller area $\pi D^2/4$ (see equation 5.21):

$$(C_P)_{wt} \equiv \frac{P_{wt}}{(1/2)\rho V_w^3(\pi D^2/4)} = \mathcal{F}_{wt}\left\{\frac{\rho \Omega D^2}{\mu}, \frac{V_w}{\Omega D}\right\} \tag{10.62}$$

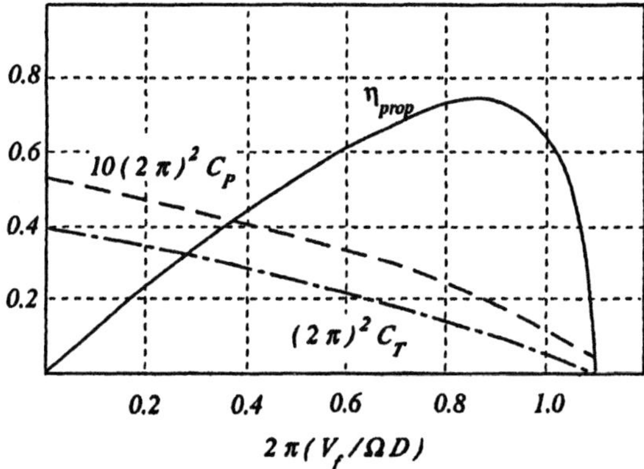

Figure 10.3 The thrust coefficient $(2\pi)^2 C_T$, power coefficient $10(2\pi)^2 C_P$, and efficiency η_{prop} for a model ship's propeller plotted as a function of the ratio $2\pi(V_f/\Omega D)$. (Reprinted, by permission, from E. V. Lewis, ed., 1986, *Principles of Naval Architecture, Second Revision*, Vol. 2, Jersey City, N.J.: Society of Naval Architects and Marine Engineers.)

thus defining the *wind turbine power coefficient* $(C_P)_{wt}$. Wind turbines usually drive an induction generator at constant speed so that the inverse Strouhal number $V_w/\Omega D$ varies only with the wind speed V_w. As for propellers, model tests of wind turbines usually do not duplicate the prototype Reynolds number, only the Strouhal number.

Example 10.4

A model of a wind turbine is tested in a wind tunnel. Its best performance occurs when $(V_w/\Omega D)_m = 0.1$, for which the power coefficient $(C_P)_m = 0.50$. For a similar flow through the prototype turbine at a wind speed $V_w = 10\,m/s$, what turbine diameter D_p and shaft speed Ω_p would you specify to produce a power $(P_{wt})_p = 100\,kW$?

Solution

Noting that the prototype has the same values of $(V_w/\Omega D)_p = 0.1$ and $(C_P)_p = 0.50$ as the model, by rearranging equation 10.62, we can solve for Ω_p,

$$P_{wt} = C_P(1/2)\rho V_w^3(\pi D^2/4) = C_P\left(\frac{\pi}{8}\right)\rho V_w^3 D^2 = C_P\frac{\pi}{8}\rho\frac{V_w^5}{\Omega^2}\left(\frac{\Omega D}{V_w}\right)^2$$

$$\Omega_p^2 = \left\{\frac{\pi C_P \rho V_w^5}{8P_{wt}}\left(\frac{\Omega D}{V_w}\right)^2\right\}_p = \frac{\pi(0.5)(1.2\,kg/m^3)(10\,m/s)^5}{8(1E(5)\,W)(0.1)^2} = 23.56\,s^{-2}$$

Figure 10.4 A model ship propeller in a tunnel showing cavitation forming on the forward face of the propeller at the tip, where the pressure is lower than the water vapor pressure. The helical trail of bubbles shows where the tip fluid moved from earlier times. (Photograph courtesy of Professor J. E. Kerwin, MIT)

$$\Omega_p = 4.854\, s^{-1} = 4.854\, s^{-1} \times \frac{60\, s}{min} \times \frac{rev}{2\pi\, rad} = 46 RPM$$

and then D_p:

$$D_p = \left\{ \left(\frac{\Omega D}{V_w} \right) \frac{V_w}{\Omega} \right\}_p = \frac{10\, m/s}{0.1(4.854\, s^{-1})} = 20.6\, m$$

Pumps and Hydroturbines

Centrifugal or axial flow pumps, fans, and hydroturbines process large volume flows of fluids utilizing the effect of high peripheral speeds of their vanes to change the pressure of the flowing fluid. In contrast with positive displacement pumps and hydraulic motors, which utilize the principle of the piston moving in a cylinder to change the fluid pressure, turbomachines alter the momentum of the working fluid to bring about a pressure change.

A pump will increase the pressure of a fluid by an amount that depends upon the volume flow rate Q, the pump physical variables diameter D and angular speed Ω, and the fluid properties density ρ and viscosity μ. If we measure the pump pressure increase by the change in head h, or more conventionally the product gh, then the dimensional relation between the dependent variable gh and the independent variables may be represented by:

$$gh = f_h\{Q, D, \Omega, \rho, \mu\} \tag{10.63}$$

To convert this relationship into dimensionless form by dimensional analysis, choose D, Ω and ρ as the selected variables to reduce the five independent variables to two:

$$C_h \equiv \frac{gh}{\Omega^2 D^2} = \mathcal{F}_h\left\{\frac{Q}{\Omega D^3}, \frac{\rho \Omega D^2}{\mu}\right\} \tag{10.64}$$

which defines the *head coefficient* C_h. The dimensionless volume flow rate $Q/\Omega D^3$ is called the *flow coefficient* C_Q:

$$C_Q \equiv \frac{Q}{\Omega D^3} \tag{10.65}$$

and the dimensionless variable $\rho\Omega D^2/\mu$ is the flow Reynolds number. At large enough Reynolds numbers ($\geq 1E(5)$), the head coefficient C_h is practically independent of Reynolds number and thus depends only upon the flow coefficient C_Q. The head coefficient C_h is generally a maximum when $C_Q = 0$ and declines toward zero with increasing C_Q.

Another dependent variable of interest is the power P required to operate a pump (or delivered by a hydroturbine). Using dimensional analysis, we can find that the dimensionless power C_P, called the *power coefficient*, is related to C_Q and the Reynolds number by:

$$C_P \equiv \frac{P}{\rho \Omega^3 D^5} = \mathcal{F}_P\left\{C_Q, \frac{\rho \Omega D^2}{\mu}\right\} \tag{10.66}$$

For high Reynolds number, the power coefficient C_P depends only upon C_Q, reaching a maximum value toward the highest value of C_Q, that at which $C_h = 0$.

Using equations 10.64–10.66, the pump efficiency η_p, defined in equation 8.18, can be shown to be related to the head, flow and power coefficients by:

$$\eta_p \equiv \frac{\rho g Q h}{P} = \frac{C_Q C_h}{C_P} \tag{10.67}$$

For a hydroturbine, the turbine efficiency η_t (see equation 8.20) is the reciprocal of 10.67:

$$\eta_t \equiv \frac{P}{\rho Q h} = \frac{C_P}{C_Q C_h} \tag{10.68}$$

The pump or turbine efficiency reaches a maximum value when the flow through the turbomachine best matches the flow for which the blades and flow passages were designed. This usually occurs in the middle of the range of C_Q.

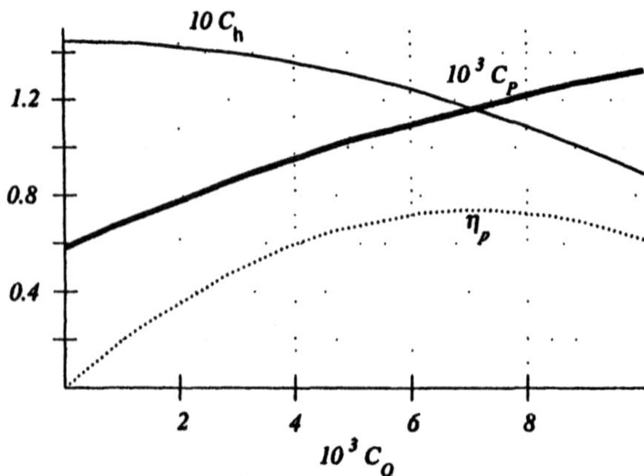

Figure 10.5 A plot of model pump performance characteristics. The ordinate is $10^3 C_P$ for the heavy solid line, $10\, C_h$ for the light solid line and η_p for the dotted line. The abscissa is $10^3 C_Q$. For this pump $N_S^* = 0.4$. (Data, reprinted, by permission, from Val S. Lobanoff and Robert R. Ross, 1992, *Centrifugal Pumps: Design and Application*, [Houston: Gulf Publishing Co.]. All rights reserved.)

A dimensionless parameter independent of D that is often used in selecting pumps for engineering systems is the *specific speed* N_S:

$$N_S \equiv \frac{Q^{1/2}\Omega}{(gh)^{3/4}} = \frac{C_Q^{1/2}}{C_h^{3/4}} \tag{10.69}$$

where the last term in 10.69 is obtained by using equations 10.64 and 10.65 to eliminate gh and Q in the definition of N_S. The square of the specific speed, $(N_S)^2$, is approximately the ratio of the average axial flow speed to the rotor tip speed. The value of N_S at the condition of maximum efficiency, denoted by N_S^*, lies between 0.3 for centrifugal machines and 4.0 for purely axial flow machines, intermediate values applying to mixed flow machines.

If a pump is to be selected to provide a given flow rate Q at a head h, then equation 10.69 may be used to find the speed Ω of the pump drive motor if the value of N_S^* is known for the type of pump being considered. Pump diameter D and power P can then be calculated from values of C_h^* and C_Q^* for this type of pump.

The head and power coefficients and efficiency of a model centrifugal pump are shown in figure 10.5 as a function of the flow coefficient. At higher flow coefficients than shown, both the head coefficient and efficiency will decline further.

Example 10.5

A pump having the characteristics shown in figure 10.5 is to be used to pump 1,000 *GPM* of water against a head of 100 *m*. Assuming the pump operates at maximum efficiency, calculate the pump speed Ω, diameter D and power P.

Solution

Reading values from figure 10.5 at the point of maximum efficiency, $C_Q = 7E(-3)$, $C_h = 0.116$ and $C_P = 1.16E(-3)$. Noting that $Q = (1,000\,gal/min)(min/60\,s)$ $(3.785E(-3)\,m^3/gal) = 6.308E(-2)\,m^3/s$, we use equation 10.69 to find Ω:

$$\Omega = \frac{(gh)^{3/4}(C_Q)^{1/2}}{Q^{1/2}C_h^{3/4}} = \frac{[(9.807\,m/s^2)(100\,m)]^{3/4}[7E(-3)]^{1/2}}{[6.308E(-2)\,m^3/s]^{1/2}[0.116]^{3/4}}$$

$$= 2.937E(2)\,s^{-1} = [2.937E(2)\,s^{-1}]\left(\frac{60\,s}{2\pi}\right) = 2805\,RPM$$

The diameter D is calculated from equation 10.65:

$$D = \left(\frac{Q}{\Omega D}\right) = \frac{6.308E(-2)\,m^3/s}{[2.937E(2)\,s^{-1}][7E(-3)]} = 0.3131\,m$$

and the power from equation 10.66:

$$P = \rho\Omega^3 D^5 C_P = [1E(3)\,kg/m^3][2.937E(2)\,s^{-1}]^3[0.3131\,m]^5[1.16E-3)] = 88.43\,kW$$

10.4 Drag

The drag force $\mathcal{D}$ required to move a body at a steady speed V through a viscous fluid is a function of the body shape, size L and speed V and the fluid properties ρ and μ. The usual dimensional analysis leads to a relationship between the drag coefficient $C_\mathcal{D}$ and the Reynolds number Re_L:

$$C_\mathcal{D} \equiv \frac{\mathcal{D}}{(1/2)\rho V^2 \mathcal{A}} = \mathcal{F}\left\{\frac{\rho V L}{\mu}\right\} \qquad (10.70)$$

where L is the body dimension in the direction of flow and where $\mathcal{A}$ is a suitable body area, either the surface area of the body or the area of the body when projected onto a plane normal to the flow (in the case of blunt bodies).[16] An immense variety of bodies has been tested in wind and water tunnels to determine their drag forces and other interesting attributes of the flow surrounding them. For each shape, the result of such tests may be portrayed by a plot of the drag coefficient as a function of the Reynolds number.

To illustrate the relationship between the drag coefficient and the Reynolds number, we will consider three simple body forms: (*i*) a sphere of diameter D, (*ii*) a circular cylinder of diameter D and length $L \gg D$, whose axis is normal to the flow direction, and (*iii*) a flat plate aligned with the flow of length D in the flow direction and length $L \gg D$ in the normal direction. The drag coefficients for these bodies are shown in figure 10.6 as a function of the Reynolds number $\rho VD/\mu$. (For the sphere, $\mathcal{A} = \pi D^2/4$ while for the cylinder $\mathcal{A} = DL$, where L equals the cylinder length. For the flat plate, $\mathcal{A} = 2DL$ is the surface area of the plate.)

For creeping or Stokes flow, when $Re_D \ll 1$, we have analytical solutions for the viscous flow. In the case of a sphere, the drag force is given by equation 6.62, and the drag coefficient is:

$$(C_D)_{sphere} = \frac{24}{Re_D} \qquad \text{if } Re_D \ll 1 \tag{10.71}$$

while for a cylinder,[17]

$$(C_D)_{cyl} = \frac{8\pi}{Re_D \ln(7.4/Re_D)} \qquad \text{if } Re_D \ll 1 \tag{10.72}$$

These values agree quite closely with the measurements when $Re_D \leq 1$. In this regime, the flow field is nearly symmetric in the flow direction and approximately equal contributions to the drag are due to the pressure and viscous stress on the body surface.

At higher Reynolds number, $1 < Re_D < 100$, a low speed flow, called the wake, develops immediately behind the body, within which mostly the same fluid circulates in a symmetric pattern. Toward the higher end of this range, the wake becomes unsteady with a well-defined frequency of motion in which portions of the wake fluid detach themselves from the wake in a regular pattern called a *vortex street*. For a cylinder, the Strouhal number formed from this frequency f is found empirically to

---

[16]For a body close to a free surface, the Froude number would also be an independent dimensionless variable, as would be the Mach number if the flow were compressible.

[17]See G. K. Batchelor, 1967, *An Introduction to Fluid Dynamics* (Cambridge: Cambridge University Press.)

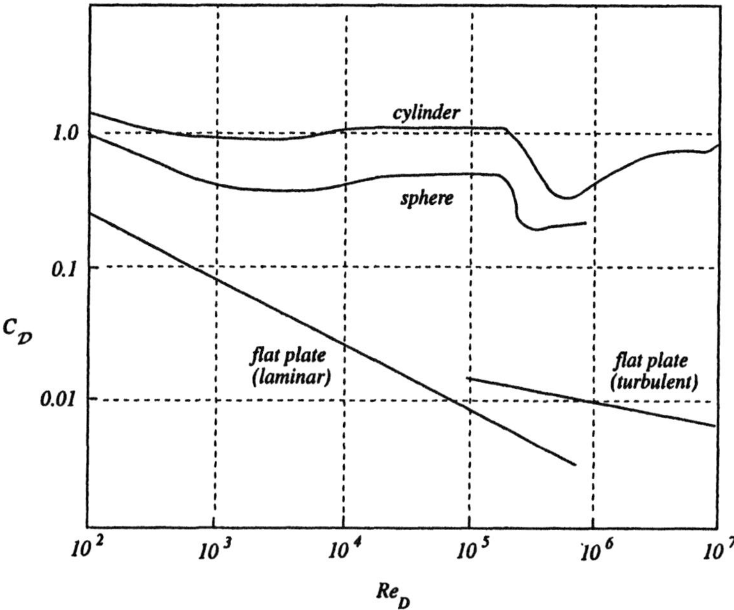

Figure 10.6 The drag coefficients C_D of a sphere, circular cylinder and a flat plate of dimension D plotted as a function of the Reynolds number $Re_D \equiv \rho VD/\mu$. (Reprinted, by permission of Macmillan Publishing Company, from Rolf H. Sabersky, Allan J. Acosta and Edward G. Hauptmann, *Fluid Flow. A First Course in Fluid Mechanics*, 3 ed. Copyright ©1989 by Macmillan Publishing Company)

depend upon the Reynolds number:

$$\left(\frac{fD}{V}\right)_{cyl} = 0.198\left(1 - \frac{19.7}{Re_D}\right) \tag{10.73}$$

that reaches a steady value of 0.198 at high Reynolds numbers. Similar, but less regular, wakes develop behind a sphere. At the upper end of this range of Reynolds number, the component of the drag due to the pressure force on the body surface, called the *form drag*, accounts for almost all the drag because the other component, due to the viscous shear stress on the surface, becomes relatively smaller with increasing Reynolds number, as it does on a flat plate (see equations 6.89–6.90). The pressure in the wake is nearly the same as that far from the body so that the average pressure difference between the front and the rear of the body is approximately $(1/2)\rho V^2$ and the drag coefficient approaches unity.

At yet higher Reynolds numbers, $100 < Re_D < 10^5$, the flow and the drag coefficient show little dependence upon Reynolds number. On the front of the body, a thin laminar boundary layer develops that eventually separates from the body, forming the outside edge of the wake region. The wake region becomes turbulent for $Re_D \geq 10^3$, remaining constant in size and pressure until about $Re_D \simeq 2E(5)$ when the boundary layer on the front of the body becomes turbulent. When this occurs, the separation point moves toward the rear of the body, and the wake size is greatly reduced. Both the drag (and drag coefficient) are significantly reduced with only a slight increase in Reynolds number. Once the boundary layer has become fully turbulent, further increases in Reynolds number produce little change in the flow properties and drag coefficient.

The marked changes in flow at $Re_D \sim 10^5$ occur on smooth, bluff bodies that have a separated boundary layer forming a wake region behind the body, where the separation process can be strongly affected by transition to turbulent flow at, or near, the separation point. It is possible to reduce the Reynolds number at which this transition occurs by roughening the body surface, thereby reducing the drag below the smooth surface value over a limited range of Reynolds numbers centered at $Re_D \sim 10^5$.[18]

Now let us return to the drag on a flat plate aligned with the flow. In Stokes flow ($Re_D \ll 1$) the drag coefficient is:[19]

$$(C_D)_{fp} = \frac{4\pi}{Re_D \ln(24.4/Re_D)} \qquad \textit{if } Re_D \ll 1 \tag{10.74}$$

This is hardly different from the drag on a circular cylinder, equation 10.72, if we take into account the difference in the reference area $\mathcal{A}$.[20] This reflects the fact that, in Stokes flow, the drag is not much influenced by the body shape. In the laminar boundary layer regime, $10 < Re_D < 10^5$, the flat plate drag coefficient is that of equation 6.93:

$$C_D = \frac{1.3282}{\sqrt{Re_D}} \tag{10.75}$$

This flat plate drag coefficient is always lower than that for the blunt shapes, the sphere and the circular cylinder, the more so at the higher Reynolds numbers. Even when the flat plate boundary layer becomes turbulent ($Re_D \gg 10^5$), $(C_D)_{fp}$ (equation 7.20 and figure 7.5) is still considerably less than that for the blunt shapes. This difference in behavior between the flat plate and the circular cylinder or sphere stems from the

---

[18] It is claimed that dimpled golf balls benefit from this effect.

[19] See H. Lamb, 1945, *Hydrodynamics*, 6 ed. (New York: Dover Publications).

[20] If we define $\mathcal{A}$ in the drag coefficient (equation 10.70) as the surface area of the body, then we should divide C_D by π and 4, respectively, for the circular cylinder and sphere in figure 10.6 and equations 10.72–10.71.

fact that the form drag of a flat plate is zero and that the entire drag is caused by the tangential shear stress on the surface of the plate. For $Re_D \gg 1$, the drag coefficient of blunt bodies is of the order of unity because the separated flow produces a wake of ambient pressure behind the body.

Separated flow can be prevented by streamlining a body, *i.e.*, by rounding sharp corners and increasing the streamwise dimension compared to the transverse dimension. A slender, airfoil-shaped body has a drag coefficient much closer to a flat plate than to a circular cylinder. At high Reynolds number, it pays to streamline bodies to take advantage of the reduced drag.

Values of C_D for bodies of various shapes at selected Reynolds number, are given by Blevens.[21]

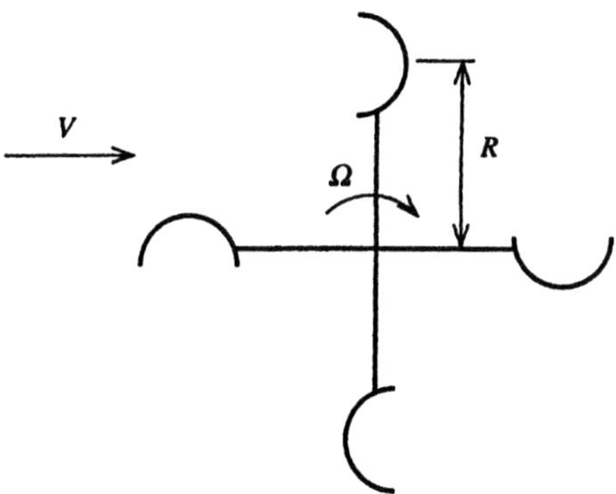

Figure E 10.6

Example 10.6

A rotating cup anemometer measures the speed of the wind. It consists of four hemispherical cups attached to the end of two crossed arms that, as sketched in figure E 10.6, rotate in the wind in the clockwise direction with an angular velocity Ω that is measured by an electronic frequency meter. When in the position shown in figure E 10.6, the upwind facing cup (at the top of the figure) has a drag coefficient $(C_D)_u = 1.4$ while the remaining three cups, which face downwind with respect to their relative flows have equal drag coefficients, $(C_D)_d = 0.4$. Noting that the net

---

[21] R. D. Blevens, 1984, *Applied Fluid Dynamics Handbook* (New York: Van Nostrand Reinhold).

torque T caused by the cup drag forces is zero, calculate the dimensionless angular speed $\Omega R/V$ at which the anemometer rotates.

Solution

The torque T due to each cup is:

$$T = R\left(\frac{1}{2}\rho V_{rel}^2 A C_D\right) = \frac{1}{2}\rho R A (C_D V_{rel}^2)$$

where V_{rel}^2 is the speed relative to the cup. Adding the torques of the four cups,

$$T = \frac{1}{2}\rho R A[(C_D)_u(V - \Omega R)^2 - (C_D)_d(V + \Omega R)^2 - 2(C_D)_d(\Omega R)^2] = 0$$

$$\frac{(C_D)_u}{(C_D)_d}(V - \Omega R)^2 - [(V + \Omega R)^2 + (\Omega R)^2] = 0$$

which may be solved for the ratio $\Omega R/V$:

$$\left(\frac{\Omega R}{V}\right)^2 = \frac{(C_D)_u - (C_D)_d}{(C_D)_u + 4(C_D)_d} = \frac{1}{3}$$

$$\frac{\Omega R}{V} = 0.5774$$

Note that the anemometer would work if the cups were replaced by any bluff asymmetrical shape having different drag coefficients for the forward and reverse flow. The hemispherical cups have a greater difference in these drag coefficients than any other shape and therefore yield a higher value for $\Omega R/V$.

10.5 Lift

We tend to think that the concept of an aerodynamic lift force was born when gliders and airplanes were first developed in the late nineteenth and early twentieth century. But long before then a practical understanding of lift was known to the designers of sailing vessels and windmills and to the engineers who developed the ship propeller. Today the quantitative characterization of lift lies at the heart of the design of airplanes, propellers, aircraft engines, sailboats, hydrofoil watercraft, pumps, hydroturbines, wind turbines, fans, and blowers.

The generation of a lift force by an airplane wing placed at an angle of attack to an oncoming flow, as illustrated in figure 10.7, is a consequence of the distortion of the streamlines of the fluid passing above and below the airfoil, both of which are shown

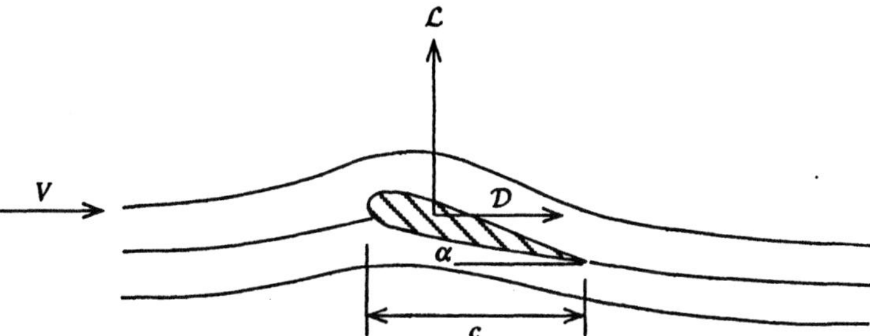

Figure 10.7 Streamlines of the flow for an airfoil at an angle of attack α in a steady flow of speed V. The components of the force on the airfoil are the lift $\mathcal{L}$ and drag $\mathcal{D}$.

to be concave downward. For a steady inviscid flow, the pressure decreases toward the center of curvature of a streamline (equation 4.19), which means that the pressure on the upper surface of the airfoil (called the suction side) is less than ambient pressure while that on the lower surface (the pressure side) is greater. This pressure difference gives rise to lift force $\mathcal{L}$, which is defined as the component of the total force on the wing in the direction normal to the oncoming flow. The other component, in the direction parallel to the oncoming flow, is the drag force $\mathcal{D}$.

Both the lift and drag forces depend upon the airfoil shape, its cord dimension c, the wing span s, the air speed V, and angle of attack α as well as the fluid properties ρ and μ. By applying dimensional analysis to these relationships, we find that the lift and drag coefficients depend upon the Reynolds number $\rho Vc/\mu$, the angle of attack α, and the ratio s/c (called the *aspect ratio*):

$$C_{\mathcal{L}} \equiv \frac{\mathcal{L}}{(1/2)\rho V^2(cs)} = \mathcal{F}_{\mathcal{L}}\left(\frac{\rho Vc}{\mu}, \alpha, \frac{s}{c}\right) \tag{10.76}$$

$$C_{\mathcal{D}} \equiv \frac{\mathcal{D}}{(1/2)\rho V^2(cs)} = \mathcal{F}_{\mathcal{D}}\left(\frac{\rho Vc}{\mu}, \alpha, \frac{s}{c}\right) \tag{10.77}$$

where the product cs is the wing area.

At any given Reynolds number and aspect ratio, both the lift and drag coefficients tend to increase with angle of attack α, as shown in figure 10.8. However, the drag coefficient is much less than the lift coefficient, usually by about a factor of ten.[22] The lift coefficient at first rises linearly with α, but then reaches a maximum, usually about $C_{\mathcal{L}} \sim 1$, implying that the average pressure difference between the lower and upper airfoil surfaces is approximately the dynamic pressure, $(1/2)\rho V^2$. This leveling off

[22]The disparity in size between $C_{\mathcal{L}}$ and $C_{\mathcal{D}}$ is essential to the economical flight of aircraft and the energy efficiency of bird flight.

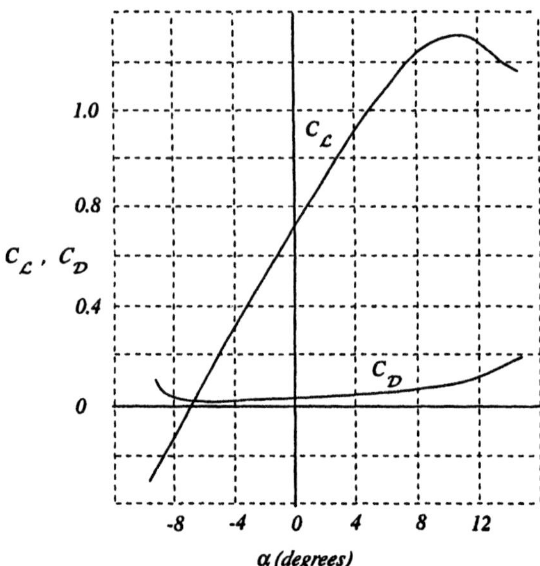

Figure 10.8 Typical lift and drag coefficients $C_\mathcal{L}$ and $C_\mathcal{D}$ versus angle of attack α for a wing. (From H. Schlicting, 1968, *Boundary-Layer Theory*, 6 ed. [New York: McGraw-Hill Book Co.]. Reprinted with permission of McGraw-Hill Book Co.)

of $C_\mathcal{L}$ is caused by the separation of the boundary layer from the upper surface of the airfoil, creating a large wake region and greatly increasing the drag. This condition is called *stall.*

If the wing aspect ratio s/c is reduced, the lift coefficient is reduced, and the drag coefficient increased. These changes are caused by the greater influence along the span of the deleterious flow at the wing tips. The corresponding increase in drag is called the *induced drag*.

Airplanes are designed to cruise under conditions where the lift-drag ratio $\mathcal{L}/\mathcal{D}$ is a maximum, *i.e.*, the drag is minimized for the amount of lift needed to overcome the force of gravity on the airplane and its cargo. However, when taking off and landing, which occur at lower speeds than cruising in level flight, ρV^2 is smaller, and $C_\mathcal{L}$ must be proportionately higher to maintain the same lift force. This is accomplished by increasing the angle of attack to increase $C_\mathcal{L}$ and applying sufficient engine thrust to overcome the effects of the much higher $C_\mathcal{D}$. In commercial jet aircraft, lift during landing and takeoff is enhanced by deploying flaps that increase both wing area and the maximum $C_\mathcal{L}$ but at the price of increasing the drag.

The same considerations govern the design of blades in an axial flow compressor. We would like to maximize the difference between the components of the lift and drag forces in the direction of motion of the blade so as to maximize the pressure rise, but because the pressure rises in a compressor stage, the blade is more susceptible to flow separation, which limits the most efficient angle of attack and the consequent pressure rise obtainable in a stage.

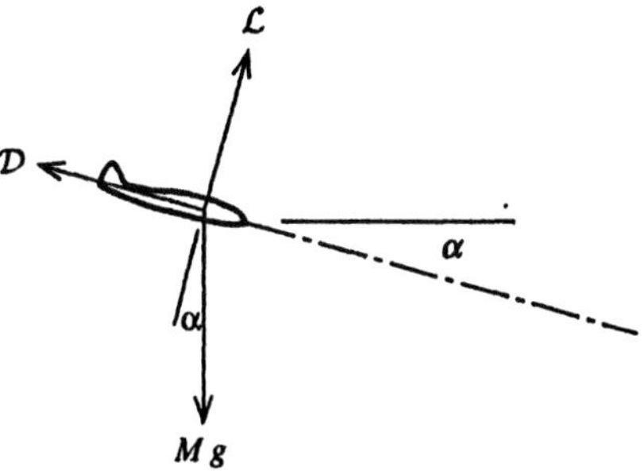

Figure E 10.7

Example 10.7

A sailplane glides steadily through still air at an angle α below the horizontal, as sketched in figure E 10.7. Prove that α depends only upon the ratio $C_{\mathcal{L}}/C_{\mathcal{D}}$ for the sailplane.

Solution

Since the sailplane is not accelerating, the aerodynamic forces must balance the gravity force on the vehicle. The force of gravity Mg has one component that balances the lift force $\mathcal{L}$ and another that balances the drag force $\mathcal{D}$:

$$\mathcal{L} = Mg \cos \alpha$$

$$\mathcal{D} = Mg \sin \alpha$$

Eliminating Mg:

$$\alpha = \cot^{-1}\left(\frac{\mathcal{L}}{\mathcal{D}}\right) = \cot^{-1}\left(\frac{C_{\mathcal{L}}}{C_{\mathcal{D}}}\right)$$

10.6 Problems

Problem 10.1

A medical researcher investigating contraceptive techniques is studying the speed with which various spermatozoa propel themselves. Since their motion is a very low Reynolds number phenomenon, the researcher recognizes that their speed V cannot depend upon the density of the fluid through which they move. What is more, since the various species are roughly similar, the major parameters which should govern the speed of propulsion of spermatozoa are the length L of the organism, the viscosity μ of the fluid through which it moves and the rate of energy expenditure $\dot{e}$ per unit volume of organism. (a) From dimensional considerations, derive the functional form of the dependence of the propulsion speed V on L, μ and $\dot{e}$. (b) If a spermatozoa starts out having available a total energy e per unit volume, derive an expression for the total distance ℓ that it can travel in terms of L, μ, e and $\dot{e}$.

Problem 10.2

If a long rectangular piece of cardboard or stiff paper is allowed to fall freely in air, it is observed to rotate at a constant frequency ω as it falls. Assuming that this frequency depends upon the the card length s, card width c, card mass per unit area Δ, air density ρ, air viscosity μ and gravity g, (a) define a dimensionless tumbling frequency and determine the dimensionless groups upon which it depends. (b) If the viscosity μ and the card length s are unimportant, determine how the results of (a) are affected. (c) We can further argue that, as the card falls at a constant vertical velocity, there exists a balance between the drag per unit area and the gravitational force per unit area. The drag depends only upon aerodynamic forces and hence upon ρ, c and f alone. Using this information, obtain an exact expression for the ratio ω_A/ω_B of tumbling frequencies in two fluids of different densities, ρ_A and ρ_B.

Problem 10.3

Rain drops are formed in clouds by condensation of water vapor around nuclei. As they fall through the air, larger drops fall faster than smaller ones, catching up and merging with the latter. As a result a larger drop becomes ever larger by this agglomeration process until the pressure force set up by the airflow around it overcomes the surface tension force which holds the drop together and causes it to break up into small drops. This scenario is repeated until the drop reaches the earth.

(a) Assuming that the large drop diameter D at which break-up begins is dependent upon its free-fall speed U, the air density ρ_a and the surface tension Υ, derive an expression that relates D with these other variables. (b) For a very small droplet, the free-fall speed V is expected to depend upon its diameter D, the air viscosity μ_a

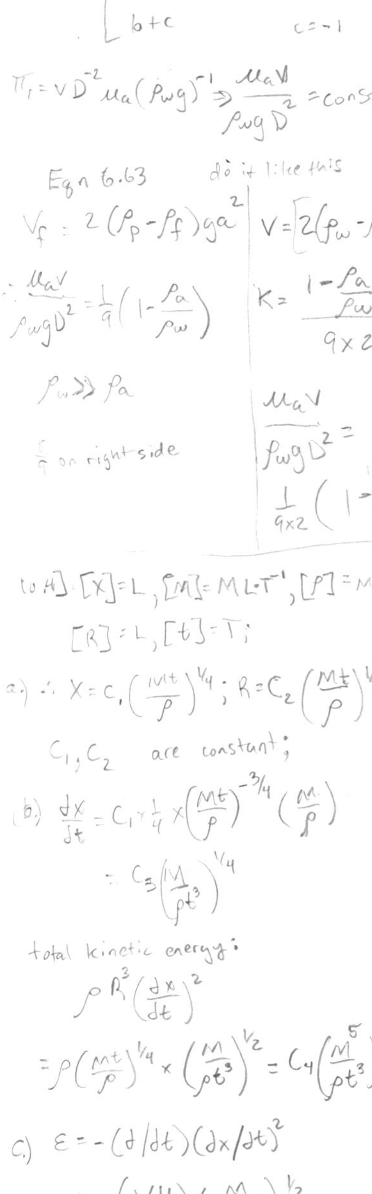

and the product $\rho_w g$, which is the gravity force per unit volume of droplet. Derive a relation between V and these other variables. Compare your answer to equation 6.63 and explain the difference.

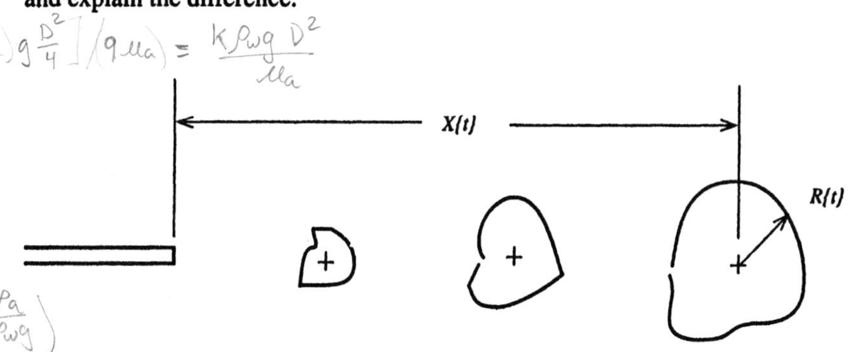

Figure P 10.4

Problem 10.4

A puff of air, marked by smoke, is suddenly ejected from the end of a small-diameter tube into a stationary, constant-pressure atmosphere. As the puff moves away from the end of the tube, it grows in volume by turbulent mixing with the surrounding air yet remains approximately spherical in shape. At various times after ejection, the distance $X\{t\}$ the puff has moved and its average radius $R\{t\}$ are measured.

The ejection of the puff imparts a total horizontal momentum M that remains constant over time as the puff mixes with the surrounding air. Since the average horizontal velocity of the puff is dX/dt, conservation of horizontal momentum for the puff may be written approximately as:

$$\rho\left(\frac{4\pi}{3}R^3\right)\frac{dX}{dt} \simeq M$$

where ρ is the air density.

The evolution of the puff is an unsteady turbulent flow in which the observed quantities X and R depend upon time t and the parameters M and ρ but not upon the viscosity μ. Employing dimensional analysis, answer the following questions. (a) Derive expressions for the dependence of X and R on t, ρ and M. (b) Assuming that the kinetic energy per unit mass in the puff is proportional to $(dX/dt)^2$, show how the kinetic energy of the puff varies as a function of time t. (c) The turbulent kinetic energy dissipation rate $\epsilon \simeq -(d/dt)(dX/dt)^2$. Determine how ϵ depends upon t, ρ and M.

Problem 10.5

A humming bird is a 1/50 linear scale model of an albatross. Both birds fly in a gravitational field g in air of density ρ_a, have the same average density ρ_b and store the same energy per unit mass of bird, ϵ. The size of each bird can be characterized by the wing span d.

(a) Determine the minimum set of independent variables that is sufficient to specify completely the mass M of either bird. Calculate the ratio of the mass M_a of an albatross to the mass M_h of a humming bird. (b) It has been suggested that the frequency f at which a bird flaps its wings should depend at most upon M, ϵ, ρ_b, d, g and ρ_a. Using dimensional analysis, express a dimensionless flapping frequency of a bird in terms of a set of other dimensionless variables. (c) The effective value of g may be changed artificially by exposing the bird to a vertical acceleration. Tests conducted with humming birds show that the flapping frequency is proportional to the square root of the effective value of g, other things being held fixed. Modify your expression in (b) to reflect this experimental fact. (d) Based upon your answer to (c), how does f depend upon ϵ? (e) Based upon your answer to (d), at what frequency does a humming bird flap if an albatross flaps at 1 beat per second?

Problem 10.6

A wind tunnel is being used to determine the drag on one-quarter scale models of automobiles ($SR = 1/4$). The model length $L_m = 1\,m$. The wind tunnel density $\rho = 1.1\,kg/m^3$, viscosity $\mu = 2E(-5)\,Pa\,s$ and pressure $p = 1E(5)\,Pa$ are the same as those of the prototype flow. The full-sized vehicle speed to be modeled by the tunnel experiments is $V_p = 25\,m/s$. (a) Calculate the wind speed V_m to be used in the model tests. (b) The model drag force measured at this speed is 600 N. Calculate the prototype vehicle drag force D_p. (c) After extensive testing, the model drag coefficient $(C_D)_m$ is found to be independent of Reynolds number. Its value, based upon a model cross-sectional area $A_m = 0.19\,m^2$ is $(C_D)_m = 0.50$. Calculate the power $\mathcal{P}_p$ required to overcome the air drag on the full-scale vehicle. (d) An improved model is tested at the speed V_m given in (a). It is found that the power required to operate the tunnel is higher by 15 kW than when the model is absent from the tunnel, all other conditions being identical. Calculate the power required to overcome the air drag of the full-scale car of this new design.

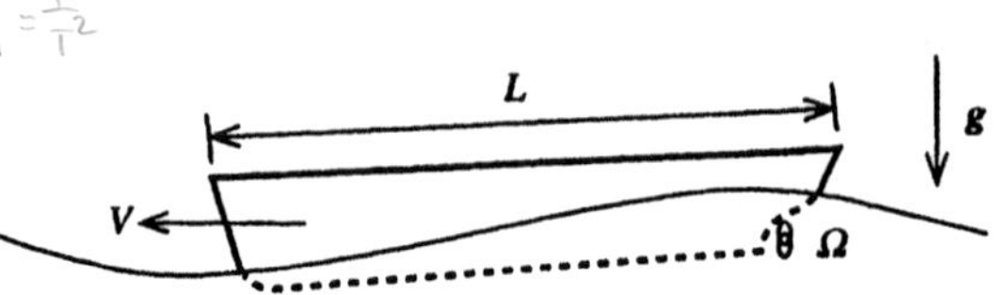

Figure P 10.7

Problem 10.7

A self-propelled model of a high speed ship is to be tested in a ship model basin. The prototype ship, whose model is to be tested, is designed to be 150 m long (L_p) and is expected to travel at 60 km/hr (V_p) when the propeller speed (Ω) is 400 RPM. The ship model length $L_m = 5\,m$. In these tests, because of the high ship speed, viscous forces on the hull are believed to be negligible compared to pressure forces. (a) At what speed (Ω_m) should the model propeller be rotated so as to model the design speed of 400 RPM? (b) When the test is run with the propeller speed given in (a), the model speed $V_m = 3\,m/s$. At what speed V_p would you predict the full-scale ship to run? (c) The ship designers are concerned about how the vessel will perform in the presence of ocean waves. Under storm conditions, it is expected that waves of length $\lambda_p = 300\,m$ and height $h_p = 10\,m$ will be encountered. What wave length λ_m and height h_m should be generated in the model basin for tests that would model the prototype performance? (d) In tests that are run according to the requirements of (c), a maximum internal bending moment (M_m) in the ship model structure of $6E(5)\,Nm$ is measured. What bending moment M_p would you predict for the prototype ship for these conditions?

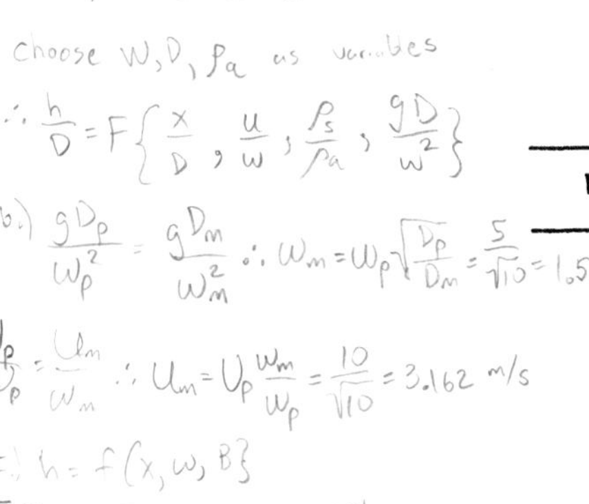

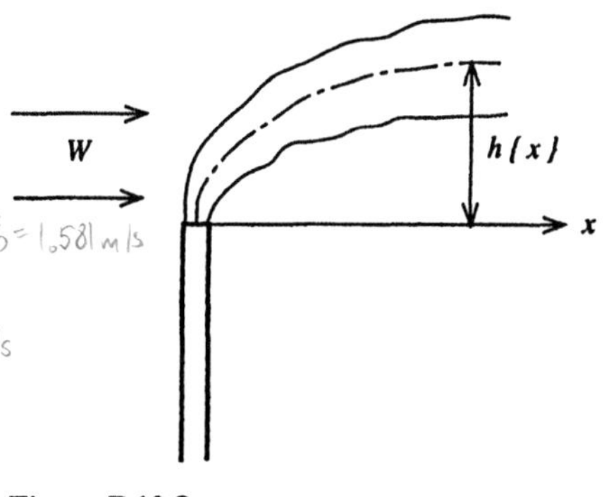

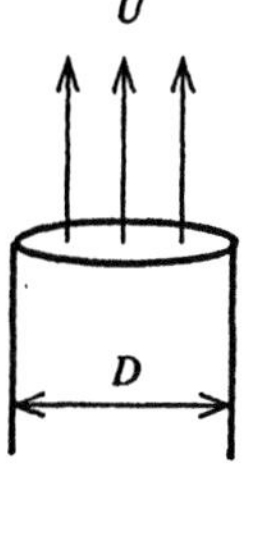

Figure P 10.8

Problem 10.8

When the wind blows past a smokestack, the hot gas leaving the stack rises vertically as it drifts downwind, as shown in figure P 10.8. We wish to model this flow in a wind tunnel by running experiments that are dynamically similar to the full-scale flow.

We are interested in determining the height $h\{x\}$ of the plume centerline as a function of the distance x downwind of the stack. From physical considerations we expect the height h to depend upon x, the wind speed W, the gas speed U at the stack exit, the stack diameter D, the air density ρ_a, the stack gas density ρ_s at the stack exit and gravity g because the flow is buoyant. (Because both the model and full-scale flows are turbulent free shear flows having high Reynolds numbers, we do not expect them to depend significantly upon the fluid viscosity.)

(a) Derive a complete set of dimensionless variables whose equality for model and full-scale flows will ensure dynamical similarity. (b) Assuming that the wind tunnel model will have the same values for the densities as the full-scale flow, calculate the tunnel wind speed W_m and exit speed U_m required to simulate a full-scale wind speed $W_p = 5\,m/s$ and exit speed $U_p = 10\,m/s$ when the model stack diameter D_m is one-tenth of the full-scale value D_p. (c) It is claimed that the height h far from the stack exit depends only upon the distance x, wind speed W and a product B (called the buoyancy flux):

$$B \equiv gUD^2\left(1 - \frac{\rho_s}{\rho_a}\right)$$

If this is true, derive a dimensionless implicit expression for the height h as a function of x. (d) It is planned to take moving pictures of the model experiment using a high framing speed of Ω_m frames per second so that when the film is shown at a standard speed of Ω_0 frames per second it will simulate the full-scale turbulent motions, *i.e.*, it will show the correct frequencies of the unsteady turbulent eddies at the full-scale conditions. What should be the ratio Ω_m/Ω_0 if the tunnel is run under the conditions described in (b)?

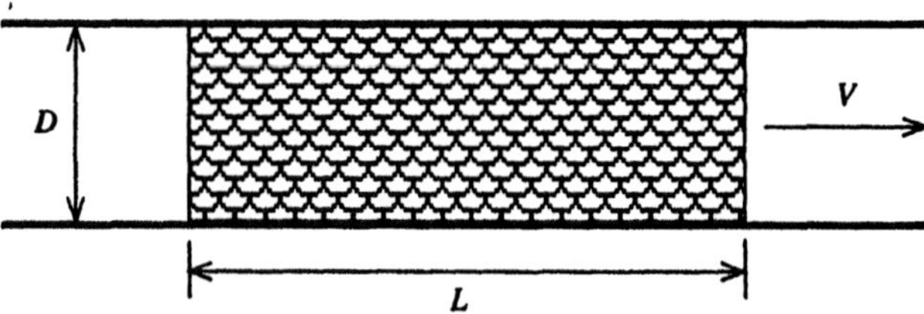

Figure P 10.9

Problem 10.9

In order to damp out large-scale turbulence, a section of a pipe of diameter D is filled with spherical balls of diameter d for a length L, as figure P 10.9 shows. The balls are kept in place by two fine-mesh screens, the effects of which can be neglected. It is desired to predict the dependence of the pressure drop Δp across this section on the fluid density ρ, viscosity μ, mean velocity V and the lengths d, D and L. Assume the fluid is incompressible.

(a) Express the functional relationship between Δp and the remaining variables in dimensionless form. (b) Simplify your answer to (a) for the case of negligible wall effects when $d \ll D$. (c) Further simplify your answer to (b) for the case when end effects are also small ($d \ll L$). (d) In operation, air will be used as the fluid. In order to determine the pressure drop, a $1/10$ scale model is constructed and tested with water as the fluid. What should be the ratio of the water velocity V_w to the air velocity V_a in order to simulate the air flow with the water model? Under this condition, what will be the ratio of the pressure drops, $\Delta p_a/\Delta p_w$?

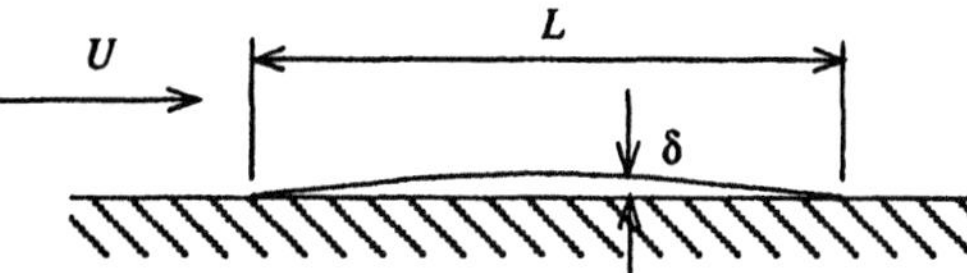

Figure P 10.10

Problem 10.10

In the blow drying of a thin liquid film, the liquid evaporates and the vapor is swept away by the air moving past the surface of the film, as depicted in figure P 10.10. We are interested in how the drying time t depends upon the film length L and film thickness δ, the liquid vapor pressure p, the air speed U, the air viscosity μ, and air density ρ. (a) Derive a set of dimensionless variables for relating the drying time t to the remaining variables. (b) We would like to design a laboratory experiment to determine the drying time of a football field where $p_p = 2E(3)\,Pa$, $L_p = 100\,m$, $\delta_p = 0.01\,m$ and $U_p = 2\,m/s$. In this experiment, the air density and viscosity will be the same as that for the football field, but $L_m = 20\,m$. Calculate the values of U_m, δ_m and p_m in order to ensure dynamic similarity with the full-scale flow. (c) If the drying time measured in the laboratory experiment of (b) is $t_m = 10\,min$, calculate the time t_p to dry the football field.

Problem 10.11

A new design of a rotary pump, which will be used to raise ground water from a well through a height H_p, is being developed. The pump impeller diameter is D_p, and it will operate at an angular speed Ω_p. The pump head increase, measured as gH, is expected to depend upon the diameter D, the angular speed Ω, the fluid density ρ, the fluid viscosity μ, and the volumetric flow rate Q. (a) Find the dimensionless groups that govern the performance of the pump. (b) To study the pump performance, a scale model of diameter $D_m = D_p/5$ is built. Calculate the ratios Q_m/Q_p and Ω_m/Ω_p that will ensure that the flows are dynamically similar when water is used in the model

experiment. (c) Calculate the values of H_m/H_p and $\mathcal{P}_m/\mathcal{P}_p$, where $\mathcal{P}$ is the pump power.

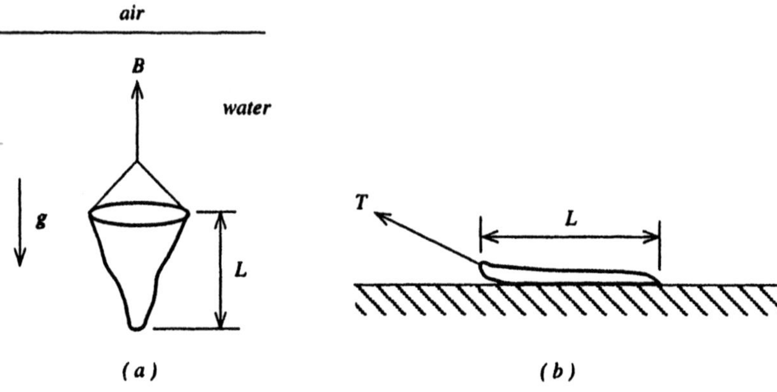

Figure P 10.12

Problem 10.12

A ship model towing tank is to be used for a model test of the towing of a fish net along the bottom of the ocean. The model net will be one-tenth the size of the full-scale (prototype) net. In order to duplicate the full-scale behavior, the same materials having the same specific gravity SG as the prototype will be used in the scale model, $(SG)_m = (SG)_p$.

(a) When the net is suspended motionless over the side of a stationary vessel, it exerts a tension B in the tow cable as shown in figure P 10.12(a). Calculate the ratio B_m/B_p of the values of B for the model and the prototype, assuming that B depends upon L, SG, g, and the water density ρ. (b) When towed along the bottom, as in figure P 10.12(b), the tension T in the towline is expected to depend upon the tow speed V, the length L of the net, the water density ρ and viscosity μ, and the submerged force B. Using dimensionless analysis, derive two dimensionless variables on which the dimensionless tension depends. (c) It is expected that the flow around the net will be turbulent for both the prototype and the model so that the tension T will not depend at all on the viscosity and hence the Reynolds number. If this is so, calculate the speed ratio V_m/V_p that relates the model and prototype speeds for dynamically similar flows. (d) After a series of model experiments, it is found that the tension T_m depends upon the towing speed V_m according to:

$$T_m = K_m V_m^2$$

where K_m is a dimensional constant having the value, $10\,Ns^2/m^2$. What value of K_p would you recommend to the fishing vessel captain so that he could determine the tension T_p as a function of the vessel speed V_p using a similar formula?

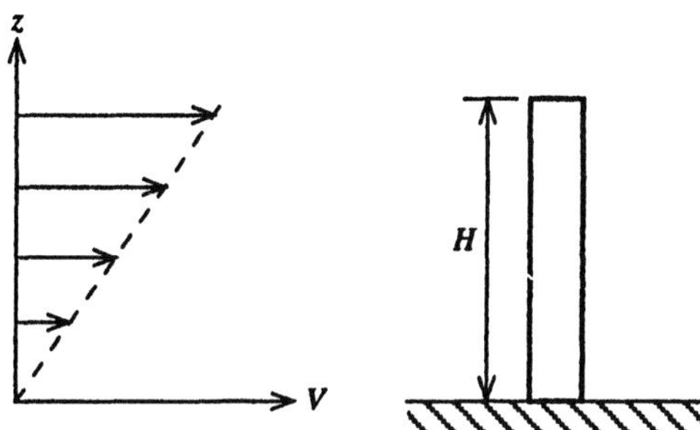

Figure P 10.13

Problem 10.13

The aerodynamic forces on a tall building are to be determined by measurements on a model in a wind tunnel. The wind field approaching the building is one having a constant velocity gradient dV/dz, as shown in figure P 10.13. The independent variables of the flow are thus dV/dz, the building height H and the air density ρ. (The test will be conducted at a high enough Reynolds number so that the air viscosity is unimportant.) Measurements will be made of the moment T of the aerodynamic drag force about the base of the building and the pressure $p\{z\}$ as a function of the height z above the base.

(a) Using dimensional analysis, express the form of the functional relationship between the dimensionless moment T and any remaining dimensionless independent variables. (b) A model test is run for $(dV/dz)_m = 1\,s^{-1}$, $H_m = 1\,m$ and $\rho_m = 1.25\,kg/m^3$. The measured value of $T_m = 0.03\,Nm$. What would be the numerical value of the prototype T_p if $H_p = 100\,m$, $\rho_p = 1.25\,kg/m^3$ and $(dV/dz)_p = 0.30\,s^{-1}$? (c) Using dimensional analysis, express the form of the functional relationship between the dimensionless pressure $p\{z\}$ and any remaining variables. (d) Under the same conditions as the test for (b), the model pressure rise $(p\{z\} - p_\infty)_m$ is measured to be $0.1\,Pa$ at $z_m = H{-}m/2$. What would be the prototype pressure rise at $z_p = H_p/2$?

Problem 10.14

A manufacturer of pumps produces a centrifugal pump of impeller diameter $D_m = 8\ in$, that the company claims will deliver a volumetric flow rate $Q_m = 300$ gallons per minute, which equals $1.893E(-2)\ m^3/s$) of water, at a pressure rise $\Delta p_m = 60\ psi = 2.482E(7)\ Pa$ when rotating at $\Omega_m = 3600$ revolutions per minute. Under these conditions, the pump efficiency η_p is said to be 85%.

(a) Calculate the power P_m required to operate the pump under these conditions. (b) The manufacturer proposes to develop a geometrically similar pump that is twice as large in all linear dimensions ($D_p = 2D_m$) and wants to determine the pressure rise and power requirement of the larger pump. For a water flow in the larger pump to be dynamically similar to that in the smaller pump, what should be the numerical values of the flow rate Q_p and rotational speed Ω_p? (c) For the conditions of (b), what will be the numerical values of the pressure rise Δp_p and the power P_p?

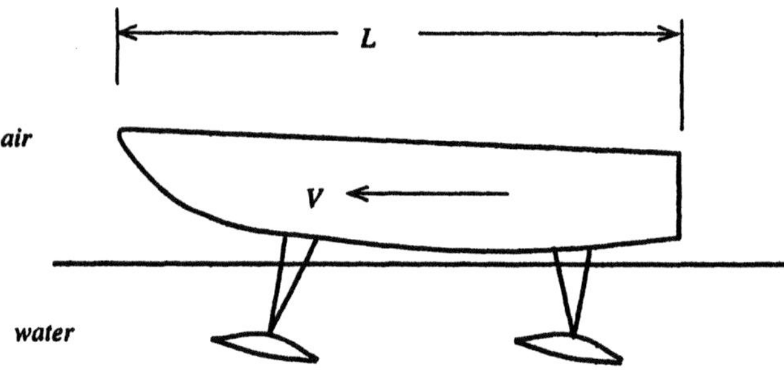

Figure P 10.15

Problem 10.15

A scale model of a hydrofoil vessel is tested in a ship model towing tank to determine the minimum speed V at which the vessel will rise clear of the water, the hydrofoil lift just balancing the weight of the vessel. The model length L_m is $1/10$ that of the prototype vessel, L_p. The model mass M_m is adjusted so that it displaces the same proportionate volume in still water as the prototype, *i.e.*, $M_m/L_m^3 = M_p/L_p^3$.

(a) The model speed V_m in the test is found to be $3\ m/s$. Assuming the lift coefficient $(C_{\mathcal{L}})_m$ of the model airfoil is the same as that of the prototype vessel under this condition, calculate the ship speed V_p at which you would predict the full scale ship to rise clear of the water. (b) In conducting model tests in a towing tank, it is usual to require that the Froude number $V/\sqrt{gL}$ be the same for both model and prototype. Is this so for the test described above? (c) Calculate the ratio Re_m/Re_p of the Reynolds numbers of the model and prototype in this test. Given this value, discuss the validity of the assumption regarding the lift coefficients made in part (a).

Problem 10.16

A design of an aircraft, of length $L_p = 50 ft$, is tested with a scale model, $L_m = 5 ft$, in a water tunnel. It is found in the water tunnel that the drag force $\mathcal{D}_m$ is related to the tunnel flow speed V_m by:

$$\mathcal{D}_m = 1.10V_m + 0.049V_m^2$$

where the unit of drag is *lbf* and that of velocity is ft/s. When the tunnel speed is reduced to $30 ft/s$, the aircraft model experiences stall.

For the model experiment $\rho_w = 1E(3)\, kg/m^3$, $\mu_w = 1E(-3)\, Pa\, s$, while for the prototype flow $\rho_a = 1.2\, kg/m^3$, $\mu_a = 1.8E(-5)\, Pa\, s$.

(a) Calculate the flight speed V_p at which the prototype would experience stall, in units of ft/s. (b) Derive an expression for the drag force $\mathcal{D}_p$ of the prototype (in units of *lbf*) as a function of the prototype flight speed V_p (in units of ft/s). (c) Calculate the power P_p (in units of horsepower) required to overcome the drag of the prototype when flying at a speed of $V_p = 300 ft/s$.

Problem 10.17

A researcher is interested in how fish propel themselves through water by oscillating their fins. The researcher considers a class of geometrically similar fish and observes that the thrust F which a fish generates depends on the length L of the fish, the density ρ of the water, the speed V with which it moves through the water and the frequency Ω with which it oscillates its tail. In a set of experiments, the researcher measures the thrust developed by a baby marlin ($L_m = 0.1\, m$) as it moves at a speed $V_m = 0.1\, m/s$ by observing its acceleration when the fish increases its tail oscillation frequency Ω, finding that:

$$F_m = [2.5E(-3)\, Ns]\Omega + [4E(-5)\, Ns^2]\Omega^2$$

where Ω is the frequency in rad/s.

(a) Using dimensional analysis, define a dimensionless thrust and determine the dimensionless variables on which it depends. (b) Using the empirical relation between F_m and Ω_m, derive an expression for the thrust F_p as a function of V_p, L_p and Ω_p for all fish which are geometrically similar to the marlin. (c) From the results of part (b), calculate the thrust of an adult marlin for which $L_p = 0.1\, m$, $V_p = 0.1\, m/s$ and $\Omega_p = 5/s$.

Problem 10.18

As is well known by baseball and golf players, a ball traveling through air will tend to "curve", *i.e.*, move transverse to its primary direction of motion if it has spin.

Consider a sphere of diameter D traveling at a speed V through an incompressible fluid of density ρ and viscosity μ. The sphere is rotating at an angular speed Ω. We are interested in the force F exerted on the sphere in the direction perpendicular to its motion and to its axis of rotation. We have available an experimental study that shows, first of all, that under the conditions of high Reynolds number, F is independent of the viscosity μ. In addition, measurements on a sphere of diameter $D_m = 0.02 ft$ moving through water ($\rho_w = 1E(3)\, kg/m^3$) at a speed $V_m = 3 ft/s$ show that the force F is proportional to Ω:

$$F = (4.4E(-6)\, lbf\, s)\Omega; \quad \Omega \leq 120\, rad/s$$

where Ω is the rotation speed in units of *rad/s*.

(a) Using dimensional analysis, define a dimensionless force F and a set of dimensionless variables upon which it depends. (b) Using the information from (a) and the water experiments, calculate the force F_p (in units of *lbf*) acting on a sphere of diameter $D_p = 0.25 ft$ moving through air ($\rho_a = 1.2\, kg/m^3$) while spinning at $\Omega_p = 10$ *revolutions per second.*

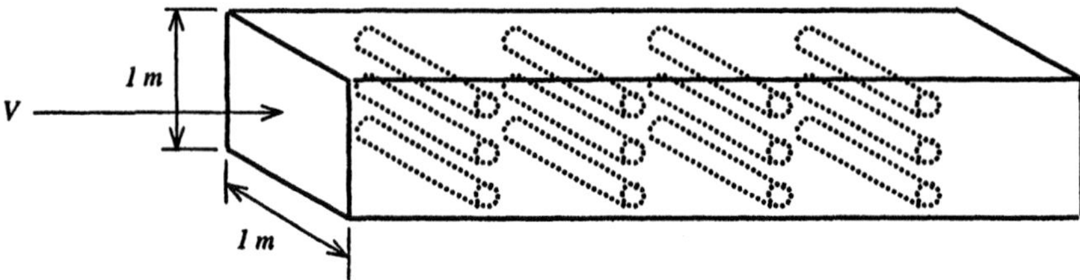

Figure P 10.19

Problem 10.19

A heat exchanger consists of 500 cylindrical copper tubes placed transverse to the axis of a square rectangular duct of $1\, m \times 1\, m$ cross-section, as shown in figure P 10.19. The air flow through the duct has an average speed $\bar{V} = 10\, m/s$, average density $\rho = 1.2\, kg/m^3$ and average viscosity $\mu = 1.8E(-5)\, Pa\, s$.

(a) Calculate the drag force on one tube, assuming it to be alone in a flow with the average flow properties. (Use the drag coefficient of figure 10.6.) (b) Assuming that the total drag force on all the tubes is 500 times that of part (a) and that the frictional forces on the tube wall can be neglected, calculate the head loss coefficient K of the heat exchanger for the flow speed $\bar{V} = 10\, m/s$.

Problem 10.20

An inventor is developing a tiny micropump that might be used to improve circulation when inserted into a small vein or artery. The pump diameter is $D_p = 0.1\,mm = 1E(-4)\,m$. It will rotate at an angular speed of $\Omega_p = 100\,s^{-1}$ in a fluid of density $\rho_p = 1E(3)\,kg/m^3$ and viscosity $\mu_p = 1E(-3)\,Pa\,s$. To predict how the micropump will perform, the inventor has made a large-scale model of the pump having a diameter $D_m = 3\,cm = 3E(-2)\,m$. This model will be tested in glycerine, whose density $\rho_m = 1.26E(3)\,kg/m^3$ and viscosity $\mu_m = 1.49\,Pa\,s$. Both the model and the prototype pump will be run under conditions where the Reynolds number $\rho\Omega D^2/\mu$ is very much less than one.

The inventor reasons correctly that the pump performance will not depend upon the fluid density, because the Reynolds number will be very small. The pump presssure rise Δp will thus be a function of μ, Ω, D and Q:

$$\Delta p = f\{\mu,\ \Omega,\ D,\ Q\}$$

(a) Using dimensional analysis, derive a dimensionless form of this relationship. (b) In running a model test, calculate what should be an upper limit for the angular speed Ω_m such that the Reynolds number will be less than one. (c) The inventor runs a model test with $\Omega_m = 0.1\,s^{-1}$. At zero pressure rise ($\Delta p_m = 0$), the flow rate is $Q_m = 5E(-8)\,m^3/s$, while at zero flow ($Q_m = 0$) the pressure rise is $\Delta p_m = 8E(4)\,Pa$. Calculate the corresponding values of the micropump flow rate Q_p and pressure rise Δp_p.

Problem 10.21

For the prototype of the model tested in figure 10.1, the product $C_D A_f = 0.329\,m^2$. Calculate the power $\mathcal{P}$ required to overcome the aerodynamic drag of the prototype at 55 *mph* when $\rho = 1.2\,kg/m^3$.

Bibliography

Batchelor, G. K. 1967. *An Introduction to Fluid Dynamics.* Cambridge: Cambridge University Press.

Betz, Albert. 1966. *Introduction to the Theory of Flow Machines.* New York: Pergamon.

Blevens, R. D. 1984. *Applied Fluid Dynamics Handbook.* New York: Van Nostrand Reinhold.

Lewis, Edward V., ed. 1988. *Principles of Naval Architecture Second Revision.* Vol. 2, *Resistance, Propulsion and Vibration.* Jersey City: Society of Naval Architects and Marine Engineers.

Howarth, L., ed. 1953. *Modern Developments in Fluid Dynamics.* Vol. 1, *High Speed Flows.* Oxford: Oxford University Press.

Lobanoff, Val S., and Robert R. Ross. 1985. *Centrifugal Pumps Design and Applications.* Houston: Gulf Publishing Co.

Lamb, H. 1945. *Hydrodynamics.* 6th ed. New York: Dover Publications.

Rouse, Hunter, and Simon Ince. 1957. *History of Hydraulics.* Ames: Iowa Institute of Hydraulic Research. State University of Iowa.

Sabersky, Rolf S., Allan J. Acosta, and Edward G. Hauptman. 1989. *Fluid Flow.* 3rd ed. New York: Macmillan Publishing Co.

White, Frank M. 1986. *Fluid Mechanics.* 2nd ed. New York: McGraw-Hill Book Co.

11 Irrotational Flow

11.1 Introduction

Many fluid flows of practical importance are ones for which the dependent variables, velocity **V** and pressure p, depend upon two or three spatial coordinates, and perhaps even time. Even if the density is constant, we may need to find as many as four scalar dependent variables (*e.g.,*, u, v, w, p) as functions of four scalar independent variables (*e.g.* x, y, z, t). This is a formidable task, even with the aid of a high speed computer and involves the assembly and display of vast amounts of numerical data in order to obtain a physical picture of the flow.

Thus far we have had to rely upon Bernoulli's integral to help us in describing some aspects of an inviscid flow for which streamlines were easily identifiable. The momentum integrals were also helpful in determining forces in flows that were constrained by boundaries. In one example of a viscous flow, that over a flat plate, we found two velocity components (u, v) as functions of two spatial coordinates (x, y), but it was no easy matter to do so. It would seem that predicting multidimensional flows would be very difficult.

There is a kind of flow, called *irrotational flow*, for which multidimensional flows may be handled quite easily. More to the point, many common flows are irrotational, at least in major parts of the flow field. Even more surprising, we often can find multidimensional flow solutions in analytic form or through use of simple numerical algorithms. Most of all, it is possible to develop good physical understanding of fluid flows by utilizing the concepts introduced in the description of irrotational flow.

In this chapter, we define the conditions under which a flow may be irrotational and describe analytical techniques that may be used for constant-density flows. As examples, we treat only two-dimensional flows involving two velocity components and two spatial dimensions. These examples are illustrative of many practical flows, such as flow over airplane wings. We place the most emphasis upon the definition of the stream function of these flows, which allows us to define the streamlines of the flow and develop an understanding of how fluids behave when they move in several spatial directions.

11.2 Vorticity

Vorticity is a kinematic property of the flow field which, at each point, measures the angular velocity of a fluid particle. It is a vector quantity, having three scalar components. Denoted by the symbol $\boldsymbol{\omega}$, it is defined by:

$$\boldsymbol{\omega} \equiv \nabla \times \mathbf{V} \tag{11.1}$$

$$= \mathbf{i}_x \left(\frac{\partial w}{\partial y} - \frac{\partial v}{\partial z} \right) + \mathbf{i}_y \left(\frac{\partial u}{\partial z} - \frac{\partial w}{\partial x} \right) + \mathbf{i}_z \left(\frac{\partial v}{\partial x} - \frac{\partial u}{\partial y} \right) \tag{11.2}$$

$$= \mathbf{i}_r \left(\frac{1}{r}\frac{\partial V_z}{\partial \theta} - \frac{\partial V_\theta}{\partial z} \right) + \mathbf{i}_\theta \left(\frac{\partial V_r}{\partial z} - \frac{\partial V_z}{\partial r} \right) + \mathbf{i}_z \left(\frac{1}{r}\frac{\partial}{\partial r}(rV_\theta) - \frac{1}{r}\frac{\partial V_r}{\partial \theta} \right) \tag{11.3}$$

where we have used the vector definitions of equations 1.35 and 1.55 to express the vorticity $\boldsymbol{\omega}$ in cartesian and cylindrical coordinates, equations 11.2 and 11.3.

To see how the vorticity $\boldsymbol{\omega}$ is related to the angular velocity of a fluid element, consider the rigid-body rotation of a fluid about a fixed axis, such as would happen if we placed water in a glass centered on a rotating turntable and waited long enough for all the fluid to rotate at the same angular velocity Ω as the turntable. Using cylindrical coordinates fixed in the laboratory inertial reference frame, with the z-axis vertical, the velocity components are $V_r = 0$, $V_\theta = \Omega r$ and $V_z = 0$. Substituting these values in equation 11.3 to find the vorticity, only one derivative is non-zero:

$$\boldsymbol{\omega} = \mathbf{i}_z \left(\frac{1}{r}\frac{\partial}{\partial r}(r^2\Omega) \right) = (2\Omega)\mathbf{i}_z \tag{11.4}$$

Because the fluid is in rigid-body rotation, the angular velocity $(\Omega)\mathbf{i}_z$ is the same for each fluid element and so is the vorticity by equation 11.4. Therefore, *the vorticity $\boldsymbol{\omega}$ is twice the angular velocity of a fluid element.*

The vorticity $\boldsymbol{\omega}$ is a vector field just like the velocity field $\mathbf{V}$. A "streamline" of the vorticity field, *i.e.*, the line whose tangent has the direction of $\boldsymbol{\omega}$ at each point, is called a *vortex line*. Vortex lines, like streamlines, do not intersect. In fact, the vorticity field is like a magnetic field in that it has zero divergence:

$$\nabla \cdot \boldsymbol{\omega} = \nabla \cdot (\nabla \times \mathbf{V}) \equiv 0 \tag{11.5}$$

because of the vector calculus identity, equation 1.37. Vortex lines cannot begin or end in the interior of a fluid, only on its boundaries. Like a magnetic field, there can be no "source" for a vortex line. Such a vector field is called *solenoidal*.[1]

What is the physical significance of the angular velocity (or vorticity) of a fluid element? If the fluid element were a rigid body, then its angular velocity would remain unchanged in the absence of a couple acting upon it. A fluid element with zero angular velocity will not acquire vorticity unless acted upon by a couple to cause it to rotate. One obvious source of rotation is unbalanced shear stresses acting on its periphery. Another, less obvious, is the effect of a pressure force acting in a direction not aligned with a density gradient in the fluid. In flows where these effects are absent, a fluid element with zero initial vorticity will not acquire any. For good reason, then, such flows are called *irrotational*, and their velocity fields are such that:

$$\nabla \times \mathbf{V} = 0 \tag{11.6}$$

For irrotational flows, the (zero) vorticity is a constant of the motion, *i.e.*, it is unchanging along any particle path.

Example 11.1

Derive an expression for the vorticity $\boldsymbol{\omega}$ in circular Poiseuille flow for which the velocity $\mathbf{V}$ is $V_z \mathbf{i}_z$, where:

$$V_z = V_m \left(1 - \frac{r^2}{a^2}\right)$$

V_m being the maximum velocity and a the tube radius. What is the shape of a vortex line?

Solution

Using equation 11.3, the only non-zero component of $\boldsymbol{\omega}$ is:

$$\boldsymbol{\omega} = -\mathbf{i}_\theta \frac{\partial V_z}{\partial r} = \mathbf{i}_\theta \frac{2rV_m}{a^2}$$

A vortex line is a circle of radius r whose center is on the z axis, lying in a plane perpendicular to that axis.

---

[1]The velocity field of an incompressible fluid is also solenoidal because, in this case, $\nabla \cdot \mathbf{V} = 0$.

Boundary Layer Vorticity

In the viscous boundary layer that develops next to a solid wall past which fluid flows, vorticity is created by an unbalanced viscous force that rotates the fluid particles. The amount of vorticity at the wall, ω_w, can be related to the magnitude of the wall shear stress τ_w by noting that, if x is the stream direction and y the normal direction at the wall, then by equation 11.2:

$$\omega_w = \mathbf{i}_z \left(-\frac{\partial u}{\partial y} \right)_w = -\frac{\tau_w}{\mu} \mathbf{i}_z \tag{11.7}$$

since v, w and their derivatives are zero at the wall. The vorticity vector ω_w at the wall is perpendicular to the stream direction and parallel to the wall. Since the fluid viscous shear stress decreases below the wall value at points within the boundary layer that are further from the wall, so will the vorticity decrease below its wall value ω_w.

Flow within a viscous boundary layer cannot be an irrotational flow since the vorticity is not zero there. But flow exterior to a boundary layer can be irrotational, provided it satisfies certain requirements discussed below. Since real flows always stick to solid surfaces, forming boundary layers, they cannot be irrotational everywhere.

11.2.1 Circulation

The *circulation* is a scalar integral of the velocity field that can be related to its vorticity. Denoted by the symbol Γ, circulation is defined by the line integral:

$$\Gamma \equiv \int_C \mathbf{V} \cdot d\mathbf{c} \tag{11.8}$$

where $\mathbf{V}$ is the velocity and $d\mathbf{c}$ the line element along a *closed* curve C. By utilizing Stokes' theorem, equation 1.45, the circulation may be expressed also as a surface integral:

$$\Gamma = \int_C \mathbf{V} \cdot d\mathbf{c} = \iint_S \mathbf{n} \cdot (\nabla \times \mathbf{V})\, dS = \iint_S \boldsymbol{\omega} \cdot \mathbf{n}\, dS \tag{11.9}$$

where the surface S is any surface whose edge is the curve C and whose normal is $\mathbf{n}$.[2] We may call Γ the vorticity flux through the area S enclosed by the curve C, in analogy with the magnetic flux enclosed by a wire loop carrying an electric current.[3]

---

[2]The positive direction of $\mathbf{n}$ is determined by the right hand rule: if the positive direction of $d\mathbf{c}$ is that of the index finger of the right hand, then the positive direction of $\mathbf{n}$ is that of the thumb.

[3]The algebraic sign of Γ depends upon the direction along the curve C for which $d\mathbf{c}$ is defined as positive. The usual convention for two-dimensional flows in the x, y or r, z plane is that

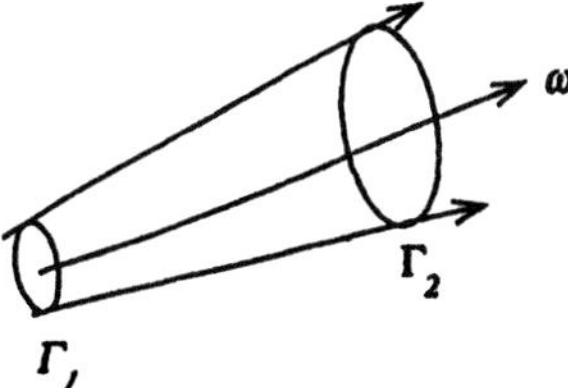

Figure 11.1 This vortex tube shows the vortex lines and two curves on the tube lateral surface at which the circulation is Γ_1 and Γ_2.

An alternate interpretation of equation 11.9 is illustrated in figure 11.1, showing a finite segment of a *vortex tube, i.e.*, a volume whose lateral surface is tangent to vortex lines and whose end surfaces is normal to them, similar to a stream tube formed about streamlines. The circulation Γ_1 enclosed by the curve C_1 may be evaluated in terms of an area integral extending over the lateral surface $\mathcal{S}_{lat}$ and the area $\mathcal{S}_2$:

$$\Gamma_1 = \int_{C_1} \mathbf{V} \cdot d\mathbf{c}_1 = \iint_S \boldsymbol{\omega} \cdot \mathbf{n}\, d\mathcal{S}_{lat} + \iint_S \boldsymbol{\omega} \cdot \mathbf{n}\, d\mathcal{S}_2 = 0 + \Gamma_2 \tag{11.10}$$

since $\boldsymbol{\omega} \cdot \mathbf{n} = 0$ on the lateral surface. Thus *the circulation about a vortex tube does not vary with distance along the vortex tube.*

Kelvin's Theorem and Irrotationality

The circulation Γ is defined for a closed curve C lying within a fluid flow. If we choose this curve to be a *material curve* C_m, *i.e.*, a line passing through a set of marked fluid particles that moves with the flow, we can calculate the time rate of change of the circulation, $D\Gamma_m/Dt$,

$$\begin{aligned} \frac{D\Gamma_m}{Dt} &= \frac{D}{Dt}\int_{C_m} \mathbf{V} \cdot d\mathbf{c} \\ &= \int_{C_m} \left(\frac{D\mathbf{V}}{Dt} \cdot d\mathbf{c} + \mathbf{V} \cdot d\left[\frac{D\mathbf{c}}{Dt}\right] \right) \\ &= \int_{C_m} \left(\frac{D\mathbf{V}}{Dt} \cdot d\mathbf{c} + \mathbf{V} \cdot d\mathbf{V} \right) \end{aligned}$$

---

the clockwise direction along C defines the sign of Γ. In this convention, the direction of $\mathbf{n}$ is $-\mathbf{i}_z$ or $-\mathbf{i}_\theta$, respectively. We will use this convention in subsequent analyses, examples and problems.

$$= \int_{C_m} \frac{D\mathbf{V}}{Dt} \cdot d\mathbf{c} \tag{11.11}$$

where the last term of the next-to-last line is zero because it is the integral of $d(V^2/2)$ around a closed curve whose limits of integration are identical.

Equation 11.11 is *Kelvin's Theorem.*[4] To understand its physical significance, it may be expressed as a surface integral by invoking Stokes' theorem, equation 1.45:

$$\frac{D\Gamma_m}{Dt} = \iint_S \left(\nabla \times \frac{D\mathbf{V}}{Dt}\right) \cdot \mathbf{n}\, dS \tag{11.12}$$

For a viscous incompressible fluid, $D\mathbf{V}/Dt$ is given by the Navier-Stokes equation 6.11 and its curl is:

$$\nabla \times \frac{D\mathbf{V}}{Dt} = \nabla \times \left(\frac{1}{\rho}\nabla p\right) + \nabla \times \mathbf{g} + \nabla \times (\mu \nabla^2 \mathbf{V})$$

$$= \nabla\left(\frac{1}{\rho}\right) \times \nabla p + \mu \nabla^2 \boldsymbol{\omega} \tag{11.13}$$

where we have used equations 1.39 and 1.36 to simplify $\nabla \times (\nabla p/\rho)$ and noted that $\nabla \times \mathbf{g} = 0$. Using equation 11.9 to express Γ_m as a surface integral and substituting 11.13 in 11.12, Kelvin's theorem for an incompressible fluid takes the form:

$$\frac{D}{Dt}\iint_S \boldsymbol{\omega} \cdot \mathbf{n}\, dS = \iint_S \left[\nabla\left(\frac{1}{\rho}\right) \times \nabla p + \mu \nabla^2 \boldsymbol{\omega}\right] \cdot \mathbf{n}\, dS \tag{11.14}$$

Equation 11.14 states that the rate of change of the vorticity flux enclosed by the material curve C is the sum of two effects, the first due to the gradient of $1/\rho$ and the second to the viscous diffusion of vorticity. The first of these arises when the direction of the pressure gradient is different from that of the density gradient, causing the fluid particle to change its angular velocity and hence vorticity $\boldsymbol{\omega}$. The second effect comes from the transfer of angular velocity between adjacent fluid particles by viscous shear stresses. If the fluid density is constant everywhere, or if it is everywhere a function of the pressure only,[5] then $\nabla(1/\rho)$ is zero or parallel to ∇p and $\nabla(1/\rho) \times \nabla p = 0$.

If a flow is irrotational ($\boldsymbol{\omega} = 0$) at some instant of time, then Kelvin's theorem 11.14 shows that it will remain irrotational if the density is everywhere a constant

[4]William Thompson [Lord Kelvin] (1824–1907), best known for his invention of the absolute temperature scale, graduated from the University of Glasgow at the age of 10. In fluid mechanics, he developed the theory of sound waves, vortex motion, capillary waves, and viscous flow stability. He introduced the word turbulence to characterize unsteady random flow.

[5]The density would be a function of the pressure only if the entropy is everywhere constant. This is termed a *homentropic flow*. When this is not the case, the flow is termed *stratified.*

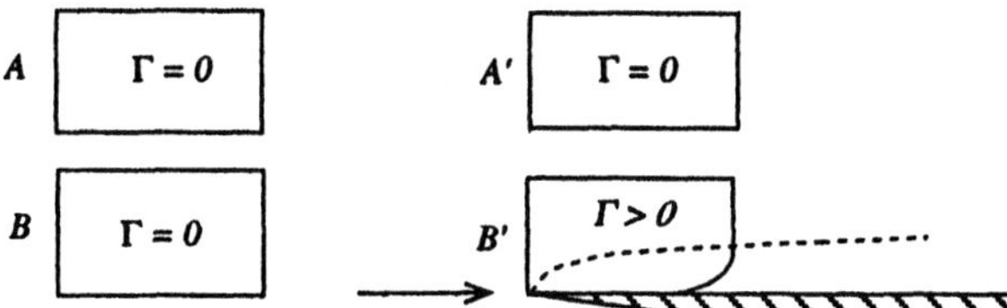

Figure 11.2 An illustration of Kelvin's theorem applied to the steady flow of a viscous fluid over a flat plate.

or a function of pressure and if the flow lies outside those regions, such as viscous boundary layers, where vorticity may diffuse from solid boundaries into the flow. A flow started from rest or having uniform conditions far upstream would be irrotational initially and would therefore remain so if these conditions were satisfied.

This principle is illustrated in figure 11.2, where two material curves, *A* and *B*, in the uniform upstream flow approaching a flat plate enclose zero vorticity and have zero circulation. When these material curves have drifted to the locations *A′* and *B′*, the former still has no circulation, but the latter has developed a positive (clockwise) circulation because it has intruded into the viscous boundary layer on the upper surface of the flat plate, where vorticity has been generated within the boundary layer.

In section 4.7, equation 4.34, we showed that Bernoulli's equation was observed along any line C in an inviscid, irrotational constant-density flow ($\mu = 0$, $\omega = 0$, $\nabla\rho = 0$):

$$\int_1^2 \frac{\partial \mathbf{V}}{\partial t} \cdot d\mathbf{c} + \left(\frac{V_2^2}{2} + \frac{p_2}{\rho} - \mathbf{g} \cdot \mathbf{R_2}\right) - \left(\frac{V_1^2}{2} + \frac{p_1}{\rho} - \mathbf{g} \cdot \mathbf{R_1}\right) = 0$$

But, in a viscous, incompressible irrotational flow, the viscous force per unit volume, $\mu\nabla^2\mathbf{V}$, is zero:

$$\nabla^2\mathbf{V} = \nabla(\nabla \cdot \mathbf{V}) - \nabla \times (\nabla \times \mathbf{V}) = 0 \tag{11.15}$$

where we have used equation 1.42 and applied the conditions of incompressibility ($\nabla \cdot \mathbf{V} = 0$) and irrotationality ($\omega = 0$). Thus a viscous, constant-density irrotational flow obeys Euler's equation, and Bernoulli's equation along any line of the flow, even though there may be viscous stresses in the fluid. Because of the irrotational nature of the flow, these viscous stresses do not provide a viscous force per unit volume and the fluid behaves as an inviscid one. It is for this reason that some fluid flows may be considered inviscid even though the fluid viscosity is not zero.

Example 11.2

The velocity **V** of a steady plane viscous flow having a constant density ρ is:

$$\mathbf{V} = \Omega x\,\mathbf{i}_x - \Omega y\,\mathbf{i}_y$$

(a) Is the flow irrotational? (b) Derive expressions for the viscous stresses τ_{xy}, τ_{xx} and τ_{yy}. (c) If the pressure p^* at $x = 0$, $y = 0$ is p_0^*, derive an expression for $p^*\{x, y\}$.

Solution

(a) Since $u = \Omega x$, $v = -\Omega y$, $w = 0$, the value of ω obtained from equation 11.2 is:

$$\omega = \mathbf{i}_x\left(0 - \frac{\partial(-\Omega y)}{\partial z}\right) + \mathbf{i}_y\left(\frac{\partial(\Omega x)}{\partial z} - 0\right) + \mathbf{i}_z\left(\frac{\partial(-\Omega y)}{\partial x} - \frac{\partial(\Omega x)}{\partial y}\right) = 0$$

so the flow is irrotational.

(b) The viscous stress components for a incompressible fluid are given by equations 6.4 and 6.6:

$$\tau_{xy} = \mu\left(\frac{\partial u}{\partial y} + \frac{\partial v}{\partial x}\right) = \mu(\Omega - \Omega) = 0$$

$$\tau xx = 2\mu\left(\frac{\partial u}{\partial x}\right) = 2\mu\Omega$$

$$\tau yy = 2\mu\left(\frac{\partial v}{\partial y}\right) = 2\mu\Omega$$

(c) At $x = 0$, $y = 0$ the velocity is zero. Since Bernoulli's equation applies to this viscous irrotational flow,

$$p^* + \frac{\rho}{2}(u^2 + v^2) = p_0^* + 0$$

$$p^* = p_0^* - \frac{\rho}{2}([\Omega x]^2 + [\Omega y]^2)$$

11.3 The Stream Function for Incompressible Flow

It is a theorem of vector calculus that any vector field can be given as the sum of the curl of a vector function and the gradient of a scalar function of space and time. For irrotational flows, we choose to determine the velocity **V** in terms of the curl of a

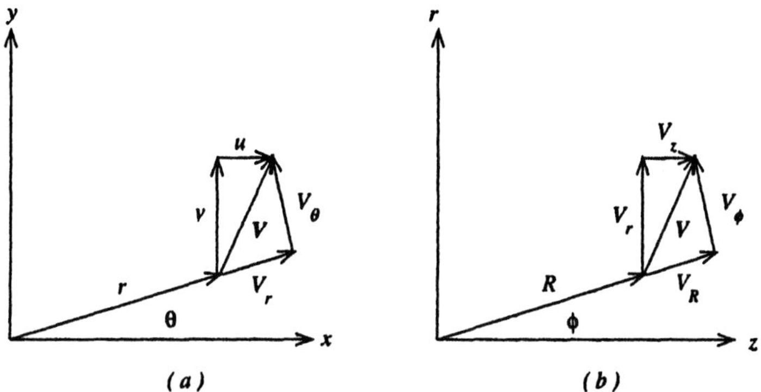

Figure 11.3 The spatial and velocity coordinates for (*a*) two-dimensional plane and (*b*) axisymmetric flow, including polar coordinates in the plane of the flow.

vector $\boldsymbol{\Psi}\{\mathbf{R}, t\}$ that is a function of position $\mathbf{R}$ and time t:

$$\mathbf{V} = \nabla \times \boldsymbol{\Psi}\{\mathbf{R}, t\} \tag{11.16}$$

Because of the vector identity 1.37, such a vector field satisfies mass conservation for an incompressible fluid:

$$\nabla \cdot \mathbf{V} = \nabla \cdot (\nabla \times \boldsymbol{\Psi}) \equiv 0 \tag{11.17}$$

We will confine our discussion of irrotational flow to two types of two-dimensional flows, plane flow in the x, y-plane where u and v are functions of x, y and t (but not z) and $w = 0$, and axisymmetric flow in the r, z-plane where V_r and V_z are functions of r, z and t (but not θ) and $V_\theta = 0$. (The coordinates and velocity components in these two-dimensional flows are shown in figure 11.3, together with the alternate polar coordinates of the respective planes.) To provide for these special two-dimensional flows, we choose $\boldsymbol{\Psi}$ to be, for plane and axisymmetric flow, respectively,

$$\boldsymbol{\Psi}\{x, y, t\} = \psi\{x, y, t\}\, \mathbf{i}_z \tag{11.18}$$

$$\boldsymbol{\Psi}\{r, z, t\} = \frac{1}{r}\psi\{r, z, t\}\, \mathbf{i}_\theta \tag{11.19}$$

where ψ is a scalar function. Using equations 1.35 and 1.55 to find $\mathbf{V} = \nabla \times \boldsymbol{\Psi}$ for 11.18 and 11.19, the velocity $\mathbf{V}$ can be found in terms of the derivatives of ψ for plane flow:

$$\mathbf{V} = \nabla \times (\psi\, \mathbf{i}_z) = \mathbf{i}_x \left(\frac{\partial \psi}{\partial y} - 0 \right) + \mathbf{i}_y \left(0 - \frac{\partial \psi}{\partial x} \right) + 0$$

$$u = \frac{\partial \psi}{\partial y}; \qquad v = -\frac{\partial \psi}{\partial x} \tag{11.20}$$

and for axisymmetric flow:

$$\mathbf{V} = \nabla \times \left(\frac{1}{r} \psi \, \mathbf{i}_\theta \right) = \mathbf{i}_r \left[0 - \frac{\partial}{\partial z} \left(\frac{\psi}{r} \right) \right] + 0 + \mathbf{i}_z \left[\frac{1}{r} \frac{\partial \psi}{\partial r} - 0 \right]$$

$$V_r = -\frac{1}{r} \frac{\partial \psi}{\partial z}; \qquad V_z = \frac{1}{r} \frac{\partial \psi}{\partial r} \tag{11.21}$$

It is sometimes desirable to use the polar coordinates r, θ in the x, y-plane or R, ϕ in the r, z-plane for the plane or axisymmetric flow, respectively, as shown in figure 11.3. With this choice of coordinates, the velocity components for plane flow become:

$$V_r = \frac{1}{r} \frac{\partial \psi}{\partial \theta}; \qquad V_\theta = -\frac{\partial \psi}{\partial r} \tag{11.22}$$

while for axisymmetric flow we find:

$$V_R = \frac{1}{R^2 \sin \phi} \frac{\partial \psi}{\partial \phi}; \qquad V_\phi = -\frac{1}{R \sin \phi} \frac{\partial \psi}{\partial R} \tag{11.23}$$

where ψ is, respectively, a function of r, θ, t and R, ϕ, t.

The scalar function ψ is called the *stream function* because it defines the streamlines of the flow. Along a line $\psi = constant$ in plane flow, we have the differential relation:

$$d\psi = \frac{\partial \psi}{\partial x} dx + \frac{\partial \psi}{\partial y} dy = 0$$

$$\frac{dy}{dx} = -\frac{\partial \psi / \partial x}{\partial \psi / \partial y} = \frac{v}{u} \tag{11.24}$$

where we have used 11.20 in 11.24 to replace the ratio of partial derivatives by the ratio of the velocity components. Thus the slope dy/dx of the curve $\psi = constant$ is the same as that (v/u) of the velocity vector, which means that $\psi = constant$ is a streamline. By a similar analysis, we find this to be true for axisymmetric flow as well.[6] A knowledge of the stream function ψ enables us to plot the streamlines of the flow, to find the velocity at any point in the flow, and by use of Bernoulli's equation, to determine the pressure field when the flow is irrotational.

The difference in the values of ψ on two different streamlines is equal to the volume flow rate of fluid between them. To prove that this is so for plane flow, connect two points, 1 and 2, on different streamlines ψ_1 and ψ_2 by a curve C lying in the x, y-plane. The volume flow rate, per unit distance normal to the x, y-plane, across an increment dc of the curve C is $\mathbf{V} \cdot \mathbf{n} \, dc$, where $\mathbf{n}$ is the normal to the curve C. However, from

---

[6] $\psi\{r, z, t\}$ is known as the *Stokes stream function*.

vector calculus, $\mathbf{n}\,dc = \mathbf{i}_x\,dy - \mathbf{i}_y\,dx$ so that:[7]

$$\mathbf{V}\cdot\mathbf{n}\,dc = \mathbf{V}\cdot(\mathbf{i}_x\,dy - \mathbf{i}_y\,dx) = u\,dy - v\,dx$$

$$= \frac{\partial\psi}{\partial y}\,dy - \left(-\frac{\partial\psi}{\partial x}\right) = d\psi$$

$$\int_1^2 \mathbf{V}\cdot\mathbf{n}\,dc = \int_1^2 d\psi = \psi_2 - \psi_1$$

where we have used equation 11.20 to replace the velocity components by the partial derivatives of ψ.

In subsequent illustrations, we will plot streamlines having equal increments in ψ so as to delineate equal volume flow rates for the fluid between the streamlines.

Irrotational Flow: Laplace's Equation

So far, we have not required that the flow be irrotational ($\omega = 0$). For a plane irrotational flow, using equations 11.2 and 11.20,

$$\omega \equiv \nabla\times\mathbf{V} = \left(\frac{\partial v}{\partial x} - \frac{\partial u}{\partial y}\right)\mathbf{i}_z = 0$$

$$\frac{\partial^2\psi}{\partial x^2} + \frac{\partial^2\psi}{\partial y^2} = 0 \tag{11.25}$$

while for axisymmetric flow, using equations 11.3 and 11.21, this condition becomes:

$$\omega \equiv \nabla\times\mathbf{V} = \mathbf{i}_\theta\left(\frac{\partial V_r}{\partial z} - \frac{\partial V_z}{\partial r}\right) = 0$$

$$\frac{\partial^2\psi}{\partial z^2} + \frac{\partial^2\psi}{\partial r^2} - \frac{1}{r}\frac{\partial\psi}{\partial r} = 0$$

$$\frac{\partial^2}{\partial z^2}\left(\frac{\psi}{r}\right) + \frac{\partial^2}{\partial r^2}\left(\frac{\psi}{r}\right) + \frac{1}{r}\frac{\partial}{\partial r}\left(\frac{\psi}{r}\right) - \frac{1}{r^2}\left(\frac{\psi}{r}\right) = 0 \tag{11.26}$$

Equations 11.25 and 11.26 are *Laplace's equation* for the plane and axisymmetric variables $\psi\mathbf{i}_z$ and $(\psi/r)\mathbf{i}_\theta$, respectively. Any function that satisfies Laplace's equation can describe an irrotational constant-density flow. However, it is impossible to find a solution to Laplace's equation that satisfies the no-slip condition ($\mathbf{V} = 0$) for a viscous fluid at a stationary solid boundary. Instead, we may only specify that the

[7]For axisymmetric flow, the volume flow rate increment is $2\pi r\mathbf{V}\cdot\mathbf{n}\,dc$, and the volume flow rate between two stream tubes is $2\pi(\psi_2 - \psi_1)$.

fluid velocity is parallel to a stationary wall, which makes $\psi = constant$ along a wall, *i.e.*, a stationary wall is a streamline of the flow.[8] Equations 11.25 and 11.26 may be solved to find the stream function of an irrotational constant-density flow whose boundary conditions are specified.

The Superposition Principle

The velocity components determined from the derivatives of the stream function, as in equations 11.20 and 11.21, are linear in the stream function. The sum of two stream functions, ψ_1 and ψ_2, is also a stream function and describes a velocity field that is the vector sum of the individual fields. For example, for a plane flow:

$$\begin{aligned} \mathbf{V} &= \nabla \times (\psi\, \mathbf{i}_z) = \nabla \times ([\psi_1 + \psi_2]\, \mathbf{i}_z) = \nabla \times (\psi_1\, \mathbf{i}_z) + \nabla \times (\psi_2\, \mathbf{i}_z) \\ &= \mathbf{V}_1 + \mathbf{V}_2 \end{aligned} \tag{11.27}$$

Thus we may superpose two or more irrotational flows by adding their stream functions, and hence their velocity fields, to find a new flow. This is called the *superposition principle.* However, the pressure fields of the superposed flow are *not* additive because the pressure, via Bernoulli's equation, is quadratic in the flow velocities $((\mathbf{V}_1 + \mathbf{V}_2)^2 \neq \mathbf{V}_1^2 + \mathbf{V}_2^2)$.

We will use the superposition principle to add together simple irrotational flows to produce more complex but equally interesting flows.

11.4 Plane Irrotational Flows

In treating plane flows, we will use both cartesian (x, y) and polar (r, θ) coordinates, for it will be found that one or the other may be more convenient in working examples, depending upon the circumstances. First we will consider four elementary flows (uniform flow, line source, line vortex and corner flow) and then somewhat more complex flows formed by superposition of two elementary flows, illustrating the principle of constructing complex flows by superposition of several elementary flows.

[8] A solution to Laplace's equation exists only when either ψ or its normal derivative is specified on a boundary.

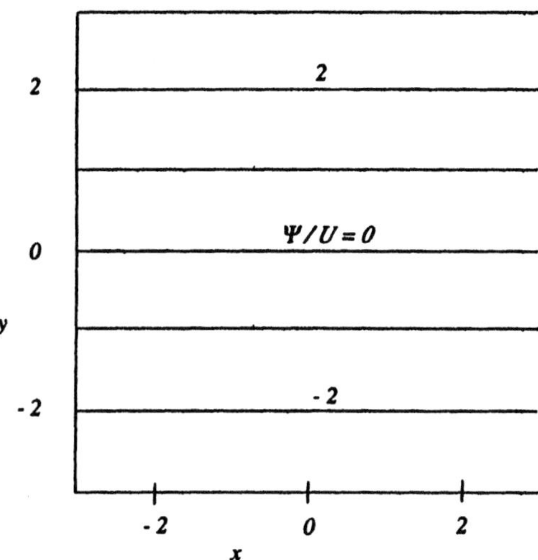

Figure 11.4 The streamlines of the uniform plane flow, $\psi = Uy$, for uniformly spaced values of ψ/U.

11.4.1 Elementary Flows

Uniform Flow

The simplest possible irrotational flow is one that moves from left to right with a uniform speed U. It would have a velocity $\mathbf{V} = U\mathbf{i}_x$ and horizontal streamlines $y =$ *constant*. The stream function that produces this velocity field is:

$$\psi = Uy = Ur \sin\theta \tag{11.28}$$

and its velocity components are found from equations 11.20 and 11.21:

$$u = \frac{\partial \psi}{\partial y} = U; \qquad v = -\frac{\partial \psi}{\partial x} = 0$$

$$V_r = \frac{1}{r}\frac{\partial \psi}{\partial \theta} = U\cos\theta; \qquad V_\theta = -\frac{\partial \psi}{\partial r} = -U\sin\theta \tag{11.29}$$

The streamlines of the flow are plotted in figure 11.4 for integral values of ψ/U. The dimensions of the stream function are those of velocity times length, or $\mathcal{L}^2\mathcal{T}^{-1}$, so the ratio ψ/U has the dimension of length.

Because the velocity is uniform in this flow, there will be no change in pressure p^*, which will have the same value everywhere.

The Line Source

Consider a plane flow that starts from the origin ($x = 0$, $y = 0$) and moves out radially and equally in all directions of θ. The velocity would be purely radial ($\mathbf{V} = \mathbf{i}_r V_r$), and the streamlines would be straight lines emanating from the origin, $\theta = constant$. To create such a flow in the laboratory, we would supply a porous pipe, extending perpendicularly between two parallel walls, separated by a distance W, with a volume flow rate Q. The volume flow rate per unit length of pipe, $q \equiv Q/W$, determines the radial flow speed V_r, the stream function being:

$$\psi = \frac{q}{2\pi}\theta = \frac{q}{2\pi}\tan^{-1}\left(\frac{y}{x}\right) \tag{11.30}$$

and the flow velocity components being:

$$V_r = \frac{1}{r}\frac{\partial\psi}{\partial\theta} = \frac{q}{2\pi r}; \qquad V_\theta = -\frac{\partial\psi}{\partial r} = 0$$

$$u = \frac{\partial\psi}{\partial y} = \frac{q}{2\pi}\left(\frac{x}{x^2+y^2}\right); \qquad v = -\frac{\partial\psi}{\partial x} = \frac{q}{2\pi}\left(\frac{y}{x^2+y^2}\right) \tag{11.31}$$

Note that this radial velocity satisfies the conservation of mass applied to a cylindrical control volume of radius r:

$$\dot{m}_{in} = \dot{m}_{out}$$

$$\rho Q = \rho V_r W(2\pi r)$$

$$Q = Wq$$

The streamlines of the line source are sketched in figure 11.5 for constant values of the dimensionless ratio ψ/q.

The stream function ψ of equation 11.30 satisfies Laplace's equation in polar coordinates:

$$\frac{\partial^2\psi}{\partial r^2} + \frac{1}{r}\frac{\partial\psi}{\partial r} + \frac{1}{r^2}\frac{\partial^2\psi}{\partial\theta^2} = 0 + 0 + 0 = 0$$

The pressure p^* may be found by applying Bernoulli's theorem between a point at a radius r, where the velocity $V_r = (q/2\pi r)$, and one at $r = \infty$, where $V_r = 0$ and p^* is assumed to be a fixed value, p^*_∞:

$$p^* + \frac{\rho}{2}\left(\frac{q}{2\pi r}\right)^2 = p^*_\infty$$

$$p^* = p^*_\infty - \frac{\rho}{2}\left(\frac{q}{2\pi r}\right)^2 \tag{11.32}$$

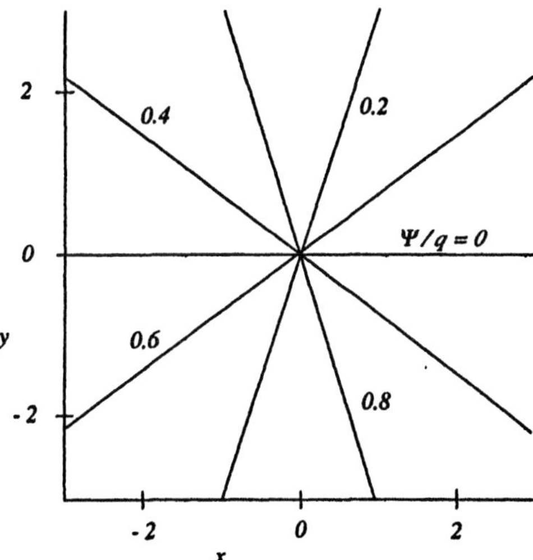

Figure 11.5 The streamlines of the line source flow, $\psi = q\theta/2\pi$, for uniformly spaced values of ψ/q.

At any radius, the pressure is *less* than its value at infinity because the outwardly moving flow is decelerating, requiring that the pressure force be acting inward (positive $\partial p/\partial r$).[9] But the pressure becomes zero when $r = \sqrt{\rho q^2/2p^*_\infty}/2\pi$, so that equation 11.32 is physically unrealistic for smaller radii than this value, where the pressure would be negative.

The distribution of pressure with radial distance is shown in figure 11.6, using dimensionless variables. The flow at the origin, $r = 0$, is *singular*, *i.e.*, the magnitudes of V and p^* become infinite and the streamlines converge to a point. Thus we have to disregard the solution at $r = 0$ because it is physically impossible for the fluid to behave this way. Nevertheless, the outer portion of the flow field describes a physically achievable flow.

The Line Vortex

A line vortex is a flow in which the fluid particles move in circles about the origin with a tangential speed V_θ that varies with radius r so as to maintain the same circulation

---

[9]If the flow is moving inward (negative q), then it is accelerating inward and $\partial p/\partial r$ is still positive.

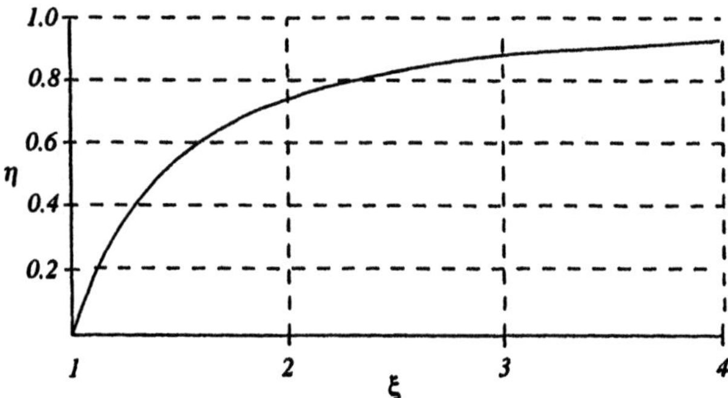

Figure 11.6 The line source pressure ratio $\eta \equiv p^*/p^*_\infty$ as a function of the dimensionless radial distance $\xi \equiv 2\pi r/q\,(\rho/2p^*_\infty)^{1/2}$ for distances where the pressure is positive.

$\Gamma = -2\pi r V_\theta$ on all streamlines. Its stream function and velocity components are:

$$\psi = \frac{\Gamma}{2\pi}\ln r = \frac{\Gamma}{2\pi}\ln\sqrt{x^2+y^2} \tag{11.33}$$

$$V_r = \frac{\partial\psi}{\partial\theta} = 0; \qquad V_\theta = -\frac{\partial\psi}{\partial r} = -\frac{\Gamma}{2\pi r}$$

$$u = \frac{\partial\psi}{\partial y} = \frac{\Gamma}{2\pi}\left(\frac{y}{x^2+y^2}\right); \qquad v = -\frac{\partial\psi}{\partial x} = -\frac{\Gamma}{2\pi}\left(\frac{x}{x^2+y^2}\right) \tag{11.34}$$

The line vortex stream function, equation 11.33, satisfies Laplace's equation in polar coordinates:

$$\frac{\partial^2\psi}{\partial r^2} + \frac{1}{r}\frac{\partial\psi}{\partial r} + \frac{1}{r^2}\frac{\partial^2\psi}{\partial\theta^2} = \frac{1}{r^2} - \frac{1}{r^2} + 0 = 0$$

so the flow is irrotational. But, if the flow is irrotational ($\omega = 0$), how can the circulation on a circular streamline be non-zero when this circulation is equal to the integral of ω over the area of the circle by equation 11.9? The answer lies in the singular behavior of the flow at the origin where the velocity becomes infinite. At this point, there is an infinitesimal region of infinite vorticity having a finite circulation Γ. As in the line source flow, the origin is a point in the flow that is physically impossible to achieve, but the remainder of the flow is irrotational.

The streamlines of the line vortex flow are shown in figure 11.7. Note how the streamlines are more closely spaced for smaller r because V_θ is larger there.

The pressure in a steady line vortex flow is obtained by applying Bernoulli's equation between any radius r, where $V = V_\theta$, and $r \to \infty$, where $V = V_\theta = 0$ and p^* has

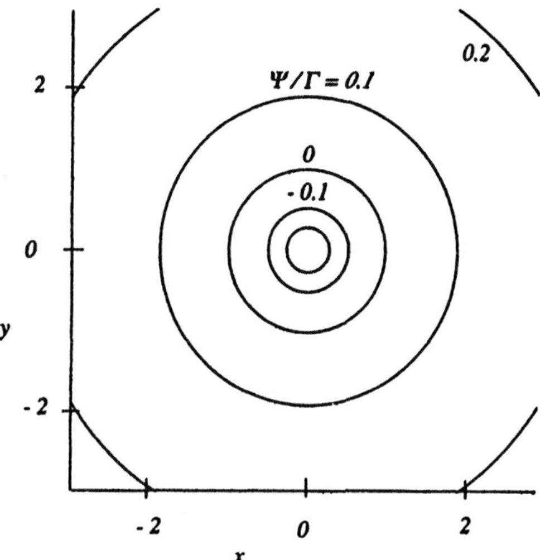

Figure 11.7 The circular streamlines of the line vortex, $\psi = -(\Gamma/2\pi)\ln r$, for uniformly spaced values of ψ/Γ.

a constant value, p^*_∞:

$$p^* + \frac{\rho}{2}\left(\frac{\Gamma}{2\pi r}\right)^2 = p^*_\infty$$

$$p^* = p^*_\infty - \frac{\rho}{2}\left(\frac{\Gamma}{2\pi r}\right)^2 \tag{11.35}$$

Like the line source flow, the line vortex flow has a pressure lower than p^*_∞ at any finite radius, the more so as the radius is smaller. This pressure p^* becomes negative at $r < r_c \equiv \sqrt{\rho\Gamma^2/2p^*_\infty}/2\pi$ so that the flow is physically impossible for $r < r_c$. The positive pressure gradient $\partial p/\partial r$ provides the inward force needed to accelerate centrifugally the fluid particles moving in a circle along a streamline. The radial pressure distribution in a line vortex is identical to that shown in figure 11.6 if q is replaced by Γ.[10]

[10]For dimensional consistency in an actual problem, we should write equation 11.33 as $\psi = (\Gamma/2\pi)\ln(r/a)$, where a is an arbitrary dimension of the flow and could be taken to be equal to r_c. This has no effect on the velocity field.

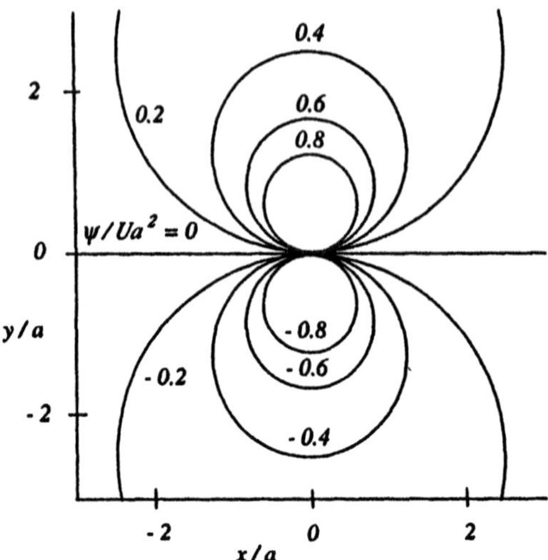

Figure 11.8 The streamlines of the line doublet, $\psi = Ua^2 \sin\theta/r$, for uniformly spaced values of ψ/Ua^2.

The Line Doublet

A line doublet is a flow formed in the limit from a line source of strength q at $x = \epsilon$ and a line sink of strength $-q$ at $x = -\epsilon$ that are brought together ($\epsilon \to 0$) while the product $q\epsilon$ is held constant. Its stream function and velocity components are:

$$\psi = Ua^2\left(\frac{\sin\theta}{r}\right) = Ua^2\left(\frac{y}{x^2+y^2}\right) \tag{11.36}$$

$$V_r = \frac{1}{r}\frac{\partial\psi}{\partial\theta} = Ua^2\left(\frac{\cos\theta}{r^2}\right); \qquad V_\theta = -\frac{\partial\psi}{\partial r} = Ua^2\left(\frac{\sin\theta}{r^2}\right)$$

$$u = \frac{\partial\psi}{\partial y} = Ua^2\left(\frac{x^2-y^2}{(x^2+y^2)^2}\right); \qquad v = -\frac{\partial\psi}{\partial x} = Ua^2\left(\frac{xy}{(x^2+y^2)^2}\right) \tag{11.37}$$

where U and a, a velocity and length scale characterizing the flow, are introduced for convenience in subsequent analyses. The streamlines of equation 11.36 for the line doublet are shown in figure 11.8.

The line doublet stream function, equation 11.36, satisfies Laplace's equation:

$$\frac{\partial^2\psi}{\partial r^2} + \frac{1}{r}\frac{\partial\psi}{\partial r} + \frac{1}{r^2}\frac{\partial^2\psi}{\partial\theta^2} = Ua^2\left(\frac{2\sin\theta}{r^3} - \frac{\sin\theta}{r^3} - \frac{1}{r^3}\sin\theta\right) = 0$$

The line doublet flow has a special property. Suppose that at a given instant of time there is a solid circular cylinder of radius a positioned with its center at $r = 0$. (This would be a circle of unit radius in figure 11.8.) If this cylinder were moving at this instant with a velocity $\mathbf{V} = U\,\mathbf{i}_x$ in the x direction, then at any point on its circumference, the component of its velocity in the radial direction would be $U\cos\theta$. However, at $r = a$, the radial velocity of the doublet flow is also $V_r = U\cos\theta$, *i.e.*, there would be no flow across the boundary of the circular cylinder if the flow external to it were the doublet flow for $r \geq a$. Thus the doublet stream function describes the instantaneous unsteady flow outside a circular cylinder centered at $r = 0$ and moving with a velocity $\mathbf{V} = U\,\mathbf{i}_x$.

The fluid set into motion by a moving circular cylinder, illustrated by the streamlines of figure 11.8 for $r/a \geq 1$, has kinetic energy imparted to the flow by the acceleration of the cylinder from rest to the speed U. We can calculate this kinetic energy per unit width of flow (ke) by integrating $\rho V^2/2$ over the plane area for $r \geq a$:

$$\begin{aligned} ke &= \frac{\rho}{2}\int_a^\infty (V_r^2 + V_\theta^2) 2\pi r\,dr = \pi\rho(Ua^2)^2 \int_a^\infty \left(\frac{\cos^2\theta}{r^4} + \frac{\sin^2\theta}{r^4}\right) r\,dr \\ &= \pi\rho(Ua^2)^2 \int_a^\infty \frac{1}{r^3}\,dr = \pi\rho(Ua^2)^2 \left| -\frac{1}{2r^2} \right|_a^\infty \\ &= \frac{U^2}{2}(\rho\pi a^2) \end{aligned} \tag{11.38}$$

This fluid kinetic energy per unit length of cylinder is thus equal to the kinetic energy per unit mass of the cylinder ($U^2/2$) times the mass of fluid displaced by the cylinder per unit length of cylinder, $\rho\pi a^2$. This mass of fluid is called the *entrained mass* because it effectively increases the mass to be accelerated when an external force is applied to a cylinder immersed in a liquid.

Example 11.3

The frequency f of vibration of a wire of length L subject to a tension T is:

$$f = \frac{1}{L}\sqrt{\frac{T}{m}}$$

where m is the mass of the wire per unit of length. If a wire of specific gravity SG is immersed in water, derive an expression for its frequency of vibration.

Solution

The effective mass per unit length of the wire is the sum of the wire mass per unit length m plus the mass of water per unit length displaced by the wire, m/SG, or a total of $(1 + 1/SG)m$. The frequency f of the submerged wire thus becomes:

$$f = \frac{1}{L}\sqrt{\frac{T(SG)}{m(1 + SG)}}$$

Flow at a Corner

A simple stream function that satisfies Laplace's equation describes a flow in the vicinity of two straight walls that meet at $r = 0$:

$$\psi = Ar^n \sin n\theta \tag{11.39}$$

where A is a constant. When $n\theta = 0$ or π, $\psi = 0$, and the radial lines $\theta = 0$ and π/n are straight streamlines of the flow that can be considered to be walls intersecting at $r = 0$ and including the angle $\phi \equiv \pi/n$. The velocity components V_r and V_θ and the magnitude V of the velocity are:

$$V_r = \frac{1}{r}\frac{\partial\psi}{\partial\theta} = Anr^{n-1}\cos n\theta; \qquad V_\theta = -\frac{\partial\psi}{\partial r} = -Anr^{n-1}\sin n\theta$$

$$V = \sqrt{V_r^2 + V_\theta^2} = Anr^{n-1} \tag{11.40}$$

Note that the flow speed V is a function of radius r only. If $n > 1$ $(\phi < \pi)$, $V = 0$ at $r = 0$, and the flow at the corner is a stagnation point while $V = \infty$ at $r = \infty$. On the other hand, if $n < 1$ $(\phi > \pi)$, then $V = \infty$ at $r = 0$, and the flow is singular at the corner. The case of $n = 1$ $(\phi = \pi)$ reduces to equations 11.28–11.29 for uniform flow with $A = U$.

The pressure p^* may be found by applying Bernoulli's equation between a point in the flow and $r = 0$, where $p^* = p_0^*$ when $\phi < \pi$, or $r = \infty$, where $p^* = p_\infty^*$ when $\phi > \pi$:

$$\begin{aligned} p^* &= p_0^* - \frac{\rho}{2}(Anr^{n-1})^2 \qquad (n > 1;\ \phi < \pi) \\ &= p_\infty^* - \frac{\rho}{2}(Anr^{n-1})^2 \qquad (n < 1;\ \phi > \pi) \end{aligned} \tag{11.41}$$

Note that the pressure p^* becomes negative when $r > (\sqrt{2p_0^*/\rho}/An)^{1/n-1}$ or $r < (\sqrt{2p_\infty^*/\rho}/An)^{1/n-1}$ for $n > 1$ or $n < 1$, respectively.

The streamlines of the flow for $\phi = \pi/4$ and $2\pi/3$ are plotted in figure 11.9, left and right respectively. For both of these flows, the origin $r = 0$ is a stagnation point. In the range $2\pi/3 > \theta > -2\pi/3$, the latter flow also describes the symmetric flow past a wedge of included angle $2\pi/3$.

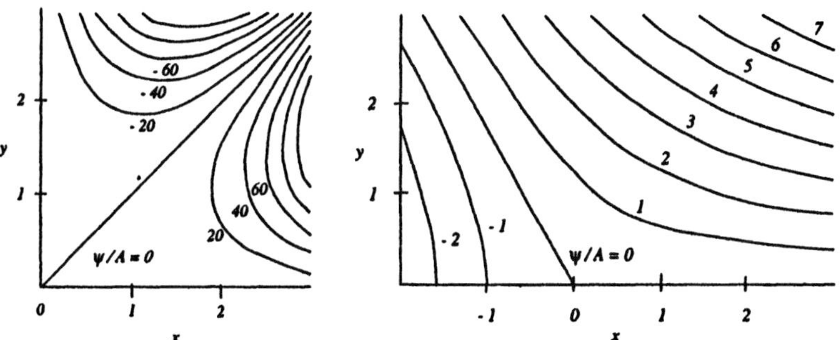

Figure 11.9 The streamlines for corner flow when (*left*) $\phi = \pi/4$ and (*right*) $\phi = 2\pi/3$.

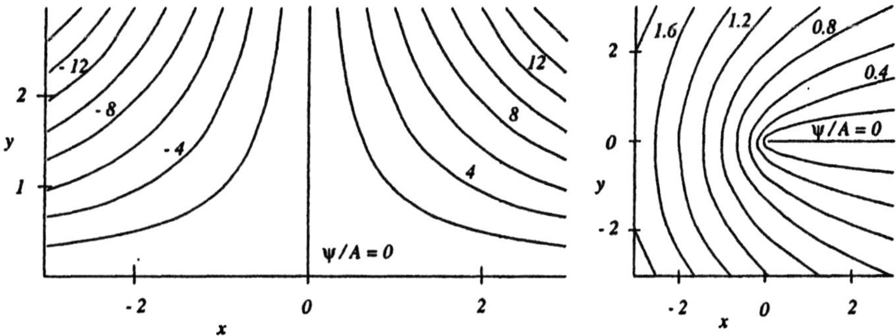

Figure 11.10 The streamlines of corner flow for (*left*) $\phi = \pi/2$ and (*right*) $\phi = 2\pi$.

The corner flow for $\phi = \pi/2$, whose streamlines are shown on the left of figure 11.10, depict a plane stagnation point flow for $\pi \geq \theta \geq 0$. For this flow, the stream function and velocities are:

$$\psi = Ar^2 \sin 2\theta = 2Ar^2 \sin\theta \cos\theta = 2Axy \tag{11.42}$$

$$V_r = 2Ar\cos 2\theta; \qquad V_\theta = -2Ar\sin 2\theta$$

$$u = \frac{\partial\psi}{\partial y} = 2Ax; \qquad v = -\frac{\partial\psi}{\partial x} = -2Ay \tag{11.43}$$

The streamlines are hyperbolas.

The streamline pattern for $\phi = 2\pi$, shown on the right of figure 11.10, depicts the flow around the end of a semi-infinite flat plate ($y = 0$; $x \geq 0$). For this flow, at $r = 0$, $V = \infty$, and $p^* = -\infty$.

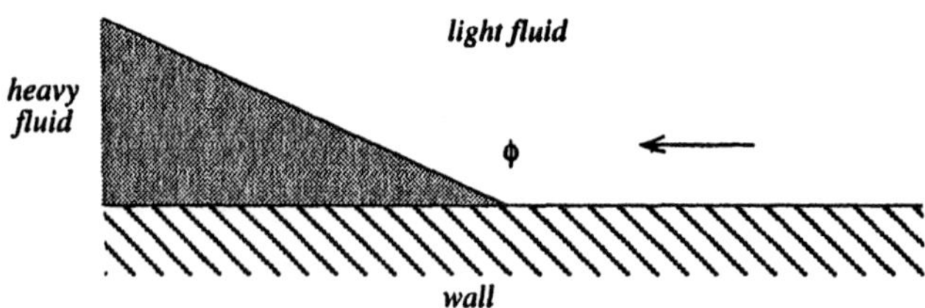

Figure E 11.4

Example 11.4

A light fluid of density ρ_l flows above a wedge-shaped heavy fluid (density ρ_h) that is stationary, as shown in figure E 11.4. Derive an expression for the angle ϕ and the constant A of the light fluid stream function in terms of the density ratio ρ_h/ρ_l.

Solution

Along the streamline of the light fluid that is in contact with the heavy fluid, Bernoulli's equation 11.41 gives:

$$p + \rho_l g r \sin\phi + \frac{\rho_l}{2}(Anr^{n-1})^2 = p_0$$

while in the stationary heavy fluid we find:

$$p + \rho_h g r \sin\phi = p_0$$

Since the pressure p is the same in each fluid at the common boundary,

$$(Anr^{n-1})^2 = \left(\frac{\rho_h}{\rho_l} - 1\right) 2gr \sin\phi$$

But the exponent of r must be the same on both sides of this equation. Thus $n = 3/2$, and $\phi = 2\pi/3$, the same value as the flow on the right side of figure 11.9. Substituting,

$$\left(\frac{3}{2}A\right)^2 = \left(\frac{\rho_h}{\rho_l} - 1\right) 2g \sin\left(\frac{2\pi}{3}\right)$$

$$A = \frac{2}{3}\sqrt{(\rho_h/\rho_l - 1)\sqrt{3}\,g}$$

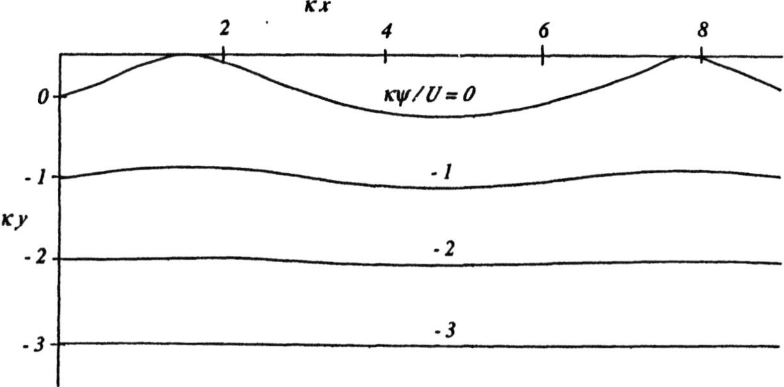

Figure 11.11 The streamlines of the flow along a corrugated wall (uppermost streamline), showing equal increments of $\kappa\psi/U$ for $\epsilon = 0.3$. The ordinate is κy and the abscissa, κx, where $\kappa = 2\pi/\lambda$, λ being the wavelength of the corrugation.

Flow along a Corrugated Wall; Gravity Waves

The function $\exp(\kappa y)\sin\kappa x$ is a solution to Laplace's equation and thus a possible stream function. If we choose a stream function which is the sum of a uniform flow plus this new function, of the form:

$$\psi = Uy - \epsilon\left(\frac{U}{\kappa}\right)\exp(\kappa y)\sin\kappa x \tag{11.44}$$

we would have a steady flow of average horizontal speed U beneath a corrugated wall of wavelength $\lambda = 2\pi/\kappa$ for $\psi \leq 0$, $y < 0$. The streamlines of the flow are shown in figure 11.11 for the case of $\epsilon = 0.3$.

The velocity components of this flow are:

$$u = \frac{\partial\psi}{\partial y} = U[1 - \epsilon\exp(\kappa y)\sin\kappa x]$$

$$v = -\frac{\partial\psi}{\partial x} = \epsilon U\exp(\kappa y)\cos\kappa x$$

$$V^2 = u^2 + v^2 = U^2[1 - 2\epsilon\exp(\kappa y)\sin\kappa x + (\epsilon\exp[\kappa y])^2] \tag{11.45}$$

The disturbance to the flow caused by the wall dies out exponentially as $y \rightarrow -\infty$.

To find the pressure in the flow, apply Bernoulli's equation between a point in the flow and $y \rightarrow -\infty$, where $p^* = p^*_\infty$ and $V = U$:

$$\frac{p^*}{\rho} + \frac{U^2}{2}[1 - 2\epsilon \exp(\kappa y) \sin \kappa x + (\epsilon \exp[\kappa y])^2] = \frac{p^*_\infty}{\rho} + \frac{U^2}{2}$$

$$p^* - p^*_\infty = \frac{\rho}{2}[2\epsilon \exp(\kappa y) \sin \kappa x - (\epsilon \exp[\kappa y])^2] \quad (11.46)$$

We can find a particular value of the flow speed U for which the pressure p becomes constant along a streamline whenever $\mathbf{g} = -g\mathbf{i}_y$ and $\epsilon \ll 1$. When this is so, we may remove the wall, replacing it by air at constant pressure. The resulting flow is a standing (stationary) gravity wave on the surface of a liquid, for which the streamlines are the same as those of figure 11.11, the liquid occupying the space where $\psi \leq 0, y < 0$. To see how to choose U for this case, consider the flow when $\epsilon \ll 1$ and when terms proportional to ϵ^2 may be dropped from equations 11.45 and 11.46. Bernoulli's equation becomes:

$$\frac{p}{\rho} + gy + \frac{U^2}{2}[1 - 2\epsilon \exp(\kappa y) \sin \kappa x] = \frac{p_0}{\rho} + \frac{U^2}{2}$$

$$p = p_0 - \rho g y + \rho \epsilon U^2 \exp(\kappa y) \sin \kappa x \quad (11.47)$$

in which p_0 is the pressure at $x = 0$, $y = 0$, where $u^2 + v^2 = U^2$. Combining 11.47 with equation 11.44 for the stream function ψ,

$$p = p_0 - \frac{\rho g \psi}{U} - \rho\left(\frac{g}{\kappa} - U^2\right) \epsilon \exp(\kappa y) \sin \kappa x \quad (11.48)$$

Thus the pressure will be constant along a streamline $\psi = constant$ if:

$$U = \sqrt{\frac{g}{\kappa}} = \sqrt{\frac{g\lambda}{2\pi}} \quad (11.49)$$

The velocity U in equation 11.49 is called the *phase velocity* of the gravity wave of wavelength λ.[11] Unlike a sound wave, the speed of the gravity wave depends strongly on its wavelength λ. The period $\mathcal{T}$ of a gravity wave, the time required for it to travel one wave length, is:

$$\mathcal{T} = \frac{\lambda}{U} = \sqrt{\frac{2\pi\lambda}{g}} \quad (11.50)$$

[11]The phase velocity U of a wave equals the ratio ϖ/κ, where $\varpi = 2\pi U/\lambda$ is the angular frequency of the wave motion as measured in a coordinate system fixed in the fluid. The *group velocity*, which is the speed of propagation of wave energy, is $d\varpi/d\kappa$. For gravity waves on the surface of a fluid, the group velocity thus becomes one-half of the phase velocity. Gravity waves undergo *dispersion, i.e.*, a wave pulse spreads out with time because the phase and group velocities are different.

This is identical to the relation between the period T and length λ of a simple pendulum.

Example 11.5

A fluid of density ρ_1 occupies the upper half-space $y > 0$ above a more dense fluid of density ρ_2, occupying the lower half-space $y < 0$. The interface between the two fluids is distorted by a sinusoidal gravity wave of wave length $\lambda = 2\pi/\kappa$. Derive a relation between the speed of propagation of this wave, U, and the parameters λ, ρ_1 and ρ_2.

Solution

For the lower half-space in fluid 2, select the stream function of equation 11.44:

$$\psi_1 = Uy - \epsilon\left(\frac{U}{\kappa}\right)\exp(\kappa y)\sin\kappa x$$

with $\epsilon \ll 1$. For the upper half-space, we need a similar stream function, except that the disturbance should die out as $y \to \infty$, but which equals ψ_1 at the interface $y = 0$. Thus we choose:

$$\psi_2 = Uy - \epsilon\left(\frac{U}{\kappa}\right)\exp(-\kappa y)\sin\kappa x$$

The pressure p_1 in fluid 1 is given by equation 11.48 with $\rho = \rho_1$:

$$p_1 = p_0 - \frac{\rho_1 g\psi_1}{U} - \rho_1\left(\frac{g}{\kappa} - U^2\right)\epsilon\exp(\kappa y)\sin\kappa x$$

Following the same steps that led to this expression, p_2 is found to be:

$$p_2 = p_0 - \frac{\rho_2 g\psi_2}{U} - \rho_2\left(\frac{g}{\kappa} + U^2\right)\epsilon\exp(\kappa y)\sin\kappa x$$

the change in the algebraic sign of the U^2 term coming from that in the exponential factor. Along the interface between the two fluids, $\psi_1 = \psi_2 = 0$, the pressures must be equal in the two fluids, *i.e.*, $p_1 = p_2$ or:

$$\rho_1\left(\frac{g}{\kappa} - U^2\right) = \rho_2\left(\frac{g}{\kappa} + U^2\right)$$

$$U = \sqrt{\left(\frac{\rho_2 - \rho_1}{\rho_2 + \rho_1}\right)\frac{g}{\kappa}}$$

Note that as $\rho_2 - \rho_1 \to 0$, the interfacial wave velocity $U \to 0$.

A wave of this type is called an *internal wave*. It is commonly encountered in river estuaries where fresh water flows above slightly more dense sea water.

11.4.2 Superposition of Elementary Flows

Flow Past a Line Source

If we add a uniform flow to that of a line source, the fluid emitted from the line source is swept away in the downstream direction. The stream function and velocity coordinates for this flow are simply the sum of those for uniform flow and the line source, equations 11.28–11.31. However, it is convenient to introduce a length $a \equiv q/2\pi U$, the radial distance from the line source at which the radial velocity equals U, in place of q. In terms of U and a, the stream function and velocity components are:

$$\psi = Ur\sin\theta + Ua\theta = Uy + Ua\tan^{-1}\left(\frac{y}{x}\right) \tag{11.51}$$

$$V_r = U\cos\theta + \frac{Ua}{r}; \qquad V_\theta = -U\sin\theta$$

$$u = U + Ua\left(\frac{x}{x^2+y^2}\right);. \qquad v = Ua\left(\frac{y}{x^2+y^2}\right) \tag{11.52}$$

The streamlines of the flow are shown in figure 11.12. At the point $x = -a$, $y = 0$, the fluid velocity is zero: this is a stagnation point of the flow where the source flow is turned around by the oncoming uniform flow. The parabolic-shaped streamline passing through the stagnation point, $\psi = \pm\pi Ua$, separates the uniform flow from the source flow. This streamline eventually becomes parallel to the x-axis at $x \to \infty$ where $y = \pm\pi a$.

To find the pressure in the flow, we apply Bernoulli's equation between a point in the flow and $R \to \infty$, where $\mathbf{V} = U\mathbf{i}_x$ and $p^* = p^*_\infty$:

$$p^* + \frac{\rho}{2}\left(V_r^2 + V_\theta^2\right) = p^*_\infty + \frac{\rho}{2}U^2$$

$$p^* = p^*_\infty + \frac{\rho}{2}\left(U^2 - V_r^2 - V_\theta^2\right)$$

$$p^* = p^*_\infty - \frac{\rho}{2}U^2\left(\frac{a}{r}\right)\left[\left(\frac{a}{r}\right) + 2\cos\theta\right] \tag{11.53}$$

For the flow of figure 11.12 to persist, there must be an external force per unit flow width, f_{ex}, acting on the line source in the positive x direction. This force provides for the increase in the source flow momentum from zero at the source to ρqU at $x = \infty$. To prove that this is indeed the case, let us apply the x-component of the

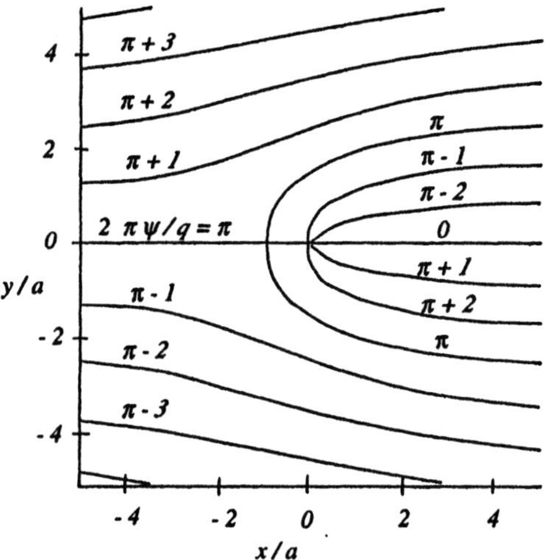

Figure 11.12 The streamlines of the flow past a line source for equal increments of $2\pi\psi/q$. The plane coordinates are x/a, y/a, where $a \equiv q/2\pi U$.

linear momentum theorem, equation 5.11, to the flow through a control volume of radius r and unit distance normal to the plane of the flow:

$$\iint_S \rho u \mathbf{V} \cdot \mathbf{n}\, dS = \iint_S (-p^*)\mathbf{n}\, dS + f_{ex}$$

$$\rho \int_0^{2\pi} u V_r r\, d\theta = -\int_0^{2\pi} (p^* \cos\theta) r\, d\theta + f_{ex}$$

$$\rho \int_0^{2\pi} U^2 (1 + \frac{a}{r}\cos\theta)(\cos\theta + \frac{a}{r}) r\, d\theta = \int_0^{2\pi} \rho U^2 \left(\frac{a}{r}\right)(\cos^2\theta) r\, d\theta + f_{ex}$$

$$\rho a U^2 \int_0^{2\pi} (1 + \cos^2\theta)\, d\theta = \rho a U^2 \int_0^{2\pi} \cos^2\theta\, d\theta + f_{ex}$$

$$f_{ex} = 2\pi \rho a U^2 = \rho q U \tag{11.54}$$

where we have used equation 11.53 to replace p^* and where in lines three and four, we have eliminated integrals that are zero because their integrands have the factor $\cos\theta$ to the first power.

If we replace the source fluid by a semi-infinite solid body having the shape of the streamline passing through the stagnation point and extending out to $x = \infty$ (called a *Rankine half-body*), the resulting residual flow would not exert a net force on the body.[12]

By superposing several line sources and sinks along the x-axis of a uniform flow, we can find closed streamlines of the flow that may be considered to form the surface of closed bodies if the total source and sink strengths add to zero. The fluid will exert no net drag force on such bodies, a consequence of irrotational flow that is called *d'Alembert's paradox.*[13]

Flow Past a Line Vortex

If we superpose a uniform flow with a line vortex, the resulting flow will be asymmetrical about the x-axis. Subtracting equations 11.33 and 11.34 from 11.30 and 11.31 so that the clockwise circulation Γ is positive, the stream function and velocities become, after introducing the length scale $a \equiv \Gamma/2\pi U$:

$$\psi = Ur\sin\theta + Ua\ln\left(\frac{r}{a}\right) = Uy + Ua\ln\left(\frac{\sqrt{x^2+y^2}}{a}\right) \tag{11.55}$$

$$V_r = U\cos\theta; \qquad V_\theta = -U\sin\theta - \frac{Ua}{r}$$

$$u = U + Ua\left(\frac{y}{x^2+y^2}\right); \qquad v = -Ua\left(\frac{x}{x^2+y^2}\right) \tag{11.56}$$

The streamlines of the flow in figure 11.13 show a stagnation point at $x/a = 0$, $y/a = -1$, the point where the vortex velocity V_θ just counteracts the uniform flow U. Above this stagnation point is a closed streamline, $\psi = -Ua = -\Gamma/2\pi$. The interior of this streamline could be replaced by a tear-shaped solid body outside of which is an irrotational flow with circulation Γ.

Applying Bernoulli's equation, the pressure in the flow may be found:

$$p^* + \frac{\rho}{2}\left(V_r^2 + V_\theta^2\right) = p_\infty^* + \frac{\rho}{2}U^2$$

---

[12]Applying the linear momentum theorem to the source flow only, the outflow momentum ρqU equals the applied force f_{ex} at the source, so that the pressure integral on the body surface must be zero.

[13]Jean le Rond d'Alembert (1717–1783) was the first to derive the expression for mass conservation in a fluid flow. He stated his paradox in a treatise on fluid resistance but could not explain why real fluids cause drag on immersed bodies. The explanation was subsequently given by Ludwig Prandtl in 1905 in his famous paper on boundary layer flow.

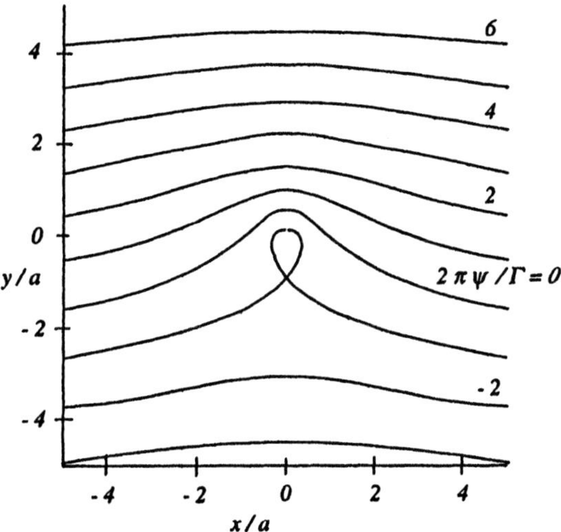

Figure 11.13 The streamlines of the flow past a line vortex of clockwise circulation Γ for equal increments of $2\pi\psi/\Gamma$. The plane coordinates are x/a, y/a, where $a \equiv \Gamma/2\pi U$.

$$p^* = p^*_\infty + \frac{\rho}{2}\left(U^2 - V_r^2 - V_\theta^2\right)$$

$$p^* = p^*_\infty - \frac{\rho}{2}U^2\left(\frac{a}{r}\right)\left[\left(\frac{a}{r}\right) + 2\sin\theta\right] \tag{11.57}$$

The body enclosed by the closed streamline, or the line vortex, experiences a lift force per unit length ℓ that is proportional to Γ. To derive the exact relationship between ℓ and Γ, we apply the y component of the linear momentum theorem to a control surface of radius r and unit thickness in the direction normal to the x, y-plane, noting that the external force applied to the control surface is $-\ell$:

$$\iint_S \rho v \mathbf{V}\cdot\mathbf{n}\,dS = \iint_S (-p^*)\mathbf{n}\,dS - \ell$$

$$\rho\int_0^{2\pi} vV_r r\,d\theta = -\int_0^{2\pi}(p^*\sin\theta)r\,d\theta - \ell$$

$$\rho\int_0^{2\pi} U^2\left(\frac{a}{r}\cos\theta\right)(\cos\theta)r\,d\theta = \int_0^{2\pi}\rho U^2\left(\frac{a}{r}\right)\sin^2\theta\, r\,d\theta - \ell$$

$$-\rho a U^2\int_0^{2\pi}\cos^2\theta\,d\theta = \rho a U^2\int_0^{2\pi}\sin^2\theta\,d\theta - \ell$$

$$\ell = 2\pi\rho a U^2 = \rho U\Gamma \tag{11.58}$$

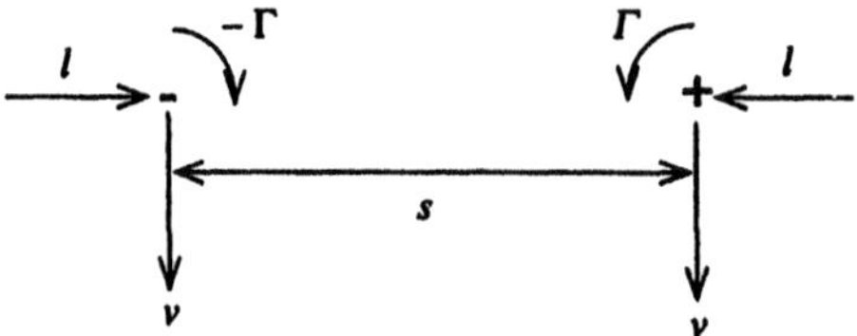

Figure 11.14 The induced velocity v and lift force l experienced by two bound vortices of opposite circulation, separated by a distance s.

where we have used 11.57 to replace p^* and have dropped terms in the integrand containing sin θ to the first power, whose integrals are zero.

Equation 11.58, which relates the lift per unit length ℓ with the circulation Γ about a cylindrical body in plane flow of speed U, is known as the *Kutta-Joukowski theorem.* To relate the directions of the lift force per unit length, the flow velocity and the circulation, it may be expressed in vector form as:

$$\boldsymbol{\ell} = \rho \mathbf{V} \times \boldsymbol{\Gamma} \tag{11.59}$$

where the circulation vector $\boldsymbol{\Gamma}$ has the magnitude of Γ and the direction of the normal to the surface enclosed by the curve C defining the circulation. (For clockwise circulation in the plane of figure 11.13, $\boldsymbol{\Gamma} = -\Gamma \mathbf{i}_z$.)

The line vortex of figure 11.13 is called a *bound vortex* because it is held in a stationary position as the oncoming fluid flows around it. A bound vortex experiences a lift force given by equation 11.59. If the vortex is free to move with the fluid, *i.e.*, has no velocity relative to the fluid surrounding it, then it experiences no lift force and is called a *free vortex.*

Two bound line vortices exert forces on each other because each induces a flow at the location of the other vortex. Figure 11.14 shows two line vortices having circulations that are equal in magnitude but opposite in direction, separated by a distance s, in a fluid that is stationary at infinity. Each of these vortices induces a downward flow of speed $v = -\Gamma/2\pi s$ at the center of the other vortex. By equation 11.59, the resulting lift per unit length, $\ell = \rho(\Gamma/2\pi s)\Gamma = \rho\Gamma^2/2\pi s$, is directed toward the other vortex. In other words, two bound vortices of opposite circulation attract each other with a force equal to $\rho\Gamma^2/2\pi s$ that must be counterbalanced by an opposite external force to maintain their separation. However, if the vortices are free to move, they will drift downward at a speed $v = -\Gamma/2\pi s$, experiencing no attractive force because they have no motion relative to the nearby fluid. If the vortices have circulations of the same direction, they will experience a repulsive force if bound and drift in a circle about their geometric center if free.

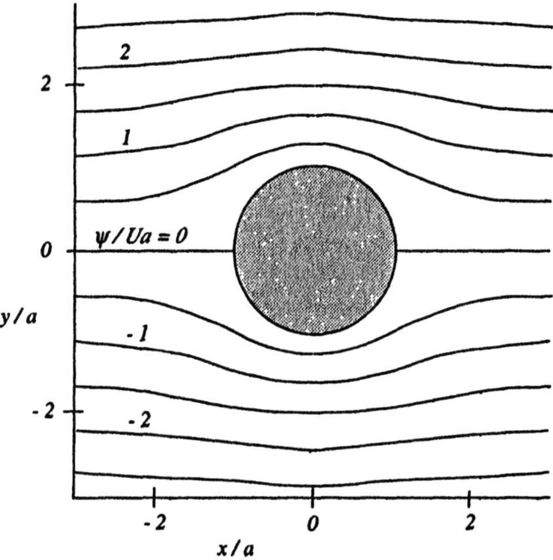

Figure 11.15 The streamlines of the flow over a circular cylinder of radius a for equal increments of ψ/Ua.

Flow Over a Circular Cylinder

The unsteady flow of a moving circular cylinder, equations 11.36 and 11.37 for $r \geq a$, may be transformed into the steady flow over the cylinder by subtracting the unsteady flow from the uniform flow, equations 11.28 and 11.29, obtaining:

$$\psi = Ur\left(1 - \frac{a^2}{r^2}\right)\sin\theta = Uy\left(1 - \frac{a^2}{x^2 + y^2}\right) \tag{11.60}$$

$$V_r = U\left(1 - \frac{a^2}{r^2}\right)\cos\theta; \qquad V_\theta = -U\left(1 + \frac{a^2}{r^2}\right)\sin\theta$$

$$u = U - Ua^2\left[\frac{x^2 - y^2}{(x^2 + y^2)^2}\right]; \qquad v = -Ua^2\left[\frac{xy}{(x^2 + y^2)^2}\right] \tag{11.61}$$

The streamlines of the flow, which is symmetric about $x = 0$ and $y = 0$, are shown in figure 11.15. The flow within the circular closed streamline ($\psi = 0$, $r = a$) may be replaced by a solid circular cylinder of radius a. There are two stagnation points ($u = v = 0$) at $x = \pm a$, $y = 0$, the front and back ends of the cylinder.

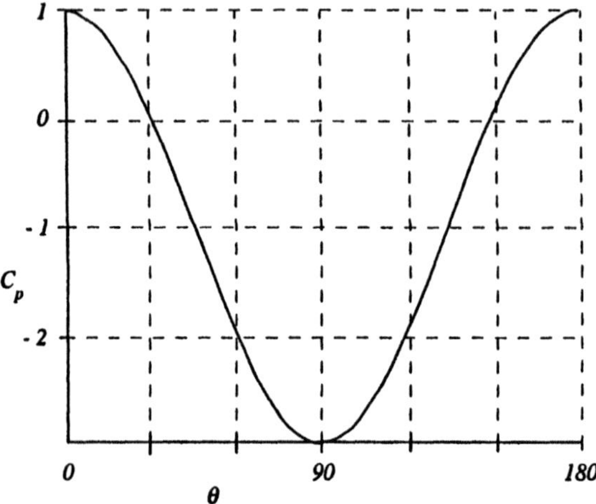

Figure 11.16 The pressure coefficient C_p on the surface of a circular cylinder in a uniform flow.

The pressure in the flow, from Bernoulli's equation, is:

$$p^* + \frac{\rho}{2}\left(V_r^2 + V_\theta^2\right) = p_\infty^* + \frac{\rho}{2}U^2$$

$$\begin{aligned} p^* &= p_\infty^* + \frac{\rho}{2}\left(U^2 - V_r^2 - V_\theta^2\right) \\ &= p_\infty^* + \frac{\rho}{2}U^2\left[1 - \left(1 - \frac{a^2}{r^2}\right)^2 \cos^2\theta - \left(1 + \frac{a^2}{r^2}\right)^2 \sin^2\theta\right] \\ &= p_\infty^* - \frac{\rho}{2}U^2\left(\frac{a^2}{r^2}\right)\left[\left(\frac{a^2}{r^2}\right) - 2\cos 2\theta\right] \end{aligned} \tag{11.62}$$

and the pressure coefficient C_p is:

$$C_p \equiv \frac{p^* - p_\infty^*}{\rho U^2/2} = \left(\frac{a^2}{r^2}\right)\left[2\cos 2\theta - \left(\frac{a^2}{r^2}\right)\right] \tag{11.63}$$

The pressure coefficient distribution on the surface of the cylinder ($r = a$) is plotted in figure 11.16. Since the pressure distribution is symmetric about $x = 0$ and $y = 0$, there is no net force on the cylinder. Notice that the pressure on most of the surface is less than p_∞^*, and the minimum value of C_p on the surface is -3 at $\theta = \pi/2$.

The irrotational flow field around a circular cylinder, shown in figure 11.15, does not describe very well the real viscous flow over a cylinder because the latter separates from the surface, usually between 80° and 120° from the forward stagnation

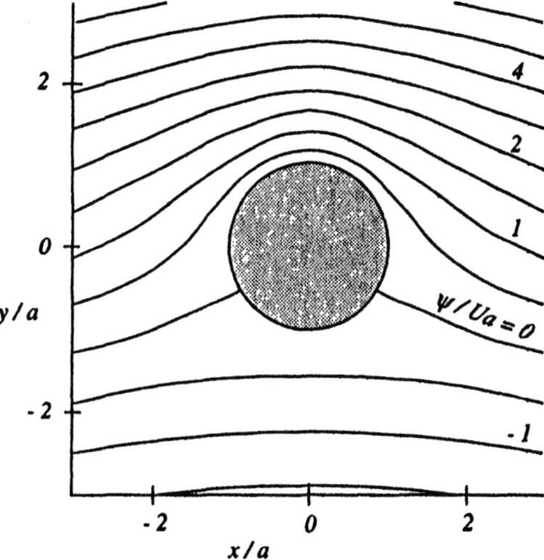

Figure 11.17 Flow over a cylinder having a circulation of $\Gamma = 2\pi Ua$. Streamlines are spaced for equal increments of ψ/Ua.

point (depending upon the Reynolds number), forming a large wake region behind the cylinder. The altered contour of this wake region substantially changes the surrounding irrotational flow and the pressure distribution on the cylinder, contributing to the cylinder drag. However, the region near the forward stagnation point is barely affected and closely resembles the irrotational flow in figure 11.15.

If a line vortex is added to the flow of figure 11.15, the resulting flow is that over a circular cylinder with lift. The appropriate stream function is:

$$\psi = Ur\left(1 - \frac{a^2}{r^2}\right)\sin\theta + \frac{\Gamma}{2\pi}\ln\left(\frac{r}{a}\right)$$

$$= Uy - Ua^2\left(\frac{y}{x^2+y^2}\right) + \frac{\Gamma}{2\pi}\ln\left(\frac{\sqrt{x^2+y^2}}{a}\right) \qquad (11.64)$$

The streamlines for the case of $\Gamma = 2\pi Ua$, corresponding to a vortex having $V_\theta = -U$ at $r = a$, are plotted in figure 11.17. The corresponding lift $\ell = \rho U\Gamma = \rho U^2(2\pi a)$, and the lift coefficient $C_L = \ell/(\rho/2)U^2(2a) = 2\pi$.

Flow similar to that of figure 11.17 can be generated by rotating a circular cylinder about its axis in the direction of the circulation. The resulting lift is called the *Magnus effect*. Lift coefficients of up to $\sim 2\pi$ may be generated in this manner, but lift/drag ratios are less than 3, and mechanical power must be supplied to the rotating cylinder.

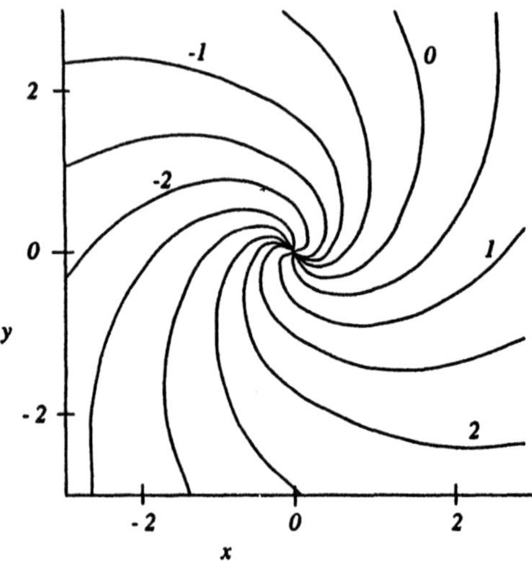

Figure 11.18 The streamlines of the flow of a line vortex and sink when $\Gamma = q$ for equal increments of $2\pi\psi/\Gamma = 2\pi\psi/q$.

Line Vortex and Sink

If we combine a line vortex with a line sink, the resulting flow swirls while it moves inward toward the origin. The stream function and velocities determined from equations 11.30–11.31 and 11.33–11.34 are:

$$\psi = \frac{\Gamma}{2\pi}\ln r - \frac{q}{2\pi}\theta = \frac{\Gamma}{2\pi}\ln\sqrt{x^2+y^2} - \frac{q}{2\pi}\tan^{-1}\left(\frac{y}{x}\right) \qquad (11.65)$$

$$V_r = -\frac{q}{2\pi r}; \qquad V_\theta = -\frac{\Gamma}{2\pi r}$$

$$u = \frac{1}{2\pi}\left(\frac{qx - \Gamma y}{x^2+y^2}\right); \qquad v = -\frac{1}{2\pi}\left(\frac{\Gamma x + qy}{x^2+y^2}\right) \qquad (11.66)$$

while the pressure is:

$$p^* = p^*_\infty - \frac{\rho}{2}\left(\frac{\Gamma^2+q^2}{(2\pi r)^2}\right) \qquad (11.67)$$

The streamlines, plotted in figure 11.18, are spirals whose tangent makes a fixed angle with the radius vector, $\tan^{-1}(\Gamma/q)$. Flows of this type occur in solid particle separators, called cyclone separators, where an inflow stream introduced tangentially at the periphery of a circular cylindrical container flows inward toward the axis. Heavy particles are centrifuged outward where they impinge on the outer wall and are collected. The flow in a tornado is also similar to figure 11.18, but it is not strictly an irrotational flow because of the earth's rotation.

The Method of Images

For inviscid irrotational flow, the boundary condition on the velocity field at a stationary solid wall requires that the component of the velocity normal to the wall be zero. This condition can be satisfied for an elementary flow, such as a line source or vortex, near a plane wall by superposing on the basic flow a flow field from an image of the elementary flow located symmetrically on the opposite side of the wall boundary. For example, suppose a line source located at $x = 0$, $y = a$ is bounded by a solid plane wall at $y = 0$. In the absence of the wall, the source flow stream function would be that of equation 11.30 with y replaced by $y - a$:[14]

$$\psi = \frac{q}{2\pi} \tan^{-1}\left(\frac{y-a}{x}\right)$$

To satisfy the wall boundary condition, we must add another source of equal strength at $x = 0$, $y = -a$:

$$\psi = \frac{q}{2\pi} \tan^{-1}\left(\frac{y-a}{x}\right) + \frac{q}{2\pi} \tan^{-1}\left(\frac{y+a}{x}\right) \tag{11.68}$$

This produces a flow that is symmetric about the x-axis, which is a streamline of the flow, as plotted in figure 11.19 for the case of $a = 1$. The streamline $\psi = 0$, which coincides with the y-axis for $0 \leq y \leq 1$, and the entire x-axis has a stagnation point at the origin.

Referring to equation 11.52, the velocity field of the flow of equation 11.68 is:

$$u = \frac{q}{2\pi}\left(\frac{x}{x^2 + (y-a)^2} + \frac{x}{x^2 + (y+a)^2}\right);$$

$$v = \frac{q}{2\pi}\left(\frac{y-a}{x^2 + (y-a)^2} + \frac{y+a}{x^2 + (y+a)^2}\right) \tag{11.69}$$

Along the wall, where $y = 0$, the velocity components of equation 11.69 are $v = 0$ and $u = (q/2\pi)[2x/(x^2 + a^2)]$ so that the wall pressure p_w^* is:

$$p_w^* = p_\infty^* - \frac{\rho}{2}\left(\frac{q}{2\pi}\right)^2 \left(\frac{2x}{x^2 + a^2}\right)^2 \tag{11.70}$$

The wall pressure p_w^* equals the stagnation pressure p_∞^* at $x = 0$ and has minima at $x = \pm a$.

---

[14]If $\psi\{x, y\}$ is the stream function of an elementary flow centered at $x = 0$, $y = 0$, then $\psi\{x-b, y-a\}$ is the stream function of the elementary flow when it is centered at $x = b$, $y = a$.

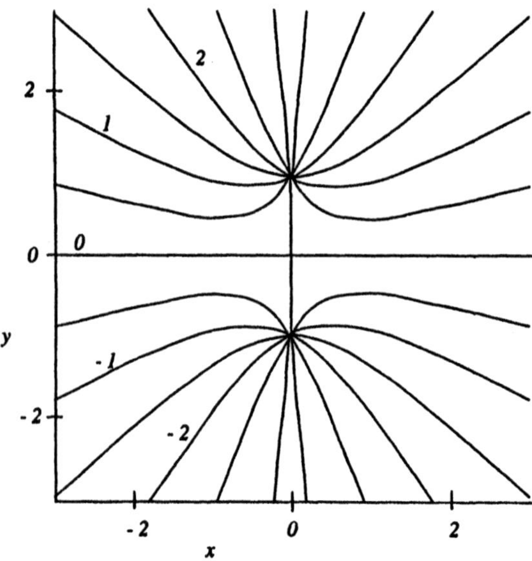

Figure 11.19 The streamlines of a flow from a line source at $x = 0$, $y = 1$ above a solid boundary at $y = 0$ are formed by placing an image line source of equal strength at $x = 0$, $y = -1$, and plotted for equal increments of $2\pi\psi/q$.

From equation 11.69, we find that the image line source induces a velocity $v_i = (q/4\pi a)$ at the location ($x = 0$, $y = a$) of the original line source. By equation 11.54, there must be an external force per unit length $\mathbf{f}_{ex}$ applied to the line source at $x = 0$, $y = a$:

$$\mathbf{f}_{ex} = \rho q(v_i \mathbf{i}_y) = \frac{\rho q^2}{4\pi a}\mathbf{i}_y \tag{11.71}$$

This external force is required to balance the attractive force that the wall, or the image source, exerts on the line source. Thus a line source is attracted by a nearby wall, and a line sink is repelled by it. There is no net force acting on the fluid in the upper half-plane ($y \geq 0$) because the external force $\mathbf{f}_{ex}$ is balanced by the pressure force $\int (p^*_\infty - p^*_w)\, dx$ exerted by the wall on the flow.

Example 11.6

A line vortex of circulation Γ is located at $x = 1$, $y = 1$. It is bounded by walls at $x = 0$ and $y = 0$, forming a right angle corner, the vortex flow being confined to the first quadrant of the x, y-plane. (a) Write the stream function of this vortex flow and sketch the streamlines. (b) Derive an expression for the lift force per unit length $\boldsymbol{\ell}$ exerted on the line vortex by its proximity to the wall.

Solution

(a) The stream function of an unconstrained line vortex, equation 11.33, when located at $x = 1,\ y = 1$ would be:

$$\psi = -\frac{\Gamma}{2\pi}\ln\sqrt{(x-1)^2+(y-1)^2} = -\frac{\Gamma}{4\pi}\ln[(x-1)^2+(y-1)^2]$$

To provide for no flow across the $y = 0$ boundary, we must locate an image line vortex of opposite circulation at $x = 1,\ y = -1$, whose stream function is:

$$\psi = \frac{\Gamma}{4\pi}\ln[(x-1)^2+(y+1)^2]$$

However, to ensure that there is no flow across the $x = 0$ boundary, we need to place two image sources at $x = -1$ that are opposite in circulation to these two. The resulting stream function is the sum of these four, which may be expressed as the logarithm of the product of the arguments:

$$\psi = -\frac{\Gamma}{4\pi}\ln\left(\frac{[(x-1)^2+(y-1)^2][(x+1)^2+(y+1)^2]}{[(x-1)^2+(y+1)^2][(x+1)^2+(y-1)^2]}\right)$$

The streamlines have the following shape:

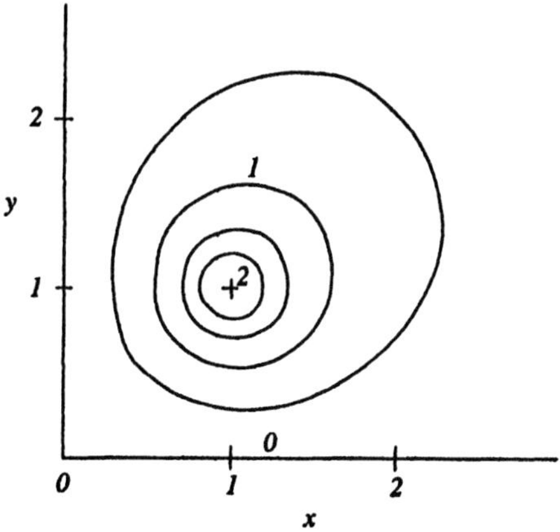

(b) According to the discussion referring to figure 11.14, the line vortex at $x = 1,\ y = 1$ is subject to a lift force $\boldsymbol{\ell} = -(\rho\Gamma^2/4\pi)\mathbf{i}_y$ due to its image of opposite circulation at $x = 1,\ y = -1$. Similarly, there will be a lift force $\boldsymbol{\ell} = -(\rho\Gamma^2/4\pi)\mathbf{i}_x$ due to the image vortex at $x = -1,\ y = 1$. The third image source of the same circulation, at $x = -1,\ y = -1$, exerts a lift force that is smaller by a factor of $\sqrt{2}$ than either of the others, having components in the opposite direction of the first two that are smaller by a factor of $(\sqrt{2})^2 = 2$. The total lift force is thus:

$$\boldsymbol{\ell} = -\frac{\rho\Gamma^2}{8\pi}(\mathbf{i}_x + \mathbf{i}_y)$$

The line vortex is attracted to the walls, as was the case of the line source.

11.5 Axisymmetric Irrotational Flows

In this section, we consider irrotational flows whose velocity vector lies in the r, z-plane (called the *meridional plane*) of a cylindrical coordinate system whose coordinates were defined in figure 11.3(b). The velocity components V_r, V_z (or V_R, V_ϕ) are expressible in terms of the derivatives of the stream function $\psi\{r, z\}$ (or $\psi\{R, \phi\}$), as given in equation 11.22 (or equation 11.23). We first consider some elementary flows analogous to those previously treated in section 11.4.1 and then use the superposition principle to treat some practical applications of these flows.

11.5.1 Elementary Flows

Uniform Flow

In uniform flow parallel to the z-axis at a speed U, the velocity $\mathbf{V}$ is $U\mathbf{i}_z$. The stream function of this velocity field is:

$$\psi = \frac{U}{2}r^2 = \frac{U}{2}(R \sin\phi)^2 \qquad (11.72)$$

and the velocity components are:

$$V_r = -\frac{1}{r}\frac{\partial\psi}{\partial z} = 0; \qquad V_z = \frac{1}{r}\frac{\partial\psi}{\partial r} = U$$

$$V_R = \frac{1}{R^2 \sin\phi}\frac{\partial\psi}{\partial\phi} = U\cos\phi; \qquad V_\phi = -\frac{1}{R\sin\phi}\frac{\partial\psi}{\partial R} = -U\sin\phi \qquad (11.73)$$

The streamlines of this flow are shown in figure 11.20 for equal increments in ψ/U. Note that the streamlines are not equally spaced, as they were for plane flow in figure 11.4, because the flow area within each streamline is proportional to r^2. The volume flow rate per unit increase in r increases with distance from the axis $r = 0$.

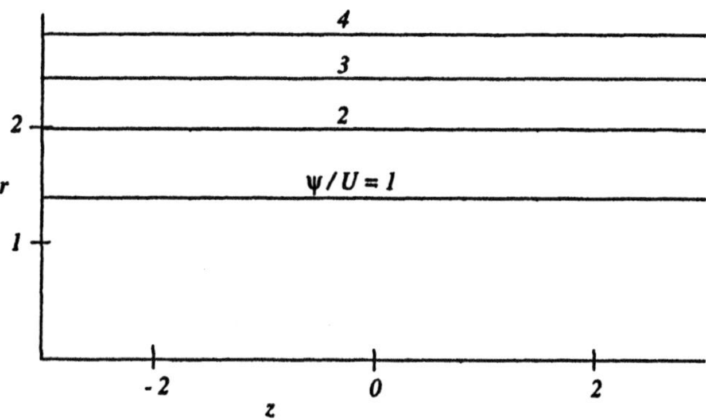

Figure 11.20 The streamlines of the uniform axisymmetric flow, $\psi = Ur^2/2$, for equal increments of ψ/U.

The Point Source

If fluid flows radially outward equally in all directions from a point at a volume flow rate Q, then for an incompressible fluid, the radial velocity V_R would have to satisfy the mass conservation principle:

$$\rho V_R(4\pi R^2) = \rho Q; \qquad V_R = \frac{Q}{4\pi R^2} \tag{11.74}$$

The stream function of the point flow velocity field and the corresponding velocity components are:

$$\psi = -\left(\frac{Q}{4\pi}\right)\cos\phi = -\left(\frac{Q}{4\pi}\right)\frac{z}{\sqrt{r^2+z^2}} \tag{11.75}$$

$$V_R = \frac{1}{R^2\sin\phi}\frac{\partial\psi}{\partial\phi} = \frac{Q}{4\pi R^2}; \qquad V_\phi = -\frac{1}{R\sin\phi}\frac{\partial\psi}{\partial R} = 0$$

$$V_r = -\frac{1}{r}\frac{\partial\psi}{\partial z} = \left(\frac{Q}{4\pi}\right)\frac{r}{(r^2+z^2)^{3/2}};$$

$$V_z = \frac{1}{r}\frac{\partial\psi}{\partial r} = \left(\frac{Q}{4\pi}\right)\frac{z}{(r^2+z^2)^{3/2}} \tag{11.76}$$

The streamlines of the point source flow are shown in figure 11.21. Compared to figure 11.5, the streamlines are more closely spaced near $\phi = \pi/2$, where the mass flow rate per unit increase in ϕ is greatest.

The pressure p^* may be found by applying Bernoulli's equation between any point

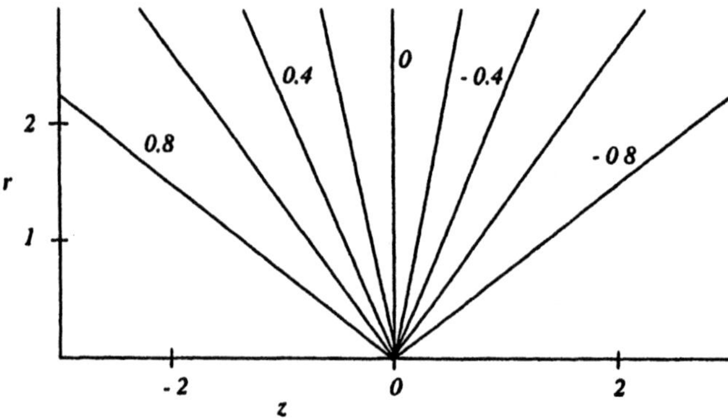

Figure 11.21 The streamlines of the point source flow plotted for equal increments of $4\pi\psi/Q$.

R and $R = \infty$, where the pressure is p^*_∞ and $V_R = 0$:

$$p^* = p^*_\infty - \frac{\rho}{2}V_R^2 = p^*_\infty - \frac{\rho}{2}\left(\frac{Q}{4\pi R^2}\right)^2 \tag{11.77}$$

Note that the pressure difference $p^*_\infty - p^*$ varies as R^{-4}. The pressure p^* becomes negative at $R < R_c \equiv \sqrt{Q/4\pi}(\rho/2p^*_\infty)^{1/4}$ so that the flow is not physically realizable for values of R less than R_c.

Example 11.7

A point source of fluid in a stationary medium has a time-dependent flow rate $Q\{t\} = Q_0 \sin \varpi t$, where Q_0 and ϖ are the constant magnitude and frequency of the flow rate. Derive expressions for the time-varying pressure $p^*\{R, t\}$, the time-averaged pressure and the amplitudes of the time-varying components of $p^*\{R, t\}$.

Solution

(a) From equation 11.75, the fluid velocity $V = V_R = (Q_0/4\pi R^2)\sin \varpi t$. Substituting in the unsteady Bernoulli integral (equation 4.14) evaluated between a radius R and $R = \infty$, where $p^* = p^*_\infty$, we find:

$$\int_R^\infty \frac{\partial V_R}{\partial t}\,dR + \frac{p^*_\infty}{\rho} - \left(\frac{p^*}{\rho} + \frac{V_R^2}{2}\right) = 0$$

$$p^* = p^*_\infty - \frac{\rho}{2}\left(\frac{Q_0}{4\pi R^2}\right)^2 \sin^2 \varpi t + \rho\int_R^\infty \left(\frac{\varpi Q_0}{4\pi R^2}\right)\cos \varpi t\,dR$$

$$= p^*_\infty - \frac{\rho}{4}\left(\frac{Q_0}{4\pi R^2}\right)^2 (1 - \cos 2\varpi t) + \frac{\rho Q_0 \varpi}{4\pi}\int_R^\infty \frac{dR}{R^2}$$

$$= \left[p^*_\infty - \frac{\rho}{4}\left(\frac{Q_0}{4\pi R^2}\right)^2\right] + \frac{\rho}{4}\left(\frac{Q_0}{4\pi R^2}\right)^2 \cos 2\varpi t + \frac{\rho}{4}\left(\frac{\varpi Q_0}{\pi R}\right)\cos \varpi t$$

The last two terms on the right side will average to zero over one cycle so that the first term on the right side is the time-average of p^*. (Note that this is *not* the same as the pressure of a steady flow at the time-average flow rate Q_0, given by equation 11.77.) The coefficients of the last two terms are the amplitudes of the sinusoidal components of p^*. The pressure includes a harmonic component at twice the fundamental frequency (2ϖ) of the source.

The Doublet

As in the case of a line doublet, we can form an axisymmetric doublet by combining a source at $z = \epsilon$ and a sink of equal strength at $z = -\epsilon$, obtaining in the limit of $\epsilon \to 0$ the doublet stream function:

$$\psi = \left(\frac{Ua^3}{2}\right)\frac{\sin^2\phi}{R} = \left(\frac{Ua^3}{2}\right)\frac{r^2}{(r^2+z^2)^{3/2}} \tag{11.78}$$

whose velocity components are:

$$V_R = \frac{1}{R^2\sin\phi}\frac{\partial\psi}{\partial\phi} = \left(\frac{Ua^3}{2}\right)\frac{2\cos\phi}{R^3};$$

$$V_\phi = -\frac{1}{R\sin\phi}\frac{\partial\psi}{\partial R} = \left(\frac{Ua^3}{2}\right)\frac{\sin\phi}{R^3}$$

$$V_r = -\frac{1}{r}\frac{\partial\psi}{\partial z} = \left(\frac{Ua^3}{2}\right)\frac{3rz}{(r^2+z^2)^{5/2}};$$

$$V_z = \frac{1}{r}\frac{\partial\psi}{\partial r} = \left(\frac{Ua^3}{2}\right)\frac{2z^2-r^2}{(r^2+z^2)^{5/2}} \tag{11.79}$$

The doublet streamlines, plotted in figure 11.22, are more elongated in the r direction than those of the line doublet, figure 11.8, reflecting the axisymmetric nature of the flow.

At $R = a$, the radial velocity $V_R = U\cos\phi$. However, $U\cos\phi$ also equals the radial component of the velocity of the surface of a sphere of radius a centered at the origin and moving in the z direction at a speed U. Thus the flow field for $R \geq a$ is an instantaneous flow external to a solid sphere of radius a moving at a speed U. For this

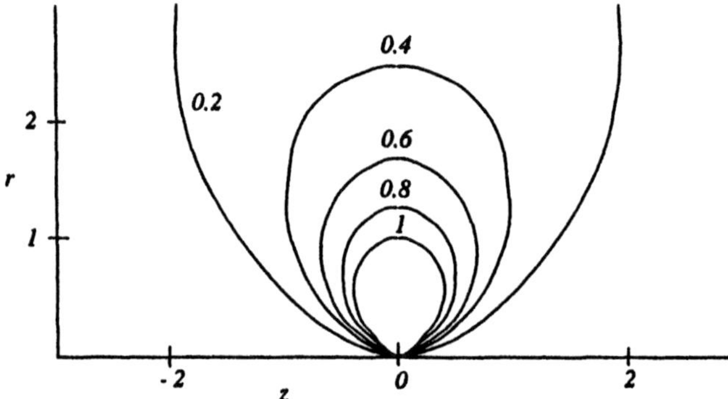

Figure 11.22 The streamlines of the axisymmetric doublet plotted for equal increments of $2\psi/Ua^3$.

flow field, we may determine the kinetic energy KE of the fluid external to the sphere:

$$KE = \iiint \frac{\rho}{2} V^2 \, d\mathcal{V} = \frac{\rho}{2}\left(\frac{Ua^3}{2}\right) \int_a^\infty \int_0^\pi \frac{(4\cos^2\phi + \sin^2\phi)}{R^6} (2\pi R \sin\phi) R \, d\phi \, dR$$

$$= \frac{\rho}{2}\left(\frac{Ua^3}{2}\right)(2\pi) \int_a^\infty \frac{dR}{R^4} \int_0^\pi (4\cos^2\phi + \sin^2\phi)\, d\phi$$

$$= \frac{U^2}{2}\left(\rho \frac{2\pi a^3}{3}\right) \tag{11.80}$$

The kinetic energy is the same as that of a fluid mass of magnitude $2\pi a^3 \rho/3$ moving at the speed U of the sphere or the same as that of a fluid mass equal to *one-half* of the displaced volume $4\pi a^3/3$ of the sphere. Thus the force required to accelerate a sphere in a fluid is equal to the acceleration times the sum of the mass of the sphere and the mass of this entrained fluid. This contrasts with the case of the circular cylinder, equation 11.38, where the entrained mass equals the mass of the displaced fluid.

Flow at a Stagnation Point

Near the front stagnation point on a blunt body of revolution whose axis is aligned with an oncoming flow, there is an axially symmetric flow that is analogous to the plane stagnation point flow illustrated on the left side of figure 11.10 and described by equations 11.42–11.43. The stream function and velocity components of the axially symmetric flow can be given as:

$$\psi = -\Omega r^2 z = -\Omega R^3 \sin\phi \cos\phi \tag{11.81}$$

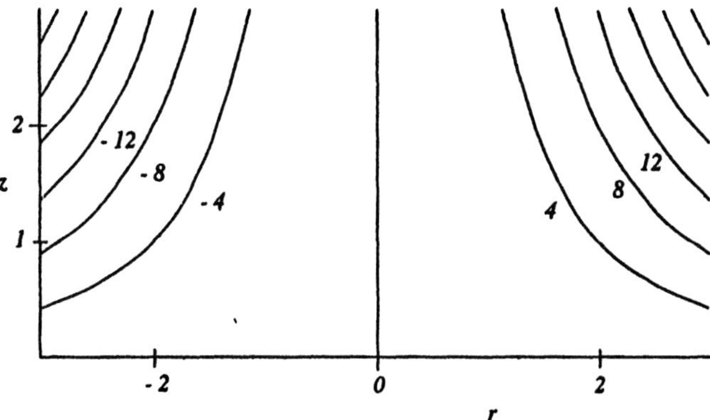

Figure 11.23 The streamlines of the flow near an axisymmetric stagnation point for equal increments in ψ/Ω with the axial dimension vertical.

$$V_r = -\frac{1}{r}\frac{\partial\psi}{\partial z} = \Omega r; \qquad V_z = \frac{1}{r}\frac{\partial\psi}{\partial r} = -2\Omega z$$

$$V_R = \frac{1}{R^2\sin\phi}\frac{\partial\psi}{\partial\phi} = -\Omega R(3\cos^2\phi - 1);$$

$$V_\phi = -\frac{1}{R\sin\phi}\frac{\partial\psi}{\partial R} = 3\Omega R\sin\phi\cos\phi \tag{11.82}$$

where the constant Ω, having the dimensions of reciprocal time, can be related to the pressure distribution near the stagnation point, as shown below.

The streamlines of this flow are plotted in figure 11.23 with the z-axis vertical for comparison with the plane stagnation point flow illustrated on the left side of figure 11.10. The stagnation point of the flow ($V_r = V_z = 0$) is located at the origin, $r = 0$, $z = 0$, where the streamline $\psi = 0$ that comes from $z = \infty$ bifurcates to run along the r-axis. The streamlines farther from the z-axis are more closely spaced in axisymmetric flow than in plane flow.

The pressure distribution in the vicinity of the stagnation point may be related to the stagnation pressure p_s^* at the stagnation point $r = 0$, $z = 0$ by utilizing Bernoulli's equation between any point r, z and the stagnation point:

$$p_s^* = p^* + \frac{\rho}{2}(V_r^2 + V_z^2)$$

$$p^* = p_s^* - \frac{\rho}{2}\Omega^2(r^2 + 4z^2) \tag{11.83}$$

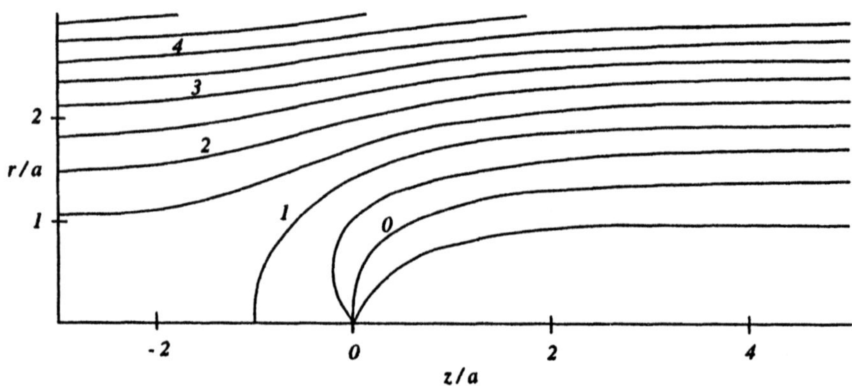

Figure 11.24 The streamlines of the uniform flow of speed U past a point source of strength Q for equal increments of $4\pi\psi/Q$. The dimensionless radial and axial coordinates are r/a and z/a, respectively, where $a \equiv (Q/4\pi U)^{1/2}$.

If we differentiate equation 11.83 twice with respect to r, we find:

$$\frac{\partial^2 p^*}{\partial r^2} = -\rho\,\Omega^2; \qquad \Omega = \sqrt{\frac{1}{\rho}\frac{\partial^2 p^*}{\partial r^2}} \tag{11.84}$$

The constant Ω is thus related to the second derivative of p^*.

11.5.2 Superposition of Axisymmetric Flows

Flow Past a Point Source

If we combine the uniform flow, equations 11.72–11.73, with the flow from a point source, equations 11.75–11.76, we obtain the flow past a point source:

$$\psi = \frac{U}{2}R^2\sin^2\phi - \left(\frac{Q}{4\pi}\right)\cos\phi = \frac{U}{2}r^2 - \left(\frac{Q}{4\pi}\right)\frac{z}{\sqrt{r^2+z^2}} \tag{11.85}$$

$$V_R = U\cos\phi + \frac{Q}{4\pi R^2}; \qquad V_\phi = -U\sin\phi$$

$$V_r = \left(\frac{Q}{4\pi}\right)\frac{r}{(r^2+z^2)^{3/2}}; \qquad V_z = U + \left(\frac{Q}{4\pi}\right)\frac{z}{(r^2+z^2)^{3/2}} \tag{11.86}$$

The streamlines of this flow are plotted in figure 11.24 as functions of the dimensionless radial distance $\xi \equiv r/a$ and axial distance $\eta \equiv z/a$, where $a \equiv \sqrt{Q/4\pi U}$, for constant values of $4\pi\psi/Q$. There is a stagnation point at $r = 0$, $z = -a$. The streamline $4\pi\psi/U = 1$ divides the oncoming flow from the source flow. As $z \to \infty$, this streamline becomes parallel to the z-axis at a radial distance r_∞ such that $\pi r_\infty^2 U = Q$ or $r_\infty = 2a$. The fluid within this dividing streamline may be replaced by a solid body of revolution, the flow external to that streamline being the flow around this semi-infinite body.

Following the argument that led to the derivation of equation 11.54 for the external force exerted on the line source in a uniform flow, there must be an external force $\mathbf{F}_{ex}$, of amount

$$\mathbf{F}_{ex} = \rho Q U \, \mathbf{i}_z \tag{11.87}$$

acting on the point source in a uniform flow so as to account for the increase in axial momentum of the source flow. It also follows that there is no net pressure force on the semi-infinite body of revolution that replaces the source flow.

Flow Past a Sphere

The flow past a sphere is formed by subtracting the doublet flow, equations 11.78–11.79, from the uniform flow, equations 11.72–11.73, obtaining:

$$\psi = \frac{UR^2 \sin^2\phi}{2}\left(1 - \frac{a^3}{R^3}\right) = \frac{Ur^2}{2}\left(1 - \frac{a^3}{(r^2+z^2)^{3/2}}\right) \tag{11.88}$$

$$V_R = U\cos\phi\left(1 - \frac{a^3}{R^3}\right); \qquad V_\phi = -U\sin\phi\left(1 + \frac{a^3}{2R^3}\right)$$

$$V_r = -\left(\frac{Ua^3}{2}\right)\frac{3rz}{(r^2+z^2)^{5/2}}; \qquad V_\phi = U - \left(\frac{Ua^3}{2}\right)\frac{2z^2 - r^2}{(r^2+z^2)^{5/2}} \tag{11.89}$$

The streamlines of this superposed flow, plotted in figure 11.25, are divided by the streamline $\psi = 0$, which forms the surface of a sphere, $R = a$. Outside this streamline, we have the flow past a sphere which is noticeably different from the plane flow past a circular cylinder shown in figure 11.15. There are two stagnation points, at $r = 0$, $z = \pm a$.

The pressure on the surface of the sphere, p^*_{sph}, may be obtained by writing Bernoulli's equation between a point on the sphere where $R = a$ and $V = V_\phi = -(3/2)U\sin\phi$ and a point at $R = \infty$, where $V = U$ and $p^* = p^*_\infty$:

$$p^*_{sph} + \frac{\rho}{2}\left(\frac{3U\sin\phi}{2}\right)^2 = p^*_\infty + \frac{\rho}{2}U^2$$

$$p^*_{sph} - p^*_\infty = \frac{\rho}{2}U^2\left(1 - \frac{9}{4}\sin^2\phi\right)$$

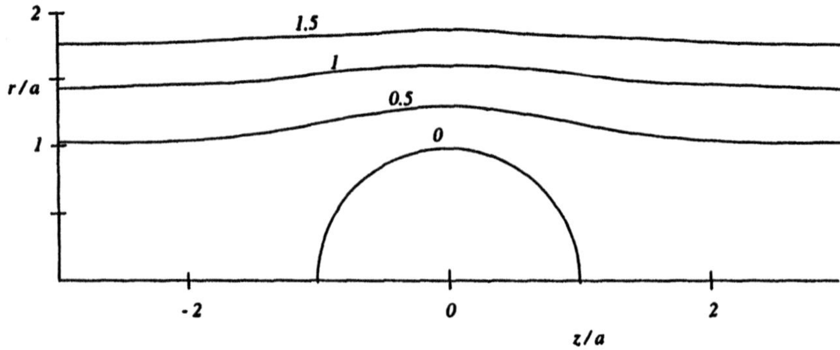

Figure 11.25 The streamlines of the uniform flow of speed U superposed on the flow of a doublet for equal increments of ψ/Ua^2.

$$(C_p)_{R=a} = 1 - \frac{9}{4}\sin^2\phi \tag{11.90}$$

Notice that the minimum pressure coefficient C_p on the surface of the sphere, at $\phi = \pi/2$, is $-5/4$, compared to the much larger negative value of -3 on the circular cylinder (equation 11.63). The disturbance of the uniform flow due to the sphere is much less at this point than it is for the circular cylinder.

The irrotational flow over a solid sphere exerts no drag on the sphere, the pressure distribution on the surface being symmetric about $\phi = \pi/2$. In a viscous flow at moderate and high Reynolds numbers, the boundary layer separates from the surface and a viscous wake region develops behind the body, where the pressure is nearly p^*_∞. Only near the forward stagnation point, $\phi = \pi$, is the flow fairly well described by the irrotational model.

11.6 Flow over Airfoils and Wings

Aircraft wings have airfoil shapes that provide for developing the requisite lift while minimizing drag. For wings that have a large aspect ratio AR, which is the ratio of wing span s to chord length c,[15] the air flow at any point along the span is very nearly a two-dimensional flow in the plane perpendicular to the span. This flow can by adequately calculated as an irrotational flow as long as the boundary layers on the airfoil surface do not separate.

[15]The wing *span* is the distance between wing tips while the *chord* is the distance between the front (*leading edge*) and rear (*trailing edge*) of the wing at any location along the span.

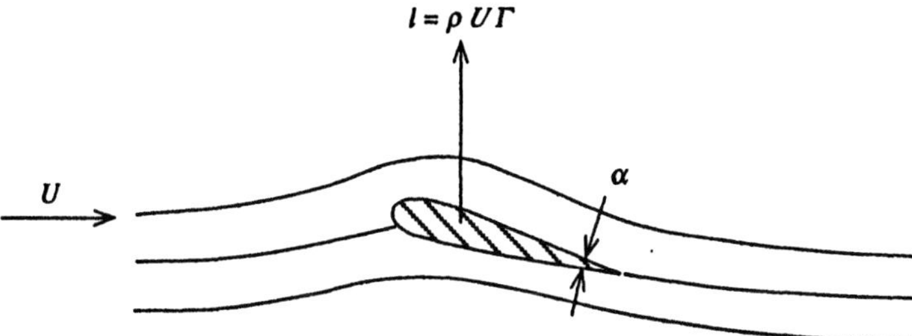

Figure 11.26 The plane irrotational flow over an airfoil at an angle of attack that develops a lift per unit span of $\ell = \rho U \Gamma$.

The flow over a typical airfoil at a small angle of attack, developing a lift force per unit span ℓ, is sketched in figure 11.26. The middle streamline of figure 11.26 is the dividing streamline between the flow above and below the airfoil. It is the streamline that touches the airfoil near the leading edge and at the trailing edge, both of which are stagnation points of the irrotational flow. At the leading edge stagnation point, this dividing streamline is normal to the airfoil surface and the local flow is that of the plane stagnation point flow shown on the left of figure 11.10. At the trailing edge, also a stagnation point, the dividing streamline bisects the angle α formed by the sharp wedge-shaped trailing edge. On either side of this streamline, the flow is locally that of a corner flow, equations 11.39–11.40, having a corner angle $\phi = \pi - \alpha/2$. Because $\phi < \pi$, the trailing edge is a stagnation point of the flow.

To calculate the irrotational flow about a given airfoil shape is no simple matter. To find the flow field, we must look for a solution to Laplace's equation for the stream function in the region surrounding the airfoil which meets the boundary conditions both on the airfoil, where $\psi_{airfoil} = constant$, and at $R = \infty$, where $\psi_\infty = Uy$, the uniform flow of equation 11.28. Whether we use analytical or numerical methods, we would find that there is no unique solution to this problem unless we specify the value of the circulation Γ on a contour C that encloses the airfoil.[16] However, we would find that there is only one value of Γ that provides a stagnation point at the trailing edge of the airfoil and a flow that leaves the trailing edge smoothly, as shown in figure 11.26. All other values of Γ make the flow at the sharp trailing edge singular, like those corner flows (equations 11.39–11.41) for which $\phi > \pi$. The requirement of selecting Γ so as to produce a stagnation point at the trailing edge is called the *Kutta condition.*

[16]Since the flow is irrotational, this circulation is the same for any contour enclosing the airfoil, including the contour of the airfoil itself.

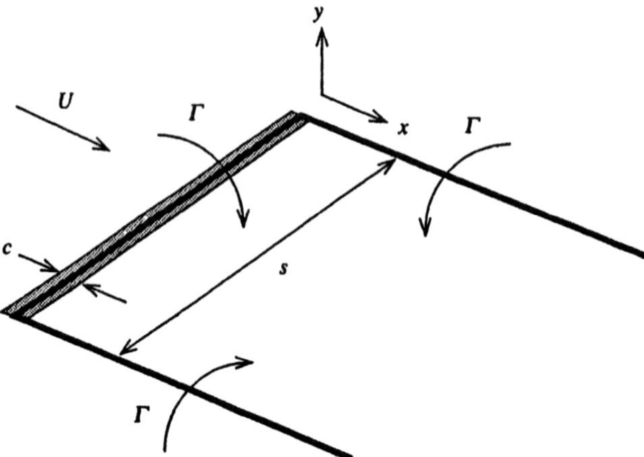

Figure 11.27 A sketch of the vortex system generated by a wing of finite span s and chord c, developing a lift force per unit length $\ell = \rho U \Gamma$. The wing tip vortices trailing downstream of the wing are also shown.

There is a physical basis for the flow at the trailing edge of an airfoil to be that of a stagnation point. Any other irrotational flow would be singular at the trailing edge, having infinite pressure and velocity gradients that would cause the viscous boundary layer to separate and the flow to readjust to the smooth trailing edge flow that attached boundary layers can tolerate.

For any given airfoil, as we calculate irrotational flows having greater angles of attack and correspondingly greater circulation and lift, we would find that the adverse pressure gradients on the upper surface of the airfoil become stronger. When they are sufficiently strong to cause boundary layer separation, the real flow departs substantially from the calculated flow, and the real lift is less than the calculated value. Irrotational flow solutions are only useful as long as boundary layer separation does not occur.

The three-dimensional flow over airplane wings can be described quite well by irrotational flow theory, which is the basis for aircraft aerodynamic design. Some of the three-dimensional effects of the flow surrounding an airplane wing of finite span are illustrated in figure 11.27, a perspective view showing a high aspect ratio wing ($s \gg c$) developing a lift per unit span, $\ell = \rho U \Gamma$, that is uniform along the wing. The circulation Γ around the wing airfoil section can be considered to be that of a bound vortex of strength Γ attached to the wing. Since this vortex cannot end at the wing tips, it trails off downstream from either wing tip as two trailing vortices, each of strength Γ but of opposite direction.

The flow induced by the trailing vortices produces a drag force on the wing, called the *induced drag*. This is not a viscous effect but one due to the wake region of the wing embodied in the trailing vortex system. The wake region has both vertical and horizontal momentum that balances the wing lift and induced drag.

The trailing vortices induce a downward velocity v_i at the wing. This downward velocity generates a force per unit wing span d_i that, by equation 11.59, acts in the flow direction and is thus a drag force. To estimate the magnitude of d_i, we first estimate the downward velocity v_i induced by the trailing vortices as:

$$v_i \sim \frac{\Gamma}{2\pi s}$$

and the resulting drag force per unit span as:

$$d_i = \rho v_i \Gamma \sim \frac{\rho \Gamma^2}{2\pi s}$$

The ratio of the induced drag d_i to the lift ℓ is:

$$\frac{d_i}{\ell} = \frac{d_i}{\rho U \Gamma} \sim \frac{\Gamma}{2\pi U s}$$

Noting that the wing lift $L = \rho U \Gamma s$, this ratio may be expressed as:

$$\frac{d_i}{\ell} \sim \frac{1}{4\pi} \frac{L}{(\rho/2)U^2[cs]} \left(\frac{c}{s}\right) = \frac{1}{4\pi} \frac{C_L}{AR} \tag{11.91}$$

where C_L is the wing lift coefficient. The ratio of the induced drag to the lift represents the losses that must be made up by propulsive power. The induced drag is lessened by using a wing of high aspect ratio AR, such as is found on sailplanes and modern racing yachts. When an airplane lands or takes off, the maximum lift coefficient is achieved and thus the maximum induced drag, requiring more power than for cruising flight.

The trailing vortices attached to the wing tips, as sketched in figure 11.27, are free vortices. Having circulation of opposite sign, they will drift downward at a speed $\Gamma/2\pi s$, as explained in the text referring to figure 11.14. Measured from the horizontal, the angle of the vortex centerline is $\tan^{-1}(2\pi s U/\Gamma) \sim \tan^{-1}(d_i/\ell)$. This downward motion of the trailing vortices and their surrounding air flow contains the vertical momentum flux caused by the external downward force applied to the wing and equal to the wing lift force. This flow field is called the *wing downwash*.

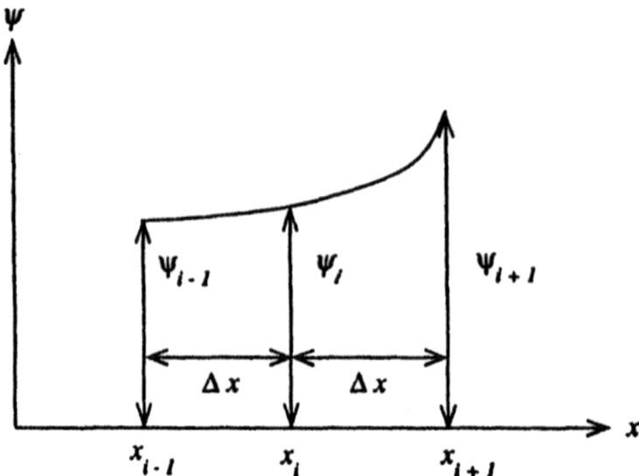

Figure 11.28 The representation of the stream function by discrete values.

11.7 Numerical Solutions

The two-dimensional irrotational flows that we have considered thus far were obtained analytically. The solution to many practical flow problems cannot be reduced to an analytical form, and we must resort to numerical techniques utilizing digital computers to obtain information about the flow field. We generally are interested in determining streamline patterns, flow velocities and pressures throughout a flow field and thus need to present the results of a numerical calculation in a convenient form, such as figures, graphs and tables rather than a large file of numbers.

In a numerical analysis of a an irrotational flow problem, we represent the solution to Laplace's equation for the flow by a set of values of the stream function ψ at the grid points of a finite mesh chosen to cover the flow field. If the mesh is sufficiently fine enough with a large number of grid points, we may have a sufficiently accurate portrayal of the flow to satisfy the requirements of an engineering design or analysis.

The basis of numerical solutions of irrotational flow is the approximate evaluation of the terms of Laplace's equation at each grid point of the mesh by means of a difference equation containing the values of the stream function at several grid points. Thus we satisfy Laplace's equation approximately at a finite number of points in the flow field. As we increase the number of grid points, we improve the accuracy of the solution but at the expense of computer processing time and file space.

To illustrate this principle, consider a stream function $\psi\{x\}$ that is a function of x only. Suppose that we want to find the value of $\psi\{x\}$ that satisfies Laplace's equation for values of x between 0 and L, with $\psi\{0\}$ and $\psi\{L\}$ having known values. We divide the x-axis into N equal intervals of length $\Delta x \equiv L/N$ and indicate the location of these points on the x-axis by x_0, x_1, $\ldots, x_{i-1}$, x_i, $x_{i+1}, \ldots, x_N$ and the corresponding values of ψ by $\psi_0, \psi_1, \ldots, \psi_i, \ldots, \psi_N$. Now consider a portion of the x-axis near x_i, as shown in figure 11.28. We will represent the stream function in this region by the first three terms of a Taylor series expansion about the point x_i:

$$\psi\{x\} = \psi_i + (x - x_i)\left(\frac{\partial\psi}{\partial x}\right)_i + \frac{1}{2}(x - x_i)^2\left(\frac{\partial^2\psi}{\partial x^2}\right)_i \tag{11.92}$$

Evaluating equation 11.92 at x_{i-1} and x_{i+1}, we find:

$$\psi_{i-1} = \psi_i - \Delta x\left(\frac{\partial\psi}{\partial x}\right)_i + + \frac{1}{2}(\Delta x)^2\left(\frac{\partial^2\psi}{\partial x^2}\right)_i$$

$$\psi_{i+1} = \psi_i + \Delta x\left(\frac{\partial\psi}{\partial x}\right)_i + + \frac{1}{2}(\Delta x)^2\left(\frac{\partial^2\psi}{\partial x^2}\right)_i \tag{11.93}$$

The two equations 11.93 can now be solved for $(\partial\psi/\partial x)_i$ and $(\partial^2\psi/\partial x^2)_i$:

$$\left(\frac{\partial\psi}{\partial x}\right)_i = \frac{-\psi_{i-1} + \psi_{i+1}}{2\Delta x} \tag{11.94}$$

$$\left(\frac{\partial^2\psi}{\partial x^2}\right)_i = \frac{\psi_{i-1} - 2\psi_i + \psi_{i+1}}{(\Delta x)^2} \tag{11.95}$$

Equation 11.94 enables us to find the velocity component v_i at x_i after we have found the ψ's:

$$v_i = \left(-\frac{\partial\psi}{\partial x}\right)_i = \frac{-\psi_{i-1} + \psi_{i+1}}{2\Delta x} \tag{11.96}$$

while equation 11.95 is used in Laplace's equation at x_i for this simple case:

$$\left(\frac{\partial^2\psi}{\partial x^2}\right)_i = 0$$

$$\psi_{i-1} - 2\psi_i + \psi_{i+1} = 0 \tag{11.97}$$

There are $N + 1$ values of ψ_i, of which $\psi_0 = 0$ and $\psi_N = 1$ are known, leaving $N - 1$ values to be determined from the $N - 1$ inhomogeneous linear equations 11.97:

$$-2\psi_1 + \psi_2 = -\psi_0$$

$$\psi_1 - 2\psi_2 + \psi_3 = 0$$

$$\ldots$$

$$\psi_{i-1} - 2\psi_i + \psi_{i+1} = 0$$

$$\ldots$$

$$\psi_{N-3} - 2\psi_{N-2} + \psi_{N-1} = 0$$

$$\psi_{N-2} - 2\psi_{N-1} = -\psi_N \tag{11.98}$$

There are two standard methods of finding the solution of 11.98. The first is to express 11.98 as a matrix equation in which the left side is an $N-1$ by $N-1$ matrix of the coefficients of the ψ_i's times the vector ψ_i and the right side is the vector of the boundary values, and then find a numerical solution by matrix inversion. The second method is to use successive approximations to ψ_i by expressing 11.98 in the form:

$$^{n+l}\psi_1 = \frac{1}{2}(\psi_0 + {}^n\psi_2)$$

$$^{n+l}\psi_2 = \frac{1}{2}({}^n\psi_1 + {}^n\psi_3)$$

$$\ldots$$

$$^{n+l}\psi_i = \frac{1}{2}({}^n\psi_{i-1} + {}^n\psi_{i+1})$$

$$\ldots$$

$$^{n+l}\psi_{N-1} = \frac{1}{2}({}^n\psi_{N-2} + \psi_N) \tag{11.99}$$

where the prefix superscript of ψ_i identifies the level of approximation. The solution to 11.99 is started by guessing an initial set of ${}^0\psi_i$ and then solving for ${}^1\psi_i$, repeating the process until there is a satisfactorily negligible change in the values of ψ_i between successive iterations. Usually about N^2 iterations may be required if the initial guess is not very good.

It is a straightforward matter to extend this derivation to include y as an independent variable. Dividing the range of y into M equal intervals Δy having values $y_0, \ldots, y_j, \ldots, y_M$ and denoting the value of $\psi\{x_i, y_j\}$ by $\psi_{i,j}$, Laplace's equation becomes:

$$\psi_{i,j-1} + \psi_{i,j+1} - 2(1+\beta^2)\psi_{i,j} + \beta^2(\psi_{i-1,j} + \psi_{i+1,j}) = 0; \qquad \beta \equiv \frac{\Delta y}{\Delta x} \tag{11.100}$$

and the velocity components are:

$$u_{i,j} = \frac{\psi_{i,j-1} - \psi_{i,j+1}}{2\Delta y}; \quad v_{i,j} = \frac{-\psi_{i-1,j} + \psi_{i+1,j}}{2\Delta x} \tag{11.101}$$

Laplace's equation 11.100 in two independent variables may be solved by matrix inversion or successive approximations. There will be as many as $(N-1)(M-1)$ values of $\psi_{i,j}$ to be found, depending upon the shape of the boundaries.

Because the solution of Laplace's equation arises in many scientific and engineering problems problems, there exist many commercial software packages for its solution.

Some of these are suitable for use on a PC or work station and are specifically designed for irrotational flow calculations.

11.8 Potential Flow

Potential flow is the name given to an incompressible irrotational flow whose velocity field $\mathbf{V}\{\mathbf{R}, t\}$ is defined in terms of a scalar potential function $\Phi\{\mathbf{R}, t\}$ called the *velocity potential*:

$$\nabla\Phi \equiv \mathbf{V} \tag{11.102}$$

Such a definition automatically satisfies the irrotationality condition, equation 11.6:

$$\nabla \times \mathbf{V} = \nabla \times (\nabla\Phi) \equiv 0 \tag{11.103}$$

because of the vector identity, equation 1.36. But, if the flow is incompressible, Φ must satisfy Laplace's equation:

$$\nabla \cdot \mathbf{V} = \nabla \cdot (\nabla\Phi) = \nabla^2\Phi = 0 \tag{11.104}$$

It is possible to solve irrotational flow problems by using Φ rather than ψ as the dependent variable of the flow field. To find such solutions, Laplace's equation 11.104 for Φ is solved subject to the boundary condition that $\nabla\Phi = \mathbf{V}$ on the boundaries of the flow field. However, such solutions do not easily provide for plotting the streamlines of the flow, which are useful to understanding the flow behavior. However, Φ can be used for three-dimensional flows while ψ, as we have used it, is restricted to two-dimensional flows.

For plane flows, the relationship between the velocity potential Φ and the stream function ψ can be derived by expressing the alternate forms of the velocity components:

$$\begin{aligned} u &= \frac{\partial\psi}{\partial y} = \frac{\partial\Phi}{\partial x} \\ v &= -\frac{\partial\psi}{\partial x} = \frac{\partial\Phi}{\partial y} \end{aligned} \tag{11.105}$$

The partial differential relations of 11.105 are called the *Cauchy-Riemann* equations, which figure prominently in the theory of complex variables. Because of the condition 11.105, it is possible to show that equipotential lines (ϕ = *constant*) and streamlines (ψ = *constant*) are everywhere perpendicular to each other. Equally useful, there is always a complex function $w\{z\}$ of the complex spatial variable $z \equiv x + iy$ such that:

$$w\{z\} = \Phi + i\psi$$

$$\frac{dw}{dz} = u - iv \qquad (11.106)$$

Thus the complex potential $w\{z\}$ generates both the stream function ψ and the velocity potential Φ. It is especially useful in determining analytical expressions for plane irrotational flows. However, it has no advantage in numerical solutions, for which either ψ or Φ is sufficient to define the flow field.

11.9 Problems

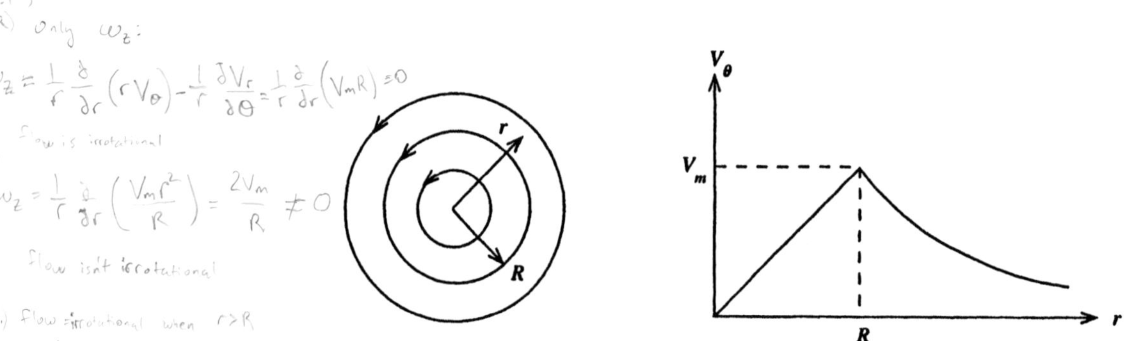

Figure P 11.1

Problem 11.1

A plane, inviscid, constant-density flow consists of circular streamlines centered about an axis normal to the plane of the flow, as sketched in figure P 11.1. There is no radial component of the velocity ($V_r = 0$) and the tangential component V_θ is proportional to the radius r when $0 < r < R$ and inversely proportional to r when $R < r < \infty$:

$$V_\theta = V_m\left(\frac{r}{R}\right) \quad \textit{if } 0 < r < R; \qquad V_\theta = V_m\left(\frac{R}{r}\right) \quad \textit{if } R < r < \infty$$

as shown in figure P 11.1, where V_m is a constant.

(a) Is the flow for $R < r < \infty$ an irrotational flow? (Prove your answer.) (b) Is the flow for $0 < r < R$ an irrotational flow? (Prove your answer.) (c) At $r = \infty$, the pressure is p_∞. Derive an expression for the pressure p_R at $r = R$ in terms of the density ρ, V_m and p_∞. (d) Derive an expression for the pressure p_0 at $r = 0$ in terms of the density ρ, V_m and p_∞.

y

(moving)

$h\{t\}$

x

(stationary)

Figure P 11.2

Problem 11.2

An unsteady, irrotational flow of constant density ρ is established between two parallel flat plates separated by a distance $h\{t\}$ that varies with time t. As shown in figure P 11.2, the upper plate ($y = h\{t\}$) moves vertically up or down but remains parallel to the lower plate $y = 0$. The stream function describing this flow is said to be:

$$\psi = -\frac{1}{h}\left(\frac{dh}{dt}\right)xy = -\left(\frac{d\ln h}{dt}\right)xy$$

(a) Derive expressions for the velocity components u and v. (b) Prove that this stream function will satisfy the velocity conditions on the surfaces of the two plates. (c) Derive an expression for the pressure difference $p\{x, y, t\} - p\{0, 0, t\}$ between any point in the fluid and the point $x = 0$, $y = 0$.

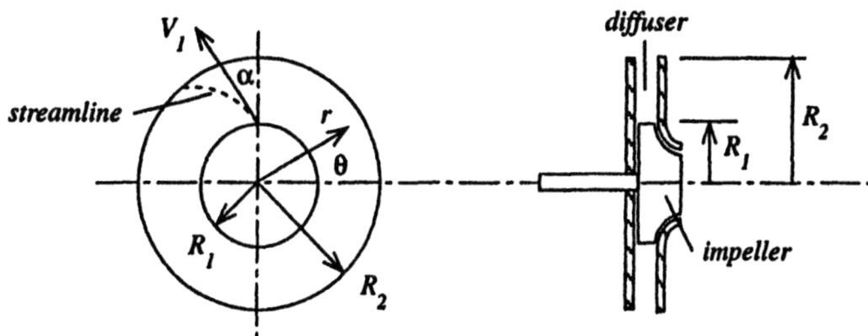

Figure P 11.3

Problem 11.3

In figure P 11.3, a radial vaneless diffuser for a centrifugal pump consists of two parallel plates which guide the flow discharged from the rotating impeller. A steady, swirling radially symmetric flow of constant density ρ exists in the diffuser, for which the velocity field is defined by the stream function $\psi\{r, \theta\}$:

$$\psi = V_1 R_1[(\cos\alpha)\theta - (\sin\alpha)\ln r]$$

where R_1 is the impeller exit radius, V_1 the speed of the flow leaving the impeller and α the angle of the streamline at R_1 (see figure P 11.3). R_2 is the outer radius of the diffuser.

(a) Prove that the given stream function satisfies the velocity conditions at the impeller exit, R_1. (b) Is the flow field irrotational? (c) Derive an expression for the pressure rise $p_2 - p_1$ across the diffuser in terms of the given parameters.

Problem 11.4

A line sources of strength q in a plane flow is located at the point $y = a$, $x = 0$ above a solid plane wall at $y = 0$. In addition to the line source, there exists a steady uniform flow $\mathbf{V} = U\mathbf{i}_x$. Derive expressions for (a) the stream function $\psi\{x, y\}$ of this flow, (b) the velocity $u\{x\}$ at the wall, the pressure $p_w^*\{x\}$ at the wall, and the external force per unit length $\mathbf{f}_{ex}$ required to hold the line source in place.

Problem 11.5

A circular cylinder of radius a is subject to a horizontal flow of time-varying speed $U\{t\}$, so that the stream function for the flow is:

$$\psi = U\{t\}r\left(1 - \frac{a^2}{r^2}\right)\sin\theta$$

Derive expressions for (a) the pressure difference $p\{\theta\} - p_0$ between a point on the cylinder surface at θ and that at $\theta = 0$ and (b) the pressure force $\mathbf{f}\{t\}$ per unit length caused by the pressure on the cylinder surface, in terms of ρ, a, and dU/dt.

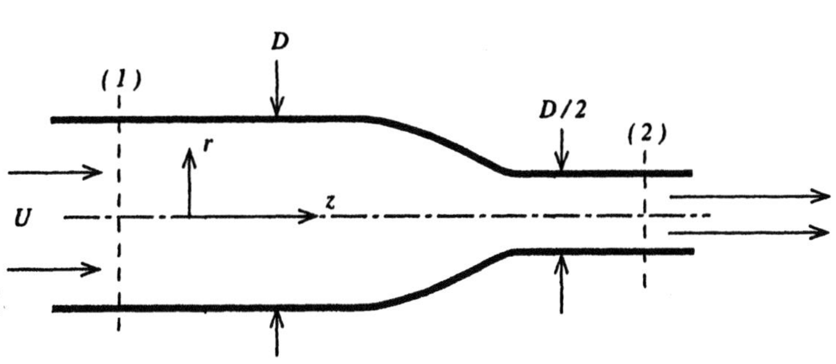

Figure P 11.6

Problem 11.6

An irrotational axisymmetric flow consists of a swirling flow in a circular pipe that necks down from a diameter D to one of diameter $D/2$, as shown in figure P 11.6. Upstream, at section 1, the axial flow speed $(V_z)_1 = U$ is uniform across the pipe cross-section, and the tangential flow speed $(V_\theta\{r\})_1 = \Gamma/2\pi r$, where Γ is the circulation measured along a contour that encloses the axis $r = 0$.

(a) Derive expressions for the velocity components $(V_z)_2$ and $(V_\theta\{r\})_2$ at the downstream section 2 in terms of U and Γ. (b) Derive an expression for the pressure difference $(p_w)_1 - (p_w)_2$ between points on the wall at 1 and 2 in terms of ρ, U and Γ.

Bibliography

Batchelor, G. K. 1967. *An Introduction to Fluid Dynamics*. Cambridge: Cambridge University Press.

Lamb, Horace. 1945. *Hydrodynamics*. 6th ed. New York: Dover Publications.

White, Frank M. 1986. *Fluid Mechanics*. 2nd ed. New York: McGraw-Hill Book Co.

12 Compressible Flow

12.1 Introduction

This chapter develops the principles of compressible flow. In contrast with the predominantly incompressible flow treated in previous chapters, compressible flow requires the integration of the equations of conservation of mass and momentum with that of energy conservation. In compressible flow, these three conservation requirements are closely coupled to each other in a way that increases the complexity of the problem of evaluating such flows. Furthermore, the energy conservation equation explicitly incorporates thermodynamic properties of the fluid. As a consequence, there is no general solution to such flows, and we resort instead to simple flows, such as unsteady one-dimensional and steady two-dimensional inviscid flows of perfect gases having constant specific heats. Nevertheless, these flows exhibit all of the important characteristics of compressible flows and even have some important engineering applications.

In this section, before proceeding with the development of some compressible flows, we explain the flow conditions that require the flow to be treated as compressible rather than as incompressible as we have done in previous chapters.

12.1.1 The Conditions for Compressible Flow

In section 3.3.3, we defined an incompressible flow as one for which the density of a fluid particle does not change with time, *i.e.* $D\rho/Dt = 0$, but we also noted that small density changes did not violate this condition provided that:

$$\frac{1}{\rho}\left|\frac{D\rho}{Dt}\right| \ll \left|\frac{\partial u}{\partial x}\right| + \left|\frac{\partial v}{\partial y}\right| + \left|\frac{\partial w}{\partial z}\right| \qquad (12.1)$$

Under the condition that the density was constant along a streamline, we were able to integrate Euler's equation 4.5,

$$\frac{D\mathbf{V}}{Dt} = -\frac{1}{\rho}\nabla p - \mathbf{g} \tag{12.2}$$

to obtain Bernoulli's integral, equation 4.13.

Let us now reconsider the condition of equation 12.1 in order to determine when it might be violated. First, let us consider a *steady flow* characterized by a velocity scale V and a length scale L. Estimating the magnitude of the terms in 12.1 while employing these scales, we find,

$$\frac{V}{L}\frac{\delta\rho}{\rho} \ll \frac{V}{L}$$

$$\frac{\delta\rho}{\rho} \ll 1 \tag{12.3}$$

The fractional change in density in the flow field, $\delta\rho/\rho$, must therefore be very small for the flow to be incompressible.

But what characteristics of the flow will ensure that 12.3 is valid? To answer this question, we must estimate the size of the terms in Euler's equation 12.2, finding (neglecting gravity):

$$\frac{V}{L}V \sim \frac{\delta p}{\rho L}$$

Now let us further assume that the fluid entropy is everywhere the same. When this is so, the changes in pressure δp and density $\delta\rho$ are related to each other by $\delta p/\delta\rho = (\partial p/\partial\rho)_s = a^2$, where a is the speed of a sound wave.[1] Substituting, we find,

$$\frac{\delta\rho}{\rho} \sim \frac{V^2}{a^2} \tag{12.4}$$

Comparing 12.4 with 12.3, we can conclude that a flow is not incompressible when the flow speed is comparable to the speed of sound.

When a flow speed V is not much smaller than the sound speed a of the fluid, we must include the conservation of energy with the conservation of mass and momentum in order to determine the flow field. In a compressible flow, some of the internal energy of the fluid is changed into kinetic energy of the flow (or vice versa), and the ratio V/a is a measure of how important this energy exchange is to the flow.

Let us return to the case of *unsteady flow*. If τ characterizes the time scale of the

[1] See equation 12.12 below.

unsteady flow, we can estimate the magnitude of the terms in 12.1 and 12.2, finding:

$$\frac{\delta\rho}{\rho\tau} \ll \frac{V}{L}$$

$$\frac{V}{\tau} \sim \frac{a^2\delta\rho}{\rho L}$$

Eliminating $\delta\rho$, we find:

$$\frac{L}{\tau} \ll a \tag{12.5}$$

Thus the ratio of the length and time scales must be much less than the speed of sound for the flow to be treated as incompressible. For a sound wave, $L \sim a\tau$, and the flow is necessarily compressible even though $\delta\rho/\rho$ may be very small.

A flow must be treated as compressible whenever either of the above conditions are present, *e.g.*, $V \not\ll a$ or $L/\tau \not\ll a$.[2]

12.2 The Speed of Sound

We all know that sound travels more slowly than light. When we see a distant carpenter strike a nail, the sound reaches us a split second later. The further we are from the source, the greater the time delay. What determines the speed of travel of a sound pulse in air?

A sound wave is a compressible flow because there is a significant exchange of energy between the internal energy of an air particle and its kinetic energy. This coupling of the random kinetic energy of individual molecules with the orderly motion of fluid particles in a sound wave provides for the propagation of sound energy from a source to a distant receiver, even though the fluid through which the wave passes has no net motion. It is remarkable that every kind of sound propagates at the same speed: a musical note, a bird's call or the crack of lightning.

Audible sounds are characterized by changes in pressure and density of the ambient atmosphere that are minute compared to the average atmospheric pressure and density. Our ears are so sensitive that only a few picowatts per square meter of sound power are sufficient to cause a neural response and the sensation of sound. Humans hear

[2]Flows with heat addition may not satisfy the condition of incompressibility, equation 12.1, but may not be compressible because they do not meet the requirements of compressible flow. These flows are usually termed *variable density flows*. In these flows, the energy conservation equation is coupled to the mass and momentum conservation equations only through the variable density.

over a frequency range of about 200 : 1 (about eight octaves), but some animals can respond to an even greater range. Within this frequency range, sound intensities that vary by about fourteen orders of magnitude can be differentiated by the human ear.

The propagation speed of a sound wave may be determined from a solution of the inviscid compressible flow equations, mass conservation (equation 3.16) and Euler's equation (equation 4.5):

$$\frac{D\rho}{Dt} + \rho\nabla\cdot\mathbf{V} = 0$$

$$\frac{D\mathbf{V}}{Dt} + \frac{1}{\rho}\nabla p = 0 \tag{12.6}$$

where we have neglected gravity, which is unimportant in the propagation of ordinary sound waves. But equations 12.6 can be simplified for the case of sound waves which have changes in the flow variables that are very small. First of all, the material derivative can be replaced by the time derivative:

$$\frac{D\rho}{Dt} \simeq \frac{\partial\rho}{\partial t}; \quad \frac{D\mathbf{V}}{Dt} \simeq \frac{\partial\mathbf{V}}{\partial t}$$

because the fluid particle moves much more slowly than the wave form itself. Secondly, the density multiplier in equations 12.6 can be replaced by its average value ρ_0 in the undisturbed air because the changes in density are so small. With these simplifications, equations 12.6 assume the form:

$$\frac{\partial\rho}{\partial t} + \rho_0\nabla\cdot\mathbf{V} = 0$$

$$\rho_0\frac{\partial\mathbf{V}}{\partial t} + \nabla p = 0 \tag{12.7}$$

The changes in pressure and density, although rapid, are slow enough that the process is a thermodynamically reversible one for which the fluid entropy $\hat{s}$ remains unchanged. Hence changes in density and pressure are related by:

$$d\rho = \left(\frac{\partial\rho}{\partial p}\right)_{\hat{s}} dp \tag{12.8}$$

and equations 12.7 may be rewritten as:

$$\left(\frac{\partial\rho}{\partial p}\right)_{\hat{s}} \frac{\partial p}{\partial t} + \rho_0\nabla\cdot\mathbf{V} = 0 \tag{12.9}$$

$$\rho_0\frac{\partial\mathbf{V}}{\partial t} + \nabla p = 0 \tag{12.10}$$

If we now subtract the divergence of equation 12.10 from the time derivative of equation 12.9, we find:

$$\left(\frac{\partial \rho}{\partial p}\right)_s \frac{\partial^2 p}{\partial t^2} + \rho_0 \frac{\partial(\nabla \cdot \mathbf{V})}{\partial t} - \rho_0 \nabla \cdot \left(\frac{\partial \mathbf{V}}{\partial t}\right) - \nabla^2 p = 0$$

$$\frac{\partial^2 p}{\partial t^2} = \left(\frac{\partial p}{\partial \rho}\right)_s \nabla^2 p \qquad (12.11)$$

Equation 12.11 is called the *wave equation*, and the coefficient $(\partial p/\partial \rho)_s$ is the square of the speed of propagation of sound waves. We use the symbol a to define the speed of sound:

$$a \equiv \sqrt{\left(\frac{\partial p}{\partial \rho}\right)_s} \qquad (12.12)$$

Employing the thermodynamic relations of equations 1.9 and 1.10, the speed of sound in a perfect gas is found to be expressible as:

$$a = \sqrt{\frac{\hat{c}_p p}{\hat{c}_v \rho}} = \sqrt{\frac{\gamma p}{\rho}} = \sqrt{\gamma R T} \qquad (12.13)$$

where the letter symbol γ is used to denote the ratio of the specific heats:

$$\gamma \equiv \frac{\hat{c}_p}{\hat{c}_v} = \frac{\hat{c}_p/R}{\hat{c}_p/R - 1} \qquad (12.14)$$

the alternate form being given by equation 1.13. In general, γ is a function of temperature T alone, which decreases as the temperature increases, but always lies in the range $5/3 \geq \gamma \geq 1$.

For liquids, the speed of sound is much higher than it is for gases because the density increase for a given pressure increase is much smaller in liquids than in gases. For a pure substance near its critical temperature and pressure, however, the sound speeds of the vapor and liquid phases are not very different.

Example 12.1

Calculate the speed of sound in (a) air and (b) helium at $T = 300\,K$.

Solution

(a) From table 1.2 for air, $R = 287.0\,J/kg\,K$ and $\hat{c}_p/R = 3.5$. Using equations 12.13–12.14,

$$a = \sqrt{(3.5/2.5)(287.0\,J/kg\,K)(300\,K)} = 347.2\,m/s$$

(b) For helium from table 1.2, $R = 2077\,J/kg\,K$ and $\hat{c}_p/R = 2.5$. Thus:

$$a = \sqrt{(2.5/1.5)(2077\,J/kg\,K)(300\,K)} = 1019\,m/s$$

Plane Sound Waves

As a simple example, consider a plane sound wave, *i.e.*, one for which the flow quantities depend only upon x and t and are thus constant in any y, z-plane. For this case, the sound wave equation 12.11 takes the form:

$$\frac{\partial^2 p}{\partial t^2} = a^2 \frac{\partial^2 p}{\partial x^2} \tag{12.15}$$

There is a very general, yet simple, solution to this equation 12.15 (and equation 12.9) for a sound wave that propagates from left to right. It is:

$$p - p_0 = \rho_0 a u = a^2(\rho - \rho_0) = f\{x - at\} \tag{12.16}$$

where $f\{x - at\}$ is any function of the argument $x - at$. For example, for a pure continuous tone, $f\{x - at\} = A\sin(x - at)$, while for a spoken word, $f\{x - at\}$ is exceedingly complex. However, for any $f\{x - at\}$, the sound signal, as a function of time, measured at x is the same as that measured at $x = 0$ at a time earlier by the amount x/a:

$$f\{x - at\} = f\{0 - a(t - x/a)\}$$

Thus a plane sound wave propagates the information in a sound pulse without change from one location to another.

Sound Pressure Level

When a loud speaker generates a plane sound wave, it does so by moving backward and forward with an alternating velocity u that generates a pressure change $p - p_0$ given by equation 12.16. The power per unit area of loud speaker, P_{sw}, required to generate this sound wave is the product of the pressure $p - p_0$ times the velocity u:

$$P_{sw} = (p - p_0)u = \frac{(p - p_0)^2}{\rho_0 a} = \rho_0 a u^2 \tag{12.17}$$

This power is proportional to the *square* of the pressure amplitude of the wave, $p - p_0$, or the velocity amplitude of the wave, u. It is transmitted from the sender to the receiver of the sound wave.

The intensity of a sound wave is measured by the square root of the time-averaged value of the square of the pressure amplitude, *i.e.*, $\sqrt{\overline{(p - p_0)^2}}$. But it is the convention of sound specialists to use a logarithmic scale of sound intensity whose unit is the *decibel (dB)* and which has a zero value when $\sqrt{\overline{(p - p_0)^2}} = 2E(-5)\,Pa$. On this scale, the *sound pressure level (SPL)*, measured in dB, has the value:

$$SPL \equiv 10 \log \left(\frac{\sqrt{\overline{(p - p_0)^2}}}{2E(-5)\,Pa} \right) dB \tag{12.18}$$

The reference pressure level of $2E(-5)\,Pa$ for zero dB on the sound pressure level scale was chosen because it is the approximate amplitude at which a sound wave is barely audible to the normal human ear. For each factor of ten increase in pressure amplitude above this level, the sound pressure level increases by $10\,dB$. Continuous exposure to sound pressure levels above $90\,dB$ (*i.e.*,$\sqrt{\overline{(p - p_0)^2}} \geq 2E(4)\,Pa$) can cause hearing loss in humans.

Example 12.2

A sound wave in air at atmospheric pressure and 20° C temperature has a sound pressure level of $20\,dB$. Calculate (a) the pressure amplitude $\sqrt{\overline{(p - p_0)^2}}$ and (b) the velocity amplitude $\sqrt{\overline{u^2}}$ as well as the power per unit area P_{sw}.

Solution

(a) Inverting equation 12.18,

$$\sqrt{\overline{(p - p_0)^2}} = [2E(-5)\,Pa] \times [1E(20/10)] = 2E(-3)\,Pa$$

(b) From equation 12.17,

$$\sqrt{\overline{u^2}} = \frac{\sqrt{\overline{(p - p_0)^2}}}{\rho_0 a}$$

From table 1.1, $\rho_0 = 1.204\,kg/m^3$. By equations 12.13–12.14 and table 1.2, $\gamma = 3.5/2.5 = 1.4$, and $a = [(1.4)(1.013E(5)\,Pa)/(1.204\,kg/m^3)]^{1/2} = 3.432E(2)\,m/s$. Thus:

$$\sqrt{\overline{u^2}} = \frac{2E(-3)\,Pa}{(1.204\,kg/m^3)(3.432E(2)\,m/s)} = 4.84E(-6)\,m/s$$

(c) From equation 12.17,

$$P_{sw} = \frac{(2E(-3)\,Pa)^2}{(1.204\,kg/m^3)(3.432\,m/s)} = 4.84E(-10)\,W/m^2$$

12.3 Steady Isentropic Flow

For an inviscid incompressible flow, the flow velocity V and the pressure p are related to each other at different points along a streamline by Bernoulli's equation 4.13. When an inviscid flow is compressible, we cannot use Bernoulli's equation because the density is no longer constant along the streamline or particle path. Instead, we must incorporate the thermodynamic principles discussed in chapter 8 that explain how the internal energy of the fluid contributes to the motion of a fluid particle. Because compressible flows are inherently more complex than incompressible flows, we shall begin by treating first a *steady* flow that is inviscid and adiabatic, *i.e.*, having negligible viscous stress $\boldsymbol{\tau}$ and heat flux $\mathbf{q}$. For such a flow, the integral form of the first law of thermodynamics, equation 8.29, reduces, to:

$$\iint_S \rho \left(\hat{h} + \frac{V^2}{2} - \mathbf{g} \cdot \mathbf{R} \right) \mathbf{V} \cdot \mathbf{n}\, dS = 0 \tag{12.19}$$

For compressible flows, where V has the same magnitude as a, the term $\mathbf{g} \cdot \mathbf{R}$ is negligible whenever the length scale L of the flow is small compared to a^2/g, which is about $1E(4)\,m$ in air. Applying equation 12.19 to the flow along a streamline that originates in a region of stationary gas (called the stagnation state), whose properties are identified by the subscript s, we find:

$$\frac{V^2}{2} + \hat{h} = \hat{h}_s \tag{12.20}$$

However, when a flow is inviscid and adiabatic, the second law of thermodynamics, equation 8.43, requires that the entropy of a fluid particle remains constant. Thus the changes in enthalpy $\hat{h}$ along a streamline in a steady flow are such that the entropy has the same value at all points along the streamline as it has at the stagnation state, namely, $\hat{s}_s$. Using p and $\hat{s}_s$ as the independent thermodynamic variables defining the enthalpy $\hat{h}$, we may rewrite equation 12.20 as:

$$\frac{V^2}{2} + \hat{h}\{p, \hat{s}_s\} = \hat{h}\{p_s, \hat{s}_s\}$$

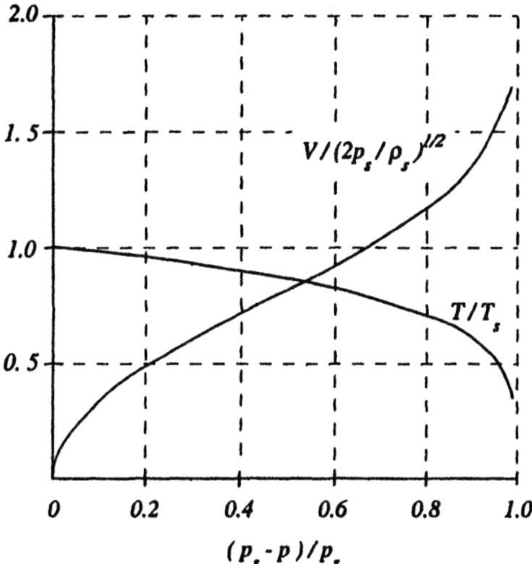

Figure 12.1 The velocity V and temperature T of superheated steam flowing isentropically from a stagnation state of $p_s = 100\,bar$, $T_s = 1073\,K$, and $\rho_s = 20.59\,kg/m^3$ plotted as a function of the pressure change $p_s - p$. (Data from Thomas F. Irvine, Jr. and James P. Hartnett, 1976, *Steam and Air tables in SI Units*, Washington DC: Hemisphere Publishing Corp.)

$$V = \sqrt{\hat{h}\{p_s, \hat{s}_s\} - \hat{h}\{p, \hat{s}_s\}} \tag{12.21}$$

Figure 12.1 shows a numerical calculation of the velocity V of superheated steam as a function of the pressure change $p_s - p$ along a streamline in isentropic flow, obtained from equation 12.21 using thermodynamic tables for superheated steam. The velocity V is made dimensionless by dividing it by $\sqrt{2p_s/\rho_s}$, the value of the velocity of an incompressible flow of the stagnation fluid when its pressure is reduced to zero. Note that the compressible flow velocity exceeds the incompressible value when the pressure change $p_s - p$ becomes comparable to p_s. The extra kinetic energy of the flow is supplied from the internal energy of the steam, whose temperature (and internal energy) decreases considerably when the pressure drop is large.

12.3.1 Isentropic Perfect Gas Relations for Constant Specific Heats

For an isentropic change in a perfect gas, the differential changes in pressure p, density ρ, temperature T and sound speed a may be found from the equations for the second law expressions (1.9–1.10), the perfect gas law (1.11) and the sound speed of a perfect gas (12.13):

$$\frac{1}{R} d\ln T = \frac{1}{\hat{c}_p} d\ln p = \frac{1}{\hat{c}_v} d\ln \rho = \frac{2}{R} d\ln a \tag{12.22}$$

These differential relations may be integrated if the specific heats are constant over the range of integration. Expressing these integrals as functions of the pressure ratio p/p_s, we find:

$$\frac{T}{T_s} = \left(\frac{p}{p_s}\right)^{(\gamma-1)/\gamma}$$

$$\frac{\rho}{\rho_s} = \left(\frac{p}{p_s}\right)^{1/\gamma}$$

$$\frac{a}{a_s} = \left(\frac{p}{p_s}\right)^{(\gamma-1)/2\gamma} \tag{12.23}$$

Thus the temperature, density and sound speed all decline with decreasing pressure but not in equal proportions.

12.3.2 Steady Isentropic Flow of a Perfect Gas

For the isentropic flow of a perfect gas, the steady flow energy equation, equation 12.20 and 12.21, has a simple form because the enthalpy $\hat{h}$ of a perfect gas and its specific heats depend only upon the temperature T. If the specific heat of the perfect gas can be considered constant over the temperature change encountered in the flow, then we may replace $\hat{h}$ by $\hat{c}_p T$ and write 12.21 as:

$$V = \sqrt{2\hat{c}_p(T_s - T)} = \left(2\hat{c}_p T_s \left[1 - \left(\frac{p}{p_s}\right)^{(\gamma-1)\gamma}\right]\right)^{1/2} \tag{12.24}$$

where we have used equation 12.23 to replace T by p in the last term of 12.24.

When the pressure in such a flow falls to zero, its lowest possible value, equation 12.24 shows that the flow speed reaches its maximum value of $\sqrt{2\hat{c}_p T_s}$. If the flow were incompressible, then the maximum velocity would be $\sqrt{2p_s/\rho_s} = \sqrt{2RT_s}$. Thus the maximum compressible flow speed is higher than the incompressible flow speed by the factor $\sqrt{\hat{c}_p/R} = \sqrt{\gamma/(\gamma-1)}$, which for air ($\gamma = 1.4$) is a factor of 1.87. The extra 87% in speed comes from the internal energy of the perfect gas, which ends up at zero temperature when its pressure falls to zero.

We may summarize the steady isentropic flow relations as a function of pressure

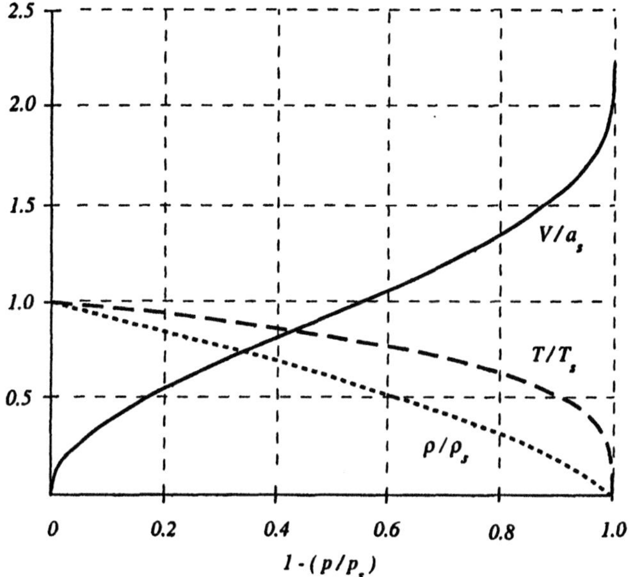

Figure 12.2 The variations of the ratios V/V_s, T/T_s and ρ/ρ_s with pressure ratio p/p_s for isentropic flow of a perfect gas having $\gamma = 1.4$.

by combining equations 12.23 and 12.24 in the form:

$$\frac{V}{a_s} = \left(\left[\frac{2}{\gamma - 1}\right]\left[1 - \left(\frac{p}{p_s}\right)^{(\gamma-1)/\gamma}\right]\right)^{1/2}$$

$$\frac{T}{T_s} = \left(\frac{p}{p_s}\right)^{(\gamma-1)/\gamma}$$

$$\frac{\rho}{\rho_s} = \left(\frac{p}{p_s}\right)^{1/\gamma}$$

$$\frac{a}{a_s} = \left(\frac{p}{p_s}\right)^{(\gamma-1)/2\gamma} \tag{12.25}$$

The variations of the velocity, temperature and density in a steady isentropic flow of a perfect gas having $\gamma = 1.4$ are plotted in figure 12.2 as functions of the dimensionless pressure drop, $(p_s - p)/p_s$. Notice how the temperature and velocity change more rapidly than the density as zero pressure is approached. For this value of γ, $V/a_s = \sqrt{5}$ when $p = 0$.

Mach Number as a Flow Variable

The Mach number M, which is the ratio of the flow speed V to the speed of sound a, is a dimensionless flow parameter whose value characterizes important features of the flow. It is sometimes more convenient to use the Mach number M rather than the pressure ratio p/p_s as a measure of the change of flow properties in an isentropic flow. For a perfect gas having constant specific heats, the relation between M and the flow variables may be found from equation 12.24 by replacing $\hat{c}_pT$ by $a^2/(\gamma - 1)$:

$$\frac{V^2}{2} + \frac{a^2}{\gamma - 1} = \frac{a_s^2}{\gamma - 1}$$

$$M^2 + \frac{2}{\gamma - 1} = \left(\frac{2}{\gamma - 1}\right)\left(\frac{a_s}{a}\right)^2$$

$$\frac{a}{a_s} = \left(1 + \frac{\gamma - 1}{2} M^2\right)^{-1/2}$$

$$\frac{V}{a_s} = \frac{Ma}{a_s} = M\left(1 + \frac{\gamma - 1}{2} M^2\right)^{-1/2}$$

$$\frac{T}{T_s} = \left(1 + \frac{\gamma - 1}{2} M^2\right)^{-1}$$

$$\frac{p}{p_s} = \left(1 + \frac{\gamma - 1}{2} M^2\right)^{-\gamma/(\gamma-1)}$$

$$\frac{\rho}{\rho_s} = \left(1 + \frac{\gamma - 1}{2} M^2\right)^{-1/(\gamma-1)} \qquad (12.26)$$

where we have used equations 12.23 in the last three lines.

As the pressure p falls in an accelerating flow, we see in figure 12.2 that the velocity V increases from zero and the sound speed $a = a_s\sqrt{T/T_s}$ decreases so that the Mach number $M = V/a$ must increase from zero. When the pressure falls to zero, V attains a finite value, but a becomes zero, while the Mach number becomes infinite. This variation of Mach number with pressure change is plotted as the upper curve in figure 12.3, showing the rapid increase in Mach number as p/p_s approaches zero.

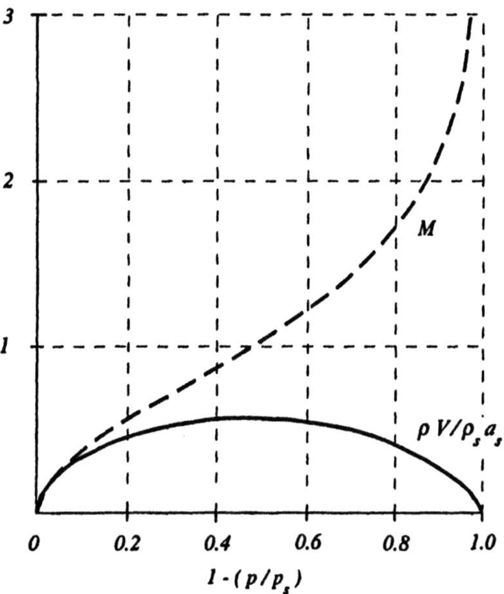

Figure 12.3 The flow mach number M and dimensionless mass flow rate per unit area, $\rho V/\rho_s a_s$, plotted as a function of pressure ratio p/p_s in an isentropic flow of a perfect gas having $\gamma = 1.4$.

Example 12.3

An experimental supersonic wind tunnel is supplied with air ($\gamma = 1.4$) stored in a large tank at a pressure $p_s = 6E(5)\,Pa$. When the tunnel runs briefly, the pressure in the test section is $p_t = 5E(4)\,Pa$. Calculate the Mach number M_t in the test section.

Solution

The relation between the pressures and Mach number in isentropic flow is given in equation 12.26:

$$\frac{p_t}{p_s} = \left(1 + \frac{\gamma - 1}{2} M_t^2\right)^{-\gamma/(\gamma-1)}$$

$$M_t^2 = \left(\frac{2}{\gamma - 1}\right)\left[\left(\frac{p_t}{p_s}\right)^{-(\gamma-1)/\gamma} - 1\right]$$

$$= \left(\frac{2}{0.4}\right)\left[\left(\frac{5E(4)}{6E(5)}\right)^{-0.4/1.4} - 1\right] = 5.17$$

$$M_t = 2.274$$

Critical Flow

In an accelerating isentropic flow, the pressure and density decrease while the velocity increases. The product ρV, which is the mass flow rate per unit area, may either increase or decrease, depending upon whether or not the increase in velocity overbalances the decrease in density. In figure 12.3, we plot the ratio $\rho V/\rho_s a_s$ as a function of the pressure ratio p/p_s in an isentropic flow of a perfect gas having $\gamma = 1.4$, using equations 12.25 to evaluate ρV:

$$\frac{\rho V}{\rho_s a_s} = \left[\left(\frac{2}{\gamma - 1}\right)\left(1 - \left(\frac{p}{p_s}\right)^{(\gamma-1)/\gamma}\right)\right]^{1/2}\left(\frac{p}{p_s}\right)^{1/\gamma} \tag{12.27}$$

The mass flow rate per unit area, ρV, has a maximum value that can be determined by setting the derivative of equation 12.27 with respect to p equal to zero. The pressure p_c at which this maximum flow rate is reached is called the *critical pressure* and, for a perfect gas with constant specific heats, has the value:

$$\frac{p_c}{p_s} = \left(\frac{2}{\gamma + 1}\right)^{\gamma/(\gamma-1)} \tag{12.28}$$

The corresponding value of the mass flow rate per unit area is:

$$\frac{\rho_c V_c}{\rho_s a_s} = \left(\frac{2}{\gamma + 1}\right)^{(\gamma+1)/2(\gamma-1)} \tag{12.29}$$

When $\gamma = 1.4$, $(p_c/p_s) = 0.5283$ and $(\rho_c V_c/\rho_s a_s) = 0.5787$.

This condition of critical flow, a maximum in ρV, is simultaneously the condition at which the Mach number is unity, as can be seen by substituting $M = 1$ in the expression for p/p_s in equation 12.26 and obtaining 12.28. Thus the critical flow separates the isentropic flow into two regimes, *subsonic flow* ($V \leq a$) when $p/p_s \geq p_c/p_s$ and *supersonic flow* ($V \geq a$) when $p/p_s \leq p_c/p_s$.

Example 12.4

A large tank contains air ($\gamma = 1.4$) at a pressure $p_s = 4E(5)\,Pa$ and sound speed $a_s = 347.2\,m/s$. A small hole in the side of the tank, having a flow area $A_c = 1\,cm^2$, permits air to flow to the atmosphere where the pressure is $p_a = 1E(5)\,Pa$. Calculate the mass flow rate of air from the tank and the external force F required to restrain the tank from moving.

Solution
Because the pressure for critical flow is higher than atmospheric pressure, from equation 12.29, the mass flow rate through the hole, $\rho_c V_c A_c$, will be:

$$\rho_c V_c A_c = \rho_s a_s \left(\frac{2}{\gamma+1}\right)^{(\gamma+1)/2(\gamma-1)} A_c$$

Since $a_s^2 = \gamma p_s/\rho_s$, $\rho_s a_s = \gamma p_s/a_s = 1.4(4E(5)\,Pa)/(347.2\,m/s) = 1.613E(3)\,kg/m^2\,s$. Thus the mass flow rate becomes:

$$\rho_c V_c A_c = (1.613E(3)\,kg/m^2\,s)(0.5787)(1E(-4)\,m^2) = 9.33E(-2)\,kg/s$$

To find V_c, we use equation 12.26 evaluated at $M = 1$:

$$V_c = a_s/\sqrt{1 + (0.4/2)} = (347.2\,m/s)(0.9129) = 317\,m/s$$

If we evaluate the linear momentum theorem for a control surface surrounding the outside of the tank, we can determine the force F:

$$(\rho_c V_c A_c)V_c = F - (p_c - p_a)A_c$$

$$F = (p_c - p_a)A_c + (9.33E(-2)\,kg/s)(317\,m/s) = (p_c - p_a)A_c + 29.59\,N$$

But we can find p_c from equation 12.28:

$$p_c = 0.5283(4E(5)\,Pa) = 2.113E(5)\,Pa$$

$$(p_c - p_a)A_c = (1.113E(5)\,Pa)(1E(-4)\,m^2) = 11.13\,N$$

Thus $F = (11.13 + 29.59)\,N = 40.72\,N$.

Flow in Variable Area Channels

For steady flow in a channel of slowly varying area A, mass conservation requires that the mass flow rate $\dot{m}$ be a constant:

$$\dot{m} = \rho V A = \textit{constant} \tag{12.30}$$

The increments in the density, velocity and flow area can be related to each other by differentiating the logarithm of equation 12.30:

$$\frac{d\rho}{\rho} + \frac{dV}{V} + \frac{dA}{A} = 0 \tag{12.31}$$

For isentropic flow, from Euler's equation (neglecting gravity) and the definition of the speed of sound,

$$V\,dV = -\frac{1}{\rho}\,dp = -\frac{a^2}{\rho}\,d\rho$$

$$\frac{d\rho}{\rho} = -\frac{V\,dV}{a^2} = -M^2\frac{dV}{V} \tag{12.32}$$

Combining with equation 12.31:

$$(1 - M^2)\frac{dV}{V} + \frac{dA}{A} = 0$$

$$\frac{dV}{V} = \frac{1}{M^2 - 1}\frac{dA}{A} \tag{12.33}$$

Equation 12.33 shows the designer how to vary the flow area A of a channel in order to accelerate or decelerate the flow. In incompressible flow ($VA = \textit{constant}$), the channel area A must decrease in order to accelerate the flow, *i.e.*, increase V. However, according to equation 12.33, for an isentropic compressible flow, the flow accelerates ($dV > 0$) in a converging channel ($dA < 0$) only if the flow is subsonic ($M^2 < 1$). To accelerate supersonic flow, the channel area must increase, not decrease. Thus to accelerate continuously a compressible flow in a nozzle, the flow area must decrease until sonic flow is achieved and then increase to further accelerate the supersonic flow.[3] For the reverse process of decelerating a supersonic flow, as in a supersonic wind tunnel, the diffuser area should first decrease until sonic flow is achieved and then increase to slow down the ensuing subsonic flow.

Figure 12.3 shows how the mass flow rate per unit area, ρV, varies in an accelerating flow, reaching its maximum value when the flow speed is locally sonic at the point of minimum flow area A_c, called the *throat*. The flow area A in a channel flow may be compared to its minimum value A_c at $M = 1$ by noting that the product ρAV is a

[3]Such converging-diverging nozzles were invented by the Swedish engineer DeLaval for use in steam turbines.

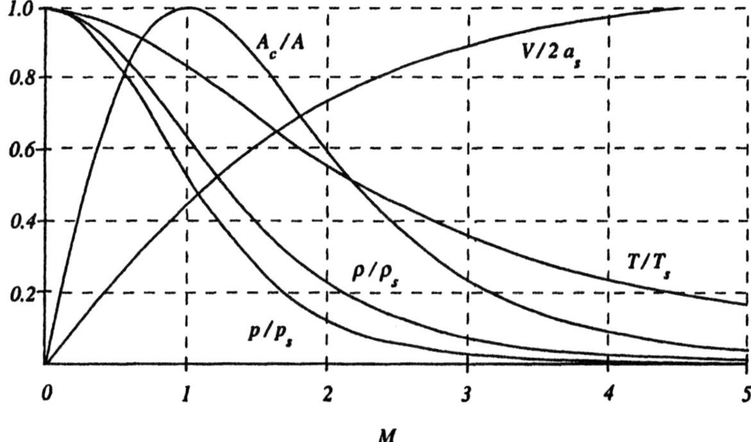

Figure 12.4 The flow variables of an isentropic steady flow as a function of Mach number M for a perfect gas with constant specific heats having $\gamma = 1.4$.

constant:

$$\begin{aligned}\frac{A_c}{A} &= \frac{\rho V}{\rho_c a_c} \\ &= M\left[\left(\frac{2}{\gamma+1}\right)\left(1+\frac{\gamma-1}{2}M^2\right)\right]^{-(\gamma+1)/2(\gamma-1)} \\ &= \left(\frac{\gamma+1}{2}\right)^{(\gamma+1)/2(\gamma-1)}\left[\left(\frac{2}{\gamma-1}\right)\left(1-\left(\frac{p}{p_s}\right)^{(\gamma-1)/\gamma}\right)\right]^{1/2}\left(\frac{p}{p_s}\right)^{1/\gamma}\end{aligned} \tag{12.34}$$

where we have used equations 12.26 and 5.30 to express the ratio A/A_c as a function of M or p/p_s.

The compressible flow variables T/T_s, p/p_s, ρ/ρ_s, and A_c/A, as given in equations 12.26 and 12.34, are plotted in figure 12.4 as functions of the flow Mach number M. The temperature, density and pressure all decrease as the Mach number increases, with the pressure declining the most rapidly. The flow area ratio A_c/A is a maximum at $M = 1$ as expected, since this is the point of critical flow.

Whether or not the fluid is a perfect gas in a steady compressible flow, the fluid cannot flow faster than the local sound speed at the location of minimum channel area A_c. To increase the mass flow rate, the density and/or the sound speed at that location must be increased by increasing ρ_s and/or a_s, but this increased flow will still be a sonic flow with $M = 1$ at that point.

Example 12.5

A rocket engine on a space vehicle expands the propellant gas to achieve an exit velocity V_e that is 90% of the maximum possible. Calculate the Mach number M_e of the exit flow and the ratio A_e/A_c of the exit area to the throat area, assuming the propellant is a perfect gas with $\gamma = 1.3$.

Solution
From equation 12.25, the maximum possible exit velocity V_{me} occurs when $p = 0$ and $V_{me}/a_s = \sqrt{2/(\gamma - 1)} = \sqrt{2/0.3} = 2.582$. The exit Mach number M_e corresponding to 90% of this value is found from equation 12.26:

$$0.9(2.582) = M_e \left(1 + \frac{0.3}{2} M_e^2\right)^{-1/2}$$

$$5.4 = \frac{M_e^2}{1 + 0.15M_e^2}$$

$$M_e = \sqrt{\frac{5.4}{1 - 0.81}} = 5.331$$

The area ratio A_e/A_c can be calculated from equation 12.34 for $M_e = 5.331$:

$$A_e/A_c = 5.331 \left[\left(\frac{2}{0.3}\right)(1 + 0.3(5.331)^2/2)\right]^{2.3/2(0.3)}$$

$$= 3.101E(3)$$

This corresponds to a diameter ratio of $\sqrt{3.101E(3)} = 55.69$.

12.4 Shock Waves

When a body moves through a fluid at a speed that exceeds the speed of sound, it produces wave fronts in which the pressure, density and temperature of the fluid change abruptly within a very small distance.[4] Such wave fronts are called *shock waves*. They can be regarded as surfaces in the flow field across which there is a discontinuous change in the flow variables. For an observer moving with the body

[4]This distance is of the order of ν/a, which for air at 20° C and one atmosphere of pressure, is about $5E(-8)\,m$. This is approximately the distance that molecules move between collisions.

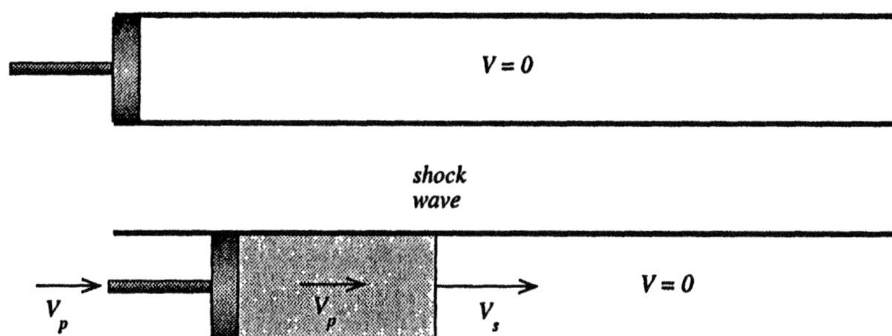

Figure 12.5 A shock wave is generated in a shock tube by the movement of a piston at speed V_p into an initially stationary gas. The fluid between the shock wave and the piston, moving at the piston speed, is shown by the shaded region.

at a constant supersonic velocity through the fluid, the oncoming flow is a steady supersonic flow and the shock wave is a surface fixed in the reference frame of the body but across which fluid flows steadily, albeit experiencing a discontinuous change in flow conditions upon crossing through the shock front. Like the gravity waves generated by a ship moving along the surface of the sea, shock waves propagate to the side and rear of a supersonic body, forming a pattern that is fixed in relation to the body when in steady flight.

A simple method of producing a shock wave is illustrated in figure 12.5. A long tube, initially filled with a uniform stationery gas, is fitted with a piston at one end. At time $t = 0$, the piston is suddenly set into motion at a constant speed V_p. This motion generates a shock wave that moves in the same direction as the piston but at a greater speed V_s that is also constant with time and greater than the speed of sound in the undisturbed gas. The fluid between the piston and the shock wave has a uniform density and pressure that are higher than in the undisturbed gas in front of the shock wave. This fluid, which has passed through the shock wave, occupies a smaller volume than it initially occupied before being processed by the shock wave.

To find the change of properties across this shock wave, it will be more convenient to describe the flow in a reference frame fixed in the shock wave, as shown in figure 12.6, for which the flow is steady. In this flow, the oncoming supersonic flow conditions in front of the shock wave are designated by the subscript 1 while those downstream are designated by the subscript 2. Select a control surface of unit area that encloses the shock wave, as shown in figure 12.6, and apply the conservation of mass, momentum and energy for an inviscid adiabatic flow:

$$\rho_2 V_2 - \rho_1 V_1 = 0$$

$$\rho_2 V_2^2 - \rho_1 V_1^2 = p_1 - p_2$$

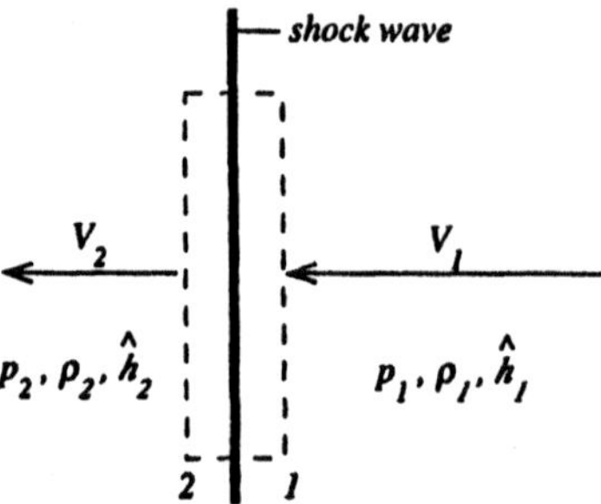

Figure 12.6 The control surface for steady flow normal to a shock wave. Velocities are measured with respect to the shock wave.

$$\rho_2 V_2 \left(\hat{h}_2 + \frac{V_2^2}{2}\right) - \rho_1 V_1 \left(\hat{h}_1 + \frac{V_1^2}{2}\right) = 0 \tag{12.35}$$

If we eliminate V_1 and V_2 among these equations, we are left with the *Rankine-Hugoniot* equation for the change of thermodynamic variables across the shock wave:

$$2(\hat{h}_2 - \hat{h}_1) = (p_2 - p_1)\left(\frac{1}{\rho_1} + \frac{1}{\rho_2}\right) \tag{12.36}$$

The thermodynamic properties of the fluid determine the enthalpy as a function of pressure and density so that equation 12.36 can be solved for the possible downstream p_2 and ρ_2 behind a shock wave moving into a fluid of initial pressure and density p_1 and ρ_1.

If we consider the case of a perfect gas with constant specific heats, for which $\hat{h} = \hat{c}_p T = \gamma p/(\gamma - 1)\rho$, the Rankine-Hugoniot equation has the form:

$$\frac{p_2}{p_1} = \frac{(\gamma + 1) - (\gamma - 1)(\rho_1/\rho_2)}{(\gamma + 1)(\rho_1/\rho_2) - (\gamma - 1)} \tag{12.37}$$

This equation is plotted in figure 12.7, showing how the possible downstream pressure p_2 is related to the density ρ_2, both expressed as the pressure ratio p_2/p_1 and density ratio ρ_1/ρ_2. Although not evident from figure 12.7, the higher pressures p_2 are associated with higher shock wave speeds. Although the pressure rise can be very high, the density ratio ρ_2/ρ_1 is limited to a maximum value of $(\gamma + 1)/(\gamma - 1)$, which equals 6 for $\gamma = 1.4$.

Also shown in figure 12.7 is a dotted curve that is an isentrope of the initial state:

$$\frac{p_2}{p_1} = \left(\frac{\rho_2}{\rho_1}\right)^{\gamma} \tag{12.38}$$

Any point p_2, ρ_2 lying above and to the right of this line has an entropy greater than

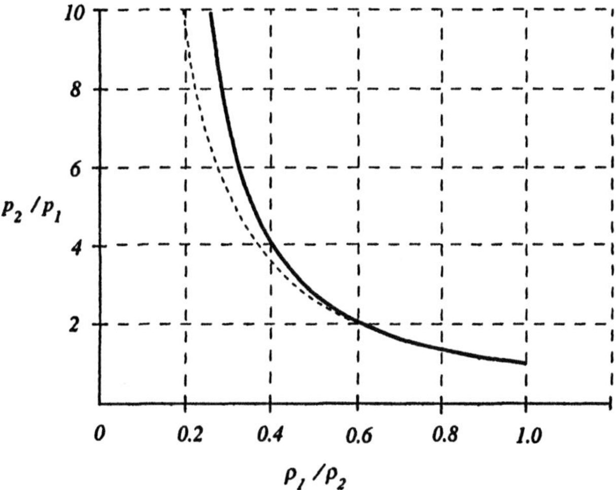

Figure 12.7 The relation between the pressure ratio p_2/p_1 across a shock wave and the density ratio ρ_1/ρ_2, as given by the Rankine-Hugoniot equation for a perfect gas with constant specific heats and having $\gamma = 1.4$, is shown as the solid line. The dotted line is the isentrope of the initial state.

$\hat{s}_1$. It is evident that the states behind a shock wave have entropies that exceed that of the upstream gas, the more so as the pressure rise is greater. Although there is a solution to the Rankine-Hugoniot equation for $p_2/p_1 \leq 1$, $\rho_1/\rho_2 \geq 1$, it is not drawn in figure 12.7 because it corresponds to states with lower entropy than $\hat{s}_1$. Such a solution violates the second law of thermodynamics, as expressed in equation 8.45, requiring that the entropy can never decrease in an adiabatic flow.

Although the flow across a shock wave is adiabatic and inviscid, the viscous stress and heat transfer are significant within the shock wave itself. The increase in entropy across the shock wave, illustrated in figure 12.7, is the result of irreversible processes within the shock front.

The flow conditions downstream of the shock wave can be expressed in terms of the shock wave Mach number, $M_1 \equiv V_1/a_1$, for the case of a perfect gas with constant specific heats, by replacing $\hat{h}$ by $\hat{c}_p T$ in equations 12.35, yielding:

$$\frac{p_2}{p_1} = \frac{2\gamma M_1^2 - (\gamma - 1)}{\gamma + 1}$$

$$\frac{\rho_2}{\rho_1} = \frac{V_1}{V_2} = \frac{(\gamma + 1)M_1^2}{(\gamma - 1)M_1^2 + 2}$$

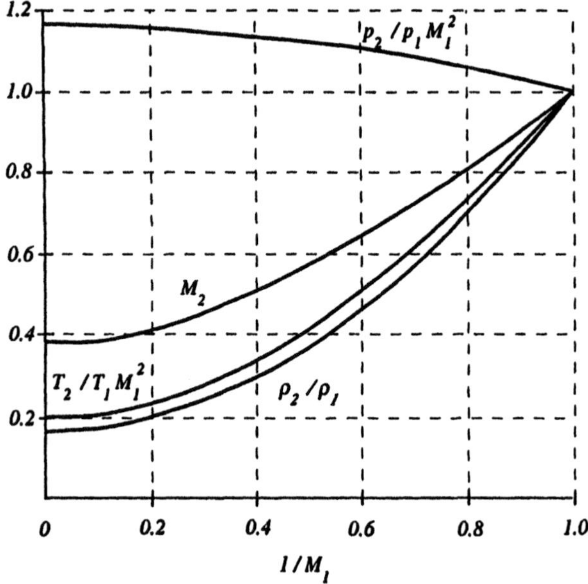

Figure 12.8 The properties of a shock wave plotted as a function of the shock wave Mach number M_1.

$$\frac{T_2}{T_1} = \frac{a_2^2}{a_1^2} = \frac{[2\gamma M_1^2 - (\gamma - 1)][(\gamma - 1)M_1^2 + 2]}{(\gamma + 1)^2 M_1^2}$$

$$M_2^2 = \frac{(\gamma - 1)M_1^2 + 2}{2\gamma M_1^2 - (\gamma - 1)} \tag{12.39}$$

These quantities are plotted in figure 12.8 as a function of $1/M_1$. Note that p_2/p_1 and T_2/T_1 are proportional to M_1^2 for very large shock Mach numbers, whereas ρ_2/ρ_1 and M_2 approach fixed values. Note also that the downstream flow is subsonic since $M_2 \leq 1$ for any value of $M_1 \geq 1$.

Example 12.6

In the flow illustrated in figure 12.5, air at rest ($a_1 = 347.2\,m/s$, $p_1 = 1E(5)\,Pa$, $\gamma = 1.4$) is accelerated by a shock wave of speed V_s that is generated by a piston moving at the speed $V_p = 100\,m/s$. Calculate the shock Mach number, the shock speed V_s and the pressure on the piston face.

Solution

Denoting the fluid state behind the shock wave by the subscript 2, mass conservation in a reference frame fixed in the shock wave is:

$$\rho_1 V_s = \rho_2 (V_s - V_p)$$

$$\frac{\rho_2}{\rho_1} = \frac{V_s}{V_s - V_p} = \frac{M_1}{M_1 - V_p/a_1}$$

since the downstream speed relative to the shock wave is $V_s - V_p$. But the density ratio ρ_2/ρ_1 is related to the shock Mach number M_1 by equation 12.39:

$$\frac{\rho_2}{\rho_1} = \frac{(\gamma+1)M_1^2}{(\gamma-1)M_1^2+2} = \frac{M_1}{M_1 - V_p/a_1}$$

$$(\gamma+1)M_1^2 - [(\gamma+1)V_p/a_1]M_1 = (\gamma-1)M_1^2 + 2$$

$$M_1^2 - [(\gamma+1)V_p/2a_1]M_1 - 1 = 0$$

Noting that $[(\gamma+1)V_p/2a_1] = 2.4(100\,m/s)/2(347.2\,m/s) = 0.3456$, we find:

$$M_1 = \frac{\sqrt{(0.3456)^2+4} + 0.3456}{2} = 1.188$$

$$V_s = M_1 a_1 = (1.188)(347.2\,m/s) = 412.5\,m/s$$

From equation 12.39, we find p_2:

$$p_2 = p_1 \frac{2(1.4)(1.188)^2 - 0.4}{2.4} = 1.48E(5)\,Pa$$

Shock Waves in Channel Flow

Shock waves can be present in supersonic channel flows, converting the supersonic to a subsonic flow whenever the downstream flow conditions require it. To illustrate the role that shock waves can play in such flows, figure 12.9 shows the pressure distribution for several possible steady compressible flows through a converging-diverging nozzle. For a very small pressure drop across the nozzle, a slow subsonic flow ensues (curve *a*) that has a minimum pressure and maximum velocity at the nozzle throat. This flow is essentially incompressible. When the pressure drop is increased somewhat (curve *b*), the flow speeds up but is still everywhere subsonic with the flow accelerating in the converging portion and decelerating in the diverging portion. When the pressure is further reduced, as in curve *c*, critical sonic flow is

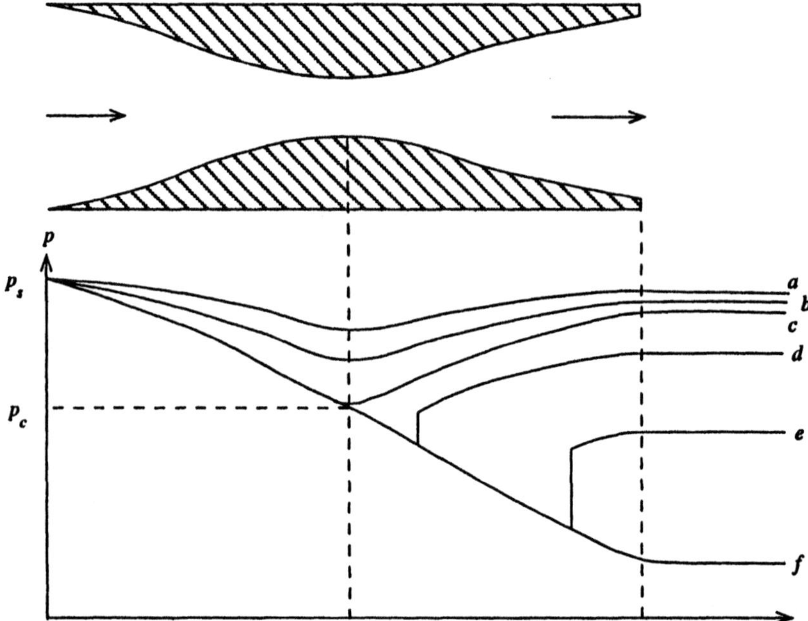

Figure 12.9 The pressure distribution in a converging-diverging nozzle depends upon the pressure change across the nozzle. Shock waves may exist for flows that are supersonic in the diverging portion.

reached at the throat at the pressure p_c, but then the flow decelerates subsonically as the pressure rises. A further lowering of the pressure, as in curve *d*, results in a supersonic flow downstream of the throat in the divergent section, terminated by a shock wave that raises the pressure abruptly and converts the subsequent flow to subsonic flow with a rising pressure. As the downstream pressure is further reduced, the region of supersonic flow is extended further along the flow axis as the shock wave moves further downstream (curve *e*). Finally, with a sharp enough reduction in pressure, the shock wave is sucked out of the nozzle and full supersonic flow prevails, as in curve *f*.

The mass flow rate through the nozzle of figure 12.9 increases with decreasing downstream pressure as long as the flow remains subsonic everywhere, as in curves *a–c*. However, once supersonic flow is established downstream of the throat, there is no further increase in mass flow rate through the nozzle, which is said to be *choked* under these conditions. For choked flows, a small change in downstream pressure cannot be propagated upstream through the throat and thus cannot affect the upstream flow and mass flow rate.

12.5 Adiabatic Viscous Pipe Flow

Gases are sometimes pumped through long pipelines, undergoing substantial changes in density as the pressure falls along the length of the pipe. This is typically the case for the long distance transmission of natural gas from the gas well to local distributors thousands of kilometers distant. Such compressible flows must be treated differently than the incompressible flows of chapter 9.

The shear stress on the wall of a pipe retards the flow, causing a pressure drop along the pipe axis. The magnitude of the pressure gradient caused by this wall shear stress is usually expressed in terms of the Darcy friction factor f by equation 6.48:

$$-\frac{\partial p}{\partial x} = \frac{f}{D}\left(\frac{\rho \bar{V}^2}{2}\right)$$

In compressible flow in a pipe, the average fluid velocity $\bar{V}$ increases as the pressure and density decline in the flow direction so that the acceleration of a fluid particle is affected by both the pressure gradient and the wall drag:

$$\bar{V}\frac{d\bar{V}}{dx} = -\frac{dp}{dx} - \frac{1}{2D}f\rho\bar{V}^2 \tag{12.40}$$

The conservation of mass in the pipe flow may be given in differential form as:

$$\frac{d\rho}{\rho} + \frac{d\bar{V}}{\bar{V}} = 0 \tag{12.41}$$

and the conservation of energy, equation 12.20, as:

$$\frac{1}{\gamma - 1}d(a^2) + \frac{1}{2}d(\bar{V}^2) = 0 \tag{12.42}$$

where we have used $d\hat{h} = d(a^2)/(\gamma - 1)$ for a perfect gas with constant specific heats.

It is convenient to introduce the Mach number $M \equiv \bar{V}/a$ as a flow variable, as we did for isentropic flow. By using the perfect gas law and equations 12.41 and 12.42, we can express the derivatives $d\bar{V}/dM$, dp/dM, *etc.*, as functions of M and then integrate these expressions from $M = 1$ to M, obtaining:

$$\frac{\bar{V}}{\bar{V}_c} = \frac{\rho_c}{\rho} = M^2\frac{p}{p_c} = M\frac{a}{a_c} = M\frac{\sqrt{T}}{\sqrt{T_c}} = M\sqrt{\frac{\gamma+1}{2+(\gamma-1)M^2}} \tag{12.43}$$

where the subscript c refers to the value of the flow variable at $M = 1$. These variables are plotted in figure 12.10 as a function of the Mach number.

The momentum equation 12.40 can also be cast in the form of a function of Mach

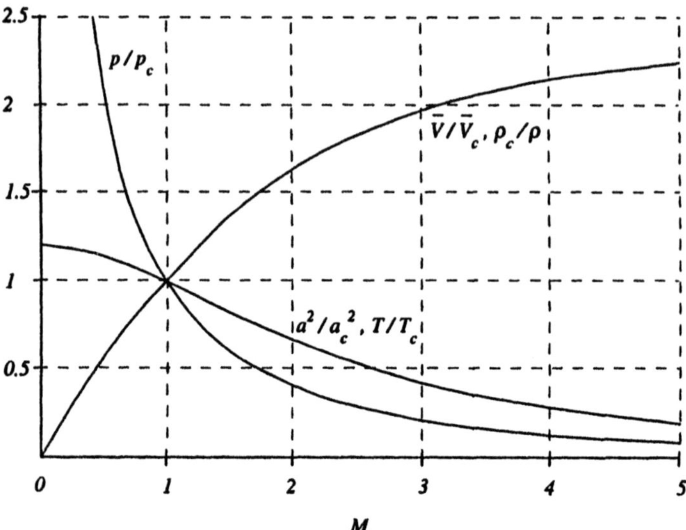

Figure 12.10 The flow variables of adiabatic pipe flow with friction as a function of Mach number $M \equiv \bar{V}/a$, for a perfect gas having $\gamma = 1.4$.

number by utilizing the relationships of 12.43:

$$\frac{f}{D}\,dx = \frac{1 - M^2}{\gamma M^4[1 + (\gamma - 1)M^2/2]}\,d(M^2) \tag{12.44}$$

The left side of this equation is always positive and so must be the right side as well. Thus $(1 - M^2)\,d(M^2)$ is always positive, and M must increase in subsonic flow and vice versa for supersonic flow. It therefore follows that the effect of friction in compressible pipe flow is to force the flow toward a sonic condition, *i.e.*, a choked flow. Once this condition is reached, no further change is possible. Thus this condition can only be reached at the end of the pipe. The upstream flow will adjust to this condition if necessary so as to maintain a steady flow.

We may integrate equation 12.44 over a length L_{max} between the entrance to the pipe and the exit when the flow there is choked ($M = 1$):

$$\frac{\bar{f}L_{max}}{D} = \frac{1 - M^2}{\gamma M^2} + \frac{\gamma + 1}{2\gamma}\ln\left[\frac{(\gamma + 1)M^2}{2[1 + (\gamma - 1)M^2/2]}\right] \tag{12.45}$$

where $\bar{f}$ is the average value of the Darcy friction factor over the pipe length L_{max}. If the inlet flow Mach number is known, then equation 12.45 defines the maximum pipe length for which steady flow may exist for that entrance condition.

The dimensionless pipe flow length $\bar{f}L_{max}/D$ is plotted in figure 12.11 as a function

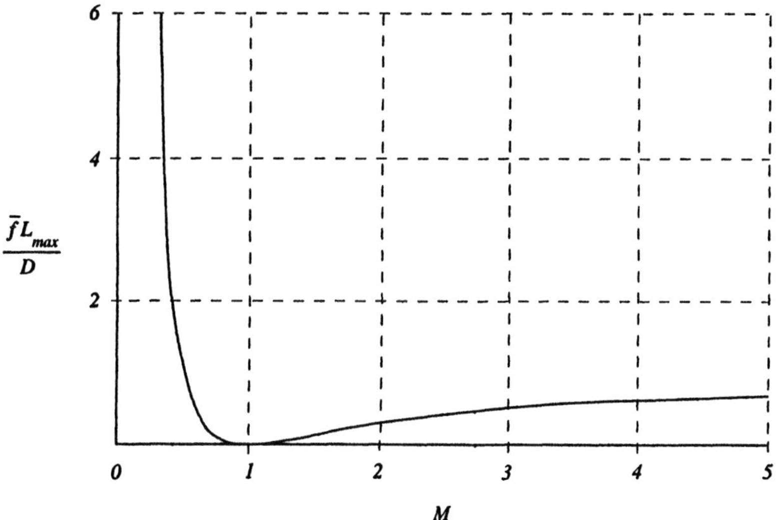

Figure 12.11 The dimensionless pipe length fL_{max}/D to reach choked flow as a function of initial Mach number M for viscous adiabatic flow in a pipe of a perfect gas having $\gamma = 1.4$.

of Mach number. Note that this length is greater for subsonic flow, the most commonly encountered case. For supersonic flow, it is generally less than unity. As can be seen from figure 12.10, the pressure falls and the velocity increases in subsonic flow while the opposite obtains in supersonic flow.

The Darcy friction factor f is a function of the flow Reynolds number $Re_D \equiv \rho\bar{V}D/\mu$, the pipe roughness ϵ/D and the flow Mach number $\bar{V}/a$. For most compressible pipe flows, the flow will be turbulent, and f can be calculated from equation 9.7. The flow Reynolds number is nearly constant along the pipe length since $\rho\bar{V}$ is constant by mass conservation and μ depends only upon temperature, which changes little. In any case, f is not sensitive to moderate changes in Reynolds number in turbulent flow so that $\bar{f}$ can be calculated for the known inflow or outflow conditions.

Example 12.7

A natural gas pipeline of diameter $D = 1\,m$ and surface roughness $\epsilon = 5E(-5)\,m$ is supplied with gas from a compressor at a pressure $p_1 = 2E(6)\,Pa$ and temperature $T_1 = 20°\,C$. At this temperature the natural gas sound speed $a_1 = 446.1\,m/s$, viscosity $\mu = 9E(-6)\,Pa\,s$ and specific heat ratio $\gamma = 1.31$ while the gas constant $R = 518.3\,J/kg\,K$.

The pipe flow speed $\bar{V}_1 = 10\,m/s$. Calculate the pipe length L_{max} at which the flow would be choked and the flow speed $\bar{V}_c$ and pressure p_c at the point of choked flow.

Solution

To find the pipe Reynolds number, first calculate the density from the perfect gas law:

$$\rho_1 = \frac{p_1}{RT_1} = \frac{2E(6)\,Pa}{(518.3\,J/kg\,K)(293.15\,K)} = 13.16\,kg/m^3$$

$$Re_D = \frac{\rho_1 \bar{V}_1 D}{\mu_1} = \frac{(13.16\,kg/m^3)(10\,m/s)(1\,m)}{9E(-6)\,Pa\,s} = 1.463E(7)$$

The Darcy friction factor f can be calculated from equation 7.17 for turbulent flow:

$$\frac{1}{\sqrt{f}} = -2.0\log\left(\frac{\epsilon/D}{3.7} + \frac{2.51}{Re_D\sqrt{f}}\right)$$

$$= -2\log\left(1.352E(-5) + \frac{1.716E(-7)}{\sqrt{f}}\right)$$

$$f = 1.076E(-2)$$

We now determine L_{max} from equation 12.45 for $M = 10/446.1 = 2.242E(-2)$:

$$\frac{\bar{f}L_{max}}{D} = \frac{1 - 5.025E(-4)}{1.31(5.025E(-4))} + \frac{2.31}{2.62}\ln\left(\frac{(2.31)5.025E(-4)}{2[1 + 0.31(5.025E(-4))/2]}\right) = 1511$$

$$L_{max} = \frac{1511(1\,m)}{1.076E(-2)} = 1.405E(5)\,m = 140.5\,km$$

To find the pressure and velocity at this point, use equations 12.43:

$$\bar{V}_c = \frac{\bar{V}}{M}\sqrt{\frac{2 + (\gamma - 1)M^2}{\gamma + 1}} = \frac{10\,m/s}{2.242E(-2)}\sqrt{\frac{2 + 0.31(2.242E(-2))^2}{2.31}} = 415.0\,m/s$$

$$p_c = \frac{p_1 \bar{V}_c M^2}{\bar{V}_1} = \frac{(2E(6)\,Pa)(415.0\,m/s)(2.242E(-2))^2}{10\,m/s} = 4.172E(4)\,Pa$$

12.6 Plane Supersonic Flow

In this section, we consider two simple supersonic flows having but two velocity components, both in the x, y-plane. The first is the flow across a plane shock wave that is oblique to the oncoming supersonic flow, such as the flow that would form

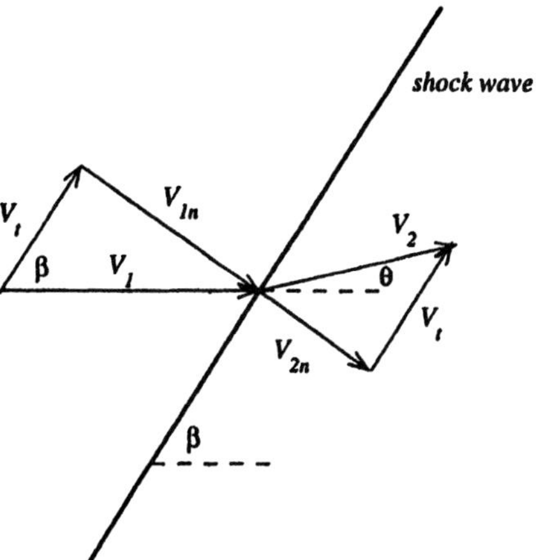

Figure 12.12 The flow across an oblique shock wave in supersonic flow, which makes an angle β with the oncoming flow, is deflected through an angle θ away from the normal direction. There is no change in the tangential velocity component V_t across the shock wave.

around a sharp-edged wedge pointing into the supersonic flow. The second is the flow along a curved wall. Both flows are inviscid and isentropic (except for the entropy increase inside the plane shock wave). These flows are the basic building blocks of two-dimensional supersonic flow and can be used to develop an approximate understanding of more complex supersonic flows.

12.6.1 Oblique Plane Shock Waves

Consider the flow sketched in figure 12.12, where a uniform horizontal supersonic flow of speed V_1 encounters an oblique plane shock wave that makes an angle β with the oncoming flow. The inflow may be decomposed into a normal component V_{1n} and a tangential component V_t that are related to V_1 by:

$$V_{1n} = V_1 \sin\beta; \quad V_t = V_1 \cos\beta \tag{12.46}$$

The flow upstream and downstream of the plane shock wave is uniform, and there is thus no pressure gradient in the direction tangent to the shock wave. Consequently, the tangential velocity component V_t is the same behind the shock wave as it is in front, as drawn in figure 12.12.

If we move along the shock wave itself at a speed V_t, we would observe a steady flow at a speed V_{ln} directed normal to the shock wave, *i.e.*, the flow would appear to be the same as that described in section 12.4. In particular, the density ratio ρ_2/ρ_1 across the shock wave could be given by equation 12.39 with M_1 replaced by

$V_{1n}/a_1 = (V_1 \sin\beta)/a_1 = M_1 \sin\beta$:

$$\frac{\rho_2}{\rho_1} = \frac{(\gamma+1)M_1^2 \sin^2\beta}{(\gamma-1)M_1^2 \sin^2\beta + 2} \tag{12.47}$$

where $M_1 = V_1/a_1$ is the Mach number of the flow upstream of the oblique shock wave of figure 12.12.

As can be seen in figure 12.12, the increase in density results in a decrease in the normal velocity V_{2n} behind the shock wave because of mass conservation:

$$\rho_1 V_{1n} = \rho_2 V_{2n}$$

$$V_{2n} = \left(\frac{\rho_1}{\rho_2}\right) V_{1n} \tag{12.48}$$

As a consequence, the downstream velocity V_2 is deflected away from the normal to the shock wave by an angle θ as compared to its upstream value. From figure 12.12, the angle between the velocity components V_2 and V_t is $\beta - \theta$ so that:

$$\tan(\beta-\theta) = \frac{V_{2n}}{V_t} = \left(\frac{\rho_1}{\rho_2}\right)\frac{V_{in}}{V_t} = \left(\frac{\rho_1}{\rho_2}\right)\tan\beta \tag{12.49}$$

where we have used equations 12.46. However, we can solve equations 12.49 and 12.47 for θ:

$$\theta = \tan^{-1}\left(\frac{2\cot\beta(M_1^2 \sin^2\beta - 1)}{M_1^2(\gamma + \cos 2\beta) + 2}\right) \tag{12.50}$$

Figure 12.13 plots the relationship between the flow deflection angle θ and the shock wave angle β for a perfect gas having $\gamma = 1.4$ for several Mach numbers. For any flow deflection angle θ, there are generally two possible shock angles β. For the smaller shock angle, the downstream flow is usually supersonic, and the shock wave is called *weak*. For the larger shock angle β, the flow downstream is subsonic and the shock wave is called *strong*, signifying a greater pressure ratio p_2/p_1 than for the weak shock wave. However, for any given Mach number, there is a maximum deflection angle θ beyond which there is no possible plane oblique shock wave with uniform flow both upstream and downstream of the shock wave. If a wedge of half-angle greater than this maximum value is inserted into a supersonic flow, the shock wave detaches from the wedge and moves upstream, becoming curved and creating a non-uniform plane flow downstream.

For very small angles of deflection θ, equation 12.50 is satisfied by $\beta = \pi/2$, a normal (strong) shock wave or by $\sin\beta = 1/M_1$, a (weak) sound wave.

The change in pressure, temperature and sound speed across the oblique shock

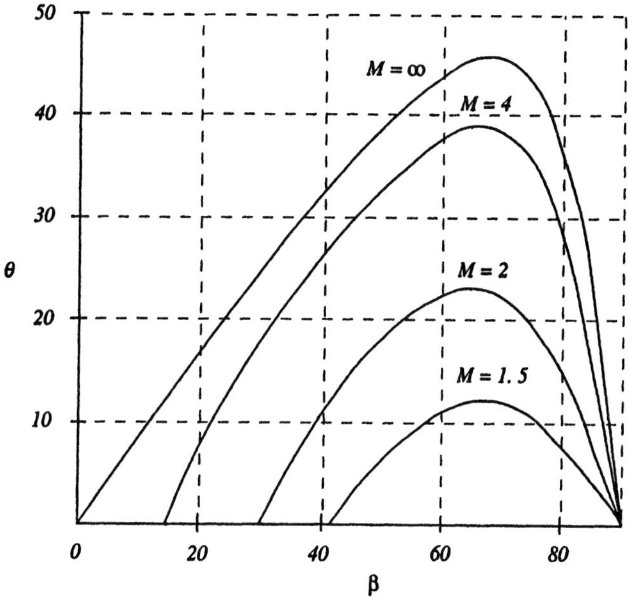

Figure 12.13 The angle θ of deflection of the flow caused by an oblique shock wave forming an angle θ with the oncoming supersonic flow for several Mach numbers M of the oncoming flow in a perfect gas; $\gamma = 1.4$.

wave may be found from the normal shock wave equations 12.39 by replacing M_1 in these equations by $V_{1n}/a_1 = (V_1 \sin\beta/a_1) = M_1 \sin\beta$:

$$\frac{p_2}{p_1} = \frac{2\gamma M_1^2 \sin^2\beta - (\gamma - 1)}{\gamma + 1}$$

$$\frac{T_2}{T_1} = \left(\frac{a_2}{a_1}\right)^2 = \frac{[2\gamma M_1^2 \sin^2\beta - (\gamma - 1)][(\gamma - 1)M_1^2 \sin^2\beta + 2]}{(\gamma + 1)^2 M_1^2 \sin^2\beta} \tag{12.51}$$

To find M_2 for an oblique shock wave, note in figure 12.12 that:

$$\cos(\beta - \theta) = \frac{V_t}{V_2}; \quad \cos\beta = \frac{V_t}{V_1} \tag{12.52}$$

so that:

$$M_2 = \frac{V_2}{a_2} = \frac{V_t}{a_2 \cos(\beta - \theta)} = \frac{V_1 \cos\beta}{a_2 \cos(\beta - \theta)} = \frac{\cos\beta}{\cos(\beta - \theta)} \left(\frac{a_1}{a_2}\right) M_1 \tag{12.53}$$

where a_1/a_2 may be obtained from 12.51.

Example 12.8

In a supersonic wind tunnel, a two-dimensional wedge of half-angle $\theta = 10°$ is inserted into the air flow, and the attached shock wave angle is observed to be $\beta = 20°$. Calculate the Mach number M of the flow in the wind tunnel.

Solution

Since we know both θ and β, we can solve equation 12.50 for the Mach number:

$$[M^2(\gamma + \cos^2\beta) + 2]\tan\theta = 2\cot\beta(M^2\sin^2\beta - 1)$$

$$M^2 = \frac{2\tan\theta + 2\cot\beta}{2\cos\beta\sin\beta - (\gamma + \cos^2\beta)\tan\theta}$$

$$= \frac{2\tan 10° + 2\cot 20°}{2\cos 20°\sin 20° - (1.4 + \cos^2 20°)\tan 10°} = 24.34$$

$$M = 4.933$$

12.6.2 Supersonic Flow along a Wall

A plane supersonic flow can abruptly change direction when it encounters an oblique shock wave, as sketched in figure 12.12, but the flow direction can also be changed gradually by a series of small amplitude sound waves in a manner that will not increase the entropy as a shock wave does. Such small amplitude sound waves are called *Mach waves*, and the change in flow direction θ caused by such waves can be much larger than that caused by an oblique shock wave.

Consider a uniform plane supersonic flow of Mach number $M = V/a$ passing over a wall that suddenly changes direction by a small angle $\delta\theta$, as sketched in figure 12.14. The change in direction is propagated into the flow along a Mach wave that makes an angle with the wall of $\beta = \sin^{-1}(1/M)$. As can be seen in figure 12.14, the component of the upstream velocity normal to the Mach wave is a because it is a sound wave so that the tangential component V_t is:

$$V_t = V\cos\beta = \sqrt{V^2 - a^2} = a\sqrt{M^2 - 1} \tag{12.54}$$

Downstream of this wave, the flow speed is increased slightly to $V + \delta V$, and the angle between $V + \delta V$ and V_t is $\beta + \delta\theta$. Consequently we have:

$$\cos(\beta + \delta\theta) = \frac{V_t}{V + \delta V} \simeq \frac{V_t}{V}\left(1 - \frac{\delta V}{V}\right) = \frac{V_t}{V} - \left(\frac{V_t}{V}\right)\frac{\delta V}{V} \tag{12.55}$$

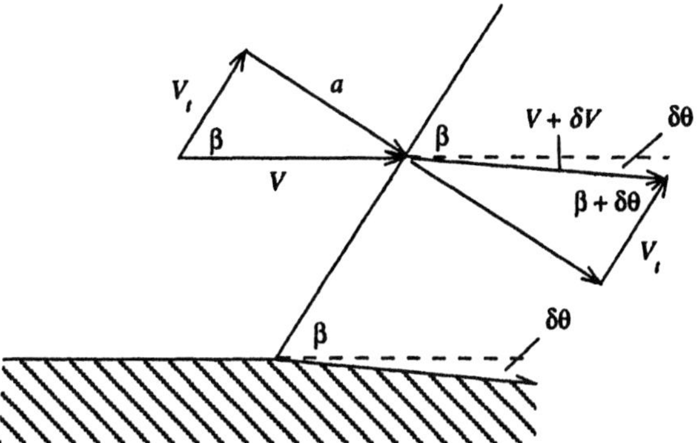

Figure 12.14 The supersonic flow past a slight bend of a wall is accomplished by a Mach wave that accelerates the flow to a slightly higher speed.

where we have assumed that $\delta V \ll V$. Noting that $\delta\theta \ll 1$ and using equation 12.54, this becomes:

$$\cos\beta - \sin\beta\,(\delta\theta) = \cos\beta - \left(\frac{V_t}{V}\right)\frac{\delta V}{V}$$

$$-\frac{1}{M}\,\delta\theta = -\frac{\sqrt{M^2-1}}{M}\,\frac{\delta V}{V}$$

$$\delta\theta = \sqrt{M^2-1}\,\frac{\delta V}{V} \tag{12.56}$$

Since $M = V/a$, small changes in these variables are related by:

$$\frac{\delta V}{V} = \frac{\delta M}{M} + \frac{\delta a}{a} \tag{12.57}$$

But, from equation 12.26 for a perfect gas with constant specific heats,

$$V\delta V + \frac{2}{\gamma - 1}\,a\,\delta a = 0 \tag{12.58}$$

Eliminating a, we find:

$$\left(1 - \frac{\gamma-1}{2}M^2\right)\frac{\delta V}{V} = \frac{\delta M}{M} \tag{12.59}$$

Substituting in equation 12.56 and using differentials for the small quantities, we find:

$$d\theta = \frac{\sqrt{M^2-1}}{M\left(1+\frac{\gamma-1}{2}M^2\right)}\,dM \tag{12.60}$$

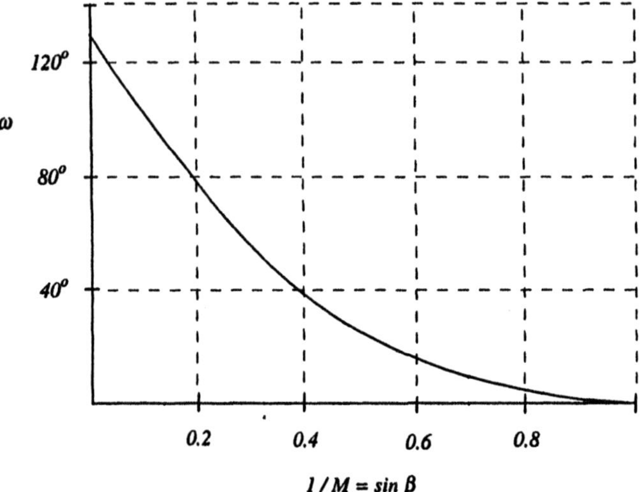

Figure 12.15 The turning angle ω of a plane supersonic flow as a function of the Mach number M for the case of $\gamma = 1.4$.

In equation 12.60, $d\theta$ is the increment of turning angle along a streamline, and dM is the increment in Mach number of the flow across the Mach wave.

If we start with a sonic flow and gradually turn the flow through an ever larger angle as the Mach number increases, the relationship between the total angle turned, denoted by ω, and the Mach number M at that angle may be found by integrating equation 12.60:

$$\omega = \sqrt{\frac{\gamma+1}{\gamma-1}} \tan^{-1}\left(\frac{(\gamma-1)(M^2-1)}{\gamma+1}\right)^{1/2} - \tan^{-1}\sqrt{M^2-1} \qquad (12.61)$$

Equation 12.61 is called the *Prandtl-Meyer function* for the turning angle of a supersonic flow. It is plotted in figure 12.15 as a function of the reciprocal of the Mach number ($1/M = \sin\beta$) for the case of $\gamma = 1.4$. Note that the turning angle ω exceeds 120° for an infinite Mach number.

Plane supersonic flow along a curved surface that is turned through a large angle θ is sketched in figure 12.16(a). The uniform inflow at Mach number M_1 and Mach wave angle β_1 is turned gradually by an *expansion wave* to a higher Mach number $M_2 > M_1$ and a smaller Mach wave angle $\beta_2 < \beta_1$. The flow downstream of the Mach wave of angle β_2 is uniform. The pressure, density and sound speed between 1 and 2 all decrease while the flow speed increases. Because θ is the difference between ω_2 and ω_1, the downstream Mach number M_2 is related to the upstream Mach number

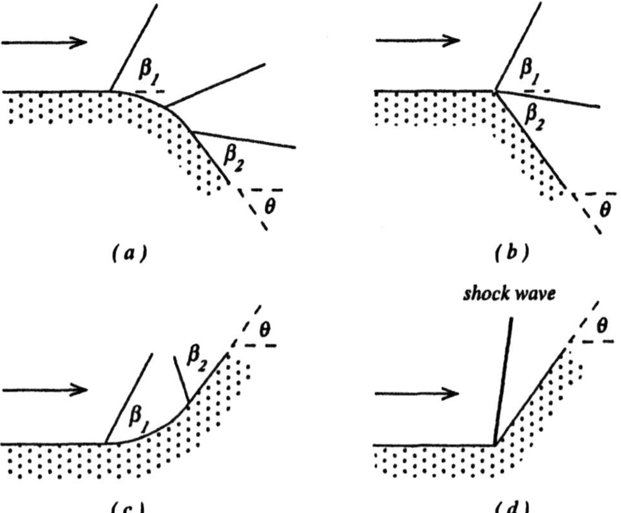

Figure 12.16 Examples of supersonic plane flow along a wall being turned through an angle θ.

M_1 by:

$$\omega_2\{M_2\} = \omega_1\{M_1\} + \theta \tag{12.62}$$

where $\omega\{M\}$ is given by equation 12.61. If the wall curvature is increased until a corner is formed, as shown in figure 12.16(b), the downstream flow is identical to that of 12.16(a), and the flow between the Mach waves 1 and 2 is said to be a *centered expansion wave*.

If the curved wall is concave toward the oncoming flow (see figure 12.16[c]) rather than convex (12.16[b]), the fluid is compressed to a higher pressure, density and sound speed but a lower velocity and Mach number in a *compression wave*. The relation between the downstream and upstream flow is determined by the turning angle θ:

$$\omega_2\{M_2\} = \omega_1\{M_1\} - \theta \tag{12.63}$$

But, as can be seen in the sketch of figure 12.16(c), the downstream Mach wave will eventually run into the upstream Mach wave, forming a shock wave at some distance from the wall. If the wall curvature is increased to form a corner, as in figure 12.16(d), the shock wave attaches to the corner, and the flow is that across an oblique shock wave as given in equations 12.50, 12.51 and 12.53. Unlike the flows of figure 12.16(a)–(c), there is an entropy increase in the downstream flow of 12.16(d).

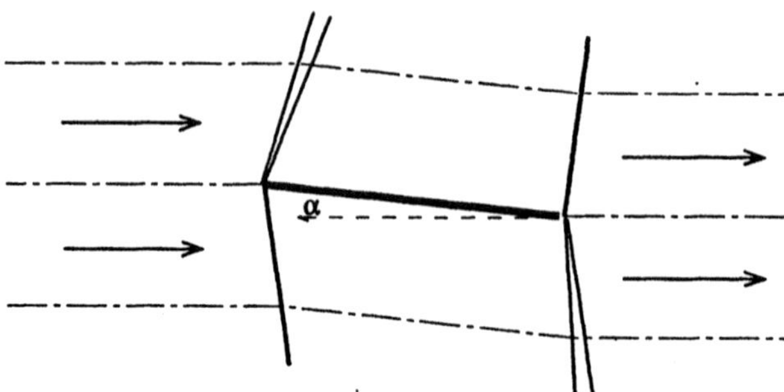

Figure 12.17 A sketch of the supersonic flow over a flat plate airfoil at an angle of attack α, showing the wave fronts generated at the leading and trailing edges, as well as some streamlines of the flow.

12.6.3 Plane Flow around Supersonic Airfoils

The principles discussed in the above sections can be applied to determine the lift and drag caused by the plane inviscid supersonic flow around an airfoil. In many cases, the flow is easier to analyze than that for incompressible flow because there is a simple relationship between the slope of the wing surface and the pressure along it.

Figure 12.17 shows the simplest airfoil in supersonic flow: a flat plate at an angle of attack α. At the leading edge, the flow above and below the airfoil encounters a corner flow, like those illustrated in figure 12.16(b) and (d). On the upper surface, a centered expansion wave turns the flow downward by the angle α with the flow Mach number M_u along the upper surface determined from equation 12.62 with $\theta = \alpha$ and $M_2 = M_u$. On this surface, the pressure difference $p_1 - p_u$ may be calculated from the isentropic flow relations of equation 12.26 for the flow between M_1 and M_u. On the lower surface, a weak shock wave increases the pressure and decreases the Mach number to M_ℓ. For this shock wave, the shock angle β may be found from equation 12.50 with $\theta = \alpha$ and then the pressure ratio p_ℓ/p_1 from equation 12.51. The pressure difference $p_\ell - p_u$ produces a force normal to the airfoil surface having a lift and drag component. The corresponding lift and drag coefficients $C_{\mathcal{L}}$ and $C_{\mathcal{D}}$ are:

$$C_{\mathcal{L}} = \frac{p_\ell - p_u}{\rho_1 V_1^2/2} \cos\alpha; \quad C_{\mathcal{D}} = \frac{p_\ell - p_u}{\rho_1 V_1^2/2} \sin\alpha \tag{12.64}$$

For small angles of attack α, the pressure difference $p_\ell - p_u$ will vary as α and so also will the lift coefficient. But the drag coefficient will vary as α^2, so the airfoil becomes inefficient at high angles of attack, having a lift/drag ratio varying approximately as $1/\alpha$. Unlike incompressible inviscid flow over airfoils, the drag force in supersonic inviscid flow unfortunately is not zero.

The flow approaching the trailing edge of the airfoil of figure 12.17 has a higher pressure below the trailing edge than above it by the amount $p_\ell - p_u$. The streamline leaving the trailing edge cannot support this pressure difference, so the flow is deflected upward by an angle very nearly equal to α, generating a shock wave going upward and an expansion wave moving downward, thus equalizing the pressures downstream of these waves. The flow below the airfoil therefore consists of a shock wave followed by an expansion wave while that above is just the reverse. In both cases, the shock wave and expansion wave will converge at large distances from the airfoil, ultimately attenuating the pressure changes across them. The downward moving shock wave is the cause of the familiar supersonic aircraft "boom."

Example 12.9

For the flat plate airfoil in supersonic flow, sketched in figure 12.17, derive an expression for the lift coefficient $C_\mathcal{L}$ in terms of the Mach number M and angle of attack α, for the case of $\alpha \ll 1$.

Solution

On the upper surface of the airfoil, for a small deflection angle $\delta\theta = \alpha$, equation 12.60 gives the corresponding increment δM of Mach number:

$$\delta\theta = \frac{\sqrt{M^2-1}}{M(1+\frac{\gamma-1}{2}M^2)}\,\delta M$$

But the flow across the expansion wave at the leading edge is isentropic, so the corresponding change in pressure δp can be found by differentiating 12.26:

$$\ln p = \ln p_s - \left(\frac{\gamma}{\gamma-1}\right)\ln\left(1+\frac{\gamma-1}{2}M^2\right)$$

$$\frac{\delta p}{p} = -\left(\frac{\gamma}{\gamma-1}\right)\frac{(\gamma-1)M\,\delta M}{1+\frac{\gamma-1}{2}M^2}$$

$$\delta p = \frac{-\gamma p M\,\delta M}{1+\frac{\gamma-1}{2}M^2}$$

Combining with the expression for $\delta\theta$:

$$\delta p = -\frac{\gamma M^2 p}{\sqrt{M^2-1}}\delta\theta$$

For a small angle of attack, the weak shock wave at the leading edge which turns the flow through the same angle $\delta\theta = \alpha$ will cause a pressure rise equal to the pressure decline on the upper surface. The lift coefficient becomes:

$$C_{\mathcal{L}} = \frac{2(-\delta p)}{\rho V^2/2} = \frac{4(-\delta p)}{\rho V^2}$$

$$= \frac{4\gamma M^2 p}{\rho V^2\sqrt{M^2-1}}\alpha = \frac{4\alpha}{\sqrt{M^2-1}}$$

since $\gamma p/\rho V^2 = a^2/V^2 = 1/M^2$.

12.7 One-Dimensional Unsteady Flow

When we investigated inviscid flow in chapter 4, we were able to treat unsteady incompressible flow along a streamline by utilizing Bernoulli's integral, equation 4.13, in which the flow unsteadiness appeared in an integral term, $\int(\partial\mathbf{V}/\partial t)\cdot d\mathbf{s}$. In cases where $\mathbf{V}$ was determinable along a streamline by applying mass conservation, flow solutions could be found. However, compressible flows are more difficult to deal with because the variable density precludes the use of Bernoulli's integral. Instead, we can treat only one-dimensional unsteady flows if we wish to obtain analytical integrals equivalent to Bernoulli's integral. Since there are practical engineering applications of such flows, we will derive the necessary equations describing them in this section.

To illustrate what might happen in an unsteady one-dimensional compressible flow, figure 12.18 shows a possible simple unsteady flow of a gas within a very long pipe. Initially, a stationary gas of uniform pressure p_0, density ρ_0 and sound speed a_0 fills the pipe to the left of a movable piston located at $x = 0$ (see figure 12.18[a)]). At time $t = 0$, the piston is moved to the right at a fixed speed V_p. The gas next to the piston moves at the same velocity as the piston, but gas further to the left of the piston moves more slowly because it has not yet experienced the full drop in pressure needed to accelerate it to the speed V_p. In fact, the gas very far to the left will not experience any movement until it is reached by a sound wave generated by the moving piston at $t = 0$. This initiating sound wave reaches $-x$ at the time $t = (-x)/a_0$. At any time t, the gas that lies in the range $-a_0 t \leq x \leq V_p t$ is in a state of unsteady flow, as shown in figure 12.18. Since this gas at time $t = 0$ occupied the length $a_0 t$, its density and pressure must have been reduced when it stretched to the length $(a_0 + V_p)t$.

For this inviscid compressible flow, there is only one velocity component, $\mathbf{V} = u\{x, t\}\mathbf{i}_x$. All flow variables depend upon only two independent variables, x and t. The flow must conform to the mass conservation law, equation 3.13, and Euler's

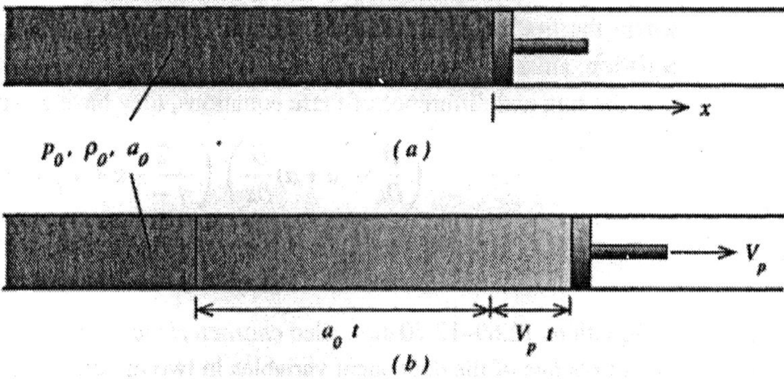

Figure 12.18 Beginning at $t = 0$, stationary gas in a pipe is set into motion by the retraction of a piston at a speed V_p.

equation 4.5, which for this case take the form:

$$\frac{\partial \rho}{\partial t} + u\frac{\partial \rho}{\partial x} + \rho\frac{\partial u}{\partial x} = 0$$

$$\frac{\partial u}{\partial t} + u\frac{\partial u}{\partial x} + \frac{1}{\rho}\frac{\partial p}{\partial x} = 0 \tag{12.65}$$

It is preferable to introduce the sound speed a as a variable in place of the density ρ. If the entropy is constant everywhere in the flow field, then by equation 12.22, we have for a perfect gas:

$$\frac{d\rho}{\rho} = \frac{2\hat{c}_v}{R}\frac{da}{a} = \frac{2}{\gamma - 1}\frac{da}{a} \tag{12.66}$$

We can replace the pressure p as a variable by using equations 12.13 and 12.66:

$$p = \frac{\rho a^2}{\gamma}$$

$$\frac{\partial p}{\partial x} = \frac{a^2}{\gamma}\frac{\partial \rho}{\partial x} + \frac{2\rho a}{\gamma}\frac{\partial a}{\partial x} = \frac{2}{\gamma - 1}a\frac{\partial a}{\partial x} \tag{12.67}$$

Substituting equations 12.67 and 12.66 into 12.65, after multiplying the first of 12.65 by a/ρ, we find:

$$\left(\frac{2}{\gamma - 1}\right)\left[\frac{\partial a}{\partial t} + u\frac{\partial a}{\partial x}\right] + a\frac{\partial u}{\partial x} = 0$$

$$\frac{\partial u}{\partial t} + u\frac{\partial u}{\partial x} + \left(\frac{2}{\gamma - 1}\right)a\frac{\partial a}{\partial x} = 0 \tag{12.68}$$

where the first of these equations is mass conservation and the second is Euler's equation, although expressed in terms of a and u as the dependent variables. If we take the sum and difference of these equations, they have a symmetrical form:

$$\left(\frac{\partial}{\partial t}+(u+a)\frac{\partial}{\partial x}\right)\left(\frac{2}{\gamma-1}a+u\right)=0 \tag{12.69}$$

$$\left(\frac{\partial}{\partial t}+(u-a)\frac{\partial}{\partial x}\right)\left(\frac{2}{\gamma-1}a-u\right)=0 \tag{12.70}$$

Equations 12.69–12.70 are called *characteristic equations*. They express the time rate of change of the dependent variables in two directions in the x, t "space" of the two independent variables. The operator $\partial/\partial t + (u + a)\partial/\partial x$ in 12.69 is the time derivative as measured by an observer moving in the $+x$ direction at a speed $u+a$, *i.e.*, at the local speed of propagation of a sound wave. Just as the time derivative following a material particle is $\partial/\partial t+\mathbf{V}\cdot\nabla$, the time derivative following a right-running sound wave is $\partial/\partial t+(u+a)\mathbf{i}_x\cdot\nabla=\partial/\partial t+(u+a)\partial/\partial x$. By the same token, the differential operator of equation 12.70 is the time derivative along a left-running sound wave. In the x, t plane, the directions $dx/dt = u \pm a$ are the local *characteristic directions*.

Equation 12.69 states that the sum $2a/(\gamma-1)+u$ does not change with time along a right-running sound wave in the flow, while 12.70 requires that $2a/(\gamma-1)-u$ does not change with time along a left-running sound wave. If, at some point x_1 of the flow, we know both the sound speed a_1 and flow speed u_1, then for all future times along the left- and right-running sound waves that passed through the point 1 at the time t_1, we have:

$$\left[\frac{2}{\gamma-1}a+u\right]_r=\frac{2}{\gamma-1}a_1+u_1 \tag{12.71}$$

$$\left[\frac{2}{\gamma-1}a-u\right]_l=\frac{2}{\gamma-1}a_1-u_1 \tag{12.72}$$

where the subscripts r and l identify the right- and left-running sound waves.

To illustrate how equations 12.72 are used to solve an unsteady compressible flow problem, let us consider the flow illustrated in figure 12.18 where a moving piston accelerates an initially stationary gas. In this case, the piston generates left-running sound waves. However, these are crossed by right-running sound waves that move from the stationary fluid where $a = a_0$ and $u = 0$. Then, by equation 12.72,

$$\left[\frac{2}{\gamma-1}a+u\right]_r=\frac{2}{\gamma-1}a_0 \tag{12.73}$$

which holds at any time in the region $-a_0t \leq x \leq V_pt$ occupied by the left-running waves between the piston and the first sound wave moving to the left, $x = a_0t$. But, at the piston, $u = V_p$, so by 12.73, the sound speed at the piston face a_p is:

$$a_p = a_0 - \left(\frac{\gamma - 1}{2}\right) V_p \tag{12.74}$$

The faster the piston moves, the lower the value of the sound speed a_p and of course the pressure and density at the piston face. When $V_p = 2a_0/(\gamma - 1)$, all these flow variables would be zero. If the piston were to be moved at an even higher velocity, it would move faster than the gas and leave a vacuum between it and the nearest gas particle, which could move no faster than $u = 2a_0/(\gamma - 1)$.

To describe the flow in more detail, it helps to plot the trajectories of sound waves in the x, t plane. Figure 12.19 shows such a plot, where we designate t as the ordinate and x as the abscissa. The first left-running wave that moves into the stationary gas moves along the line $x = -a_0 t$ that has the slope $1/(-a_0)$. The piston trajectory is the straight line $x = V_p t$ and has the slope $1/V_p$. A left-running wave that leaves the piston face where $u = V_p$ and $a = a_0 - (\gamma - 1)V_p/2$ has the slope $1/(u - a) = 1/(V_p - [a_0 - (\gamma - 1)V_p/2]) = -1/[a_0 - (\gamma - 1)V_p/2]$. Its trajectory, $x = -[a_0 - (\gamma - 1)V_p/2]t$, is plotted in figure 12.19.

There are other left-running waves between these two, for which the trajectories are:

$$x = -(a - u)t \tag{12.75}$$

However, u and a are also related by equation 12.73. Combining 12.75 and 12.73, we find $u\{x, t\}$ and $a\{x, t\}$ for the region $-a_0 t \leq x \leq -[a_0 - (\gamma - 1)V_p/2]t$:

$$a = \frac{2}{\gamma + 1} a_0 - \left(\frac{\gamma - 1}{\gamma + 1}\right) \frac{x}{t}$$

$$u = \frac{2}{\gamma - 1} a_0 + \left(\frac{2}{\gamma + 1}\right) \frac{x}{t} \tag{12.76}$$

Equation 12.76 is a *centered expansion wave* in which all the left-running sound waves in the expansion wave originate at the point $x = 0$, $t = 0$. If the piston were accelerated to the speed V_p over a finite time interval, these sound waves would originate at different points along the piston trajectory, $x_p = \int V_p \, dt$.

Because this flow is isentropic, the pressure, density, and temperature of the gas are related to the sound speed a through the isentropic relations of equation 12.23, which may be put in the form:

$$\frac{p}{p_0} = \left(\frac{a}{a_0}\right)^{2\gamma/(\gamma - 1)}$$

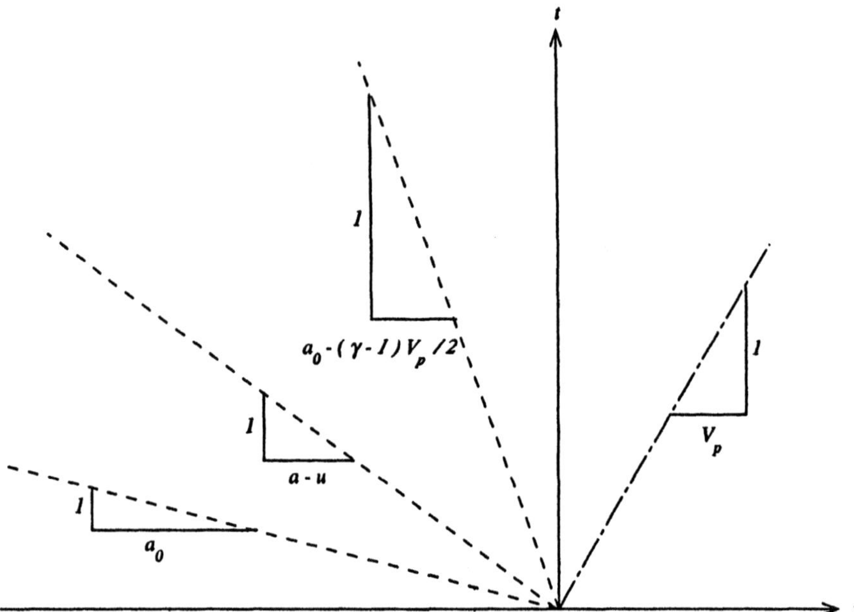

Figure 12.19 The unsteady flow of a stationary gas set into motion by the retraction of a piston moving at the speed V_p, depicted in the x, t plane. Left-running waves are indicated by dashed lines, and the piston trajectory by a dash-dot line.

$$\frac{\rho}{\rho_0} = \left(\frac{a}{a_0}\right)^{2/(\gamma-1)}$$

$$\frac{T}{T_0} = \left(\frac{a}{a_0}\right)^2 \tag{12.77}$$

Example 12.10

A pipeline carries natural gas under high pressure at an average speed $u_0 = 100 m/s$. Under the flow conditions in the pipeline, $\gamma = 1.31$ and $a_0 = 446.1\, m/s$. In an accident, the pipeline is suddenly severed and exposed to atmospheric pressure so that a sonic outflow is experienced from both the upstream ($-x$) and downstream ($+x$) sections of the pipe. Calculate (a) the outflow speeds from the two sections and (b) the ratios of the outflow mass flow rates to that in the undisturbed pipe flow.

Solution

(a) The unsteady flow out of the upstream pipe section would be similar to that illustrated in figure 12.18 except that the undisturbed flow would be moving at a speed $u_0 = 100\,m/s$. A right-running wave moving from this region to the point of the break would have the property (equation 12.72):

$$\frac{2}{\gamma-1}a + u = \frac{2}{\gamma-1}a_0 + u_0$$

But the outflow is sonic, so that $a_{out} = u_{out}$:

$$\left(\frac{2}{\gamma-1}+1\right)u_{out} = \frac{2}{\gamma-1}a_0 + u_0$$

$$u_{out} = \frac{2}{\gamma+1}a_0 + \frac{\gamma-1}{\gamma+1}u_0$$

$$= \frac{2(446.1\,m/s)}{2.31} + \frac{0.31(100\,m/s}{2.31} = 399.7\,m/s$$

For the flow out of the downstream section, start with a left-running wave from the undisturbed flow:

$$\frac{2}{\gamma-1}a - u = \frac{2}{\gamma-1}a_0 - u_0$$

In this case, the outflow speed is in the $-x$ direction, *i.e.*, is negative so that $a_0 = -u_{out}$:

$$\left(\frac{2}{\gamma-1}+1\right)(-u_{out}) = \frac{2}{\gamma-1}a_0 - u_0$$

$$u_{out} = -\left(\frac{2}{\gamma+1}a_0 - \frac{\gamma-1}{\gamma+1}u_0\right)$$

$$= -\frac{2(446.1\,m/s)}{2.31} + \frac{0.31(100\,m/s}{2.31} = -372.8\,m/s$$

The direction of the downstream flow must be reversed to flow out of the broken pipe opening so that its outflow speed is less than that from the upstream section.

(b) The mass flow rate ratio is:

$$\frac{\rho_{out}u_{out}}{\rho_0 u_o} = \left(\frac{u_{out}}{a_0}\right)^{2/(\gamma-1)}\frac{u_{out}}{u_0}$$

where we have used equation 12.77 to determine the density ratio ρ_{out}/ρ_0. For the upstream and downstream flows, this becomes, respectively:

$$\frac{\rho_{out}u_{out}}{\rho_o u_o} = \left(\frac{399.7}{446.1}\right)^{2/0.31}\left(\frac{399.7}{100}\right) = 1.968$$

$$= \left(\frac{372.8}{446.1}\right)^{2/0.31}\left(\frac{372.8}{100}\right) = 1.171$$

The total mass flow rate from the pipe break is about three times the mass flow rate in the unbroken pipeline.

12.8 Problems

Problem 12.1

Derive an expression for the speed of sound a_m in a mixture of water and small air bubbles, in which the bubbles occupy a fraction ϵ of the volume of the mixture.

Problem 12.2

Calculate the average amplitude $\sqrt{\overline{x^2}}$ of the motion of a loudspeaker blasting out a 80 dB sound wave at a frequency of $f = 400\,Hz$.

Problem 12.3

In a closed-cycle gas turbine power plant, helium enters the adiabatic turbine at a pressure $p_1 = 8\,bar$ and temperature $T_1 = 1100\,K$ and leaves at a pressure $p_2 = 1\,bar$ and temperature $T_2 = 620\,K$. Calculate (a) the adiabatic turbine efficiency η_t and (b) the work delivered by the turbine per kilogram of helium flowing through it.

Problem 12.4

In a supersonic wind tunnel, air stored at a pressure $p_s = 10\,bar$ and a temperature $T_s = 290\,K$ is passed adiabatically through a convergent-divergent nozzle and exhausts as a supersonic stream at a pressure $p_e = 1\,bar$. Assuming a constant $\gamma = 1.4$, calculate (a) the Mach number M_e of the flow at the nozzle exit, (b) the temperature T_e of the exit flow, (c) the ratio A_e/A_c of the exit area to the throat area and (d) the mass flow rate $\dot{m}$ of air if the throat area $A_e = 10\,cm^2$.

Problem 12.5

It is proposed to design a hypersonic wind tunnel using helium ($\gamma = 5/3$) as the working fluid, in which the test section will operate at a Mach number $M_t = 20$. Calculate (a) the stagnation temperature T_s if the test section temperature $T_t = 10\,K$, (b) the stagnation pressure required if the test section pressure $p_t = 1E(-4)\,bar = 10\,Pa$, and (c) the ratio A_t/A_c of the test section area to the throat area.

Problem 12.6

For a shock wave in a perfect gas with constant specific heats, derive expressions for the ratios $p_2/\rho_1 V_1^2$ and $2\hat{c}_p T_2/V_1^2$ in terms of γ alone for the limit of $M_1 = \infty$.

Problem 12.7

The entropy increase across a shock wave in a perfect gas with constant specific heats is related to the pressure and density changes by:

$$\frac{\hat{s}_2 - \hat{s}_1}{\hat{c}_v} = \ln\left(\frac{p_2}{p_1}\right) - \gamma \ln\left(\frac{\rho_2}{\rho_1}\right)$$

A weak shock wave may be specified by a small parameter $\epsilon \equiv M_1^2 - 1 \ll 1$. Derive an expression for $(\hat{s}_2 - \hat{s}_1)/\hat{c}_v$ as a power series in the small parameter ϵ, and show that the first non-zero term is $2\gamma(\gamma-1)\epsilon^3/3(\gamma+1)^3$, which is third order in the pressure rise $(p_2 - p_1)/p_1$).

Problem 12.8

Natural gas (methane), compressed to a pressure $p_1 = 60\,bar$ and a temperature $T_1 = 300\,K$, flows through a pipeline of length $L = 7\,km$ and diameter D, emerging at the other end at a pressure $p_2 = 10\,bar$. The inflow velocity $V_1 = 45\,m/s$. Assuming the flow is adiabatic, calculate (a) the temperature T_2 of the outflow, (b) the Mach numbers M_1 and M_2 of the entering and leaving flow and (c) the value of the Darcy friction factor $\bar{f}$ for this flow.

Problem 12.9

In the limit of $M_1 = \infty$ for an oblique shock wave in a perfect gas, derive an expression for the maximum deflection angle θ_m and the shock angle β_m at which it is encountered. Evaluate your answer for $\gamma = 1.4$ and compare with figure 12.13.

Problem 12.10

A large tank filled with a perfect gas having a sound speed a_s and a constant γ is connected to a very long pipe ($x \geq 0$) by a quick-opening valve at the entrance to the pipe. Initially, the pipe is completely evacuated ($p = 0$ when $t = 0$; $x \geq 0$). At time $t = 0$, the valve is suddenly opened, and the gas flows into the pipe. The flow from the tank at the pipe entrance is a steady choked flow; the flow in the pipe is an unsteady flow. (a) Derive an expression for the maximum velocity u_m of the gas in the pipe, and calculate the ratio u_m/V_m for $\gamma = 1.4$, where V_m is the maximum steady flow velocity for an isentropic flow from the tank to a region of zero pressure. (b) Derive an expression for the pipe velocity $u\{x, t\}$ as a function of x and t.

Bibliography

Sabersky, Rolf H., Allan J. Acosta, and Edward G. Hauptman. 1989. *Fluid Flow*. 3rd ed. New York: Macmillan Publishing Co.

Thompson, Philip A. 1972. *Compressible-Fluid Dynamics*. New York: McGraw-Hill Book Co.

White, Frank M. 1986. *Fluid Mechanics*. 2nd ed. New York: McGraw-Hill Book Co.

Index